Biomechanics of Active Movement and Division of Cells

AF788511

NATO ASI Series

Advanced Science Institutes Series

A series presenting the results of activities sponsored by the NATO Science Committee, which aims at the dissemination of advanced scientific and technological knowledge, with a view to strengthening links between scientific communities.

The Series is published by an international board of publishers in conjunction with the NATO Scientific Affairs Division

A Life Sciences B Physics	Plenum Publishing Corporation London and New York
C Mathematical and Physical Sciences D Behavioural and Social Sciences E Applied Sciences	Kluwer Academic Publishers Dordrecht, Boston and London
F Computer and Systems Sciences G Ecological Sciences H Cell Biology I Global Environmental Change	Springer-Verlag Berlin Heidelberg New York London Paris Tokyo Hong Kong Barcelona Budapest

NATO-PCO DATABASE

The electronic index to the NATO ASI Series provides full bibliographical references (with keywords and/or abstracts) to more than 30000 contributions from international scientists published in all sections of the NATO ASI Series. Access to the NATO-PCO DATABASE compiled by the NATO Publication Coordination Office is possible in two ways:

- via online FILE 128 (NATO-PCO DATABASE) hosted by ESRIN, Via Galileo Galilei, I-00044 Frascati, Italy.

- via CD-ROM "NATO Science & Technology Disk" with user-friendly retrieval software in English, French and German (© WTV GmbH and DATAWARE Technologies Inc. 1992).

The CD-ROM can be ordered through any member of the Board of Publishers or through NATO-PCO, Overijse, Belgium.

Series H: Cell Biology, Vol. 84

Biomechanics of Active Movement and Division of Cells

Edited by

Nuri Akkaş

Department of Engineering Sciences
Middle East Technical University
06531 Ankara, Turkey

Springer-Verlag
Berlin Heidelberg New York London Paris Tokyo
Hong Kong Barcelona Budapest
Published in cooperation with NATO Scientific Affairs Division

Proceedings of the NATO Advanced Study Institute on Biomechanics of Active Movement and Division of Cells, held in Istanbul, Turkey, September 19–29, 1993

ISBN-13: 978-3-642-78977-9 e-ISBN-13: 978-3-642-78975-5
DOI: 10.1007/978-3-642-78975-5

CIP data applied for

This work is subject to copyright. All rights are reserved, whether the whole or part of the material is concerned, specifically the rights of translation, reprinting, reuse of illustrations, recitation, broadcasting, reproduction on microfilm or in any other way, and storage in data banks. Duplication of this publication or parts thereof is permitted only under the provisions of the German Copyright Law of September 9, 1965, in its current version, and permission for use must always be obtained from Springer-Verlag. Violations are liable for prosecution under the German Copyright Law.

© Springer-Verlag Berlin Heidelberg 1994
Softcover reprint of the hardcover 1st edition 1994

Typesetting: Camera ready by authors
SPIN 10101701 31/3130 - 5 4 3 2 1 0 - Printed on acid-free paper

PREFACE

The NATO Advanced Study Institute on Biomechanics of Active Movement and Division of Cells was held September 19-29, 1993 in Istanbul and the Proceedings are presented in this volume. Sixty-eight scientists from sixteen countries attended. Prof. J. Bereiter-Hahn of Goethe-Universitat, Frankfurt, Germany, Prof. A.K. Harris of the University of North Carolina, Chapel Hill, USA, Prof. R.M. Nerem of Georgia Institute of Technology, Atlanta, USA and Prof. R. Skalak of the University of California, San Diego, USA were the members of the International Organizing Committee. As the Scientific Director of the Institute, I wish to express my sincere appreciation for their assistance without which the Institute could not have taken place.

This Institute is the third one of the meetings which are now called "the NATO Istanbul Meetings on Cytomechanics". The first one was the NATO Advanced Research Workshop on Biomechanics of Cell Division which was held October 12-17, 1986 in Istanbul. The Proceedings were published as NATO ASI Series A : Life Sciences Vol. 132 by Plenum Press in 1987.
The second one was the NATO Advanced Study Institute on Biomechanics of Active Movement and Deformation of Cells which was held September 3-13, 1989 in Istanbul. The Proceedings were published as NATO ASI Series H : Cell Biology Vol. 42 by Springer-Verlag in 1990.

The idea of having NATO Advanced Study Institutes on cytomechanics at convenient intervals was proposed and overwhelmingly supported during the previous meetings. The objective of these cytomechanics meetings is to continue the process of crossdisciplinary and international cooperation among biologists, biomechanicians, biophysicists, biochemists and biomathematicians in solving problems of active cell behaviour (division, locomotion, shape change, force generation) and to explore the applicability of such knowledge in physiology, diagnosis, therapy and similar. The meetings have, thus far, served very well to establish a dialogue and an interchange of views between participants from the several disciplines represented.

Cytomechanics is the application of the principles of classical mechanics in cytology. It is an applied science which is concerned with the description and evaluation of mechanical properties of cells and their organelles as well as of the forces exerted by them, thus being responsible for generation and maintenance of cell and tissue architecture.

The third cytomechanics meeting has shown that very considerable progress has been achieved by interdisciplinary cooperation of scientists from the diverse fields of cell biology, biomechanics, biophysics, biochemistry, engineering, mathematics and computational science in revealing the detailed and appropriate mechanical explanations of the causes and consequences of active motions of cells, and is on the verge of exciting application in biomedical field. Such progress is equally relevant to both animal and plant cell development, although the work on plant cells is much less developed and do not generally have motion, and represents an area of potential rapid progress and benefit to the peoples of the world dependent on agricultural products.

The qualitative and biochemical identification of molecular species involved in various aspects of biomechanics of active movement and division of cells has advanced to a stage that it is now possible and crucial to further progress in such biomechanics to develop molecular level models of the processes and reactions which take place in the various stages of division and active motion of cells. Elucidating the underlying molecular dynamics (such computations are now feasible because of the rapid increase in computer capacity, in particular the development of parallel processing) and events which are the basis of macroscopic, continuum mechanics modelling has proved to be fruitful. The emphasis should be on relating models at a molecular level to the macroscopic events of interest at the cellular, tissue and organ level because, although great progress has been made in molecular biology and protein folding mechanics, this information has not been utilized or incorporated in the function models of interest.

Many different kinds of shape changes and rearrangements are undergone by the cells of the body. At the sub-cellular level, these include such phenomena as mitosis, cleavage, capping, cell polarization, cell locomotion and the exertion of traction forces. At the super-cellular level there are various morphogenetic movements and wound healing phenomena. All of these are mechanical processes in which physical forces of various kinds are exerted by one component of a cell on other parts of that same or other cell. Thus, these processes are microscopic versions of the same general kinds of phenomena which it is the professional concern of mechanicians and applied mathematicians to analyze. Developmental and cell biologists are thus in considerable need of help and advice from mechanicians and mathematicians; while the mechanicians and mathematicians stand to gain access to some very stimulating intellectual problems from their contacts with the biologists, not to mention the NOT unrealistic chance of making useful contributions to the most fundamental aspects of medicine. There are many biological processes which depend on the active movement and division of cells, both in normal growth and in diseases, such as cancer. Cell activity forms the basis of life and this Advanced Study Institute was on active cellular processes which involve mechanical forces and motions.

Significance of mechanical forces in the control of cell proliferation and differentiation has now become more accepted. The mechanical interaction of a cell with its environment may be an important mechanism in the control of gene activity also. It is, therefore, vital to understand how cells develop forces. Bringing together experts from so many diverse fields and of such different but, in this case, complementary backgrounds created an excellent interdisciplinary forum for a discussion of the mentioned force generation and the related control mechanism.

At the end of the third NATO Istanbul Meeting on Cytomechanics, the participants resolved that a further conference be held on "Cytomechanics: Molecular, Cellular and Developmental Aspects", bringing together the scientists of cell biology, medicine, engineering, mathematics and computational science represented at the previous NATO Istanbul Meetings on Cytomechanics, in addition some representation of scientists specializing in molecular biology, molecular modeling of protein folding and larger molecular assemblies. Inclusion of such would enhance the conference by elucidating the underlying molecular dynamics (such computations are now feasible because of the rapid increase in computer capacity, in particular the development of parallel processing) and events which are the basis of macroscopic, continuum mechanics modelling which has proved to be fruitful in the past Cytomechanics conferences. The emphasis would be on relating models at a molecular level to the macroscopic events of interest at the cellular, tissue and organ level because, although great progress has been made in molecular biology and protein folding mechanics, this information has not been utilized or incorporated in the function models of interest. Examples of topics may include:

a. Chromosomal movement in cell division;
b. Forces generated by polymerization at the front of a pseudopod;
c. Molecular models of cell adhesion;
d. Cell junctional formation and disruption of permeability;
e. Mechanics of RNA, DNA and protein transcription;
f. Uptake of extracellular materials and their intracellular transport;
g. Active cell response to external mechanical stresses;
h. Progresses in the topics covered in the previous Cytomechanics conferences.

It has further been resolved that the continuity, momentum and expanding interdisciplinary communication and research established by the past and present NATO Istanbul Meetings on Cytomechanics be continued and expanded as described above because it represents a needed and potentially very potent cross-fertilization, scientific advancement and growth in ultimate practical application of mechanics and analysis of great benefit and value to humanity.

In conclusion, it should be reiterated that the objective of the Advanced Study Institute was to provide the forum for an exchange of views at the frontiers of knowledge and for the formulation of recommendations for new research directions. The expected output is to make the authorities' views on the necessity of interdisciplinary collaboration in cytomechanics available in a book form so that an interdisciplinary cooperation will be activated and stimulated.

January 1994, Ankara

Nuri Akkas

CONTENTS

Modelling of Biological Polymers: Discrete and Continuum Mechanics Formulations
A. ASKAR 1

Cytokinesis: The Mechanism of Formation of the Contractile Ring in Animal Cell Division
A.K. HARRIS 37

Effect of Cortex Stiffness Variation on Cleavage in Animal Cells
N. AKKAS and M. KERMANIAN 67

Multicellular Mechanics in the Creation of Anatomical Structures
A.K. HARRIS 87

The Interplay of Active Forces and Passive Mechanical Stresses in Animal Morphogenesis
L.V. BELOUSSOV 131

The Role of Elasticity in the Motile Behaviour of Cells
J. BEREITER-HAHN and H. LÜERS 181

On Free Boundary Problems and Amoeboid Motion
M. DEMBO 231

A Model of the Leukocyte Migration Through Solid Tissue
B.A. SKIERCZYNSKI, S. USAMI, and R. SKALAK 285

Adhesion and Traction Forces in Migration: Insights from Mathematical Models and Experiments
P.A. DIMILLA 329

The Active Response of an Endothelial Cell to the Onset of Flow
G. HELMLINGER, O. THOUMINE, T.F. WIESNER, and R.M. NEREM 369

Mechanical Aspects of Ciliary Propulsion
M.E.J. HOLWILL 393

Local Bending Fluctuations of the Cell Membrane
R. KORENSTEIN, S. TUVIA, L. MITTELMAN, and S. LEVIN 415

Simulation of Morphogenetic Shape Changes Using the Finite Element Method
G.W. BRODLAND and D.A. CLAUSI 425

Locomoting Blebbing Cells: A New Model to Test Whether Formation of Protrusions is Primarily due to Hydrostatic Pressure or to Actin Elongation
H.U. KELLER 431

Dynamic Morphology of Leukocytes: Statistical Analysis and a Stochastic Model for Receptor-Mediated Cell Motion and Orientation
R.T. TRANQUILLO, O. BROSTEANU, and W. ALT 437

Observation, Modeling and Simulation of Keratinocyte Movement
W. ALT and H.W. KAISER 445

Strain Transduction in Non-Excitable Cells
D.B. JONES and G. LEIVSETH 453

Self-Organization in Biological Systems as a Result of Interaction Between Active and Passive Mechanical Stresses: Mathematical Model
A.A. STEIN 459

Biomechanical Events in the Production of Ciliary Movement
P. SATIR 465

Microtubule Converging Centers: Implications for Microtubule Dynamics in Higher Plants
A.S. BAJER, E.A. SMIRNOVA, K.A. WAWROWSKY, R. WOLF, and J. MOLE-BAJER 471

Elastic Properties of Layered Membranes and Their Role in Transformations of Cellular Shapes
S. SVETINA and B. ZEKS 479

Cortical Flow in Free-Living Amoebae
A. GREBECKI 487

A Mechanical Model for Growth and Control of the Cell Wall of Rod-Shaped Bacteria
J.J. THWAITES 493

Measurement of Passive and Active Force in Single Isolated Smooth Muscle Cells
R. VAN MASTRIGT, J.J. GLERUM, and R. SCHOT 499

Molecular Mechanisms of Neutrophil Adhesion Studied by Inducing a High Avidity State at β2-Integrin
S.I. SIMON 511

Mechanical Tension Regulates the Phenotype of Cells Cultured in a Collagen Gel
C.A. LAMBERT, P.Y. LEFEBVRE, C. DEROANNE, B.V. NUSGENS, and C.M. LAPIERE 519

The Mechanism of Actomyosin and Microtubule-Kinesin Systems
Y.Z. MA and E.W. TAYLOR 525

Immunolocalization of Acetylated Microtubules in Germ Cells of Insects
K.W. WOLF 533

Mechanisms of α_1-Adrenergic Induced Positive Inotropism in Cardiac Cells
G. VASSORT, C. VANNIER, V. LAKOMKINE, M. PUCEAT, and O. CLEMENT-CHOMIENNE 539

Kinetochore Directional Instability in Vertebrate Mitotic Cells
R.V. SKIBBENS and E.D. SALMON 545

Signal Transduction and Myofibrillogenesis in Isolated Neonatal Heart Myocytes in Vitro
W. SHARP, D.G. SIMPSON, L. TERRACIO, and T.K. BORG 551

Quantitative EMG and Biomechanics
G.V. DIMITROV and N.A. DIMITROVA 557

The Anaphase Movement and Its Relationship with Cell Cleavage
M. MOTA 563

Mechanism of Induction of Microspore Embryogenesis in Brassica napus: Significance of the Preprophase Band of Microtubules in the First Sporophytic Division
D.H. SIMMONDS 569

Subject Index 575

MODELLING OF BIOLOGICAL POLYMERS : Discrete and Continuum Mechanics Formulations

Attila Askar
Koç University
Istinye, 80860 Istanbul
Turkey

1. Introduction

Proteins and chromosomes are the essential polymers of life. The proteins are the primary matter for sustaining life through constituting most of the living matter along with storing and using energy. The DNA is at the origin of life by carrying the genetic code and performing the functions of replication and transcription for the production of protein. Their dynamics are thought to determine their biological functions as well as their malfunction which have vital consequences. Similarly, and stability statics studies are also important in determining the material properties and structure of these polymers. The information provided by statics can be used either for its own sake such as for understanding the strength and deformation of muscles or provides the necessary data to be used for dynamical studies. Stability studies on the other hand would be useful for instance in understanding the formation of beta sheets and the folding of proteins.

The goal of this article is to present a detailed model for biological polymers [Saenger (84), Freifelder (87), Mc Cammon, and Harwey (87), Dickerson (83), Stryer, Freeman (81)]. Of primary interest is the need of a thorough understanding of the mechanics of helices which form the most intricate component in the biopolymers. For the molecular structures in

NATO ASI Series, Vol. H 84
Biomechanics of Active Movement
and Division of Cells
Edited by N. Akkaş
© Springer-Verlag Berlin Heidelberg 1994

consideration, the helices are of a finite pitch such that the elementary strength of materials theory of helices is not acceptable. The internal energy expressions for a protein and DNA are obtained by adding the contribution of the relevant hydrogen bonds to that of the backbond helices. In this presentation, the general formulation is followed by illustrations of the theory for finite static deformations, dynamics of a discrete helical chain, dynamics of a super coiled DNA and various soliton solutions as model for proton migration in hydrogen bonds, DNA denaturation and rotatory dynamics of the DNA bases.

2. A brief historical note

The history of the discovery of the DNA's structure as a double helix has been narrated in a very interesting little book that bears the same title by the two researchers that were awarded the Nobel prize for their discovery. There is however more to tell then just the story of the DNA. Here a brief sketch is presented indicating. The important land mark discoveries in the quest for understanding life. Rigor is of course not to be expected in the following story. Some of the dates appearing with the items are merely to locate a particular idea in time in a rough manner.

1. Hook, the famous elasticity theory pioneer, appears as an early proponent of the idea that all plants are constituted of cells in 1665. It took about two centuries for Schneider to propose in 1838 that all living matter is constituted of cells.
2. An important realisation is due to Fleming when he noticed in 1879 a colored element in the cell which he named as "chromatine". This observation is the forrunner of the idea of the chromosome.

3. The discovery of the foundations of the idea of heredity must certainly be associated with the Austrian monk Mendel in 1856. Partly perhaps due to his being in the clergy, Mendel did not push forward very vigorously some of the ideas about heredity that might have been judged as offensive to prevailing doctrines of the time. He was forgotten for about a half century and his experiments made a come back only in the 1900' s.
4. The pace of progress towards understanding the material elements of life accelerated starting in 1930 with the discovery of the chemistry of the proteins by Fisher and that of DNA by Avery in 1944.
5. Although the discoveries by Fisher and Avery revealed the elements constituting these polymers, their structure which is essential to understanding their function was not known. The structure of the proteins as an alpha helix is provided by Pauling in 1951 and that of the DNA as a double helix by Watson and Crick in 1953.
6. Three physicists should also be credited for their indirect contribution to the saga leading to the modern bio-sciences. Bohr was perhaps the first person to suggest in 1932, at least in the modern times and in a scientific context, that life does not necessitate a separate science and all can be understood within the framework of physics and chemistry. Schrodinger suggested in 1944 that the basic buldingstones of life had to have a regularity and had to carry a genetic message. He used the contradictory term of "aperiodic crystal" for describing the regularity and the term "code script" which is the predessor of the term genetic code. Schrodinger made however an error: he thought that the proteins with their much richer variety had to carry the genetic code in order to describe the extreme variety in living species. It was Gamov who suggested in 1953 that the DNA was the carrier of the hereditary information.

3. Structure of biopolymers

The basic structural element of biopolymers is a helix. The proteins involve a single helix, while the DNA has two helices. In the following, the chemical composition as well as the basic structure of these polymers are described [Schrega (84), Van Gunsteren, Karplus (82), Schlick (82), Karplus, Mc Cammon (86)].

3.1 Alpha helices and proteins

The structure of proteins is determined by a single helix called as the "alpha helix". This helical chain is a polypeptide polymer. For the alpha helices, the helical backbone is further strengthened through hydrogen bonds along the axis of the helix. The basic constituants of a protein are amino-acids. Figure 1 illustrates the basic structure of the protein. Although each type of amino acid is chemically distinct, there is a strong resemblance between these molecules. They all have an acidic end, the carboxyl group COOH, and a basic end, the amino group NH_2. Since acids and bases tend to come together, chain structures can be formed the chain has both carbon and nitrogen atoms in its backbone by joining amino acids in a head-to-tail fashion. An amino acid residue is the groups of atoms that remains after a molecule of water has been removed: H from the amino end and OH from the amino end and OH from the carboxyl end. The formation of a polypeptide is an example of condensation polymerization. It is an important feature of polypeptides that they never exhibit branching. They are always simple linear structures, although the chains can become cross-linked by the formation of secondary bonds.Together with Robert Corey, Pauling discovered a polypeptide structure that is consistent with all the experimental evidence and is attractively simple.

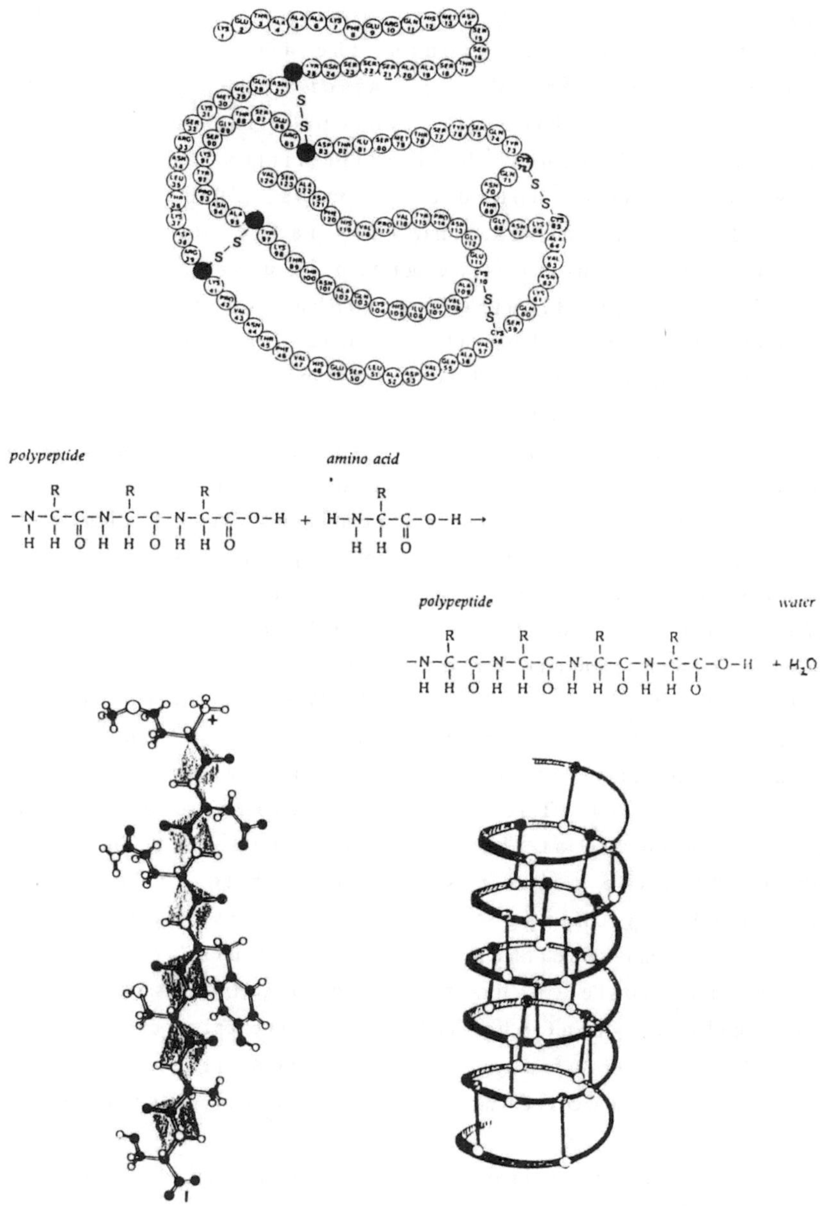

Fig. 1 The structure of Proteins (a) in a globular configuration; (b) the formation of the polypeptide chain; c) the three dimensional structure of an alpha helix; d) a model for an alpha helix.

There exist twenty different common amino acids are divided into two main groups, the distinction being whether the side group is polar or non-polar. The non-polar examples, which are hydrophobic, can be readily distinguished because their extremities usually consist of hydrogen atoms bound to carbons. There are nine non-polar amino acids, and they range from simple glycine, in which the side group is a single hydrogen atom, to the complicated tryptophan with a side group containing nine carbon atoms, one nitrogen atom, and eight hydrogen atoms. The other main group, those containing polar side groups, can be further subdivided into three classes. The first contains aspartic acid and glutamic acid. Which being acidic, both develop negative charges at their extremities. As a consequence of the charge, they are also hydrophilic. A second class counsist of lysine and arginine develop a positive charge and are thus basic and hydrophabic. The final class contains amino acids such as glutamine and cysteine, which are all neutral. Whether a particular residue is polar or non-polar is of crucial importance to the conformation of the polypeptide, when it is in aqueous solution. The non-polar residues, avoid water by tending to lie at the interior of the protein. Conversely, the polar residues prefer to be located on the outside. A change of solvent causes some of the residues to move either towards the outside or inside, leading to subtle changes of confirmation that are frequently of great physiological importance. The catalytic action of some enzymes appears to be influenced by such changes of chemical environment.

The sequence of amino acids in a protein defines what is known as its primary structure. The question arises as to how it transforms from a linear polymer molecule to a three-dimensional structure. There are several factors. One comes from steric effects due to the side groups, and another is the establishment of various bonds, the strongest type being the disulfide bridge. The weaker kind of secondary link that binds polypeptide chains together is the hydrogen band in which the hydrojen atom attached to the nitrogen atom atom on one chain forms a bond with the carbonyl oxygen of another chain. In practice, since neighbouring polypeptide chains often lie parallel to each other, hydrogen bonds frequently occur in groups. A third type of link that can stabilize a protein is the salt bridge. This can be established when amino and carboxyl side groups are in close proximity. The stabilizing force is the electrical attraction between the two opposing charges. It is not a bond in the usual sense in that there is no sharing of electrons, and the attraction survives in spite of considerable variation in the charge-charge distance.

A mean of unambiguously predicting the three-dimensional conformation of a protein, starting only with its primary structure and a knowledge of the interatomic forces is not yet possible.One must still resort to structure analysis by X-ray scattering. Another approach that has scored several spectacular successes involves building scale models. It requires knowledge of atomic sizes, interatomic bonding, a grasp of relevant biophysical and biochemical concepts and, not the least, intuition and inspired guess work.

3.2 Double helix and DNA

DNA stands for Deoxyribo Nucleic Acid. The DNA has a double helical structure where each one of the helices consists of alternating sugar and phosphate molecules. The sugar element is of a special kind called as "deoxyribose" which is essentially a deoxidized sugar with a pentagonal shape. The two helical strands are

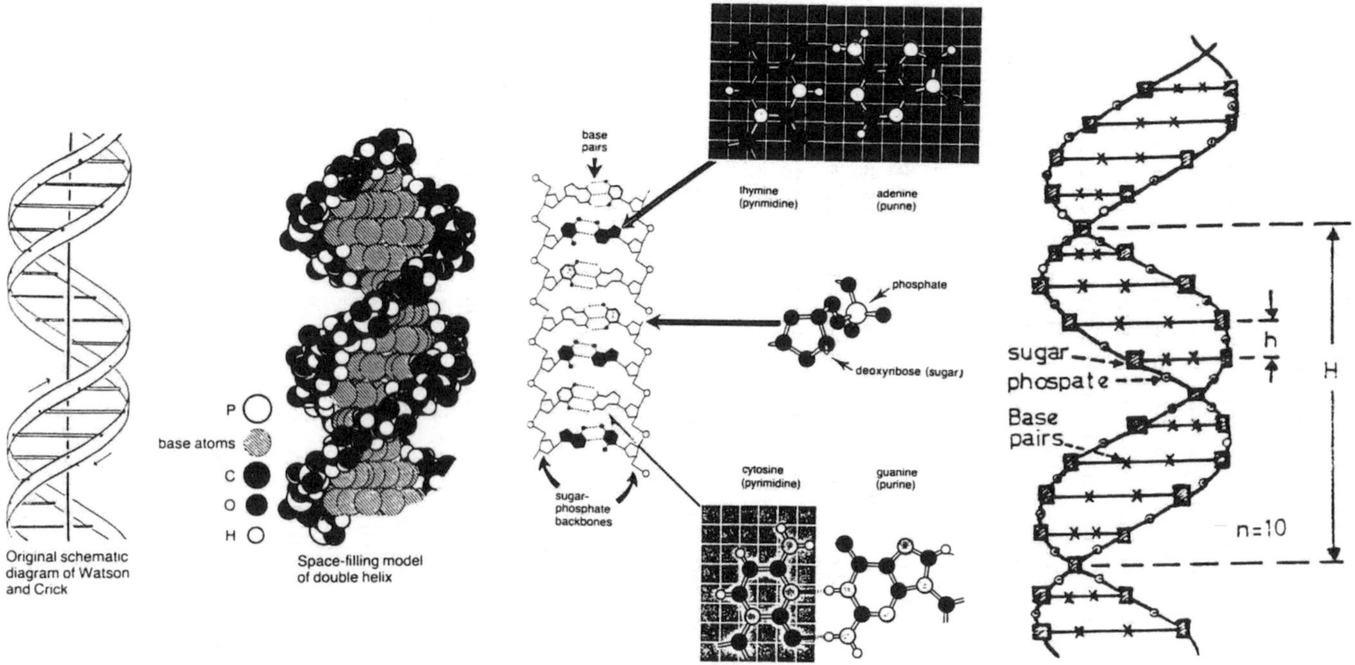

Fig. 2 The DNA molecule. (a) the three dimensional view; b) the structure of the DNA; c) a model for of DNA.

joined by base pairs which are themselves held together by hydrogen bonds. Figure 2 illustrates the basic structure of DNA. Four types of nitrogenous bases are found in DNA. These are the pyrimidine bases consisting of cytosine (C) and thymine (T) and the purine bases which consist of adenine (A) and guanine (G). The experiments that were crucial in determining the structure of DNA are those that showed that the amounts of deoxyribose and phostphate were equal. Similarly, these experiments also revealed that the amount of adenine was always the same as that of thymine, while the amounts of guanine and cytosine were also the same. This lead to the conclusion that A, T and C, G had to come in pairs. Moreover, a coplanar hydrogen-bonded arrangement of adenine and thymine has exactly the same overall length as similar combination of guanine and cytosine. These observations were the two key results that lead to the discovery of the structure of the DNA. Such base pairs can thus act as the rungs of the ladder, the sides of which are composed of sugar and phostphate units in regular alternation.Building a scale model of their arrangement, Watson and Crick found that steric hindrance causes the ladder to twist into a helical conformation: the now famous double helix. The spacing between adjacent base pairs is a few tenths of a nanometer, as expected, but two successive bases are offset from each other by an angle of just over 30°. The pitch of the double helix is such that there is parallelism between the bases spaced eleven base pairs apart, and this corresponds to 3.4 nanometers. The beauty of the double helix lies in its eminent suitability as a hereditary perpetuator. If the two strands are separated, each can act as a template for the construction of a new strand, which, because of the rigid control imposed by the base pairing, cannot help being perfectly complementary to the original strand. Because each of the original strands can function in this way, the result of such division is the production of two new double helices, each of which is a perfect copy of the original. This process of reproduction is called as replication. The genetic message was clearly encoded in the arrangement of base pairs, and this must vary from one species to another. The information stored in the DNA specifies the type of amino acids occuring in proteins, and the sequences in their primary structures. This is the process of producing the prescribed proteins i.e. transcription. The order of DNA instructions is identical to the order of amino-acid side groups on the corresponding protein. The four bases, denoted by A, T, C and G, and their pairings as AT and CG are the genetic alphabet. Figure 3 illustrates the processes of replication and transcription.

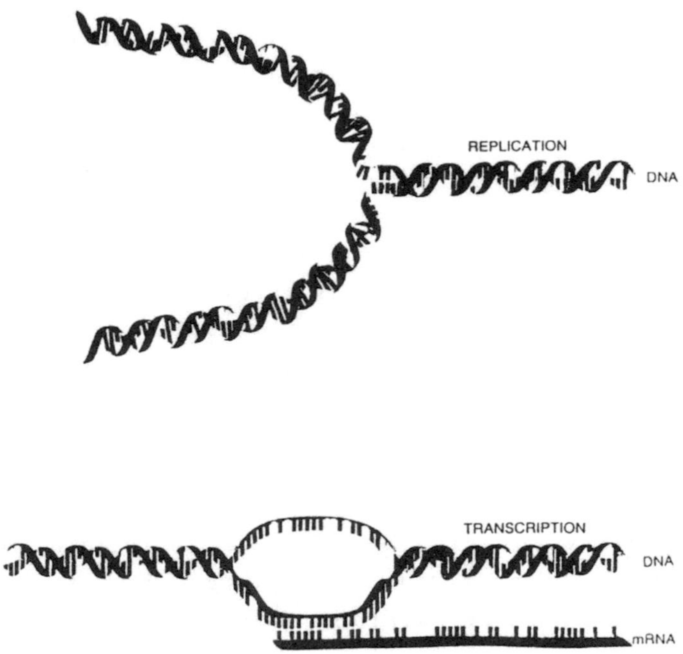

Fig. 3 Schematic description of the functions of DNA (a) replication; b) transcription.

4. Helical coil with a finite pitch

4.1 The helix geometry

Various kinematic derivations in the next sections necessitate some results about the helix geometry. The necessary tools are provided by the Frenet formulae of differential geometry. With the prime marking the derivative with respect to the arc length, $1/\rho$ and $1/\tau$ being the curvature and torsion of a general space curve; **t**, **n** and **b** indicating the unit tangent, normal and binormal vectors, the Frenet formulae read:

$$\mathbf{x}' = \mathbf{t} \qquad \mathbf{t}' = \mathbf{n}/\rho$$
$$\mathbf{n}' = -\mathbf{t}/\rho + \mathbf{b}/\tau \qquad \mathbf{b}' = \mathbf{n}/\tau \tag{4.1}$$

The next step is the derivation of the specific expressions for ρ and τ for a right handed helix. In the cartesian co-ordinates, the position vector of a helix point is defined as:

$$\mathbf{x} = (a\cos\theta,\ a\sin\theta,\ b\theta) \tag{4.2}$$

where a is the radius of the circular cylinder, $2\pi b$ the step of the helix and θ the circular angle. The arc length is related to the polar angle change $d\theta$ in the base plane as $ds = c d\theta$.

The Frenet formulae in (4.1) yield the natural basis vectors as:

$$\mathbf{t} = -\alpha\sin\theta\mathbf{i} + \alpha\cos\theta\mathbf{j} + \beta\mathbf{k}$$
$$\mathbf{n} = -\cos\theta\mathbf{i} - \sin\theta\mathbf{j} \tag{4.3}$$
$$\mathbf{b} = \beta\sin\theta\mathbf{i} - \beta\cos\theta\mathbf{j} + \alpha\mathbf{k}$$

where α and β are related to the helix parameters and the pitch angle as:

$$\alpha=\cos\gamma=a/c \qquad \beta=\sin\gamma=b/c \qquad \tan\gamma = b/a \qquad (4.4)$$

The relation between the moving trihedre of natural basis vectors and the fixed cartesian basis vectors is:

$$\mathbf{i} = -\alpha\sin\theta\,\mathbf{t} - \cos\theta\,\mathbf{n} + \beta\sin\theta\,\mathbf{b} \qquad (4.5)$$

$$\mathbf{j} = \alpha\cos\theta\,\mathbf{t} - \sin\theta\,\mathbf{n} - \beta\cos\theta\,\mathbf{b}$$

$$\mathbf{k} = \beta\,\mathbf{t} + \alpha\,\mathbf{b}$$

Finally, the curvature and torsion are obtained as:

$$\rho^{-1} = a/c^2 \qquad \tau^{-1} = b/c^2 \qquad c^2=a^2+b^2 \qquad (4.6)$$

The helix, as seen above, has both its curvature and torsion as constants throughout the space. In fact, the helix is the unique curve to have both its curvature and torsion as constants. Hence, with ρ and τ being constants for a helix, it follows that:

$$\mathbf{x}'=\mathbf{t} \qquad \mathbf{x}''=\mathbf{n}/\rho \qquad \mathbf{x}'''=-\mathbf{t}/\rho^2+\mathbf{b}/\rho\tau \qquad (4.7)$$

These geometric results will be used in the next sections.

4.2 Deformation of a helix: discrete analysis

The method for deriving the relevant equations starts with a set of discrete points connected together with interaction forces. The continuum equations are derived via the method of affine deformations or equivalently, the method of long waves [Born, Hwang (54), Askar

(88), Thacher, Rabitz (88), Thacher, Rabitz, Askar (90), Askar, Maugin (91), Thacher, Ganesan, Askar, Rabitz (86)]. Towards this goal, consider a hierarchy of interactions between the material points along the helix. first, the intramalecular interactions finding together the atoms formeng the amino-acid resiues in protein and sugar, phosphate and base molecular groups in DNA are by for the strongest. Consequently, for a model aiming the study of global interactions the molecular groups can be taken as rigid masses in the first weaker intermolecular interactions are operatue for the global modes of increst in this formulation. The first three levels of intermolecular interactions are i) the stretch interaction due to the change of the equilibrium distance between two consecutive molecular groups; ii) the bending interaction which involves the change in the angle between three consecutive points; and iii) the torsional interaction which involves the changes in the dihedral angle between four consecutive points along the helical backbone [Schlick (82)]. The dihedral angle is defined as the angle between the two planes formed respectively by the first, second, third and second, third, fourth particles in a succession. The dihedral or equivalently the torsion angle can equivalently be defined as the angle between the normals of these two planes as shown in Figure 4.

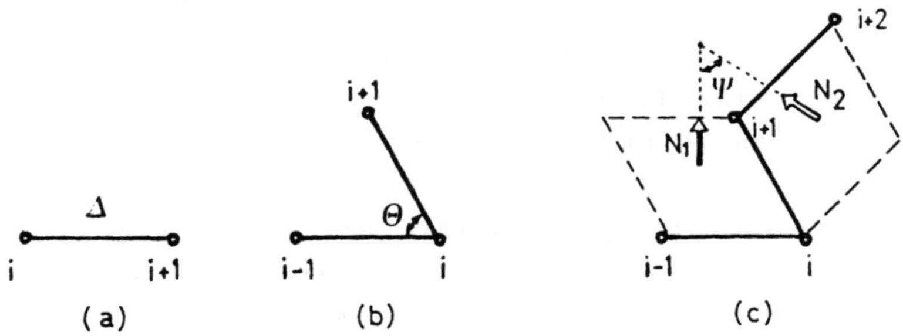

Fig. 4 The stretch, the bening and dihedral angle.

The molecular groups, represented by the material points here, are much more tightly bound against the stretching of the intermolecular distances than against the two angular changes; it is a reasonable assumption to take the intermolecular distances as unchanged during the deformations of the helix. As is common to many polymer models. The bond length, labelled as Δ is taken to be the same throughout the chain. The actual positions of the molecules are denoted by the lower case letters as $\mathbf{x}_k$ while their equilibrium positions by the capitalized letters as $\mathbf{X}_k$. The bending angles and the dihedral angles at equilibrium and after the deformation are likewise denoted respectively as B , Ψ and β, ψ. In terms of the particle positions, bond length and the kinematics variables are defined as:

$$\Delta = |\mathbf{x}_{k+1} - \mathbf{x}_k|$$

$$\cos\beta = (\mathbf{x}_{k+1} - \mathbf{x}_k).(\mathbf{x}_k - \mathbf{x}_{k-1})/\Delta^2 \qquad (4.8)$$

$$\cos\psi = \mathbf{n}_1 . \mathbf{n}_2$$

where

$$\mathbf{n}_1 = [(\mathbf{x}_k - \mathbf{x}_{k-1}) \mathrm{x} (\mathbf{x}_{k+1} - \mathbf{x}_k)]/|(\mathbf{x}_k - \mathbf{x}_{k-1}) \mathrm{x} (\mathbf{x}_{k+1} - \mathbf{x}_k)|$$
$$(4.9)$$
$$\mathbf{n}_2 = [(\mathbf{x}_{k+1} - \mathbf{x}_k) \mathrm{x} (\mathbf{x}_{k+2} - \mathbf{x}_{k+1})]/|(\mathbf{x}_{k+1} - \mathbf{x}_k) \mathrm{x} (\mathbf{x}_{k+2} - \mathbf{x}_{k+1})|$$

Similar expressions at equilibrium are given for the capitalized quantities in terms of the $\mathbf{X}$'s. Above, in evaluating the dihedral angles, first the unit normals of the two consecutive planes are evaluated through cross products and subsequently the cosine of the angles between these two normal vectors is evaluated by their inner product. The helix energy is evaluated in terms of the above quantities by the addition of the bending and torsional contributions as [Schlick (82)]:

$$V=1/2[K_s(\Delta s-\Delta S)^2+K_b(\beta-B)^2]+K_t[1-\mathrm{Cos}3(\psi-\Psi)] \quad (4.10)$$

The values of the bending and torsion interaction coefficients are known with a reasonable accuracy for all of the possible bonds found in polymers.

4.3 Deformation of a helix: continuum analysis

The continuum representation is obtained from the expressions in Eq.(4.8) and (4.9) by following the standard procedure of the affine transformations [Born, Hwang (54), Askar (88)]. The procedure essentially defines a field from the discrete variables and considers the Taylor series expansions between neighbouring lattice points. The justification for the procedure comes from the consideration of slow variations in the field values . A detail in the calculations concerns the expansions that have to be carried out at the centers of symmetry in each group of particles. The same expressions with the capital letters indicate the corresponding curvature and the torsion in the undeformed initial reference state. Considering that each step of the helix has n lattice points per cycle, in view of (4.3), the arc length between two consecutive chain points is measured as:

$$\Delta S = 2\pi C/n \quad (4.11)$$

In view of the above preparation, the secant and the arc lengths between two consecutive chain points, respectively Δ and Δs, are related in the leading order by the Taylor series expansion in (4.8) as:

$$\Delta^2=(\mathbf{x}'.\mathbf{x}'+(1/12)\mathbf{x}'.\mathbf{x}'''\Delta s^2)\Delta s^2 \quad (4.12)$$

Similarly, the bending angle expression in (4.8) yields:

$$\cos\beta = [\mathbf{x}'.\mathbf{x}' + ((1/3)\mathbf{x}'.\mathbf{x}''' - (1/4)\mathbf{x}''.\mathbf{x}'')\Delta s^2](\Delta s/\Delta)^2 \quad (4.13)$$

Above, the use of the relation in between the secant and arc lengths derived in (4.8) yields:

$$\cos\ \beta = \mathbf{x}'.\mathbf{x}' + (1/4)(\mathbf{x}'.\mathbf{x}''' - \mathbf{x}''.\mathbf{x}'')\Delta s^2 \quad (4.14)$$

Finally, the normals $\mathbf{n}_1$ and $\mathbf{n}_2$ for the torsion angle expression in (4.9) reduce to the simple formula below:

$$\mathbf{n}_1 = [\mathbf{x}'\mathrm{x}\mathbf{x}'' - \mathbf{x}'\mathrm{x}\mathbf{x}'''\Delta s/2]/|\mathbf{x}'\mathrm{x}\mathbf{x}'' - \mathbf{x}'\mathrm{x}\mathbf{x}'''\Delta s/2|$$

$$(4.15)$$

$$\mathbf{n}_2 = [\mathbf{x}'\mathrm{x}\mathbf{x}'' + \mathbf{x}'\mathrm{x}\mathbf{x}'''\Delta s/2]/|\mathbf{x}'\mathrm{x}\mathbf{x}'' + \mathbf{x}'\mathrm{x}\mathbf{x}'''\Delta s/2|$$

The substitution of the various derivatives in (4.12), (4.15) and (4.16) in terms of the Fresnet formulae in (4.8) yields:

$$\cos\beta = 1 - (1/2)(\Delta/\rho)^2 \qquad \cos\psi = 1 - (1/2)(\Delta/\tau)^2 \quad (4.16)$$

Hence for the limit of small arc lengths inherent in the continuum formulation, one finds the elegant results:

$$\beta = \Delta/\rho \qquad \psi = \Delta/\tau \quad (4.17)$$

For the helix in its undeformed configuration, similar to the above, the corresponding angles are:

$$B = \Delta/P \qquad \Psi = \Delta/T \quad (4.18)$$

The helix deformation energy in (4.10) is now written by substituting the expressions for β, ψ and their analogs in the undeformed state above followed by replacing the summation over the lattice points by an integral over the continuum field. The result is obtained as:

$$V = (1/2) \int [D_b(\rho^{-1} - P^{-1})^2 + D_t(\tau^{-1} - T^{-1})^2] ds \qquad (4.19)$$

where

$$D_b = \Delta K_b \qquad D_t = \Delta K_t \qquad (4.20)$$

D_b and D_t are the usual stiffness of the bar section for bending and torsion

5. Energy expressions for biopolymers

In the preceding sections, the deformation energy for a helix is derived. Using this basic contribution, the two specific cases of importance, proteins and DNA, are studied.

5.1 Proteins

For a simplified model, by lumping the amino acid radicals into mass points, the total energy is obtained by adding the contributions of the hydrogen bond between the helix points that are a step apart, to the helix energy of the preceding sections. It should be noted that the hydrogen bonds are not parallel to the axis of the helix. This configuration leads to a stiffer helix in its torsion. For n denoting the index difference between two connected points of the helix we have:

$$V_{Hyd} = \sum \nu(|\mathbf{x}_{k+n} - \mathbf{x}_k|) \qquad (5.1)$$

The form of the pair intersection energy ν expression is left unspecified. In many applications a Morse potential as well as its quartic approximation provide an adequate description [Schlick (82)]. Clearly, for a linear modal analysis, the quadratic approximation of the hydrogen bond energy will suffice. For the kinetic energy, when the molecular groups are represented by mass identical points having no rototary inertia, we have:

$$K.E. = \sum m_k \mathbf{x}_k \cdot \mathbf{x}_k \tag{5.2}$$

The dynamical equations are derived from the Langrangian

$$L = K.E - (V_{hel} + V_{Hyd}) \tag{5.3}$$

Here, the fact that the hydrogen bonds are not along the helix axis has to be accounted for as this introduces a coupling between the axial streching and the torsinal modes of the helix. Finally, the field equations are obtained by the well-established procedures of calculus of variatious.

An improvement on the energy expression above can be obtained by adding the contribution of the amino acid radicals that have been lumped into a mass point along the helix. In particular, their rotations as a rigid whole happen at slow enough rate that this would lead to an energy exchange with the global modes.

5.2 DNA

For this polymer, there exist two helices of the type described aboveform the backbone. The simple model is obtained by bumping the sugar, phaspate and base groups into rigid mass points.Denoting by **x** and **y** respectively, the positions on the two helices, the helix energy contributions are obtained from the addition of the energies evaluated from (4.20) once with the **x**'s and once with the **y**'s. With the **x**'s and the **y**'s denoting the bonded sugar corresponding particles on the two different helices, the hydrogen bond energies become

$$V_{Hyd}=\sum v(|x_k-y_k|) \qquad (5.4)$$

The form of v is left unspecified. It may be taken as a Morse potential, a quartic or a quadratic function. The kinetic energies associated with the two helices are obtained as in (5.2) using the **x**'s as well as the **y**'s simultaneously.

For the DNA, the rotations of the base pairs in subsequent stacks along the helices involve an energy commensurate with those entering into the helix deformations and the hydrogen bond stretchings between these. This particular energy is expressed as (Schlick 87].

$$V_{Base}=\sum(1/2)R_1(\phi_j-\phi_{j-1})^2+R_2(1-\cos\phi_j) \qquad (5.5)$$

Above ϕ_j enotes the rotation of the base pairs in the jth step of the helix with respect to its equilibrium configuration about the helix axis. Above, the first term is the interaction between the base pairs for the stacking along the helix and the second term the interaction of a base pair taken as a dipale with the dipolar electriz average cal field. A suitable approximation is to suppress the independent rotations of the base pairs from the helices and evaluate ϕ_j from the positions of the helix points they join as

$$\phi_j=|[(\mathbf{x}_j-\mathbf{y}_j)-(\mathbf{X}_j-\mathbf{Y}_j)]\times\mathbf{k}|/|(\mathbf{X}_j-\mathbf{Y}_j)| \qquad (5.6)$$

Above **k** denotes the unit vector along the helix axis. With the rotation of the base pairs taken into account, it also becomes necessary to account for their librational energy L.E. For I denoting the rotatory inertia of the base pairs about the helix axis, we have

$$L.E.=(1/2)\sum I_j\dot{\phi}_j{}^2 \qquad (5.7)$$

In this work, the rotations as propeller-twist of the base pairs are neglected. The appropriate Langrangian and the accompanying dynamical equations are obtained straight forwardly following the above descriptions. Similarly, the continuum representation follows in a from the above equations in terms of two displacements in connection with the two position vectors **x** and **y** and the base pair rotationsϕ.

6.Numerical studies

6.1 Finite axial deformations of a helix

A helix that has initially the radius A, the step B and comprises n cycles is considered. The initial heigth of a cycle of the helix is $H=2\pi B$ and thus its total height is nH. The continuum theory derived in the preceeding section assumes that the arc length remains unchanged. Therefore, the length of a cycle of the helix is $L=2\pi C$ so that its total length is nL where $C^2=A^2+B^2$. We consider deformation modes that map the initial helix onto a helix of radius a and step b under external axial forces. In addition, the deformation is considered to be uniform throughout the helix. As a result, the integral in (4.19) may be evaluat to obtain [Askar, Maugin (91), Thacher, Ganesan, Askar, Rabitz (86), Askar, Akoz(94)]:

$$V = (nL/C^2)[D_b(\rho^{-1}-P^{-1})^2+D_t(\tau^{-1}-T^{-1})^2] \qquad (6.1)$$

The above internal deformation energy is balanced by the work done by the external forces and torques and together they determine the deformed shape of the helix.

For the helix having initially the height per cycle, a convenient kinematics variable is its height h after the deformation. Similarly a full cycle of 2π radians becomes $2\pi(1+\ \phi)$ after a uniform twist by an angle of ϕ per each radian of arc along the helix. Two canonical deformation modes are pure length change (stretch or compression) and pure twist (in the winding or unwinding directions). The pure length change is defined in terms of h with $\phi = 0$ and the pure twist is defined in terms of ϕ with h = H. In the former case, an external torque is needed to keep the helix from rotating and in the second case an external axial load is needed to keep the height of the helix unchanged. Hence these two modes of deformation are defined through the conditions :

$$\begin{aligned} &\text{Mode 1; No Twist:} \quad &\phi = 0 \\ &\text{Mode 2; No Lorque:} \quad &T = 0 \end{aligned} \tag{6.2}$$

The axial extension force F and torque T are derived from V in (6.1) as:

$$F = (\ V/\ h)/n \qquad T = (\ V/\ \phi\ )/2\pi n \tag{6.3}$$

The number of cycles n in the derivatives comes in since h and $2\pi\ \phi$ are the height and twist per cycle of the helix, and the derivatives have to be evaluated with respect to the total height and twist.

In studying the deformations defined in (6.2), it should be kept in mind that the helix height is bounded by the (unchanged) length of the helix. This places the restriction $0 \le h \le L$ for h. In principle the amount of twist is unbounded, while physically meaningful ranges are of course finite.

The calculation procedure is simple: it is based on determining the mapping of the initial helix with the parameters A, B and C to the deformed helix with the parameters a, b and c through the deformation variables h and ϕ. An immediate result is provided by the inextensibility of the arc length. The arc length of a full cycle at the initial configuration is $2\pi C$. The same arc length in the deformed helix is tended by the angle $2\pi(1+\phi)$ and hence is $2\pi(1+\phi)c$. The equality of these two arc lengths yields:

$$c = C/(1+\phi) \tag{6.4}$$

Similarly, the heigth h of a loop of the helix after the deformation is related to the kinematic variables as:

$$h = 2\pi(1+\phi)b \tag{6.5}$$

For the pure stretch mode, the results are obtained by setting $\phi = 0$ in the relevant expressions. For the second mode i.e. pure twist, h and ϕ are related by requiring $T=0$. In this case $(1+\phi)$ may be solved explicitly in terms of h from $T=0$.

The results thus obtained for the the loading defined in the case 1 are displayed in Figures 5. The figure 5a compares the experimental and theoretical results according to the present formulation. In the experiments a comercially available toy, a "Slinky" is used. The radius of the helix is 3.4 cm and the rigidities for torsion and bending are obtained from a linear experimental calibration to be respectively equal to $D_t = 13.61\ kg.cm^2$ and $D_b = 203.09\ kg.cm^2$. The theoretical results could be plotted in dimensionless units for more general use. However the dimensional plots are preferred for direct comparison with experimental measurements. The Figure 5b presents the

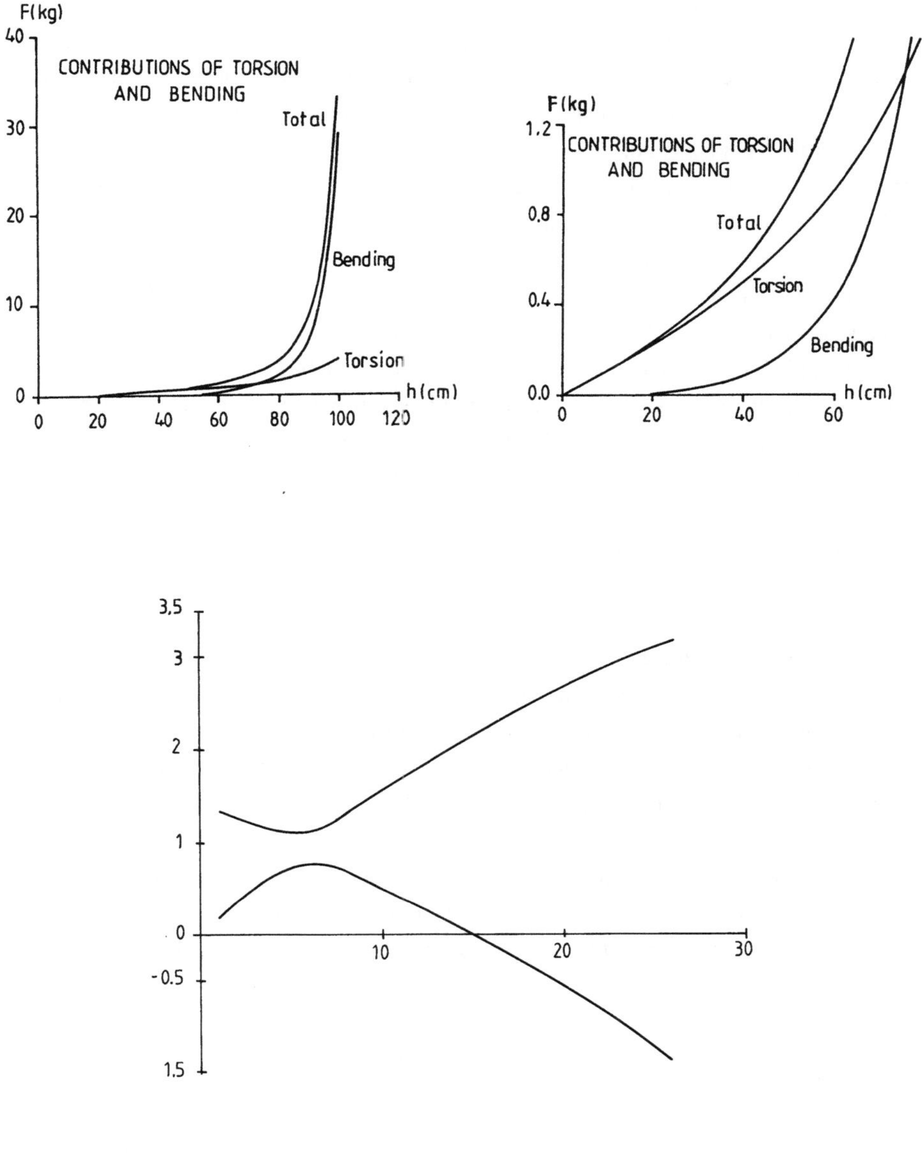

Fig. 5 Axial stretch of a helix. (a) comparison of the experimental and theoretical results; b) respective contributions of torsion and bending to the force-stretch relation; c) the eigenvalues of a helix versus the stretch.

respective contributions of the bending and the torsion type of deformations.Finally, Figure 5c shows a representive result from a linear the amounts of guanine and cytosine were also the same. This lead to the conclusion that A, T and C, G had to come in pairs. Moreover, a coplanar hydrogen-bonded arrangement of adenine and thymine has exactly the same stability analysis is based on linearizing the internal energy for infinitesimal stretches and twist about a stretched configuration. The two eigenvalues of the resulting quadratic provide the criteria for stability: any one being negative shows instability, whereas both have to be positive for stability. The parameters are for a single helix of an inital pitch of 45°.

The following conclusions follow from the observations in the figures.

1. For even moderately large deformations significant nonlinearities exist in the force-displacement relations. The elementary strength of materials theory of springs corresponds only to the initial slope of the non-linear force-displacement plots.

2. For small stretches, the mechanics is governed by torsion while bending is predominant for large displacements and the torsional contribution becomes minimal.

3. An axially strech helix may become unstable for twists about a whole range of stretching.

It is hoped that the derivations and results here should prove useful for a better understanding of the helix mechanics and provide a means to study the dynamics and the stability of the helical structures of biopolymers both in micro and macro scales.

6.2 Dynamical studies: discrete analysis

In order to give an idea on the extreme deformations that a helical polymer undergoes, the results for a simple helix are shown in Figure 6.

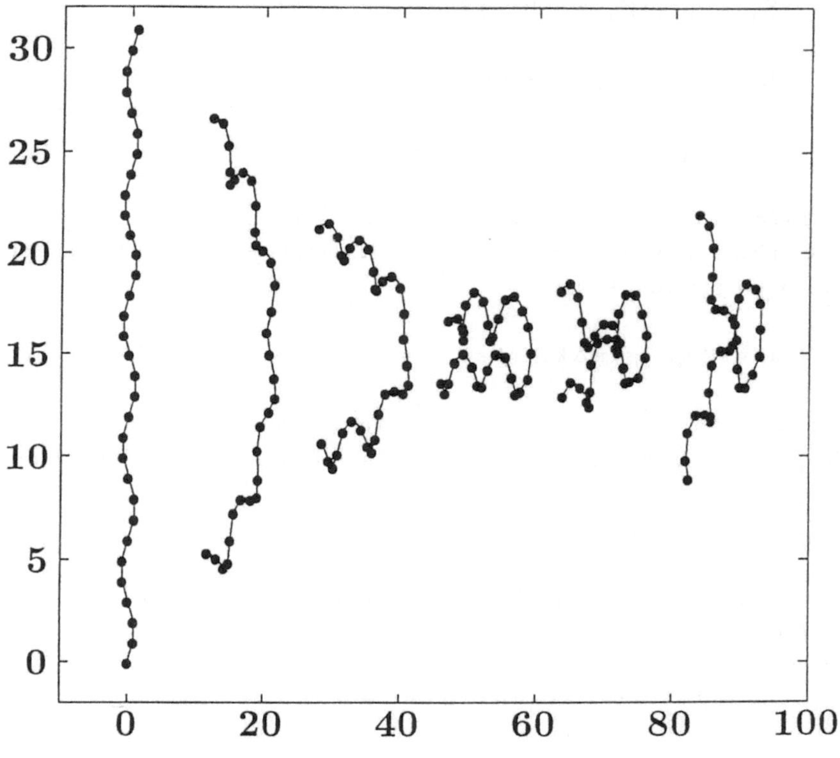

Fig. 6 Dynamics of a single helix with 32 discrete particles

Such a study contains the Elements of observing the deformations of an alpha helix leading to the beta sheets and constitutes an elementary step in the formidable protein folding problem. The potential energy is taken to be the sum of the stretch, bending and torsional contributions as:

$$V=\Sigma S_j(b_j)+\Sigma B_j(\beta_j)+\Sigma T_j(\psi_j) \qquad (6.6)$$

with

$$S_i = (1/2)K_j(b_j - B^o{}_j)^2$$

$$B_j = (1/2)K_{Bj}(\beta_j - \beta^o{}_j)^2 \qquad (6.7)$$

$$T_j = K_{Tj}[1-\cos 3(\psi - \psi^o{}_j)]$$

In our calculations, the masses in the polymer are taken to be all Carbon atoms and the initial equilibrium configuration is a helix defined as

$$\mathbf{x}_k=(a\cos\Delta k,\ a\sin\Delta k,\ h\Delta k) \quad k=1,\ 2,\ \ldots,32 \qquad (6.8)$$

For a=h and $\Delta = 60^o$, the equilibrium values for the bond length and the bending and torsion angles are found to be:

$$b^o{}_j = 2a \qquad \beta^o{}_j = 41^o \qquad \psi^o{}_j = 44^o \qquad (6.9)$$

The numerical values of the parameters used in the calculations are:

$$a = h = 1\text{Å} \qquad m=20*10^{-14}\ \text{g}$$

$$K_i = 500\ \text{kcal/molÅ}^2 = 35.0*10^{-12}\ \text{erg/cm}^2$$

$$K_{Bi} = 50\ \text{kcal/mol} = 3.50*10^{-12}\ \text{erg} \qquad (6.10)$$

$$K_{Ti} = 0.5\ \text{kcal/mol} = 3.50*10^{-14}\ \text{erg}$$

Details on these calculations as well as more results can be found in the work by [Askar, Space, Rabitz].

6.3 Dynamical studies: continuum analysis

In many organisms the DNA is found in a circular form. Cyclization or the formation of a closed ring from a helix is an important configuration. Freeing the two ends of the closed ring and resealing them after introducing an winding induces an internal stress. The inital internal enrcgy in the coil causes it to undergo important dynamics. The DNA relaxes to a new configuration by twisting its axis. This behaviour is termed as "super coiling". In traditional continuum mechanics, the super coiling is a type of dislocation. The time evolution of the dynamics of a super coiled DNA is shown in Figure 7. The calculations for obtaining this set of figures are based on a continuum representation of the internal energy as in Eq.(4.20).

6.4 Non-linear waves: solitons

Various modes of deformations are of interest for both the single and the double helices. The complications ofthe real system has lead to various models trying to capture a particular feature [Scott (85), Lomdahl, Layne, Bigio (85), Krumhansl, Alexander (83), Davydov (85), Remoissenet (90)].

(i) ϕ^4 solitons in proteins [Weiner, Askar (70), Antonchenko, Davidov, Zolataryuk (88), Peyrard, Pnevmatikos, Flytzanis (87)]. Consider a simplified one-dimensional chain cutting along an alpha-helix by following the hydrogen bonds. The hydrogen is subjected to a double well potential, called as the substrate potential which may be represented by a quartic $V_{sub} = \gamma(u_j^2 - a^2)^2$. Here u_k denotes the position of the jth hydrogen atom with respect to the mid-point between two consecutive amino-acid radicals which are in this model lumped to a point and a the distance of the two

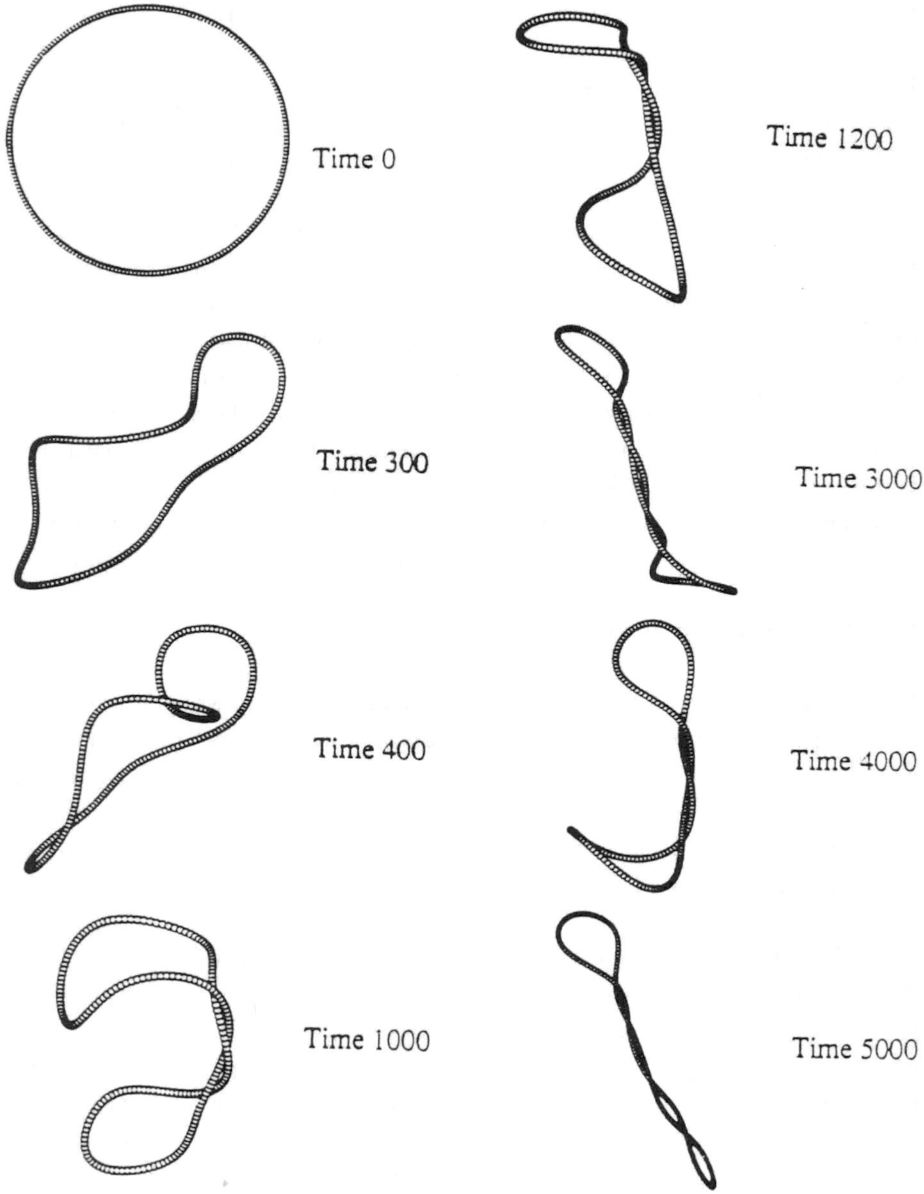

Fig. 7 The dynamics of a supercoiled DNA, t indicates the various instances of time in nondimensional units

minima on either side of the midpoint. In addition to thesubstrate potential which is the action of the amino acids on the hydrogen, these latter interact with each other. The amino acid radicals are denoted as A and the hydrogens as H in the model chain depicted in Figure 8.

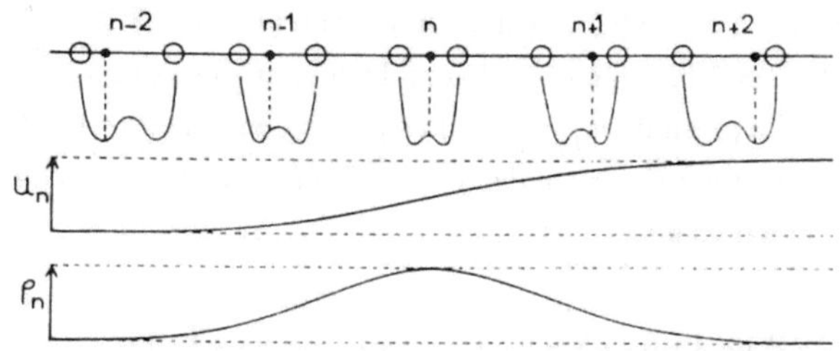

Fig. 8 Model for a hydrogen bonded chain and the generation of a soliton.

The corresponding potential may be taken as a quadratic in the simplest approximation. Hence, the total potential may be taken as:

$$V=\Sigma(1/2)K(u_{j+1}-u_j)^2+\gamma(u_j{}^2-a^2)^2 \qquad (6.11)$$

The kinetic energy is simply $K.E.=\Sigma(1/2)mu_j{}^2$. The resulting dynamical equations in the continuum limit, when nondimensionalized, read:

$$u_{tt}=u_{xx}+u(1-u^2) \qquad (6.12)$$

This equation is called as the ϕ^4 wave equation and possesses the soliton solution as a kink in the form $u = A\tanh(x-vt)$. Further refinements on the model include the motion of the A particles as well as two dimensional considerations.

(ii) DNA denaturation. In a DNA chain consider the stretching of the hydrogen bonds connecting the two strands together. When the relative distance between the two strands, i. e. the hydrogen bond length grows unboundedly, the DNA is said to be denaturated [Peyrard, Bishop (89), Peyrard, Dauvxois, Bishop, Peyrard, Bishop (89), Takeno, Homma (87), Yomosa (84), Zhang (87), Moto, Lamdahl, Christiansen (90)]. Denoting by u_j and v_j the displacements of the two base points on the two strands, the interaction potential may be represented as:

$$V = \Sigma (1/2) K[(u_{j+1} - u_j)^2 + (v_{j+1} - v_j)^2]) +$$

$$D\{\exp[-a(u_j - v_j)] - 1\}^2 \qquad (6.13)$$

Above, the first term is the coupling due to stacking of the neighbouring bases along the backbone and the second term is the hydrogen bonding interaction. Figure 9 shows the simplified model.

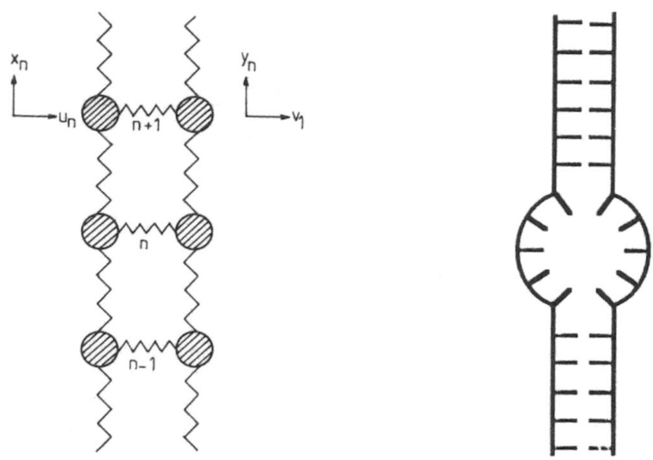

Fig. 9 Model for the DNA double helix for the formation of a bubble opening as a representation of denaturation.

With the canonical coordinates

$$w = 2(u_j - v_j) \qquad\qquad z = 2(u_j + v_j) \tag{6.14}$$

the potential and kinetic energies the two degrees of freedom are decoupled:

$$V = \Sigma\ (1/2)K(z_{j+1} - z_j)^2 +$$

$$\Sigma(1/2)K(w_{j+1} - w_j)^2 + D\{\exp(-a2w_j) - 1\}^2 \tag{6.15}$$

$$KE = \Sigma(1/2)m(z_j{}^2 + w_j{}^2)$$

The equation for the relative displacements of the two strands in the continuum limit and in a nondimensional variable is obtained as:

$$w_{tt} = w_{xx} - Ae^{-w}(e^{-w} - 1) \tag{6.16}$$

A simplified version of the above equation due to the expansion of the exponential term is also of interest:

$$w_{tt} = w_{xx} - A(w - w^2 + w^3/2) \tag{6.17}$$

The above equations possess soliton type of solutions.

(iii) Rotor model for DNA bases. The base pairs along the helical backbone of the DNA carry a nonnegligible electrostatic dipole. These electrostatic dipoles are subject to dipolar interactions with their neighbors as well as are subjected to a substrate potential field. The resulting interaction potential was given inEq.(5.5). The kinetic energy consisting of the rotatory inertia was likewise given in Eq.(5.6). Figure 10 shows the model schematically.

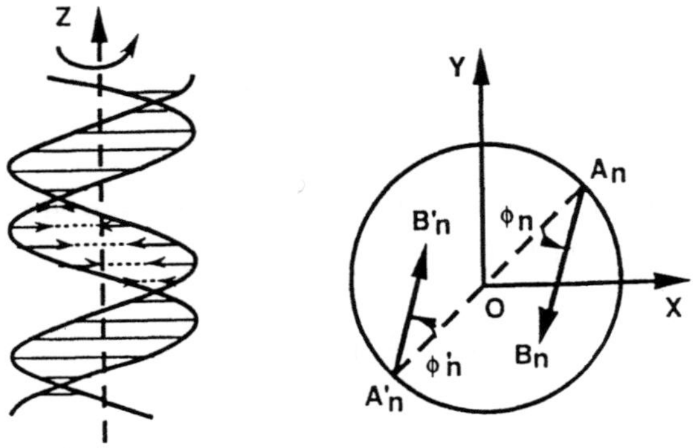

Fig. 10 Model for the rotatory motion of the base pairs

The corresponding equation in the continuum limit is that of Sine-Gordon's which, in nondimensional form, reads: [Yomosa (83), Yomosa (84)].

$$\phi_{tt} = \phi_{xx} + \sin\phi \tag{6.18}$$

This latter is known to possess soliton type of solutions.

7. Concluding remarks

As described above, a single position variable or equivalently a single displacement vector defines the dynmics of the alpha-helix. This model is complemented in the leading order by adding the hydrogen bonds between amino-acids that are taken as mass points. A more refined theory may be obtained by introducing the rotations of the amino-acid molecular groups. For the DNA, the derivations may likewise be taken with various levels of sophistication. In the basic approximation, the sugar, phosphate and base groups are lumped into mass points along the two helices that are joined with stretchable hyrogen bonds.In the next level, the rotations of the base pairs may be taken as independent dynamical variables or be related to the displacements of the surrounding sugar groups they connect. Similarly, the dynamics of the two helices may be introduced through the symmetric and the antisymmetric variables **x + y** and **x – y**. This particular decomposition permits an independent study of the phenomena involving or excluding the hydrogen bond streching . A further level of sophistication can be achieved by considering the rotations of the sugar and phosphate groups along with the propeller twist motion of the base pairs in the DNA. Finally, with the rich dynamics involving, simpler uni-modal studies may be of value by separating the axial extensional, rotational and transversal flexural modes as well as concentrating on the non-linear dynamics of the hydrogen bonds. Similarly, stability analyses that involve a delicate interplay of the bending and torsional modes are expected to provide useful insight into the folding problems. Studies conducted in the discrete as well as continuum forms are both of interest. The continuum formulation obtained in this presentation relates the material coefficients to the basic parameters in the interaction potentials at the discete level.

REFERENCES:

Askar A., "Lattice Dynamical Foundations of Continuum Theories of Solids: Elasticity, Piezoelectricity, Viscoelasticity, Plasticity", World Scientific, Singapore (1988)

Askar A., Dynamics of Single alpha and Double DNA helices, in "Continuum Models and Discrete Systems" Vol II, ed, G. A. Maugin, Longman Scientific and Technical and John Wiley & Sons, NY (1991)

Askar A., Akoz Y., Bull. Istanbul Technical University, to appear
Askar A., Space B., Rabitz H. Long time Molecular Dynamics; Pico second time steps and the subspace method (to be published)
Born M., Hwang K., " Dynamical Theories of Crystal Lathces", Oxford University Pres (1954)
Davydov A. S., (1985). "Solitons in Molecular Systems", Reidel, Dordrecht
Dickerson R. E., The DNA helix and how it is read, Sci. Amer. 249, 94 (1983)
Freifelder D., "Molecular Biology", Jones and Barlett, Boston 1987
Karplus J. A. M. and Mc Cammon "The Dynamics of Proteins", Sci. Amer. 254, 42 (1986)
Krumhansl J. A., and Alexander D. M., (1983) Nonlinear dynamics and conformal excitations in biomolecular materials. In "Structure and Dynamics :Nuclear acids and proteins". ed. Clemente E. and
Sarma R. H.. Adenine Press, New York
Lomdahl P.S., Layne S. P., and Bigio I. J., Solitons in biology (1985). Los Alamos Sciences, Spring Issue, 4-21
Mc Cammon J. A., and Harwey S. C., "Dynamics of Proteins and Nucleic Acids", Cambridge University Press, Cambridge 1987
Muto V., Lamdahl P. S., Christiansen P. L., Phys Rew A, 42, 7452 (1990)
Peyrard M., Phevmatikes S. T., Flytzanis N., Phys-Rev. A 36, 903 (1987)
Peyrard M., and Bishop A. R., (1989) Statistical mechanics of a nonlinear model for DNA denaturation. Preprint
Peyrard M., Dauvxois T., Bishop A. R., Dynamics of thermal denaturation of DNA
Peyrard M., and Bishop A. R., in Proceedings of the 6th Interdisiplinary Workshop on Nonlinear Coherent Structures in Physics, Mechanics and Biological Systems (1989), Bartes M., J. Leon eds, Springer (1990)
Remoissenet M., "Nonlinear Evolution Equations, Quasi-Solitons and Their Experimental Manifestation", Proceedings of the Winter School in Theoretical Physics on Partially integrable nonlinear equations and their in physical applications, March 21-29,
Les Houches, France. Kluwer Academic Publishers, Dordrecht, The Netherlands

Saenger W., "Principles of Nucleic Acid Structure", Springer Verlag, NY (1984)
Schlick T., Olsan W. K., J. Mol. Biol, (1987)
Scott A. C., (1985a). Solitons in biological molecules. Comments Moll. Cell. Biophys., 3, 15-37
Stryer L., Biochemistry", Freeman W. F. and Co., N.Y. (1981)
Schrega H. A., Ad. Phys Org. Chem 6, 103 (1984)
Schlick T., Modelling and Minimization for Predicting Three-Dimensional Structures of Large Biological Molecules, PhD Thesis, Courant Institute, NewYork University, NY (1982)
Takeno S., and Homma S., Prog. Theor. Phys. 77, 548 (1987)
Thacher T., Rabitz H., Biophys J., Soc, 54, G95 (1988)
Thacher T., Rabitz H., Askar A., J. Chem Phys. 93, 4673 (1990)
Thacher T., Ganesan S., Askar A., Rabitz H. R., J. Chem Phys, 85, 3655 (1986)
Van Gunsteren W. F. and Karplus M., Macromolecules 1528, 15 (1982)
Weiner J. H. and Askar A., Nature, 226, 842 (1970)
Ya Antonchende V., Davidov A. S., zolotar Yuk A. V., Phys. Stat, Solidi b 146, 487 (1988)
Yomosa S., Phys. Rev. A27, 2120 (1983) and A30, 474 (1984)
Yomosa S., Soliton excitations in deoxyribonucleic acid (DNA) double helices. Phys. Rev. A- 27, 2120-2125, (1983)
Yomosa S., (1984). Solitary excitations in deoxyribonucleic acid (DNA) double helices. Phys. Rev. A, 30,474-480
Zhang Chun-Ting , Phys. Rev. A35, 886 (1987)

Acknowledgement

The author acknowledges support from the US Air Force Office of Scientific Research and the Office of Naval Research and the Research Fund at Bogazici University.

CYTOKINESIS: THE MECHANISM OF FORMATION OF THE CONTRACTILE RING IN ANIMAL CELL DIVISION

Albert K. Harris
Department of Biology
Coker Hall CB#3280
University of North Carolina
Chapel Hill, North Carolina 27599-3280 USA

Electronic Mail: akharris@med.unc.edu

INTRODUCTION

When an animal cell divides itself in two, it accomplishes this by the active contraction of a belt-like ring of cytoplasmic actin and myosin, usually encircling its equator. That much has been established with certainty (Schroeder, 1973; Mabuchi, 1986). On the other hand, no one yet understands the mechanisms responsible for creating these contractile rings and positioning them in the correct locations. The purpose of this paper is to review the different possible mechanisms for contractile ring formation, together with the available evidence supporting and opposing each of the alternative theories. Four main questions arise about this phenomenon. Three of these questions concern the signals that are believed to control the locations where the rings will form: first, where do these signals come from; second, what is the chemical or physical nature of the signals; and, third, what local effects do the signals produce at the cell surface? In particular, is the proximate effect of these signals to weaken cortical contractility, or to strengthen it? The fourth question is whether the actin and myosin are polymerized directly at the site of ring formation, as opposed to being collected together from pre-existing fibers that had been parts of other regions of the cell cortex. As we will see, these

NATO ASI Series, Vol. H 84
Biomechanics of Active Movement
and Division of Cells
Edited by N. Akkaş
© Springer-Verlag Berlin Heidelberg 1994

questions are interrelated, with the answers to any one of them having implications for each of the others.

Some background information and terminology are now needed. The site where the cell constricts itself is called the "cleavage furrow". The process of constriction is called by either of two synonymous names: either "cytokinesis" or, more colloquially, "cleavage". These should not, however, be confused with mitosis, which consists of the partitioning of the chromosomes into the two daughter nuclei. Thus, you could say that mitosis is nuclear division, while cytokinesis is cytoplasmic division. It is also important to realize that these two processes are distinct from one another, with each having its own physical mechanism (albeit neither of them is completely understood). For example, it is not unusual for cells to undergo mitosis without accompanying cytokinesis, thereby becoming multinucleate. Conversely, there is at least one case in which, in the total absence of mitosis, cells pinch themselves in two by active constrictions very similar to those in cytokinesis (Rappaport, 1960a).

The mitotic apparatus (MA), of which the central part is more colloquially called the "mitotic spindle" (or often just the spindle), is a complex, self-organizing array of microtubules and certain other materials, including components of the endoplasmic reticulum. The MA serves to pull and push the chromosomes around during cell division, with forces being applied to a special part of each chromosome called its "kinetochore". Forces are exerted by microtubules that attach to the kinetochores and extend toward the poles of the MA. At first, each chromosome remains mechanically attached to its paired sister chromosome (i.e. the one of which it is an exact copy). Normally, one sister kinetochore becomes attached to microtubules that run to one pole, while its sister kinetochore becomes attached to microtubules running to the opposite pole. This results in what amounts to a tug-of-war, with each conjoined pair of chromosomes being suspended midway between the two poles.

Some (not yet understood) mechanism seems to make the net strengths of these pulling forces proportional to the current distance separating each kinetochore from its pole. In other words, the further the distance between pole and kinetochore, the stronger the pulling force (Ostergren, 1950, 1961; Hays, Wise and Salmon, 1982), with the result being a negative feedback cycle which causes kinetochore pairs to gravitate to the metaphase plate. This would be a good example of an asymptotically stable structure. The most recent evidence raises other possibilities, including combinations of both pushing and pulling forces and the control of net displacement by the frequencies and durations of force exertion, rather than just the relative strengths of the forces (Skibbens, Skeen and Salmon, 1993; also see the chapter by Skibbens in this volume). Comparable analyses of force balances at near-molecular levels will probably be needed before we understand cytokinesis. In the case of contractile ring formation, we should probably be looking for comparable sorts of feedback cycles at molecular levels, except that these should presumably involve positive feedback instead of negative feedback, in that what we are trying to explain are nearly spontaneous rearrangements of cortical materials into new patterns.

The accumulation of the chromosomes at the equatorial position is called "congression", and its completion marks the beginning of the phase of mitosis known as "metaphase". A further piece of nomenclature is that the plane in which the chromosomes collect is called the "metaphase plate". During the next phase of mitosis (called "anaphase") the chromosome pairs uncouple from one another, so that the poleward pulling forces are free to move the kinetochores (and thus the chromosomes themselves) toward the poles of the MA. In addition, at least in some species, the MA itself elongates at this stage, in the sense that its two poles move further away from each other. The movement of the chromosomes toward the poles is called "anaphase I", while the elongation of the MA is called "anaphase II". This terminology will later be useful in considering specific theories and experimental evidence about the mechanisms responsible.

It was long assumed that the forces responsible for cytokinesis should also be exerted by the mitotic apparatus itself. Since the MA pulls chromosomes around, then why shouldn't it also be what is pulling the cell surface inward? It would certainly seem simpler that way. Despite their plausibility, however, such ideas were disproven conclusively by the rather drastic experiment of sucking the MA completely out of dividing sea urchin egg cells, and showing that cytokinesis still continued to completion. The first report of such an experiment was by Yatsu (1912), but it was Hiramoto's (1956) performance of this experiment that conclusively proved the point. Even though cleavage continues in the absence of the MA, however, it is equally clear that the MA serves to control the location where the contractile ring will form. Rappaport (1985) proved this by using a micromanipulator to move the MA around inside an artificially elongated echinoderm egg, showing that cleavage furrows could be initiated at one site after another, always adjacent to the most recent position of the mitotic apparatus. Apparently, some sort of signal must be emitted from one or another part of the MA, with this signal somehow causing actin and myosin to accumulate into a ring.

Does the furrow-inducing signal come from the poles of the MA, or from its metaphase plate?

Simply because the typical site of furrow formation is directly adjacent to the equatorial region of the MA spindle (i.e. the metaphase plate), one tends to assume that the furrow-inducing signals should be emitted from the metaphase plate, rather than from the poles of the MA. Lending support to this very plausible idea are a series of careful observations by Earnshaw and his colleagues showing that certain special proteins (which they call INCENPs, standing for inner centromere proteins) move from the chromosomes in the metaphase plate outward into the cortex at the site of furrow formation (Cooke, Heck and Earnshaw, 1987; Earnshaw and Cooke, 1991). These proteins' behavior can be taken to indicate some causal role in the induction of the furrow, although it is not yet known whether the contractile ring would either fail to form or to function correctly were cells to be deprived of their INCENPs.

On the other hand, there are a good many cases in which contractile rings and cleavage furrows form half-way between the poles of two different mitotic apparatuses (i.e. even when there is not a metaphase plate located there). The most direct and elegant demonstration of this was Rappaport's wonderful "doughnut experiment" (Rappaport, 1961) in which a cleaving echinoderm egg was compressed by a glass bead into a ring-like shape.

The crucial aspect of this experiment was that, at the time of the second mitosis and its cytokinesis, three furrows were induced to form, not just the two adjacent to the metaphase plates of the two mitotic apparatuses, but also a third furrow half-way between a pole of one of these and a pole of the other.

Diagram of Ray Rappaport's classic "doughnut experiment" in which an echinoderm egg is squeezed into the shape of a torus by downward pressure from the rounded tip of a microneedle. The crucial point is that at the second cleavage, 3 furrows form, one at the metaphase plate of each mitotic apparatus, and a third one (black arrows) between the poles of one mitotic apparatus and the other (without a metaphase plate).

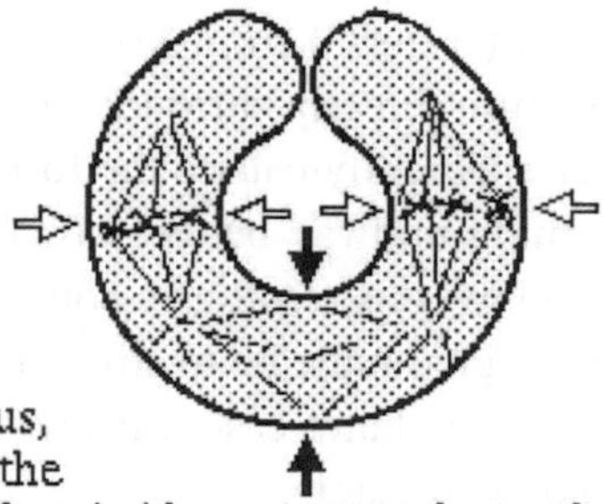

If the signals had come from the metaphase plates, or at least if they had come from them alone, then there would be no reason to expect this third cleavage furrow. It will be important to determine if echinoderm eggs have any equivalent to Earnshaw's INCENP proteins, and if so whether these proteins also accumulate in this additional (third) furrow, or only in the two furrows that are adjacent to the metaphase plates. Other examples of extra cleavage furrows forming between nearby poles of different MAs include all cases in which multinucleate cells undergo simultaneous formation of interlocking networks of cleavage furrows. This is often seen when one blocks one or more sequential divisions of, say, a sea urchin egg but then lets further cleavages proceed without disturbance. Simultaneous mass cleavage is also a normal part of development in certain species, including the coelenterate known as the "sea pansy" and the much better-known fly Drosophila (Fullilove and Jacobson, 1971). In this latter case, the cleavage furrows take the form of an interlocking hexagonal network, beneath which there is a corresponding network of myosin-staining material. (Warn, 1990).

Insect embryos undergo a simultaneous mass cleavage into several thousand cells, forming a hexagonal pattern.

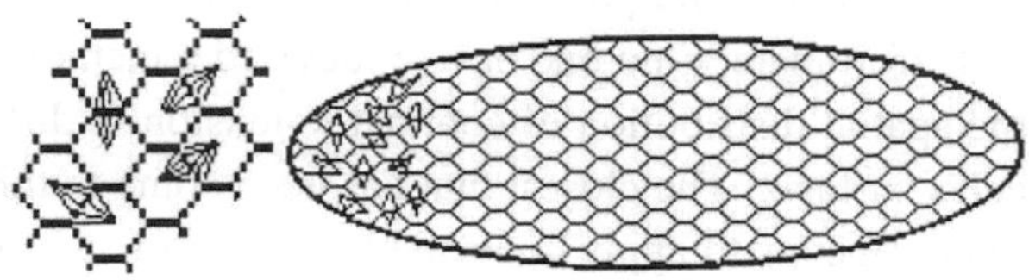

A different kind of evidence that the signals come from the poles was provided by some other experiments by Rappaport in which he separated the polar regions ("asters") from MAs and found that rudimentary furrows would sometimes form directly surrounding such an individual aster. In this case, a single aster apparently causes the formation of its own furrow, rather than a furrow forming midway between two different asters. As will be further discussed below, this result also constitutes part of the evidence that the effect of the polar signal must be to stimulate cortical contractility, rather than to weaken it. In the latter case, furrows should only form between two or more different poles or asters.

How much more strongly contractile does the furrow cortex have to become, as compared with other parts of the cell cortex?

Although the different alternative theories of cleavage have been in agreement that the cortex somehow becomes more strongly contractile at the furrow site than elsewhere, there has been surprisingly little discussion or debate about exactly (or even approximately) how much stronger it would need to be. Would twice as strong be sufficient, or would the furrow perhaps need to be ten times as strong, etc.? As we will see, the answer to this question can place significant constraints on the workability of most of the different alternative theories about how the contractility becomes localized and aligned. At least in their present forms, these theories of signalling can only produce ratios of about two to one (or less) in relative signal strengths at the furrow site relative to the poles of the cell. Yet there are simple geometric and logical reasons to conclude that the actual ratios of contractile strength must be at least twenty to one, and probably considerably greater than that.

Unfortunately, no one has yet devised a method for measuring and mapping the spatial distribution of contractile tension in different parts of a dividing cell's surface. The closest that has been come to this goal is the use of micropipettes to apply controlled amounts of suction to the surface, so as either to measure how far a bulge of membrane and cortical cytoplasm can be pulled outward by a given amount of suction, or conversely how strong you have to make the suction force in order to pull out a cortical bulge of a given size.

Application of this method has confirmed the expected major difference between the polar and furrow regions, but cannot tell exactly how great it is.

Another approach is essentially geometric, in that it depends on the variations in amount of surface curvature in the furrow site as a function of direction. To understand this reasoning, you first need to realize that the curvature of a line is defined as the rate of angular change of direction per unit length along the line (i.e. the sharper the turn, the larger its curvature). Secondly, the amount of lateral force that something like a rope will exert if it is stretched tight along a curved path turns out to be equal to its curvature times the tension acting along its length. For example, picture a flexible rope stretched around tree trunks with the different curvatures shown in the diagram below. The greater the curvature (sharper the turn), the larger the lateral force per amount of longitudinal force. This lateral pushing force is always exerted toward the concave side, and becomes twice as strong where the curvature is twice as sharp, three times as strong where the curvature is three times as sharp, etc. Thus, if you had two ropes, looped around each other so that their lateral pushing forces were balanced against each other, then you could deduce the ratio of the tensions in the two ropes by measuring the ratios of the angle changes (equivalent to curvatures) at their intersection. The same argument applies if the two ropes were perpendicular, or if you had many ropes looped together into something like a fishing net, and even if you had a continuous sheet of cloth, or a soap film, or a cell membrane.

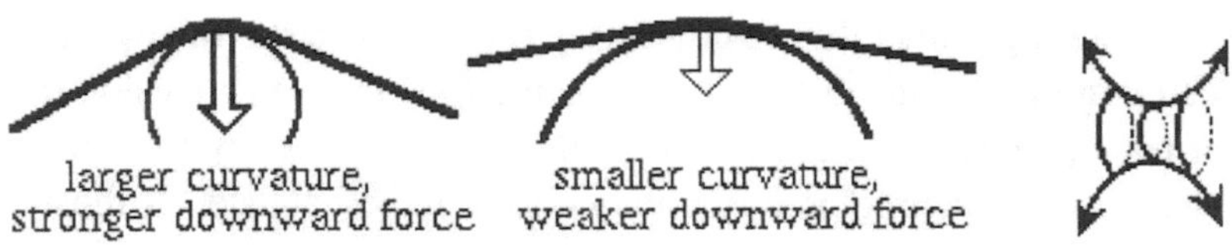

The situation at the cleavage furrow is that some contractile elements are pulling along the length of the furrow (i.e. circumferentially around the cell), while others are pulling in other directions, including the perpendicular direction, as in the drawing at the extreme right end in the figure above. Initially (assuming the cell is approximately spherical, or otherwise convex everywhere) all these forces contribute to an inward pull on the plasma membrane. But as the furrow deepens, then that part of the surface becomes convex as viewed in the pole-to-pole axis, across the width of the furrow. This creates a state of mechanical opposition between those pulling forces that are oriented in this direction, as compared with those that are oriented along the length of the furrow. Those forces that are directed along the furrow length have the effect of further deepening the furrow; but those pulling forces directed in the perpendicular direction have the opposite effect, tending to flatten the furrow out again. Since the furrow does normally continue to deepen, it is apparent that the former forces, those directed along the furrow, must be "winning". What is more, they are winning despite a large and increasing disparity in the curvatures in these two directions, a disparity that should have hindered the deepening of the furrow. Although it is true that the surface curvature along the furrow axis does keep getting larger and larger (in inverse proportion to the decreasing diameter of the constricted cell), the curvature in the perpendicular direction becomes larger still, at least until the cell has been constricted nearly all the way through.

Thus, the mechanical situation is equivalent to that of our two stretched ropes, discussed above, that were looped around each other so as to push in opposed directions. Just as the ratios of the tensions in the two ropes could be deduced from the ratio of their angles of intersection, one can also deduce the ratios of amounts of cell surface tension that would need to be acting in the two perpendicular directions. If the curvature in the direction across the furrow were about 20 times that in the direction along the furrow, then it would follow that the tension in the direction along the furrow would have to be <u>at least</u> 20 times stronger than the tension acting transverse to the furrow axis. The final step in this argument is that of equating this transverse tension to the amount of tension existing in other parts of the cell cortex, away from the furrow. The basic reasoning here is simply one of force balance. If these transverse forces were stronger than the tensions in the areas off to the sides of the furrows, then unlimited amounts of surface material from these areas would be pulled down into the furrow, and so on.

This basic line of reasoning seems to be the basis of Hiramoto's application of the equation known as "Laplace's Law" to calculate relative amounts of tension based on observations of directional surface curvatures in cleaving echinoderm eggs (Hiramoto, 1990). It also underlies the results of some of the computer simulation studies carried out by Sally Gewalt and myself in which cell surfaces are idealized as if they were networks of intersecting ropes or springs (Harris, 1990). The computer programs allow you to change the relative strengths of contractility or tension in the different ropes to see what ratios are needed to mimic cleavage. These "computer experiments" showed that ratios of 20:1 barely produced complete cleavage, whereas when the ratios were set to be only 18:1 the result was a furrow that cleaved only about half way through, and then became stuck in a static balance of forces. If the ratio of forces was instead set to only two or three to one, then the simulated cell would develop only a shallow furrow, for example resembling those observed by Bluemink when he injected amphibian eggs with cytochalasin (a drug that inhibits actin assembly). One of these simulations was also used to replicate the situation in something like a cleaving egg of a teleost fish, where the cells are hemispheres on top of a large spherical yolk. In this case, the furrow constricts inward from only the upper side of each cell, with the two ends of the furrows remaining the same distance apart. This means that the curvature along the furrow axis actually decreases during cleavage, rather than increasing as it does in something like an echinoderm or an amphibian egg. The result is that the ratio of contractile strengths has to be made even greater in order that the balance of forces be tipped in favor of constriction; in the simulations, ratios of more than 50:1 up to 100:1 were needed, depending on details of shape and proportions.

The drawings on the following page were generated by a simple computer program running on a Macintosh and designed to simulate the shape changes that would be undergone by sheets of flexible and actively contractile materials in response to patterns of differing contractility specified by the experimenter. Imagine that each line above is a stretched spring, and that these springs are attached to each other in a square grid pattern so as to enclose an approximately hemispherical volume, within which a pressure pushes outward and upward. The left drawing is a perspective view, while the one on the right is face on. The idea is to idealize the contractility of a cell's cortical acto-myosin, specifically to determine what shapes this "cell" would adopt for any arbitrary distribution of contractile forces in its surface. The furrow that has developed in this one resulted from specifying that those imaginary springs lying along that particular circumference would exert contractile forces ten times as strong as any of the other springs. This program was an amateur effort, but is available from the author in the hope of encouraging professionals to write improved versions.

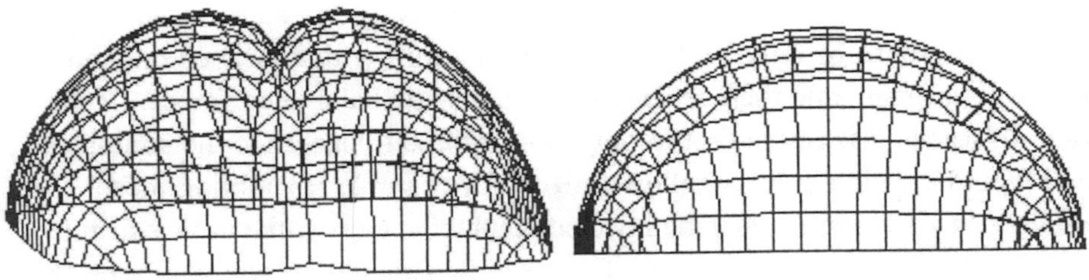

More "experimental-simulation" work of this kind seems to be needed. One of the virtues of such computer experimentation is that it allows one to try out the effects of all different possible combinations of surface flexibility, viscosity and other variables that Laplace's law simply idealizes out of existence. This approach is somewhat the opposite to many previous simulations and other mathematical models. Instead of plugging in some particular combinations of quantitative measurements taken from the biological system, combining these with a specific theoretical hypothesis, and letting the computer produce a facsimile of the observed phenomenon, our approach was meant to be more open-ended. Our programs themselves made as few assumptions as possible, thus allowing biologists to change the values of the variables, as much as possible in an experimental fashion, and watch what happens. Biologists can then ask what combinations of assumptions are capable of mimicking reality, and which ones are not. It is especially important to know which parts of the simulation (i.e. which particular assumptions) control which aspects of the output, even if only in a qualitative fashion.

Historically, most of the progress of physics and chemistry has depended upon making very precise quantitative measurements, combined with the formulation of explanatory hypotheses that make equally precise quantitative predictions. Many people extrapolate from this to conclude that equivalent progress in biology will depend on equally precise measurements and predictions. In the case of genetics, one might even regard such predictions as having been confirmed. But when it comes to the mechanics of cells and tissues, precise measurements are neither very practical nor do the contending theories usually make particularly quantitative predictions. "More versus less" and "two-fold versus twenty-fold" are more often the relevant questions.

Do the furrow-inducing signals weaken cortical contraction, or do they strengthen it?

This question has a long history of debate. One popular idea has been that the signal's effect is to weaken cortical contractility, with this weakening somehow being concentrated nearer the poles of the cell. The end result of weakening at the poles would be to leave the equatorial region comparatively greater in contractile strength. This would not be because it has become stronger, but merely because rest of the cell surface has become weaker. The majority of scientists who have discussed this issue in print seem to have espoused some form of this idea (Greenspan, 1977, 1978; Mitchison and Swann, 1954a,1954b; Schroeder, 1981; White and Borisy, 1983; Wolpert, 1960). The opposite theory is that signals from the MA have the effect of strengthening cortical contractility, with this stimulatory effect being strongest at or near the equatorial region, where the contractile ring forms. This latter hypothesis has long been championed by Rappaport (1971, 1983, 1987, 1990; see also Rappaport and Rappaport, 1985b, 1988, 1993), providing much evidence in its favor, although some troubling loose ends still remain.

The strongest evidence in favor of stimulatory signals comes from two of the characteristically elegant experiments of Ray and Barbara Rappaport (Rappaport and Rappaport, 1985b). In one set of experiments, they squeezed echinoderm eggs into cylindrical tubes and mechanically broke up some of their MAs so that their poles became separate asters. In these cases, the cylindrical cells formed cleavage furrows directly around the individual asters. In other words, when you have two poles or asters close to one another (making up the poles of a normal MA) then the furrow forms half way between them; but when you have only a single aster, then the furrow forms directly above it. This is exactly what ought to occur if these asters emit a stimulatory signal, but the idea of an inhibitory signal does not predict the formation of a furrow around a single aster.

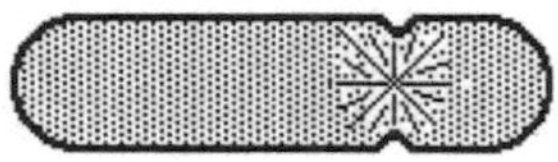

The second set of experiments supporting stimulatory signals involved cells that were pinched into hour-glass or "dumb-bell" shapes by pairs of fine glass hooks that were looped around the cell and pulled in opposite directions (Rappaport and Rappaport, 1988). This was done prior to mitosis, and sometimes the nucleus and developing MA found themselves completely on one side of the constriction, or completely on the other (rather than being situated inside the isthmus between the two sides, as I am told had been the original idea). The unexpected result was that whichever side contained the MA began to constrict itself and to squeeze cytoplasm to the other side. Eventually, the MA would likewise be squeezed over to the other side. At that point, this other side would itself begin to contract and to squeeze cytoplasm back in the other direction. The force differentials seemed to be fairly large, in that the MA was often frazzled by being pushed through the constriction, especially after being pushed back and forth 2 or 3 times.

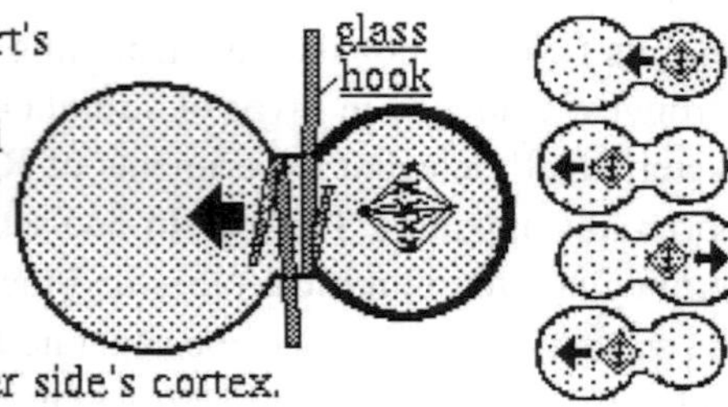

Barbara and Ray Rappaport's "dumb-bell" experiment, in which a pre-cleavage sand dollar egg is constricted by two glass hooks, with the result that whichever side contains the MA contracts more strongly that the other side's cortex.

This alternating contraction and back and forth flow makes sense if the MA were sending some stimulatory signal to nearby parts of the cell cortex. When the whole MA is on one side of the constriction, then only the cortex on that side will be stimulated to increase its contractile strength. The squeezing effect of this increased contractility will propel cytoplasm over to the other side. When the MA itself is squeezed to the other side, then the cortex on that side is induced to become more strongly contractile, thereby reversing the pressure differential. I have been entirely unable to think of any way to explain these results in terms of signals from the MA whose effect would be to weaken cortical contractility. Advocates of inhibitory signals would seem to be left only with the option of claiming that this phenomenon (back and forth flow in externally constricted cells) has some entirely different cause than does the normal formation of the cleavage furrow. But that would seem very unlikely, even if not logically impossible.

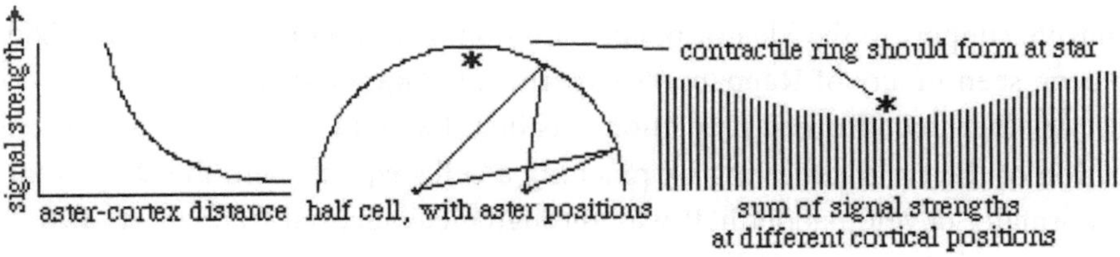

The issue of inhibitory versus stimulatory signals has also been argued at the level of computer simulations. White and Borisy (1983) showed convincingly that if each of the poles of a MA were to emit a signal that inhibited the strength of cortical contractility, with these inhibitory effects of the two poles being additive, and also with the effectiveness of the signals declining in proportion to pole-cortex distance, then the net amount of inhibition at different parts of a spherical cell's surface would indeed be a minimum along the equator. Their simulations showed that this was true whether this decline in effect obeyed an inverse square law (like gravity) or an inverse fourth power law. The point was to show the workability of a polar inhibition type of mechanism. It was later shown by Harris and Gewalt (1989) that an equatorial minimum would result for any inverse power law from inverse first power to inverse tenth power.

On the other hand, Harris and Gewalt's computer simulation also showed that this equatorial minimum would disappear in cells that had been constricted even slightly around their equators, as for example they had been in many of

Rappaport's experiments where echinoderm eggs were sucked into cylindrical tubes (1985) or stretched out by hanging weights on them (Rappaport, 1960b). In other words, White and Borisy's simulation would have predicted, on the basis of furrow location being controlled by an inverse-power inhibitory signal coming from the MA poles, that cells would become unable to form furrows in the correct equatorial position if they were even slightly constricted around the middle. In fact, our adaptation of their simulation predicts that cylindrical cells should have formed two contractile rings, parallel to each other and at a distance apart that should have become progressively greater the more constricted the cells were. It is fair to say that theoretical predictions become more convincing the more counterintuitive they are; so if cylindrical cells could be shown to form two parallel contractile rings, this would lend persuasive support to the idea of polar inhibition. But nothing of that kind had ever been seen in any of Rappaport's experiments on cylindrically distorted cells. Another unconfirmed prediction was that it should be possible to prevent induction of the contractile ring by pinching a cell's equator into the shape that cells normally develop about half way through cleavage, but Rappaport (1964), Rappaport and Rappaport (1984), not to mention the dumb-bell (reversing flow) experiment described above, all showed that furrow induction is not prevented by distorting cells into such shapes. The conclusion seems inescapable that inhibitory signals, varying inversely as any power of the pole-cortex distance, cannot after all explain the control of furrow location.

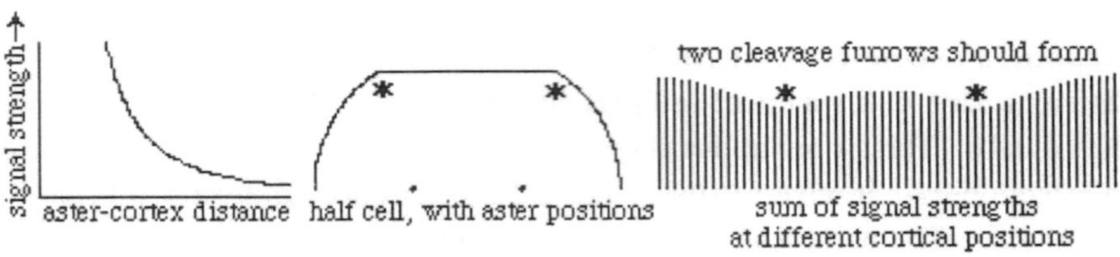

Harris and Gewalt also used these computer programs to explore the possibilities of stimulatory signals coming from the poles, specifically whether there were any "laws of variation with distance" (comparable to the inverse power laws proposed by White and Borisy) that would be capable of predicting accurately where cells would form their cleavage furrows, even in cells distorted into the various odd shapes produced in the experiments of Rappaport and others. Note that when one is hypothesizing a stimulatory signal, then the sum of the signals should be a maximum at the furrow site. After trying curves of a wide variety of shapes, it was concluded that the most accurate predictions resulted from stimulatory signals the effects of which were constant out to a distance slightly exceeding the radius of the cell, but then declined rapidly beyond this distance. Variation according to such a "law" results in having the stimuli from the two poles overlap at the equatorial furrow site, but nowhere else. In this case, a single maximum of cortical contractility should still continue to form even if the cell's equator were constricted into a

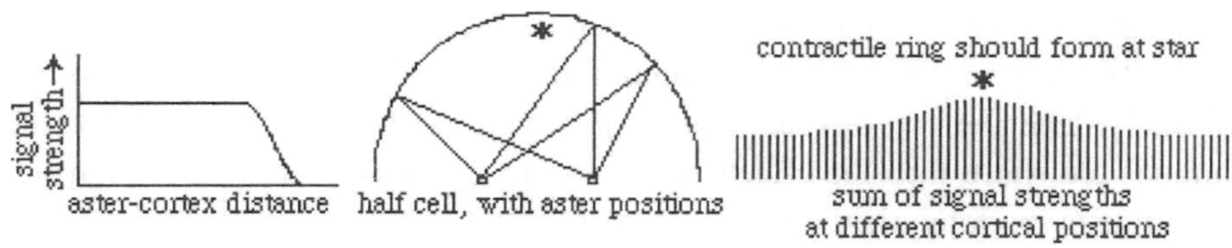

cylindrical shape, or even that of a dumb-bell. On the other hand, such a law of variation would predict that producing a large outward bulge in the equatorial region should prevent formation there of a contractility maximum, so that the contractile ring should either fail to form or should instead form just above or below the bulge. When Rappaport (1992) subsequently tried this experiment, the latter was what occurred, thus lending some support to the idea of a stimulatory signal whose effectiveness drops off rapidly beyond a certain

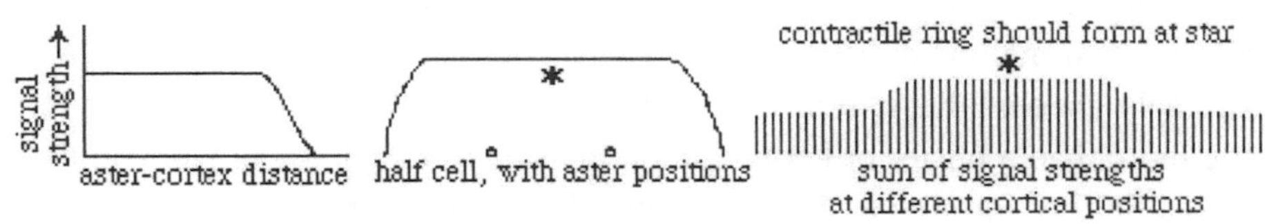

range. Harris and Gewalt (1993) also used their computer programs to show that such a stimulatory signal can also produce the hexagonal pattern of furrowing that is observed at the remarkable fourteenth cleavage of developing fly embryos, when their blastoderm cleaves simultaneously into several thousand cells. Devore, Conrad and Rappaport (1989) developed a more specific computer simulation in which the locations of furrow formation can be predicted on the basis of a hypothetical diffusion of some material, possibly calcium ions, outward from the ends of microtubules that radiate from the poles of the MA. This accurately predicts furrow locations in constricted and cylindrical cells.

On the other hand, it must also be admitted that there are some disconcerting problems with this hypothetical stimulatory signal and its rapid decline beyond a certain distance. One problem is the control of this distance or range beyond which the stimulatory distance declines so rapidly; the formation of the contractility maximum at the proper locations turns out to be critically dependent on this range being just right for the size of the cell. If the range is too short, then the effects of the two poles won't overlap at all; but if the range is too long, then the overlap will extend over most or all of the cell surface. Even worse, this range has to be made smaller if the cell is smaller; and remember that in a cleaving egg, each sequential division involves cells that are half the size of the previous one. Are we to suppose that the range of the stimulatory effects somehow declines automatically so as to match what is needed to control the formation of contractile rings in the ever-smaller daughter cells? It is not yet clear whether such considerations will lead to a disproof of this whole type of theory—or whether they will perhaps lead to the prediction of otherwise entirely unexpected phenomena, the experimental confirmation of which will then place all these theories on a new footing.

Another difficulty that arises when we carefully consider hypotheses based on stimulatory signals concerns the formation of polar bodies when oocytes undergo their meiotic divisions. The problem here is that the cleavage is so extremely unequal that it almost amounts to budding; the future polar body is little more than a blister or herniation on the side of the large oocyte, with the contractile ring forming at the rim of this blister and then pinching it off. Of course, if your goal is to form a herniation of part of a cell's surface, than the obvious way to achieve this would be to weaken the cortical contractility in the area where you want the bulge to occur. In other words, it would be much easier to explain polar body formation in terms of inhibitory, rather than stimulatory signals from the poles of the MA. That is because weakening the contractility of the cortex would be an excellent means for causing it to bulge outward at just the right place. On the other hand, it is likely that the bulging is actually a side effect of constriction by the contractile ring that encircles it, as indicated by Rappaport and Rappaport's (1985a) observation that increasing pressure in the cell does not enhance bulging at the site of polar body formation. The distinction between herniation and constriction seems to offer some prospects for critical testing of stimulatory versus inhibitory mechanisms of signalling, if only we had better criteria for distinguishing whether the surface of the bulge is being pushed out by lateral constriction or by herniation, these alternative forces being oriented at 90 degrees perpendicular to one another, and potentially being susceptible to interference by different factors that an experimenter could hope to control. It is rather embarrassing not to be able to tell the difference between lateral compression and outward herniation.

In meiotic polar body formation, and other very unequal cleavages, why does the outward bulge form unless the upper pole of the mitotic apparatus is somehow weakening the contractile strength of the adjacent cortex?

At present, however, the greater weight of the evidence is on the side of stimulatory signals (from the poles of the MA) that increase rather than weaken the local contractility of the cell's actomyosin cortex.

What is the chemical or physical basis of the furrow-inducing signals?

There is surprisingly little direct evidence on this question. The two hypotheses most frequently encountered are (1) that some chemical or ion diffuses out from the poles of the MA, and (2) that the astral microtubules themselves either carry or are the signal. A third type of possibility is that the INCENP centromere proteins discovered by Earnshaw's group serve to stimulate furrow formation in the part of the cell cortex closest to the metaphase plate (Cooke, Heck and Earnshaw, 1987; Earnshaw and Cooke, 1991) . In the context of their computer simulation, Devore, Rappaport and Conrad (1990) proposed a mechanism that combines elements of numbers one and two above. Their specific proposal was that calcium ions might diffuse outward from the ends of the astral micortubules; but their basic model would work equally well whether the diffusing material was calcium or something else, even something like the INCENP proteins. Asnes and Schroeder (1979) carried out a careful transmission electron microscope study of the spatial distributions of astral microtubules, relative to the furrow location, in cleaving echinoderm eggs. Their results indicated that these microtubules didn't even reach the cortex at the furrow site, let alone overlapping there as some theories of furrow induction seemed to require.

In the sections above that discussed computer simulations designed to predict the patterns of relative signal strengths at different parts of a cell's surfaces, a crucial question was the type of mathematical formula by which the signals varied with distance (for example whether the effects decreased with the square or some other power of distance). If we knew for sure the mathematical rule by which these signals vary in effectiveness as a function of distance, this could be an important clue as to the nature of the physical or chemical mechanism of the signals. Unfortunately, lots of different mechanisms would obey inverse power laws. For example, the cross sectional density of microtubules in a radiating array (such as an aster) would decline as a function of outward distance according to some inverse power (an inverse square, if all microtubules were to reach the surface). Likewise, the expected variation with distance of the concentration of a substance diffusing out from a center would also approximate an inverse power function. On the other hand, the combination of diffusion and microtubule radiation, as suggested by Devore, Rappaport and Conrad (1990), could produce a variation with distance by a sigmoidal rule much like the one found by Harris and Gewalt to predict most accurately where cells will form their cleavage furrows after their shapes have been experimentally distorted, for example by being sucked into cylindrical tubes. If the signals diffuse out from the ends of the microtubules, then this could help to explain why the microtubules don't seem to need to reach all the way to the surface.

The observations of Salmon and Wolniak (1990) on furrow induction in taxol treated cells are also highly relevant to these questions. They found that the stabilization of microtubules by taxol caused the asters to become highly compact. The effects of taxol are also such as to reduce the usual dynamic alternation of elongation and shortening that is typical of normal microtubules. Yet these cells still formed cleavage furrows in the correct locations midway between these somewhat abnormal asters.

Another important question is whether the signals from the two poles of the MA are merely additive, as was assumed by all the computer simulations mentioned above, or whether these signals might be more multiplicative or synergistic. For the latter to occur, it would seem to me necessary for the signals to interact with one another. This seems possible, except insofar as one wonders why the signals from one pole would not, in the same way, then be able to interact with themselves. Unfortunately, it is hard to think of any experimental evidence that either does or could bear upon this question. When we recall that at least a 20-fold increase in contractile strength seems to be necessary to produce successful cleavage, and contrast this to the paltry ratios of two-fold or less that the various simulation programs were able to generate, then the idea of some kind of multiplicative interaction seems more and more appealing. A simpler solution might involve thresholds or other highly nonlinear responses on the part of the cortex; for example, one unit of signal from the MA might produce one unit of increase in cortical contraction, but 2 units of signal might produce 20 units of increased contraction.

Do actin and myosin become concentrated at the furrow by assembly there, or by lateral flow of pre-assembled fibers?

An entirely different kind of solution to this problem of getting the needed degree of concentration of contractility was proposed by White and Borisy (1983) as part of their computer model. This was the idea of having the actin and myosin slide laterally along just underneath the plasma membrane, being pulled along by graded variations in the strength of their own contractility. In this way, it would be sufficient for the signals from the MA to produce modest gradients in the strengths of cortical contractility. These gradients would essentially magnify their own steepness, by pulling in cortical actin and myosin from the surrounding areas of the cell. This idea is equally consistent with either theories of polar relaxation or of polar stimulation, even though it was originally proposed in the context of the former. All it needs is for the signals from the MA to generate a gradient of cortical contractility, with its maximum near the cell's equator.

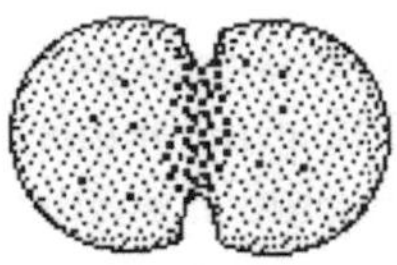

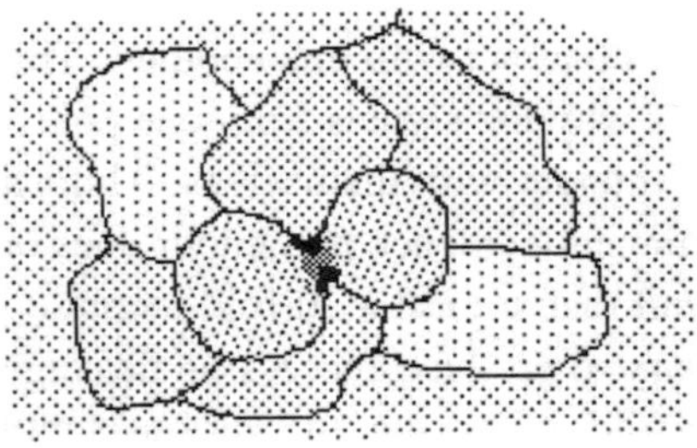

Such a lateral flow of cortical acto-myosin toward the furrow site might reasonably be expected to reveal itself by sweeping other cortical components along with it. In fact, just such an accumulation of pigment granules has been discovered. Inoue (1990) demonstrated an accumulation of pigment in the cleavage furrows of sea urchin eggs (drawing above, left), and Scott (1960) had reported comparable observations. Having seen Inoue's video recordings of this phenomenon, I subsequently noticed comparable events in video recordings of dividing cells at the surfaces of amphibian embryos at the blastula and gastrula stages (right side of previous drawing).

Cao and Wang (1990a; 1990b) have followed the movements of fluorescently labeled actin that had been microinjected into cultured cells. They report that this actin assembles into cortical fibers, and conclude that these fibers move laterally toward the furrow site, both before and during cleavage, and accumulate as part of the contractile ring. Both this and the observations on cortical granule movement support the idea of ring formation by lateral flow, although neither bears directly on the related issue of whether the cortical fibers are being pulled along by gradients of their own contractility.

A possible way to test this latter idea directly would be to relax the tensile stress in the cell cortex at the time when the lateral flow should occur. If the cortex is not stretched tight, then no gradient of contractile tension can exist, at least that is the idea. In some very preliminary experiments, Sally Gewalt and I found that perfusing sea urchin eggs with sea water that had been made hypertonic (by dissolving the inert sugar sorbitol in it) made these cells shrink slightly in volume, with a slight wrinkling of the membrane and cortex (diagrammed in the drawings below). While in this slightly shriveled state, the cells failed to form cleavage furrows at the times when their nuclear materials underwent mitoses. Later, when these (now multinucleated) eggs were returned to normal sea water, they swelled up again. At the time of the next synchronous mitoses, the cell cortexes formed networks of multiple intersecting sets of cleavage furrows, somewhat reminiscent of the synchronous mass cleavages of fly eggs.

Echinoderm eggs will shrink when perfused with sea water made hypertonic with sorbitol. The loss in tension in their cortex and membrane is accompanied by a temporary inability to cleave, perhaps because the formation of the contractile ring requires that the cortex be stretched tight.

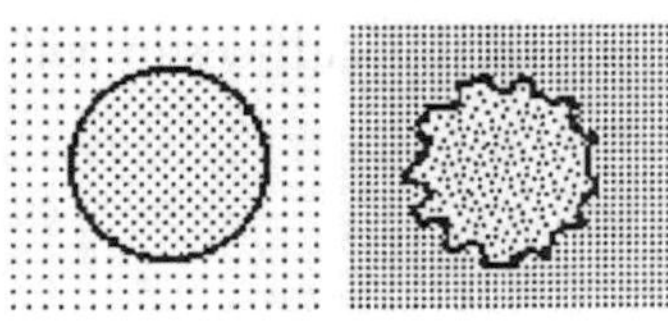

This whole question of the lateral movements and interactions of cortical fibers has yet to be adequately analyzed. A particularly interesting example arises when a furrow penetrates into a cell from one side only. This is the norm in the earlier stages of cleavage of the eggs of coelenterates, and it can sometimes be induced in other species by pushing the MA over toward one side of the cell. The interesting thing is that the contractile fibers must be concentrated together at the furrow site, but then spread out broadly toward the other side of the cell. Otherwise, that other side would itself be pulled in by the same force of constriction. Comparable situations arise in the cleavage of the yolkier parts of amphibian eggs. Such events seem to me to imply that the progression of the furrow toward the opposite side of the cell probably involves convergent lateral flow of these previously splayed out actin fibers.

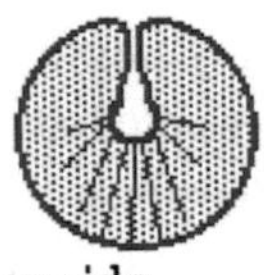

When a cleavage furrow cuts into a cell from one side only (as is normal in many coelenterate eggs, and can be induced) then the acto-myosin fibers of the contractile ring must splay out on the other side.

Why do cells round up during mitosis?

When a weight is placed on top of an echinoderm egg, this weight flattens the cell by an amount from which it is possible to calculate the tensile stress of the cell's surface and cortex, much as when one estimates the inflation of a bicycle tire by pinching it between your fingers. Another way to measure this cortical tension is with a micropipette, sucking a small blister in the cell surface, and using a manometer or other pressure device to determine how much suction suffices to pull the cell surface a given distance into the mouth

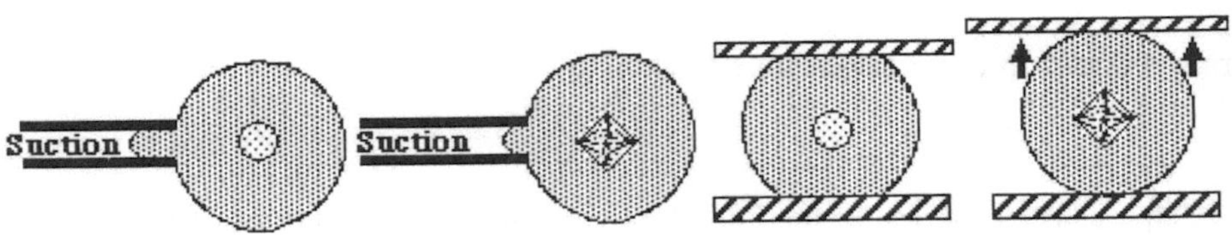

of the pipette. I think it is unfortunate that both these are commonly said to be measures of the cell's "stiffness"; it is not rigidity that we are concerned with, but rather tensile stress; in other words, how tightly the cortex is stretched, and how strongly contractile it is.

The motivation for using this word, "stiffness", was as a neutral term for cases in which one did not know whether the property being measured was rigidity (Young's modulus of elasticity) or contractility. It would be unfortunate, however, if the continued use of this word led people to confound and confuse these two very different sorts of properties. It may even be that some of the observed changes in the "stiffness" of mitotic cells are really changes in rigidity, while other changes are really changes in cortical contractility and tension. Sorting out which is which may help to resolve some of the remaining paradoxes of this subject.

Based on both types of measurement, suction as well as compression, it is believed that the tension in the egg surface increases markedly around to the time of the onset of mitosis. The same occurs in amphibian eggs, to judge from a lessening of the degree of their gravitational flattening. This globally increased contraction is a crucial element in polar relaxation hypotheses such as that of Schroeder (1981), the idea being that the contractility first increases everywhere, so that its subsequent weakening at the poles will leave the equator with enough strength to produce cleavage. Another interpretation would be that signals from the MA poles stimulate increased contractility, but at first do so all over the cell before later shortening the range of these stimulatory effects enough that they only overlap near the cell equator.

Tissue culture cells round up in mitosis **1**
but remain connected to the substratum by long retraction fibers **2**
which the daughter cells refill with cytoplasm after division. **3**

Increased rounding during mitosis is not confined to egg cells. Tissue culture cells, such as fibroblasts also round up when they divide. In this case, however, the cause seems not to be an increase in cortical contractility. When such cells are grown attached to flexible substrata such as silicone rubber or clotted blood, the contractility of each cell is made visible by the degree of mechanical distortion (wrinkling, etc.) of these substrata. Based on the degree of these distortions in the substrata, the contractile strengths of fibroblasts actually seems to decrease (greatly so, in fact) when they round up on mitosis (Harris, 1973b). Furthermore, the retracting cell margins leave behind long strands of plasma membrane, called retraction fibers. Following the completion of mitosis and cleavage, the cytoplasms of the daughter cells flow back into these retraction fibers and reinflate them (Harris, 1973a). Similar changes can be produced by perfusion with chelators of calcium ions, and by exposure to very high hydrostatic pressures. What can be the mechanical cause of these changes, can they be equivalent to those that cause increased rounding in echinoderm and amphibian eggs, and what can be the function of these changes? At present, my best guess would be that it has something to do with the ability of actin and myosin to flow laterally up gradients of their own contractility (as discussed in the preceding section). During interphase, between cell divisions, it may be that the cortex is not susceptible to such flow, only becoming capable of it during the time of mitosis, with the rounding up phenomena being side effects of this susceptibility. It will be important to determine if such changes in the mechanical properties of cells are produced by the sorts of molecular changes recently discovered by Satterwhite et al. (1992).

What further experiments are most needed?

Decisive answers could be given to nearly all of the questions discussed above if only some method could be developed to measure and map the relative strengths of cell contractility during cytokinesis. This method would not even need to be very quantitative; it would just need to distinguish between increases and decreases, and to map their locations without disturbing the division process. Almost as valuable, or perhaps more so, would be the development of methods able to induce localized increases and/or decreases in contractility at desired locations. This would allow you to compel a certain part of the cortex to become weaker than another. Such a method could test whether artificial furrows can be induced to develop, for example half-way between two bands along which cortical contractility had been artificially weakened (as diagrammed in the figure below).

Although all my attempts in this direction have been completely unsuccessful, they may help someone else. They were based on projecting narrow beams of visible and near-ultraviolet light directly onto sea urchin eggs and other cleaving cells. This was done by epiillumination through the objective, using an ordinary fluorescence microscope that had been slightly modified by substituting aluminized coverslips for the usual dichroic mirrors. Sheets of aluminum foil were placed across the light path at the proper focal planes so as to image slit-shaped holes in the foil onto the cell surfaces. The hope was that light from a mercury arc would be able to inhibit the formation of the contractile ring and cleavage furrow, thus indicating an ability to weaken cortical contractility. Had that been the result, attempts would have been made to identify the wavelengths and intensities responsible, and then to use these as the basis for experiments on the induction of artificial furrows on enucleated cells. Unfortunately, this goal of finding something equivalent to Superman's "Kryptonite rays" has not been even slightly successful. Further attempts are planned using various "caged" compounds (i.e. capable of being converted to active forms by exposure to certain wavelengths of light), and by dyes that may be able to photosensitize either the actin or the myosin components of the cortex.

One specific question to be asked is whether it is possible to induce furrows artificially, for example if cortical contraction is stimulated along a line. Or if contractility is inhibited along two parallel lines, does a furrow then form half-way in between. Likewise, could you induced furrow formation half-way between one pole of an MA and one of these bands of optically-induced cortical weakening. A further prediction would be that furrows would be optically inducible only in mitotic cells, not in interphase ones. Possible exceptions would include cells that had been artificially caused to develop mitotic behavior patterns by exposure to high hydrostatic pressures or by chelation of calcium. This is based on the hypothesis that their rounding up reflects a shift of their cortical actomyosin to a state of instability, in which these proteins become able to pull themselves up gradients of their own contractile strength.

Dedication:

This paper is respectfully dedicated to Barbara and Ray Rappaport, who have contributed more than anyone else to our understanding of cytokinesis. The consistent elegance of their experiments stands as an example to us all.

Literature Cited:

Asnes C F Schroeder T E (1979) Cell Cleavage; ultrastructural evidence against equatorial stimulation by aster microtubules. Exp. Cell Res. 122:327-338

Beams H W Kessel R G (1976) Cytokinesis: A comparative study of cytoplasmic division in animal cells. American Scientist 64: 279-290

Bluemink J G (1971) Cytokinesis and cytochalasin-induced furrow regression in the first cleavage zygote of *Xenopus laevis*. Z Zellforsch Mikrosk Anat 121:102-126

Cao L G Wang Y L (1990a) Mechanism of the formation of contractile ring in dividing cultured animal cells. I. Recruitment of preexisting actin filaments into the cleavage furrow. J Cell Biol 110:1089-95

Cao L G Wang Y L (1990b) Mechanism of the formation of contractile ring in dividing cultured animal cells. II. Cortical movement of microinjected actin filaments. J Cell Biol 111:1905-11

Cooke C A Heck M M Earnshaw W C (1987) The inner centromere protein (INCENP) antigens: movement from inner centromere to midbody during mitosis. J Cell Biol 105:2053-67.

Devore J J Conrad G W Rappaport R (1989) A model for astral stimulation of cytokinesis in animal cells. J Cell Biol 109:2225-2232

Earnshaw W C Cooke C A (1991) Analysis of the distribution of the INCENPs throughout mitosis reveals the existence of a pathway of structural changes in the chromosomes during metaphase and early events in cleavage furrow formation. J Cell Sci 98:443-61

Fullilove S L Jacobson A G (1971) Nuclear elongation and cytokinesis in *Drosophila montana*. Dev Biol 26: 560-577

Greenspan H P (1977) On the dynamics of cell cleavage. J Theor Biol 65:79-99

Greenspan H P (1978) On fluid-mechanical simulations of cell division and movement. J Theor Biol 70:125-134.

Harris A K (1973a) Location of cellular adhesions to solid substrata. Dev Biol 35: 97-114

Harris A K (1973b) Cell surface movements related to cell locomotion. in "Locomotion of Tissue Cells" Ciba Foundation Symposium, Elsevier, Amsterdam pp. 3-26

Harris A K (1990) Testing cleavage mechanisms by comparing computer simulations to actual experimental results. in "Cytokinesis: Mechanisms of Furrow Formation During Cell Division" (G Conrad and T E Schroeder Eds.) Ann N Y Acad Sci 582: 60-77

Harris A K Gewalt S L (1989) Simulation testing of mechanisms for inducing the formation of the contractile ring in cytokinesis. J Cell Biol 109:2215-2224

Hays T S Wise D Salmon E D (1982) Traction forces on a kinetochore at metaphase act as a linear function of kinetochore length. J. Cell Biol 93:347-382

Hiramoto Y (1956) Cell division without mitotic apparatus in sea urchins. Exp. Cell Res. 11: 630-636

Hiramoto Y (1990) Mechanical properties of the cortex before and during cleavage. in "Cytokinesis: Mechanisms of Furrow Formation During Cell Division" (G Conrad and T E Schroeder Eds.) Ann N Y Acad Sci 582: 22-30

Inoue S (1990) Dynamics of mitosis and cleavage. in "Cytokinesis: Mechanisms of Furrow Formation During Cell Division" (G Conrad and T E Schroeder Eds.) Ann N Y Acad Sci 582: 1-14

Mabuchi I (1986) Biochemical aspects of cytokinesis. Int Rev Cytol 101: 175-213

Mabuchi I (1990) Cleavage furrow formation and actin-modulating proteins. in "Cytokinesis: Mechanisms of Furrow Formation During Cell Division" (G Conrad and T E Schroeder Eds.) Ann N Y Acad Sci 582: 131-146

Mitchison J M, Swann M M (1954a) The mechanical properties of the cell surface I. The cell elastometer. J Exp Biol 31:443-460

Mitchison J M Swann M M (1954b) The mechanical properties of the cell surface II. The unfertilized sea-urchin egg. J Exp Biol 31:461-472

Ostergren G (1950) Considerations on some elementary features of mitosis. Hereditas 36:1-18

Ostergren G (1961) Mitosis with undivided chromosomes II Some theoretical aspects of the problem. Chromosoma 12:80-96

Rappaport R (1960a) The origin and formation of blastoderm cells in gammarid crustacea. J Exp Zool 144:43-59

Rappaport R (1960b) Cleavage of sand dollar eggs under constant tensile stress. J Exp Zool 144:225-231

Rappaport R (1961) Experiments concerning the cleavage stimulus in sand dollar eggs. J Exp Zool 148:81-89

Rappaport R (1964) Geometrical relations of the cleavage stimulus in constricted sand dollar eggs. J Exp Zool 155:225-230

Rappaport R (1971) Cytokinesis in animal cells. Int Rev Cytol 31:169-213

Rappaport R (1983) Cytokinesis: effects of blocks between the mitotic apparatus and the surface on furrow establishment in flattened echinoderm eggs. J Exp Zool 227:213-227

Rappaport R (1985) Repeated furrow formation from a single mitotic apparatus in cylindrical sand dollar eggs. J Exp Zool 234:167-171

Rappaport, R. (1987) Location of the physical mechanism in the cell. In "Biomechanics of Cell Division" Ed. Nuri Akkas, Plenum Pub. pp 1-12

Rappaport R (1990) Role of the mitotic apparatus in furrow initiation. in "Cytokinesis: Mechanisms of Furrow Formation During Cell Division" (G Conrad and T E Schroeder Eds.) Ann N Y Acad Sci 582: 15-21

Rappaport R (1991) Enhancement of aster-induced furrowing activity by a factor associated with the nucleus. J Exp Zool 257:87-95

Rappaport R (1992) Cleavage inhibition by cell shape change. J Exp Zool 264:26-31

Rappaport R Rappaport B A (1984) Division of constricted and urethane-treated sand dollar eggs: a test of the polar stimulation hypothesis. J Exp Zool 231:81-92

Rappaport R Rappaport B A (1985a) Experimental analysis of polar body formation in starfish eggs. J Exp Zool 235:87-103

Rappaport R Rappaport B A (1985b) Surface contractile activity associated with isolated asters in cylindrical sand dollar eggs. J Exp Zool 235:217-226

Rappaport R Rappaport B A (1988) Reversing cytoplasmic flow in nucleated, constricted sand dollar eggs. J Exp Zool 247:92-98

Rappaport R Rappaport B A (1993) Duration of division-related events in cleaving sand dollar eggs. Dev Biol 158:265-273

Satterwhite L L Lohka M J Wilson K L Scherson T Y Cisek L J Cordon J L Pollard T D (1992) Phosphorylation of myosin-II regulatory light chain by cyclin-p34^{cdc2}: A mechanism for the timing of cytokinesis. J. Cell Biol 118:595-605

Salmon E D Wolniak (1990) Role of microtubules in stimulating cytokinesis in animal cells. in "Cytokinesis: Mechanisms of Furrow Formation During Cell Division" (G Conrad and T E Schroeder Eds.) Ann N Y Acad Sci 582: 88-98

Schroeder T E (1973) Cleavage in HeLa cells: contractile ring filaments bind heavy meromyosin. Proc Nat Acad Sci USA 70:1688-1692

Schroeder T E (1981) The origin of cleavage forces in dividing eggs; a mechanism in two steps. Exp Cell Res 134: 231-240

Scott A (1960) Surface changes during cell division. Biol Bull 119: 260-272

Skibbens R V Skeen V P Salmon E D (1993) Directional instability of kinetochore motility during chromosome congression and segregation in mitotic newt lung cells; a push-pull mechanism. J Cell Biol 122:859-875

Warn R M Warn A Planques Robert-Nicoud M (1990) Cytokinesis in the early *Drosophila* embryo. in "Cytokinesis: Mechanisms of Furrow Formation During Cell Division" (G Conrad and T E Schroeder Eds.) Ann N Y Acad Sci 582: 222-232

White J G Borisy G G (1983) On the mechanism of cytokinesis in animal cells. J Theor Biol 101:289-316

Wolpert L (1960) The mechanics and mechanism of cleavage. Int Rev Cytol 10:163-216

Yatsu N (1912) Observations and experiments on the Ctenophore egg. I. The structure of the egg and experiments on cell-division. J Coll Sci Imp Univ Tokyo 32: 1-21

Effect of Cortex Stiffness Variation on Cleavage in Animal Cells

Nuri Akkas and Mina Kermanian
Department of Engineering Sciences
Middle East Technical University
06531 Ankara, Turkey

INTRODUCTION

Cytokinesis

Cytokinesis is the division of the cytoplasm and the plasma membrane into two. Nuclear division and cytokinesis take a relatively short time to happen in animal cells in which cytokinesis takes place by means of furrowing. Here we consider the cleavage of the animal cells only. Although the basic division mechanisms appear to be similar in all animal cells, the divisions that take place during early embryonic development, which are termed as cleavages, are not exactly the same as those that occur later. In early embryos cell divisions merely partition the existing cytoplasm without any apparent change in the total volume. During the cleavage period there is no growth. Animal cells attain a roughly spherical configuration before division. In these cells, cell division is concluded by the formation of a cleavage furrow in the equatorial plane that constricts the entire cell and finally squeezes it into two parts.

The contractile ring's structural organization is such that it exerts active tensile forces against the cell membrane of the cleavage furrow. When this ring is established at the right time and place, it continues to exist and function until the cell is completely constricted (Schroeder, 1987). Electron microscopic observations of well-preserved cells show

NATO ASI Series, Vol. H 84
Biomechanics of Active Movement
and Division of Cells
Edited by N. Akkaş
© Springer-Verlag Berlin Heidelberg 1994

that directly under the plasma membrane of the cleavage furrow a thin, unique dense layer exists which does not occur anywhere else. Schroeder (1972) was the first to take its micrograph. This dense layer is called the contractile ring which is a band or ring of about 0.1 µm thick and 10 µm wide. These dimensions are almost the same in most cells regardless of size or stage of cleavage (Pickette-Heaps et al., 1978; Scott, 1986). By electron microscopic inspections, it has been observed that the contractile ring is composed of parallel microfilaments which are oriented roughly circumferentially in the plane of the cleavage furrow. These microfilaments are fibrils with a diameter of 40-70 A and have an indeterminate length. They can slide along each other. Cleavage is caused by the muscle-like contraction of this ring. The microfilaments of the ring are attached to the plasma membrane, just like the ones in other regions of the cortical layer. As the equatorial constriction continues, this ring pulls the attached plasma membrane down in order to displace the cytoplasm to either side (Arnold 1976). Since the total cell volume is commonly assumed to remain constant during cleavage, then the surface area must increase during this process. Study of the dimension changes in contractile ring of sea urchin eggs during cleavage shows that neither the thickness nor the width of the contractile ring seems to change appreciably according to the stage of cleavage (Schroeder, 1972). Dimensional changes in this case are restricted to changes in length. Contractile ring is actually characterized by a high degree of dynamic assembly and disassembly (Schroeder, 1976).

The animal cell is a discrete mass of cytoplasm enclosed in a so-called plasma membrane or the lipid bilayer. This membrane is structuraly independent of the contained material. The plasma membrane (lipid bilayer) has a thickness of about 7.5 nm. According to the fluid mosaic model of Singer and Nicolson (1972), the plasma membrane is composed of lipids, glycolipids, proteins, and glycoproteins. The apparently elastic membrane studied by Hiramoto (1970) has a thickness of

1-3 µm. Obviously this is not the thickness of the lipid bilayer alone but it includes both the lipid bilayer and the submembraneous region containing cytoskeletal components. This compound layer is called the cortex or the cortical layer. According to Akkas (1981), it is this cortical layer which is the membrane to be used in mathematical models.

Biomechanical Models of Cleavage in Animal Cells

Greenspan (1977,1978) considered the cell as a viscous, homogeneous fluid. He introduced the concept of an effective surface tension for the approximation of the tension in the membrane of the cell, and then considered that the magnitude of the effective surface tension depends on the local surface concentration of tension elements. The cell is considered to be in equilibrium and the tension elements are assumed to be uniformly distributed over the surface before the onset of cleavage. Therefore, the tension of the surface elements is constant and is counterbalanced by the internal pressure. Cleavage is initiated as a result of the disturbance of the surface tension equilibrium which may be caused due to some chemical activity within the cell and the movements of the tension elements toward the equatorial surface. Greenspan's theory predicts that the division process is dynamically unstable and, once initiated, it is completed without further stimulus. This theory can be useful only in the case of a cell membrane with a very fluid characteristic. Cells having stiffer membranes need an equatorial constriction action. The microfilaments of the cortical layer which are interconnected and attached to the membrane protein may stimulate the surface tension. These interconnections must be broken in order to let the individual filament to flow toward the equator. According to Greenspan (1977), it is not clear whether the cytoplasm streaming can cause the breaking. This model indicates that the stirring should affect the furrowing, but this case is not observed according to Rappaport's experimental results (1971).

Doerner (1976) considered a model which assumes that the shape of the cell at the beginning of the division changes from a sphere to a short central cylinder with ellipsoidal ends. A discontinuity between the radial displacement of the cylinder and the ellipsoidal ends is created as a result of the uniform internal pressure. In this case the shearing forces and moment need to be uniformly distributed along the circumference of the joined sections and their magnitude should be such that it eliminates this discontinuity. Doerner's studies did not give any results which could be compared with experimental results.

Pujara and Lardner (1979) established a model in which the cell membrane is studied by a mathematical model which considers both the geometrical and the material nonlinearities. A review by Lardner (1987) is also available. These investigators considered an initially spherical membrane and examined its deformation pattern for the displacement at the equator, assuming that the enclosed volume is constant. Akkas (1980) extended the work of Pujara and Lardner (1979). Akkas, using Hiramoto's experimental results, considered that there is an initial stretch in the cell membrane. In this model the Mooney material is used for the nonlinearity of the material (Mooney, 1940). According to his numerical results, Akkas (1980) showed that: neither the Mooney representation nor any other exponential representation, in their classical form, is appropriate for the description of the material properties of the cell membrane during cytokinesis.

Pujara and Lardner (1979) showed that the intracellular pressure increases monotonically with the stage of the division. This result is in contrast with the experimental results of Hiramoto (1968) who found that the intracellular pressure increases at the beginning of the cleavage. During the early stages of cleavage the pressure reaches a peak and during the second half of the cleavage it decreases. Akkas (1980) studied the case in which the cell volume may change slightly during cleavage. The results obtained show that the discrepancy mentioned exists whether the cell volume is kept constant or not. It was also shown that using a different material rather than the Mooney material does not eliminate this discrepancy (Akkas and Engin, 1980).

Akkas (1981) modeled the process of cytokinesis in animal cells as the equatorial constriction of a spherical membrane made of a viscoelastic material. The assumptions used for the development of the mathematical model are generally the same as those of his elastic model (Akkas, 1980). In accordance with his numerical results, Akkas (1981) concluded that the results of the viscoelastic model were in good agreement with the results of the previous elastic model and the experimental results as far as the geometric variables of the problem are concerned. However, as was the case for the elastic model, the results on the stiffness related parameters, such as the intracellular pressure and the equatorial ring force, do not agree with the corresponding experimental results, the disagreement being qualitative in nature. According to Akkas (1981), this means that the incorporation of viscoelastic constitutive models does not resolve the differences between the predicted values and the experimental values. Akkas suggested that this discrepancy can be resolved if the stiffness is allowed to change during cleavage.

White and Borisy (1983) considered a model which proposed two separate phases of cytokinesis. The first is not dependent on the presence of the mitotic apparatus and involves a general activation of cortical contractile elements resulting in the development of a surface tension. In the second phase the asters of the mitotic apparatus interact and modulate the activities of the generating elements in the cortex to produce gradients of surface tension with the highest value being at the equator. Tension generating elements are assumed to be free to move in the plane of the cortex so that they will consequently move up the gradient of tension and accumulate as an equatorial belt of oriented elements, i.e. the contractile ring (White and Borisy, 1982). The main aspects of this model are the cortical tension generating elements, relaxing influence of asters and redistribution of tension elements.

The effect of passive surface traction on the cleavage of cells is incorporated in the hydrodynamic model of cytokinesis by Zinemanas and Nir (1987). Different rheological behaviors were examined to model the surface tensions which arise due to the passive deformation of the cortex: the Mooney material, the STZC material and a viscoelastic material (Zinemanas and Nir, 1990). The calculations show that passive surface tensions play an important role in determining the local surface deformations as well as in the modulation of the surface forces. Varying the rheological model has limited effect on the overall deformation (Zinemanas and Nir, 1987). According to the results obtained, it was observed that the overall geometric dimensions of the cell differ slightly from those observed experimently. The polar length was found to be larger by 20 % than those reported by Hiramoto (1958) and Yoneda and Dan (1972). The values reported by the experimental results of Rappaport (1976) correspond to surface contractions at the equatorial region in the range of 25-50 %, and this agrees with the hydrodynamical model presented by Zinemanas and Nir.

A fluid-mechanical approach for the cleavage of cells was also presented by Zinemanas and Nir (1987). In this study the equations of motion were combined with concentration and orientation distribution balances, for active contractile filaments on the cell surface, to provide a dynamic evolution of interfacial forces and deformation. The resulting flow and the simultaneously developing surface-tension anisotropy provide a mechanism that facilitates the generation of a contractile ring at the cell equator. The moving-boundary problem was solved numerically using boundary integral representation for the Stokes equation which was modified to incorporate the anisotropic interfacial tension (Zinemanas and Nir, 1988). According to this study it was observed that the tension increases initially but later it will be diminished.

This is caused by the combined effect of contractile filament production by the biochemical reaction, muscle dynamics and surface motions. In general the behavior of these tensions agrees with Hiramoto's (1968) observations. However, according to Hiramoto, the measured values of polar tension are higher than those of the meridional equatorial component at all times. This is not the same conclusion reached in the study explained. The resulting instantaneous cell shape found by these investigators resembles the reported experimental observation of Hiramoto (1968) and Karasiewickz (1981).

According to Bray and White (1988), during cell division in very large cells, such as sea urchin eggs, there is initially a global increase of tension, which may be related to the tendency of even small vertebrate cells to become round before division, followed by a relaxation at the two poles. Surface stiffness remains high in the equator of the cell, leading to an indentation of the surface in a circumferential furrow. This cleavage furrow becomes progressively deeper, pushing organelles to one side or the other by mechanical force, until it eventually divides the cell in two. The changes in surface stiffness that take place during cell division can be shown experimentaly to be controlled from the cytoplasm by the mitotic spindle which may either introduce contraction of the cortex at the equatorial region or cause relaxation in the vicinity of the mitotic poles. In either case, a gradient of cortical tension will develop which could, in principle, drive the movement of cortical components into equatorial regions. As localized changes in surface tension occur, the cortical elements move from regions of lowest tension (polar) to the regions of highest tension (equatorial), changing as they do so, from a random orientation to alignment with equator of the cell. According to Bray and White (1988), a suitable physical analogy is that of an elastic net stocking, drawn tightly around the cell, which is loosened at either pole by cutting some of the threads. The tension of the remaining network then draws the remaining strands together around the middle of the cell while at the same time changing their orientation until they are circumferential. Computer simulation of these properties generates a sequence of cleavage morphologies that are extremely close to those observed in living cells. A detailed discussion can be found in Akkas (1987,1990).

The Purpose of the Present Work

In 1975 Hiramoto tried to determine the equatorial constriction force during cleavage in sea urchin eggs. He injected an oil drop into the cell and observed the progress of the cleavage furrow as the furrow divided not only the cell but also the oil drop into two. Using the surface tension concept, Hiramoto (1975) determined the ring force necessary to divide the oil drop into two. Yoneda and Dan (1972) also calculated the equatorial ring force in a cleaving cell assuming that the surface tension in the cell membrane remains constant. Therefore, we now have the estimation of the contractile ring force during cleavage by two investigators. Their results are qualitatively similar and the ring force is shown in Figure 1 as a function of the state of cleavage (Hiramoto, 1975). As seen in this figure, the ring force increases during the early stages of cleavage, reaches a maximum and decreases during the later stages of cleavage. Akkas (1980) obtained a monotonically increasing ring force versus cleavage curve. Later, Akkas (1981) modified his model by including viscoelasticity but the ring force versus cleavage curve was still monotonically increasing. In both of these works the membrane stiffness was uniform all over the cell. The purpose of the present work is to modify the elastic model of Akkas (1980) by assuming that the cell membrane stiffness varies along the meridian. In one case, the polar region is taken weaker than the equatorial region. In the second case considered, the equatorial region is weaker. We want to see if the stiffness variation along the meridian will

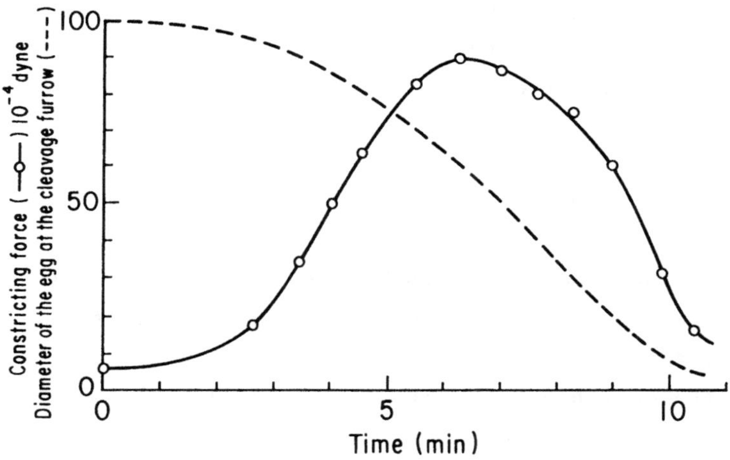

Figure 1. Constricting force developed in the furrow of the sea urchin egg during cleavage. Open circles indicate the constricting force and broken line indicates the diameter of the cleavage furrow (percent of the diameter of the egg before the onset of cleavage) (Hiramoto, 1975).

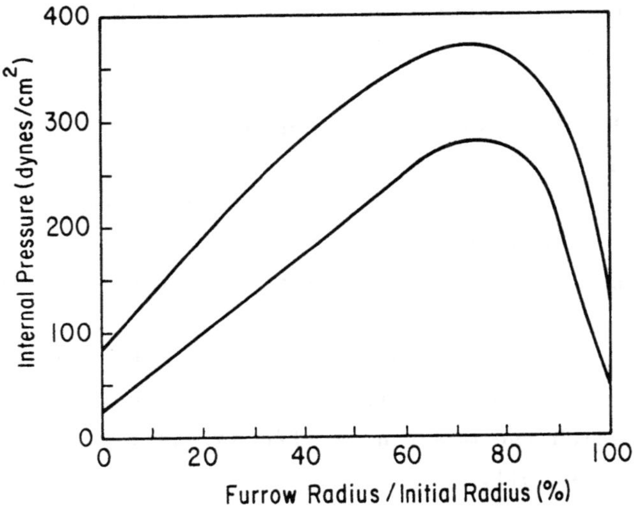

Figure 2. Envelope of the change in the intracellular pressure during cleavage. Experimental results of Hiramoto (1968).

affect the ring force versus cleavage curve qualitatively in such a way that the numerical results agree with the experimental observations.

MATHEMATICAL MODEL

The mathematical model used in this study for the investigation of cytokinesis in animal cells is based upon the following assumptions:

a. At the beginning of cleavage, the cell membrane is spherical in shape.
b. The membrane is initially slightly inflated.
c. The cell division is caused by the contraction of a ring about the equator (equatorial constriction theory).
d. The constitutive relations used to describe the behavior of the membrane material are nonlinear.
e. The viscous properties of the cell cytoplasm are neglected, so it is assumed that the intracellular pressure is uniform.
f. The enclosed volume is considered to be constant during cleavage.
g. Cell division is a quasi-static large deformation problem of an elastic membrane.
h. The membrane stiffness is a function of the meridional angle.

Assumption (h) states the only difference between the present model and the model studied in detail previously by Akkas (1980) in which the membrane stiffness was taken to be uniform all around the cell.

Through the use of the mentioned assumptions, the mathematical model of cleavage in animal cells is reduced to the large deformation analysis of a spherical membrane made of a nonlinearly elastic material subjected to an equatorial ring force. The governing equations and the method of numerical solution will not be given since they are available in Akkas (1980). The notation will be described when needed.

θ and β are the circumferential angle and the angle between the normal to the middle surface and the axis of revolution of the deformed membrane, respectively. ϕ corresponds to β in the undeformed configuration. N_ϕ and N_θ are the membrane forces per unit length of the undeformed membrane. The stretch ratios in the meridional and circumferential directions are denoted by Γ_ϕ and Γ_θ. The problem considered is a materially nonlinear and geometrically large deformation problem. As the nonlinear material, we choose a modified Mooney material. The constitutive relations used are:

$$N_\theta = (2hC_1 \Gamma_\theta^{-1}) (\Gamma_\theta^2 - \Gamma_\theta^{-2} \Gamma_\phi^{-2}) (1 + \alpha \Gamma_\phi^2) \quad (1)$$

$$N_\phi = (2hC_1 \Gamma_\phi^{-1}) (\Gamma_\phi^2 - \Gamma_\theta^{-2} \Gamma_\phi^{-2}) (1 + \alpha \Gamma_\phi^2) \quad (2)$$

where h is the initial uniform thickness of the membrane and $\alpha = C_2 / C_1$. C_1 and C_2 are the material constants which describe the stiffness properties of the cell membrane. Pujara and Lardner (1979) and Akkas (1980) took these stiffness parameters C_1 and C_2 as constants. In the present study the cortex stiffness is considered to be a function of the meridional angle ϕ. Thus, C_1 and C_2 of Equations (1) and (2) are assumed to be

$$C_1(\phi) = C_1^0 \; f_c(\phi) \quad (3)$$

$$C_2(\phi) = C_2^0 \; f_c(\phi) \quad (4)$$

Here $f_c(\phi)$ is the function describing the dependence of the membrane stiffness on ϕ. C_1^0 and C_2^0 are two constants. When we take $f_c(\phi) = 1.0$, the material becomes the classical Mooney material again. Two cases for the cortex stiffness will considered: (a) the cortex stiffness around the equatorial region is lower than that around the polar region; (b) the cortex stiffness around the equatorial region is higher than that around the polar region. A constant cortex stiffness will also be considered for comparison purposes.

The problem is axisymmetric. The unknowns of the deformed configuration of the shell are the stretch ratios Γ_ϕ, Γ_θ and the corresponding angle β. The configuration of the deformed state is obtained when these unknowns are determined. The total surface area of the two sides of the deformed membrane and the total enclosed volume of both sides of the deformed configuration can be obtained through proper integration of the variables. The formulas for these and also the axial force in the ring are available in the literature (Akkas, 1980).

NUMERICAL RESULTS AND DISCUSSION

The cell is assumed to be spherical in shape in its uninflated reference configuration. In this configuration, the diameter and the effective thickness of the cell membrane are taken to be 100 μ and 1 μ, respectively. All calculations are carried out for a cell membrane with $r_1 = 50$ and $a = 0.1$. In the following, all quantities are in nondimensional form. The stiffness of the cortex is considered to be a function of the meridional angle ϕ. We consider two cases as given below:

$$f_c(\phi) = 1.0 - \epsilon \sin(\pi/2 + \phi) \tag{5}$$

$$f_c(\phi) = 1.0 - \epsilon \sin(\phi) \tag{6}$$

Equation (5) describes the case in which the stiffness around the equatorial region is higher whereas Equation (6) describes the case in which the stiffness around the polar region is higher. Here the factor ϵ is called the weakening factor for the cortex stiffness. In this work we consider this parameter to be 0.0, 0.2 and 0.4. The results obtained for each case will be discussed. $\epsilon = 0.0$ corresponds to a uniform stiffness. The initial apex stretch ratio is a variable and it takes the values $\Gamma_0 = 1.1, 1.2, \ldots, 2.0$.

For the study of cleavage in animal cells we consider a membrane initially inflated by 20% corresponding to the apex stretch ratio $\Gamma_0 = 1.2$. Hiramoto (1968) calculated the intracellular pressure of the sea urchin egg before and during cleavage by compressing the egg in the direction of the spindle axis between a rod and a plate. The force applied to the egg was controlled by observing the displacement of the rod during the application of the force as a result of the bending of a fine glass needle fixed to the rod at the tip. The intracellular pressure of the compressed egg was calculated from the formula $P=F/(\pi r^2)$, in which P is the intracellular pressure, F is the applied force and r is the distance between the spindle axis and the point of the cell surface where the tangent is perpendicular to the axis. Hiramoto observed that the pressure increases monotonically at the beginning of cleavage and it reaches a peak during cleavage followed by a decrease during the second half of cleavage (Figure 2). The pressure increases by about tenfold during the first half of cleavage and decreases by a similar degree during the second half. On the other hand, Figure 3 shows the change in the intracellular pressure as cleavage proceeds (with constant volume constraint) obtained from the numerical results of the present model. Our results indicate that the pressure rises monotonically as the division is completed. Therefore, our mathematical model, similar to those presented before, does not predict a drop in the intracellular pressure back to zero as cleavage is completed. This is in contradiction with the experimental results. We would expect the pressure to be close to zero at the end of cleavage when the daughter cells are formed. According to our numerical results, although the pressure decreases relatively more when $\epsilon = 0.4$ and the equatorial region is stiffer than the polar region, it still does not drop to zero. In conclusion, our results do not agree with Hiramoto's experimental results, either.

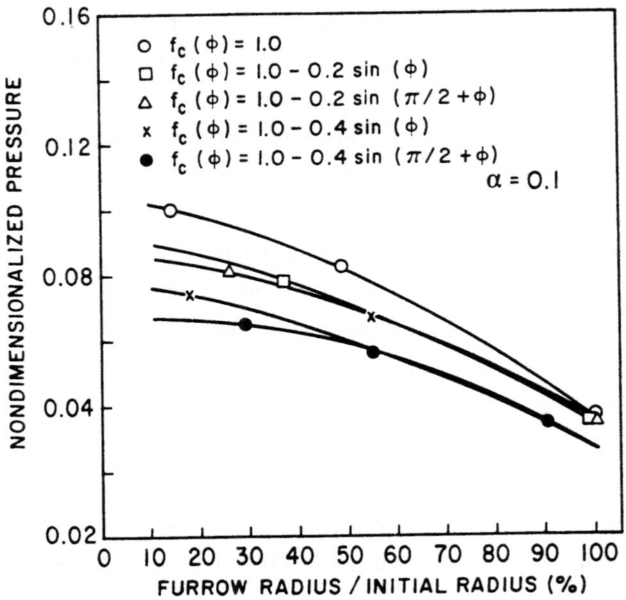

Figure 3. Change in the intracellular pressure during cleavage for the different cases considered. Numerical results of the present model.

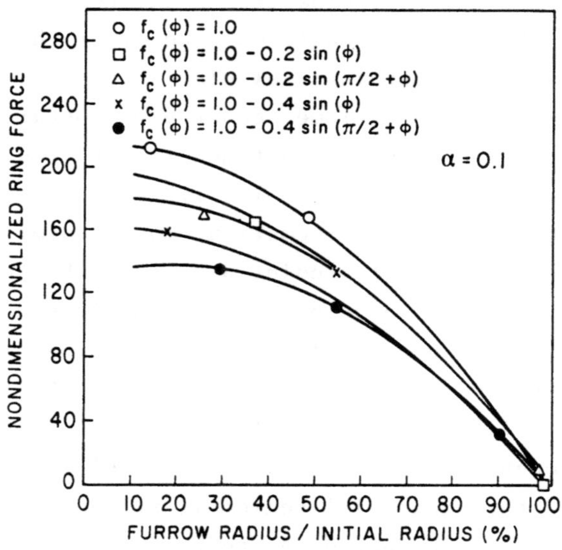

Figure 4. Change in the axial ring force during cleavage for the different cases considered. Numerical results of the present model.

As stated previously, Hiramoto (1975) calculated the force exerted by the cleavage furrow of a sea urchin egg by injecting a drop of ferrofluid into the center of a dividing sea urchin egg. The drop which is initially spherical gradually deformed into the shape of an hourglass as the cleavage advanced. The force applied to the drop was determined from the deformation of the drop and the interfacial tension between the ferrofluid and the protoplasm. Hiramoto (1975) concluded that the force applied to the drop in the dividing egg increased during the early stages of cleavage and decreased during the later stages. His result was already shown in Figure 1. The variation of the ring force as a function of the furrow radius obtained from the present work is plotted in Figure 4. The ring force increases monotonically throughout cleavage. The fact that the ring force does not drop back to zero is in contrast with the experimental results of Hiramoto (1975). Our conclusion is the same as those obtained from the elastic and viscoelastic models of Akkas (1980, 1981) who concluded that, for his numerical results to agree with Hiramoto's experimental results, it is necessary that C_1 change during cleavage. The required change in the material constant C_1 during cleavage was given as a function of both time and the furrow radius. Of course, the material corresponding to the modified stiffness coefficient C_1 which is no longer a constant is not the classical Mooney material anymore. In light of the numerical results obtained from the use of this modified Mooney material, Akkas (1980) concluded the following: the cell membrane becomes stiffer during early stages of cleavage and, after reaching a limit point, the cell membrane deteriorates as cleavage continues. This is the only way that the numerical predictions of the present elastic model can be brought to a qualitative (and even quantitative) agreement with the available experimental results.

SUMMARY AND CONCLUSIONS

According to Akkas (1980), if the material constant describing the elastic cell membrane is assumed to be a function of the stage of cleavage the numerical results obtained from the mathematical model agree with the experimental data obtained by Hiramoto (1958). The cortex stiffness must decrease during the second half of cleavage in order to obtain qualitatively the same results as found experimentally. This change in stiffness is feasible because of the fact that the meridionally-oriented microfilaments break as the furrow ring contracts. This breakage would cause the stiffness of the membrane to decrease. The meridional stretch is larger near the equatorial plane. It is reasonable that the change in stiffness is more significant near the equatorial plane and that the stiffness remains constant near the poles. In order to see if any other assumption can affect the numerical results qualitatively, we modified the model considered by Akkas (1980) by assuming that the cortex stiffness is a function of the meridional angle, rather than the stage of cleavage.

The numerical results showing the variation of the various quantities considered were obtained. As far as the geometric variables of the problem are concerned, the results of the present model are in good agreement with the results of the previous elastic models. However, as was the case for the previous models, the results on the so-called stiffness-related parameters such as the intracellular pressure and the equatorial ring force do not agree with the corresponding experimental results, the disagreement being qualitative in nature. This means that the reason behind the disagreement between the numerical and the experimental results was not the uniformity of the membrane's cortex stiffness along the meridian. According to the experimental results, we expect the pressure inside the cell and the ring force to drop back to zero towards the final stage of cleavage. However, our numerical results, and those of the previous models, indicate that both the intracellular pressure and the axial ring force increase monotonically during cleavage. Therefore, taking the cortex stiffness as a function of the meridional angle does not solve the discrepancy mentioned. At this stage, then, we must again go back to the initial suggestion of Akkas (1980, 1981) that the cortex stiffness must be a function of the stage of cleavage in order to have qualitatively agreeing numerical and experimental results. Whether this change in cortex stiffness is an active phenomenon programmed in the cell cycle already or it is a passive phenomenon resulting from the relative increase in the surface stretch leading to the break of the microfilaments is still to be answered.

REFERENCES

Akkas N (1980) On the biomechanics of cytokinesis in animal cells. J Biomechanics 13:977-988

Akkas N (1981) Viscoelastic model for cytokinesis in animal cells. J Biomechanics 14:621-631

Akkas N (1987) A mechanician's view of the biomechanics of cytokinesis. In: Akkas N (ed) Biomechanics of cell division. NATO ASI Series A: Life Sciences 132:347-369 Plenum New York

Akkas N (1990) Some basic principles of mechanics as applied to cytology. In: Akkas N (ed) Biomechanics of active movement and deformation of cells. NATO ASI Series H: Cell Biology 42:109-154 Springer Heidelberg

Akkas N, Engin AE (1980) Nonlinear mechanics and cell division. 1980 Adv Bioengng ASME 271-272

Arnold JM (1976) Cytokinesis in animal cells: New answers to old questions. In: Poste G, Nicolson GL(eds). The cell surface in animal embryogenesis and development. 55-80 Biomedical Press Elsevier

Bray D, White JG (1988) Cortical flow in animal cells. Science 239:883-888

Doerner K (1976) Some stress-strain relationship of cellular cleavage and noncleavage. J Theor Biol 61:205-209

Greenspan HP (1977) On the dynamics of cell cleavage. J Theor Biol 65:79-99

Greenspan HP (1978) On fluid-mechanical simulation of cell division and movement. J Theor Biol 70:125-134

Hiramoto Y (1958) A quantitative description of protoplasm movement during cleavage in the sea urchin egg. Exp Biol 35:407-424

Hiramoto Y (1968) The mechanics and mechanisms of cleavage in the sea urchin egg. Symp Soc Exp Biol 22:311-327

Hiramoto Y (1970) Rheological properties of sea urchin eggs. Biorheology 6:201-234

Hiramoto Y (1975) Force exerted by the cleavage furrow of sea urchin eggs. Development, Growth and Differentiation 17:27-37

Karasiewickz J (1981) Electron microscopic studies of cytokinesis in metazoan cells. In: Zimmermann AM, Forer A (eds). Mitosis - cytokinesis. 419-436 Academic Press

Lardner T J (1987) Elastic models of cytokinesis. In: Akkas N (ed). Biomechanics of cell division. NATO ASI Series A: Life Sciences 132:247-279 Plenum Press

Mooney M (1940) A theory of large elastic deformations. J Appl Phys 11:582-592

Pickette-Heaps JD, Tippit DM, Anreozzi JA (1982) Cell division in the pennate diatom Pinnularia. 2. Later stages of mitosis. Biol Cell 33:79-90

Pujara P, Lardner TJ (1979) A model for cell division. J Biomech 12:293-299

Rappaport R (1971) Cytokinesis in animal cells. Int Rev Cyt 31:169-213

Rappaport R (1976) Furrowing in altered cell surfaces. J Exp Zool 195:271-290

Schroeder TE (1972) The contactile ring. 3. Determining its brief existence, volumetric changes, and vital role in cleaving Arabacia eggs. J Cell Biol 53:419-430

Schroeder TE (1976) Actin in dividing cells : Evidence for its role in cleavage but not mitosis. In: Goldman R, Rosenbaum J (eds). Cell motility. Cold Spring Harbour

Schroeder TE (1987) The origin and action of the contractile ring. In: Akkas N (ed). Biomechanics of cell division. NATO ASI Series A: Life Sciences 132:209-230 Plenum Press

Scott RS (1986) Ultrastructure of cell division in the unicellular Red Alga Flintiella Sanguinara. Can J Bot 64:516-530

Singer SJ, Nicolson GL (1972) The fluid mosaic model of the structure of the cell membranes. Science NY 175:720-731

White JG, Borisy GG (1983) On the implications of laterally mobile cortical tension elements for cytokinesis. J Theor Biol 109:289-302

Yoneda M, Dan K (1972) Tension at the surface of the sea urchin egg. J Exp Biol 57:575-587

Zinemanas D, Nir A (1987) Fluid mechanical simulations of cell furrowing due to anisotropic surface forces. In: Akkas N (ed) Biomechanics of cell division. NATO ASI Series A: Life Sciences 132:281-305 Plenum New York

Zinemanas D, Nir A (1988) On the viscous deformation of biological cells under anisotropic surface tension. J Fluid Mech 193:217-241

Zinemanas D, Nir A (1990) Surface viscoelastic effects in cell cleavage. J Biomech 23:417-424

MULTICELLULAR MECHANICS IN THE CREATION OF ANATOMICAL STRUCTURES

Albert K. Harris
Department of Biology
Coker Hall CB#3280
University of North Carolina
Chapel Hill, North Carolina 27599-3280 USA

Electronic Mail: akharris@med.unc.edu

Introduction: The problems of getting differentiated cells to their correct anatomical positions.

Embryonic development is largely a problem of applied engineering. Molecular mechanisms combine with physical forces to create intricate networks of tubes, rods, cables and bearings, among many other types of structures. The illustrations in any textbook of anatomy will show you hundreds of structures at least as intricate (and seemingly well designed) as any ever built by human engineers. Yet the builders of anatomy are not conscious beings, much less intelligent ones, but merely cells and molecules, ultimately controlled by genes. Our problem is to trace the causal chains by which genes bring anatomy into existence.

One level of causation is that of molecular genetics, specifically the control of gene expression so as to cause each cell to differentiate into one or another specialized cell type. This is done by "turning on" the expression of appropriate subsets of genes. One cell acquires the combination of properties and behaviors that make it, say, a skin cell, while another cell turns on a different subset of genes, the result of which is to turn it into a muscle cell. Human bodies are made up of a little over two hundred of these differentiated cell types, each expressing a different subset of the genes they all share, and each cell type with its own special combination of chemical, mechanical and

NATO ASI Series, Vol. H 84
Biomechanics of Active Movement
and Division of Cells
Edited by N. Akkaş
© Springer-Verlag Berlin Heidelberg 1994

behavioral properties. Simpler animals have fewer kinds of differentiated cells, with sponges and Hydra having only a dozen or so. But no animal's body is just a random jumble of different kinds of cells, and many or most cells begin their differentiation at some location different from the one they will eventually occupy. So you also need mechanisms for arranging the different cell types into correct geometric patterns: skin cells on the outside, bone cells on the inside, muscle cells attached to the bone cells at certain places, and so on.

This problem of getting differentiated cells into their correct geometric locations can be solved by either of two extreme alternative types of method, both of which are used in actual development. The first alternative is simply to have differentiation be controlled by position. For example, any cell finding itself on the outside would therefore turn on the set of genes appropriate for becoming a skin cell, and so on for the other cell types, each according to its position. Attention has recently concentrated on this aspect of development, with various theories of "positional information" having been proposed (Wolpert, 1969).

The second alternative is the exact reverse — have position be controlled by differentiation. Include among the special characteristics of each cell type some properties or combinations of properties that will cause those cells preferentially to move to whatever locations are appropriate for them. For example, a skin cell would have properties that cause it to move to the outside surface, even if it started out somewhere else.

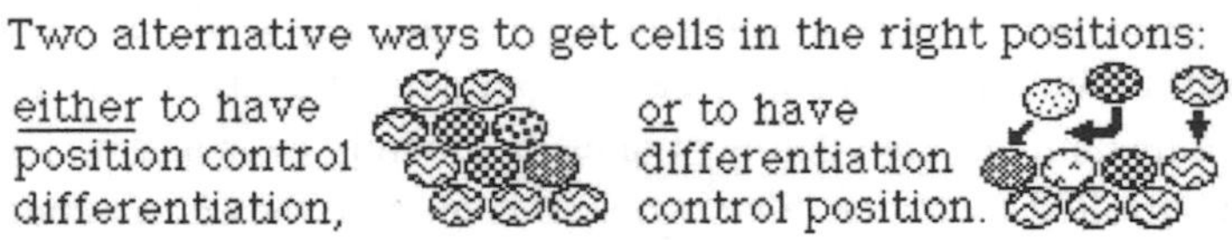

Embryological research has come to be focused so narrowly on the former of these two classes of mechanisms (in which position controls differentiation) that one could easily get the idea that animal development depends mostly if not entirely on mechanisms of that type. Because higher plants lack any capacity for locomotion, they actually are restricted to this former type of mechanism. But animal development is very different and uses mixtures of both approaches. For example, many cell types begin their differentiation at one location in the body, but then migrate long distances to some other site before acquiring their eventual geometry. Skeletal muscle cells are an example. In sponges, as we will see, cellular rearrangements are so extreme that anatomical pattern becomes completely independent of where cells had originally undergone their differentiation. Although much excellent research is now being concentrated on particular kinds of animals (nematodes and *Drosophila*), and particular stages of development, where patterns happen to be mostly or entirely produced by the positional control of differentiation, the development of vertebrates, and mammals in particular, is rather different. In our own development, there is so much cell migration and rearrangement that few structures can be explained without understanding the forces that move cells to their correct locations.

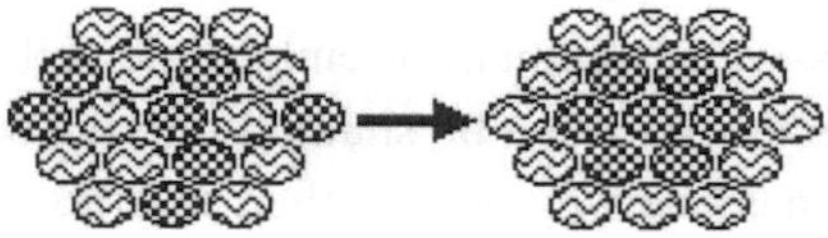

If cells have properties that move them to their correct anatomical positions, then it is not so surprising that these cells will also have some ability to sort out if dissociated and jumbled together randomly. Such an ability makes perfect sense if differentiation is controlling position. Such cell sorting was first discovered in sponges, and later shown to occur when embryonic cells of vertebrates are dissociated and mixed randomly. As will be discussed below, several very different explanations for such phenomena have been accepted and then discarded over the years. Currently, the dominant view is that tissue-specific differences in cell-cell adhesion bonds are responsible, and there is conclusive evidence that adhesive specificity does play a major role. But this should not prevent us from from asking what other cellular properties contribute to the correct positioning of cells, including mechanical and behavioral properties. Simply to equate sorting with adhesion differences is naive.

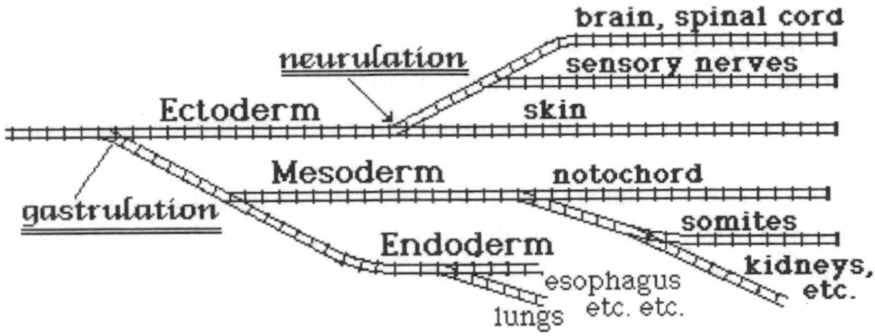

A further relationship between cell differentiation and cell rearrangement is that each can cause the other, at least sometimes. The great British embryologist C. H. Waddington liked the analogy of a railway switching yard for describing the progressive assignment of embryonic cells (as in the preceding figure). Just as box cars can be shunted into different eventual sidings by throwing switches one way or the other, a series of genetic switches (still only partially understood) control whether a given cell turns on the set of genes appropriate for being a skin cell, or those appropriate for being a nerve cell, etc. Notice how often these epigenetic switch points correspond to physical rearrangements of cells. Gastrulation and neurulation are both good examples. In the case of neurulation, the boundary between the future skin cells and the future nerve cells corresponds exactly to the separation between two different mechanical behaviors of the cells: the ones that fold become the nerve cells, the ones that don't fold become the skin cells. To ask which causes the other is to miss the point. We seem to be dealing with a positive feedback loop that combines genetic and mechanical elements. Richard Gordon and his colleagues have coined the term "cell state switcher" to refer to such situations in which mechanical properties and movements of cells are part of the control system that switches cell fate between alternate differentiated states (Bjorklund and Gordon, 1993)

Humans would probably never have designed such a system to have its genetic switch mechanisms directly coupled to its mechanical rearrangements, so we tend to miss the clear implications of what we are looking at when an embryo gastrulates or neurulates. If you just wanted to turn on different sets of genes in different cells, so that each cell type differentiated at its correct location, then you wouldn't need any such mass rearrangements. These processes make perfect sense, however, if mechanical forces are part of the system for controlling differentiation.

Embryologists need the collaboration of mechanicians

Are embryological mechanisms something we can hope to understand and control? We don't need to duplicate development, but merely to make sense of it in terms of chemistry and mechanics. But this is still a difficult task, and will require the active collaboration of those professionally trained in engineering, mechanics and applied mathematics. Their accumulated understanding of the interactions of forces and matter are the basis of modern building methods, and the mechanisms of anatomical construction cannot be understood without them.

Part of our task is to identify the physical forces that serve to move cells to their proper positions. Even more important is explaining how genes control the strengths and directions of these morphogenetic forces, such that they produce correct geometric patterns of cells and molecules, not just random arrangements. Other questions include whether we should regard wound repair as a reactivation of the same developmental mechanisms that were responsible for creating structures in the first place. We also need a better understanding of the day-to-day maintenance of anatomy. Skin, bones, and digestive tract, among many other tissues, have been shown to undergo constant and rapid turnovers of molecules and even of whole cells. As the mechanisms become better understood, we may come to regard maintenance as a special case of repair, or as an extension of construction. Such questions challenge us intellectually, and their answers hold out prospects of enormous medical benefits. Consider how much of modern medicine's power is based on understanding physiological and immunological mechanisms well enough to manipulate them for the patient's benefit. Then imagine what will become possible when we gain equivalent control over the structure-building mechanisms. Orthopedics and plastic surgery, orthodontics and periodontics are only some of the fields that can be revolutionized.

Progress in this direction requires engineers and biologists to collaborate, and for both to change some basic assumptions. Biologists have to start thinking of cells as building materials, subject to stresses and strains, and possessing the sorts of physical properties on which engineering calculations are based. Over the past 400 years, going back to Galileo and Newton, mathematicians and physicists have accumulated concepts, terminology, formulae and other tools for predicting how forces will distribute themselves through inert matter. It is past time for biologists to familiarize themselves with at least the rudiments of this knowledge. We must, for example, learn the distinctions between stress and strain, between scalar, vector and tensor properties, among many other aspects of continuum mechanics (Fung, 1977;.Gordon, 1978; Houwink, 1958; Ruoff, 1973; Shames, 1964).

Conversely, engineers need to imagine what combinations of properties would make materials "want" to adopt one shape or arrangement in preference to others. They must realize that embryological development differs from human construction techniques at levels far more fundamental than merely the building materials used (collagen instead of steel, cartilage instead of plastic, bone instead of concrete, etc.). One difference is that most of the engineers' hard-won equations assume homogenetity, isotropy and linear proportionality between stress and strain, and often become wildly inaccurate when applied to materials that disobey these idealizations. Biological morphogenesis, on the other hand, can depend precisely on deviations from linearity, etc.. An even greater difference is that human engineering is based almost entirely upon the external imposition of shape upon inert matter, whereas anatomical structures arise spontaneously, "from the inside out", by active interactions between component cells and molecules.

Construction by Self-Organization.

Human technology sometimes does use self-organization. One example is the old method for making lead buckshot spherical: by dropping molten metal down the inside of a tall tower, the metal droplets' own surface tension makes them round up. A comparable example is the manufacture of "latex microspheres" by polymerization of methacrylate (and sometimes polystyrene or other plastics) in water solutions, to which small concentrations of surfactants have also been added. Not only do surface tension effects impose a spherical shape on each tiny mass of polymerizing plastic, these forces also turn out to regulate the individual diameters of these masses with such astonishing precision that (nearly) all the products of a given batch will have diameters identical to many decimal places. They are used as size standards in electron microscopy.

Similar ideas have been developed by Elsdale (1968, 1973) and by Bard (1990). Elsdale used the analogy of grinding of spherical mirrors and lenses with abrasives, calling this "an inherently precise process" because of the automatic tendency of imperfections to be repaired. He proposed that the alignment of dermal fibroblasts is another such automatically self-correcting process. In another paper I have drawn attention to comparable properties in "quartz halogen" lamps. These are an elegant kind of incandescent light bulb, within which halogen gases not only serve to scavenge atoms of tungsten that have evaporated from the filament, but can also redeposit this tungsten wherever the filament is hottest. Since the filament becomes hottest where it is thinnest, this redeposition serves as a negative feedback loop, automatically adding metal where it is most needed. This is an analogy to the sort of geometrical homeostasis that we need to look for to explain the maintenance of anatomical structures. Self-sharpening tools and self-sealing gas tanks might also provide useful analogies.

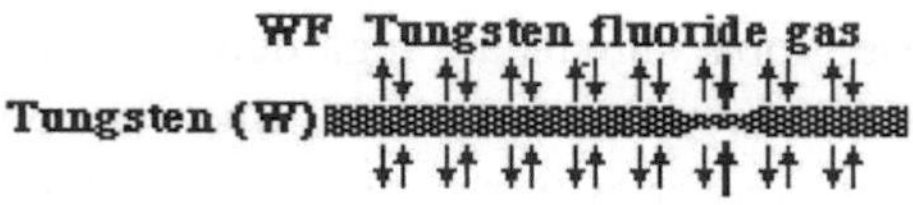

Other examples of spontaneous morphogenesis occur in the manufacture of metal alloys. As it happens, the desirable properties (hardness etc.) of alloys are largely consequences of crystalline or grain structures, the dimensions of which are far above the level of interatomic bonding. Formations of such microstructural patterns are strongly influenced by physical treatments such as tempering and forging. Ordinary cooking is another field in which small-scale textures, for example those of pastries and sauces, are caused to form spontaneously as consequences of the properties of the raw materials. Unfortunately, the development of almost all such techniques, from cooking to metallurgy, has been largely empirical, essentially by trial and error. Theoretical understanding is lagging far behind practice, so that we have little knowledge of how to make materials spontaneously adopt a given shape. It may even be that engineering technology will learn some useful lessons from biology. For example, if the dendritic condensation of polymers could be controlled at will, then you could cheaply make artificial insulating materials equal or superior to the most expensive goose down.

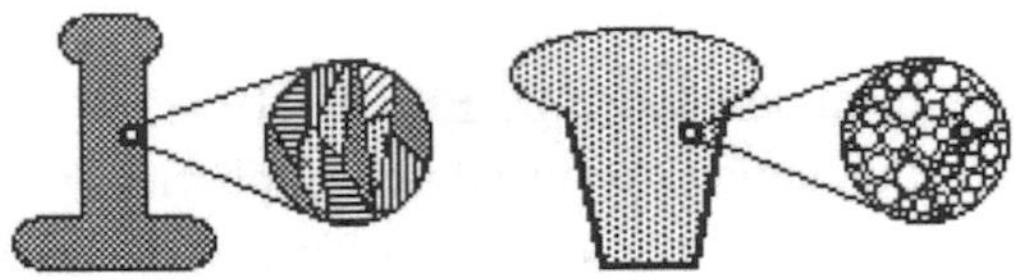

Meanwhile, nearly everybody finds it more natural to approach embryological mechanisms in terms of the imposition of shape from the outside in. People usually try to analyze pattern formation in terms of signals external to the process itself. Instead of looking for feedback loops or other internal interactions among component materials, they try to externalize causality; the more external a proposed control mechanism, the more plausible it will seem. In particular, I suggest that the much of the overwhelming recent popularity of "positional information" theories is traceable to this intuitive bias. Whenever cells at different locations behave differently, it is concluded that the cells in some sense "know where they are", with development being regarded as a series of local responses to global systems of x, y, z coordinates. These global signals are regarded as having been created separately from the cells' local responses to them. Such an approach allows one to regard development as a series of reactions, rather than interactions. It must be frankly admitted that this approach has proven very useful for making sense of certain aspects of development, such as early fly embryogenesis (Lawrence, 1992). For many other developmental systems, however, it is inadequate and misleading. This is especially true when anatomical patterns are produced by mass movements and rearrangements of cells that have already begun their differentiation somewhere else. Several such systems will be considered below.

Sponge reaggregation, and the discovery of histotypic cell sorting.

The most primitive multicellular animals are the sponges, a phylum consisting of more than 5,000 species, nearly all of which are marine, with a few freshwater species. Sponges have only three main cell types: (a) flagellated cells called choanocytes, which pump water and filter food out of it; (b) pinacocytes, which serve as epithelial cells, forming the outer surface as well as the linings of the excurrent water canals; and (c) mesenchymal cells, which are subdivided into many subcategories. Some of these secrete the important extracellular protein collagen (the form of this protein made by sponges being called "spongin"). Others make spicules, which are rods made of silicone oxide (in most species, but made instead of calcium carbonate in a few species). Spicules serve as skeletal supports, come in a wide variety of (species-specific) shapes, and are physically carried from place to place inside the sponge by mesenchymal cells. Because of this relatively small number of cell types, as well as their extremely simple anatomy, sponges are good organisms in which to begin analyzing the most basic principles of tissue construction.

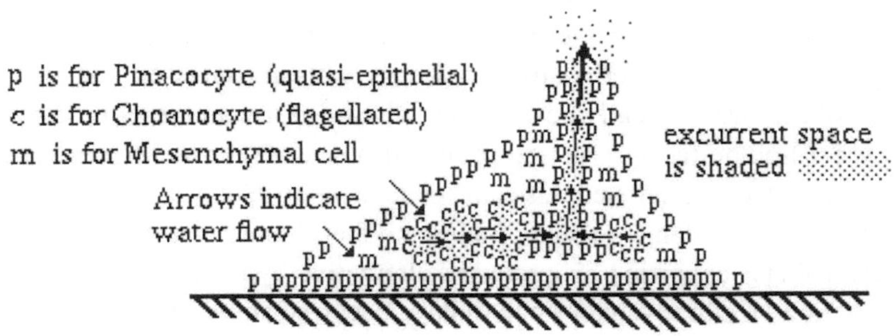

In 1907, H. V. Wilson (my predecessor as embryologist in this department) reported astonishing observations on the ability of dissociated sponge cells to reconstitute functional organization. His method was to squeeze fragments of sponge tissue through a fine cloth mesh, so that the individual cells were mechanically separated from one another. When the resulting suspensions of cells were allowed to settle out in dishes of clean sea water, he found that they aggregated together and formed small functioning sponges within two or three days (Wilson, 1907). This is scarcely less surprising than if one were to have bulldozed a building down to the level of individual bricks, and then found small bungalows organizing themselves spontaneously out of the rubble. A few years later, he reported similar results from studies of dissociated coelenterates, the next most primitive phylum after sponges (Wilson, 1911).

These findings were soon confirmed by others, and comparable reconstitution from dissociated cells has been reported for members of higher phyla, from platyhelminthes (flatworms) up to tunicates (which, as chordates, are members of the same phylum as ourselves!).

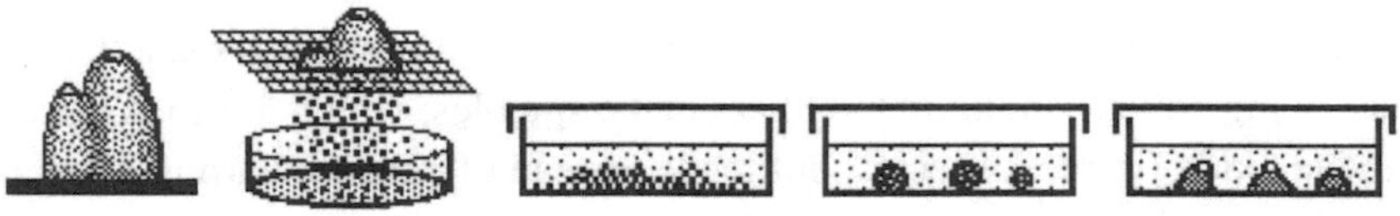

Wilson's discovery was ahead of his own time, and even ahead of ours. The idea of cells as the building blocks of tissues and organs was already well established; but few were ready for cells to be the active builders of anatomical structures. Wilson himself resisted the idea. In his paper on the reconstitution of coelenterates, he proposed two alternative explanations: either that the cells must be changing their differentiated states to match whatever new positions they found themselves in; or, alternatively, that the cells must be rearranging themselves according to differentiated cell type, moving back to their correct relative positions. Wilson confessed that the latter possibility was logically possible, but discarded it with these words: "However, I find no facts in support of this idea, and it seems to me extremely improbable." He then applied this same conclusion to sponges.

As regards this statement about "finding no facts": note that there were even fewer facts supporting his preferred alternative hypothesis of cellular re-differentiation. Cell movements had to have occurred simply to get the cells back together again. Cell rearrangements were also required to reshape these cell masses into their proper external contours, not to mention forming internal cavities that are essential parts of the functional anatomies in both sponges and coelentrates. Lastly, Wilson himself had shown that cells dissociated from different species of sponges would sort out from one another into separate masses. In retrospect, it seems obvious that even the evidence then available, while less than entirely conclusive, actually favored the discarded hypothesis (cell rearrangement), as opposed to redifferentiation. Then as now, however, certain ideas enjoyed the status of "default conditions", always to be assumed true until disproven.

Julian Huxley (1911, 1921) repeated Wilson's dissociation experiments with syconoid sponges (instead of the leucanoid sponges that Wilson had used, and which make up the vast majority of species in this phylum). Among the differences of the syconoid sponges is that their choanocytes happen to be much larger than in leucanoid sponges (where they are incredibly tiny!), and were large enough to remain distinguishable from other cell types inside the masses of reorganizing cells. Huxley tentatively concluded that cells seemed to be rearranging rather than redifferentiating. Similarly, observations by Galtsoff (1925) led that author to conclude that pinacocytes were moving back to their correct locations. Wilson himself later made made extensive histological studies or reaggregating sponge cells (Wilson and Penny,1930), but moved even further away from the cell rearrangement interpretation. Based on observations of what seemed to be intermediate stages between undifferentiated and differentiated cells, they concluded that the reconstitution of sponges was mostly a matter of regeneration, with previously undifferentiated archeocytes changing into the various cell different cell types, according to position.

It was really Johannes Holtfreter (1939), in his work on the sorting-out of dissociated amphibian embryos, who pushed the idea of differentiated cells being able to find their way back more or less to their correct relative locations. To this idea, he added the corollary that those properties that allow the cells to find their ways back after artificial dissociation are essentially the same properties that cause them to move to their correct locations during normal embryonic development. This would mean that cell sorting is a way to study the normal mechanisms that create anatomical structure. In1939, Holtfreter proposed the concept of "tissue affinity" (Gewebaffinität), to refer to the collection of cellular properties (whatever these might turn out to be) that are responsible for correct positioning, whether in normal development or among dissociated cells. He suggested that selective adhesiveness was <u>among</u> these properties, and his tissue affinity concept has been systematically mis-remembered as a proposal that cell sorting results <u>only</u> from differentiated cells being more adhesive to others of their own cell type. Nevertheless, his idea of selective adhesiveness led eventually to the identification of cadherins, N-CAMs and other cell surface adhesion proteins.

Holtfreter had used species-specific differences in cell pigmentation as a means of tracing cell fate during sorting out. For example, when he combined neural cells from embryos of a darkly pigmented species, and combined these with (say) skin cells from embryos of a species whose eggs happen to be more lightly pigmented, then he could determine whether the subsequent formation of a mass of neural cells surrounded by a layer of skin cells had been caused by the neural cells sorting out to the internal position, as opposed to those cells that happened to be on the inside switching to a neural cell type. His observations consistently indicated the former: that cells were choosing position according to differentiated cell type, as opposed to the latter possibility (which nevertheless continued to be dogma well into the 1950s). Not until Moscona (1957) made equivalent use of differences in nuclear morphology between chicken and mouse cells was it finally accepted that cells sort out according to differentiated cell type.

This belated acceptance of cell sorting also owed much to the availability of a molecular hypothesis capable of explaining it. I refer to the idea of selective adhesion: some kind of molecular "glue" specific for each cell type. This idea was championed by Moscona, although its eventual confirmation, in the form of the identification and isolation of N-CAM, cadherins, etc. was accomplished largely by others. This topic has developed vigorously, is the subject of many reviews, and will not receive much further consideration here. Instead, my goal will be to emphasize that many other cellular properties, in addition to selective adhesiveness, also contribute to the correct positioning of cells in development.

The Continuous Morphogenesis of Sponges

Although large amounts of information have been accumulated concerning sponges, there seems to be much yet to be learned. For example, the textbooks assure us that sponges are sessile, unable to move, yet in the course of studying cell movements inside sponges, Bond and I discovered that our sponges were actively crawling around the inside walls of their aquaria (Bond and Harris, 1988). This could be seen simply by tracing each sponge's outline with a pen on the outside surface of the glass, and dating each outline (as diagrammed in the following sketch).

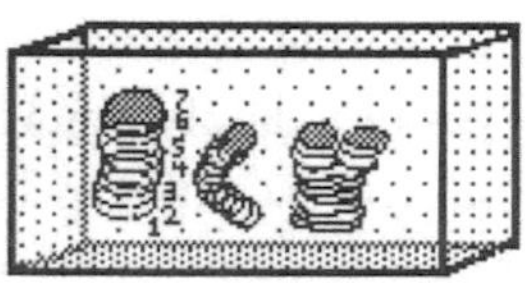

Both freshwater and marine sponges can crawl slowly over solid surfaces.

Ironically, considering that it was first thought impossible for sponge cells to rearrange themselves following dissociation, we found that sponges constantly rearrange their cells during their day to day life, and do so at least as dynamically as when dissociated. Calhoun Bond (1992) demonstrated these constant rearrangements by making time lapse films and video recordings through transparent parts of sponges of many different species, both marine and fresh water. In all of them, the time lapse records show whirlwinds of cellular movements and tissue reorganization. Of course, he had to find or create situations where the tissues were thin and transparent enough that one could see the interior cells. This was accomplished in three different ways: by looking through the thin margins of large sponges; by observing very small sponges; and by using "sandwich preparations" in which living sponges have spontaneously extended tissues into thin spaces between slides and cover slips. These sandwich preparations have been pioneered by Ankel and Eigenbrodt (1950), Weissenfels (1981) and other German students of sponge biology, and are analogous to "ant farms" in which the tunnel networks made by ants are made visible by culturing them in a thin space between sheets of glass. It is not unreasonable to worry that cell rearrangements might somehow be stimulated or accelerated by the thinness of these parts of sponges; but the patterns and speeds of the movements were the same for all three methods. Thus, pending cell labeling and tissue grafting within thick interior areas of large sponges, the burden of proof would seem to have been shifted onto those who believe in static tissue structures.

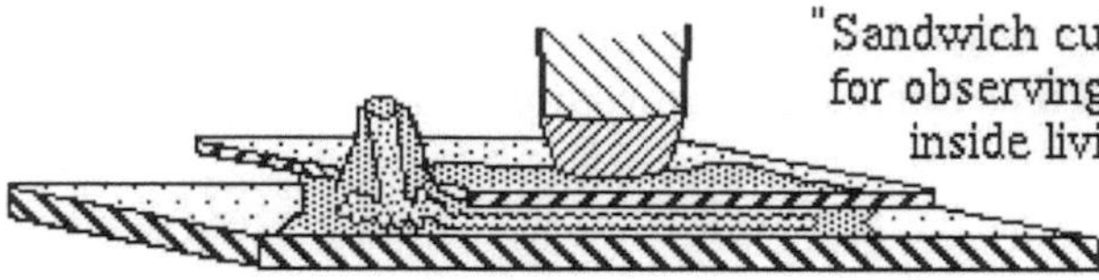

"Sandwich culture" method for observing cell movements inside living sponges.

Lessons to be learned from Bond's observations on sponges:

1) Tissues can sometimes continue to function even while undergoing massive reorganizations.

2) The phenomenon discovered by H. V. Wilson is not just something that happens in response to disaggregation.

3) Cell sorting in sponges seems to be governed by combinations of many cellular properties and behaviors, including (but not confined to) selective adhesiveness between like cell types.

4) Cell sorting should be understood as a means of studying the abilities of differentiated cells to find their correct locations and spatial arrangements, whatever the mechanisms used for this purpose (selective adhesiveness or anything else).

5) Even when you have an almost perfect view of cell behavior during morphogenesis, it can be almost impossible to make sense of what you are seeing.

The cells' apparent rules of behavior, in so far as these rules can be deduced from Bond's time lapse films and videos, are summarized in the following list. His work, much of it still unpublished, includes videos of how sponge cells react when their correct anatomical arrangements are mechanically disrupted by microneedles. These same rules seem to apply both to marine and freshwater species.

a) Choanocytes stick tightly together side by side with other choanocytes to form hollow spheres and hemispheres. These spheres and hemispheres are called "choanocyte chambers", with each having a diameter on the order of 25-50 micrometers.

b) Each choanocyte's single flagellum becomes directed to the concave side of the cell sheet, and thus into the hollow cavity of the nest.

c) Pinacocytes form continuous sheets, mostly on the outside surface, but also making up interior canals. Among the more fundamental things we do not yet know about sponges is the degree of interchangeability of these two populations of pinacocytes. Are those making up the outer surface a different cell type than those making up the canals, or should we think of them as members of one cell type doing slightly different jobs? Similar questions arise concerning those pinacocytes that form the lower surface of the sponge.

d) Those pinacocytes that form the lower surface of the sponge adhere to whatever solid substratum the cell is living on; but they also crawl actively across these substrata, with the result that the sponge as a whole is propelled along at speeds of up to a millimeter or two per day (Bond and Harris, 1988).

e) Both choanocytes and mesenchymal cells confine themselves to the interior space enclosed by the pinacocytes. Conversely, one could regard this as a tendency on the part of the pinacocyte sheets to enclose cells of the other two other types.

f) Choanocyte nests spontaneously integrate themselves into the walls of the interior canals. If separated from these canals, choanocyte nests close up like little soap bubbles. When two or more of these come together, they form openings to one another. They also form openings where they come in contact with sheets of pinacocytes making up the wall of a canal. In that case, the canal wall also opens, with the fusion of the two kinds of cell sheets serving to incorporate that choanocyte nest into the canal wall.

g) Mesenchymal cells confine themselves to the spaces inside the outer sheet of pinacocytes, but outside the interior canals. Mesenchymal cells crawl about rapidly, traversing their own lengths every two or three minutes, so that in a small sponge it is not unusual to see them move from one extreme end of the body to the other in only a few hours. There seem to be streams of at least temporary directionality within which most of the mesenchymal cells are moving in the same general direction. Spicules are carried along these streams by groups of mesenchymal cells.

h) Spicules are positioned and attached together by mesenchymal cells so as to form supports equivalent to tent poles holding up the sheet of pinacocytes that form the roof of the sponge.

i) Waves of active contraction propagate through the sponge tissues at speeds around 50 micrometers a minute. Each choanocyte chamber shrinks slightly in diameter as one of these waves passes it. Neither the driving force nor the functional significance of these waves is known.

j) Much more dramatic contractions also occur, consisting of an extreme shrinkage of the spaces where the mesenchymal cells are located, combined with huge dilations of the excurrent canals, both lasting 25-20 minutes or more. These events occur spontaneously, but they can also be induced by exposure of sponges to specific inhibitors of cyclic AMP breakdown (Bond, unpublished).

To what extent is this combination of cellular behaviors adequate to generate, maintain and restore the correct architectural arrangements of sponge anatomy? What additional rules are needed? Part of the problem is that our unaided mechanical intuition cannot predict net geometrical consequences from a given list of rules. Computer simulations could be a great help. Without something of that kind, we won't know what other sorts of behavior to look for, nor whether our list is already adequate to predict (and thereby explain) the formation of anatomical patterns.

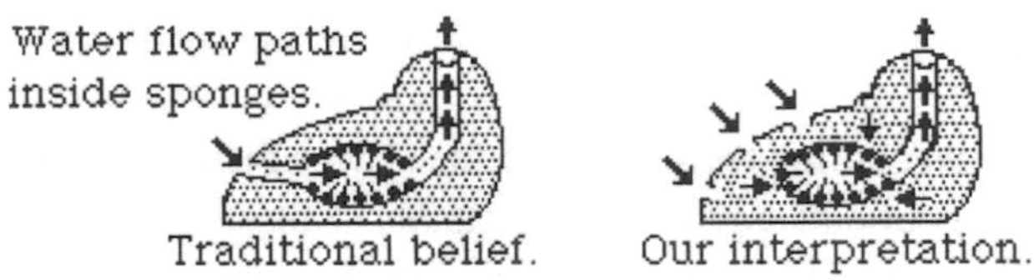

An important complicating factor is that the textbooks are (I believe) radically mistaken about sponge "plumbing", i.e. about the paths followed by the water as it move through the sponge body. All books on the subject show one set of pinacocyte-lined canals running from the sponge surface to the choanocyte chambers, and a second set of pinacocyte-lined canals then carrying the water from the choanocyte chambers back to the outside. The former are called "incurrent canals" while the latter are called "excurrent canals", and both are supposed to be separate from the space where the mesenchymal cells are. Ongoing studies in which living sponges are observed while ingesting fluorescently labeled beads and water have suggested to Bond and myself that the so-called "incurrent canals" are essentially fictional and that the water flow reaches the choanocyte chambers by flowing directly through the space where the mesenchymal cells are located. We also believe that the pumping action of the choanocytes is mechanically equivalent to that in nephridia, in that the water is pulled up through the spaces between one choanocyte and another. This is as opposed to the water being driven laterally past the flagellated ends of the choanocytes, as textbook diagrams indicate. To the extent that this new interpretation is correct, local differences in water pressure could provide important morphogenetic clues to control sponge cell behavior. The excurrent space, inside the choanocyte nests and the excurrent canals, would have a pressure higher than ouside the sponge; but the mesohyl space, where the mesenchyme cells and spicules are located, would have a lower pressure, because of the suction of water into the choanocyte nests. Sponge cells could thus detect gaps in the pinacocyte sheets from the resulting decreases in water pressure differentials. This would be one way in which cells could respond globally to local inadequacies in anatomical pattern. Such mechanisms are the sort of thing we need to be looking for as elements of each organisms' mechanisms of self-repair and geometrical homeostasis.

Hydra

The Coelenterata are considered to be, after the sponges, the second simplest of all the phyla of multicellular animals. Of these, the genus Hydra and their close relatives seem to be the ones whose anatomy is least complicated. A hydra's body consists of little more than two epithelial layers, the ectoderm on the outside and the endoderm on the inside, together forming a hollow cylinder, closed at one end and open at the other (at the mouth, with its surrounding "hypostome"). The mouth is surrounded by a whorl of tentacles, each of which is also a hollow cylinder, made of these same two epithelial layers and continuous with them. The basal surfaces of these epithelia are pressed together, with the apical surface of the ectoderm facing the outside world, while the apical surface of the endoderm faces the interior cavity, into which food is engulfed through the mouth. This interior cavity is continuous with hollow spaces running down the center of each tentacle. A network of nerve cells lies between the basal surfaces of these two epithelia; but these do not seem to be very important to structure formation, since hydras have been found to be able to live and maintain their correct morphological structures even after all the nerve cells have been killed.

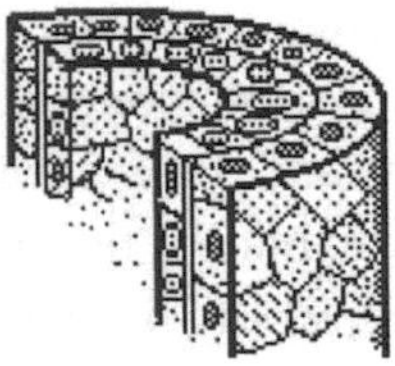

The body and tentacles of Hydra consist of 2-layered tubes.

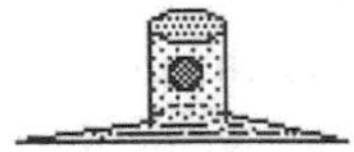

Both the inner and outer layer are made of myo-epithelial cells.

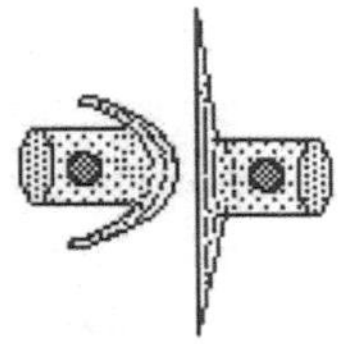

The muscle-like feet of the outer cells are oriented longitudinally; those of the inner cells are oriented circumferentially.

It was once believed that one or more layers of muscle cells were sandwiched between the two epithelia. But it is now known that both the ectodermal and the endodermal cells are myoepithelial cells, with each cell having a muscle-like portion at its basal end. These muscle-like regions are elongated in a direction transverse to the long axis of the epithelial portion. In the case of the ectodermal cells, the muscle-like part of each cell extends longitudinally, parallel to the long axis of the hydra's body. In contrast, the muscle-like part of each endodermal cell extends circumferentially, exactly perpendicular to the long axis. The net effect is as if you had two mutually-perpendicular sheets of muscle cells stuck together, with all the muscles of each sheet parallel to one another, but exactly perpendicular to those of the other sheet. This is one of many interesting cases in which layers of muscles, and sometimes of collagen fibers (or both), become aligned exactly perpendicular to adjacent layers. For example, the interior muscles of the human tongue somehow form themselves into dozens of alternating layers, each perpendicular to the adjacent ones. It will be interesting to find out what physical mechanism is responsible for producing such perpendicular layers, and whether it is basically the same mechanism in human tongues as in Hydra body walls.

Scientific studies of Hydra morphogenesis began 250 years ago, way back in the 1740s, with the classic regeneration studies of Abraham Trembly. The fascinating series of memoirs in which he described his experiments have now been translated into English and collected into book form by Lenhoff and Lenhoff (1986), and are highly recommended. Trembly began with the idea of distinguishing whether hydras were plants or animals based on whether they could regenerate parts that had been cut off. He showed that hydras could regenerate any of their parts, and also that they reproduced asexually by budding off new individuals from the sides of their trunk regions.

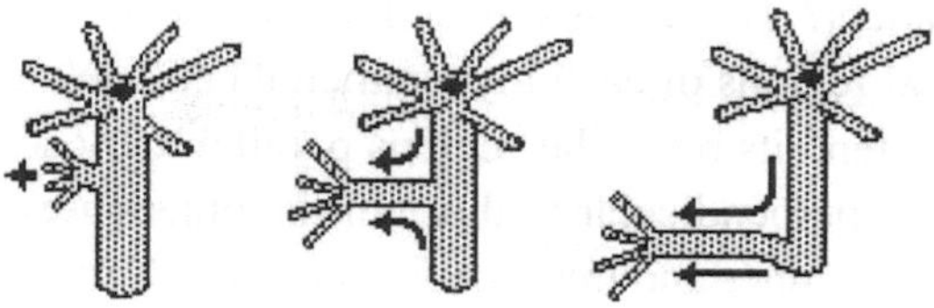

In recent years, our understanding of this budding process has been revolutionized, primarily by Richard Campbell (1967a, 1967b). He showed that budding is not so much a matter of local growth and differentiation, as had been naively supposed, but more of a case of geometrical remodeling and reorientation of the two juxtaposed epithelia. When a new bud forms and grows, this is because portions of the pre-existing or parental trunk have folded outward and reshaped themselves into a cylinder running perpendicular to the one that they had been part of. His student Joanne Otto showed that this is accompanied by (and presumably caused by, in a mechanical sense) rotations in the axes of extension of the muscle-like basal portions of both the ectodermal and endodermal myoepithelia (Otto, 1977). This rotation process is not easy to visualize, but it turns out that the angle by which these muscle-like extensions have to rotate varies as a function of position around the circumference of the bud. No rotation is necessary on the side of the bud facing up the trunk toward the mouth and tentacles; nor does rotation occur in those myoepithelial cells making up the side of the bud facing down toward the base or foot of the parental stalk. On the lateral sides of the bud stalk, however, these muscle-like bases rotate their axes of extension by 90 degrees. Intermediate amounts of basal rotation are required in the parts of the bud stalk in between.

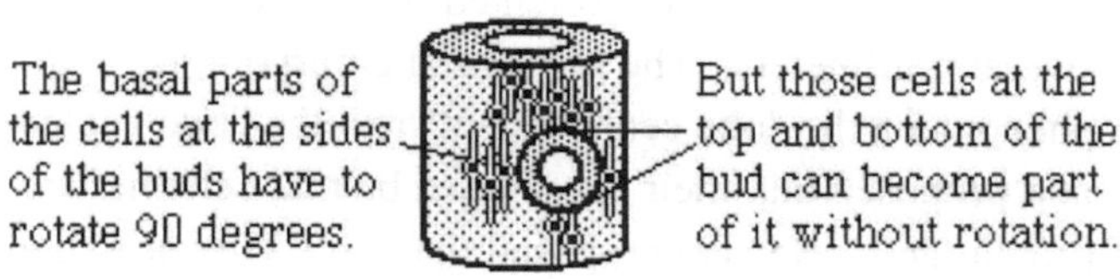

There is much work left to be done on the mechanics of pattern and structure formation in these and other lower animals. What properties of the individual myoepithelial cells cause them to tend to become arranged into double-layered cylinders? How is this related to the elongation and alignment of the muscular, basal regions of each myoepithelial cell? What mechanism causes each cell to extend its muscular regions parallel to those of its lateral neighbors, but exactly perpendicular to those of the other sheet? What differences between the inner and outer cells makes the former extend these processes circumferentially, but makes the latter extend them longitudinally? During the process of budding, what is the nature of the propagated change that makes these myoepithelial cells switch from being part of the old cylinder to become part of the new, buddling cylinder? Could this process be made to run backward? How is this process related to the gradual transfer or flow process by which myoepithelial cells of the trunk are remodeled into the narrower tubular arrangement of the tentacles? Is this process basically the same as the one that occurs in budding? What properties of the cells determine the cross-sectional diameters of the cylinders that they will form? In particular, when myoepithelial cells of the trunk become part of a tentacle, what mechanical or behavioral change in the cell's properties makes the tentacle diameter so much less?

Fibroblast traction and alignment of collagen to form tendons and skeletal muscles.

For many years, the standard substrata for tissue culture consisted of clotted lymph or blood plasma. It was frequently observed that the fibers of these gels became radially distorted in areas surrounding explants or other concentrations of cells. Likewise, in regions between nearby concentrations of cells, the gel fibers tended to become oriented into parallel arrays, even over distances of several millimeters. The influential embryologist Paul Weiss named these phenomena the "one center effect" and the "two center effect", respectively, and proposed that their cause must be that cells induce local shrinkage of protein gels. He also proposed that similar events should occur within the developing embryo, where the alignment of matrix proteins could serve important morphogenetic functions (reviewed in Weiss, 1955).

His idea was that biochemical activities of the cells induced the protein gels to shrink in volume, supposedly because cells absorbed water from the gels in proportion to the cells' growth rates. This idea of chemically induced shrinkage remained dominant for many decades. Any alternative interpretations, especially any suggestion that the cells might themselves be physically pulling on the gels, were dealt with quite severely (see Katzberg, 1951; Weiss, 1952). Weiss' main supporting evidence was that nerve cells produce little or no distortion of plasma clots, in contrast to glial cells (the idea being that glial cells grow and divide, while nerve cells do not). He concluded that the ability to divide must determine the capability to distort substrata.

Assuming that these alignment phenomena also occur inside developing embryos, with fibers of extracellular matrix proteins such as collagen being distorted and aligned in much the same patterns as had been observed for blood clot fibers in tissue culture, Weiss suggested that clusters of growing cells should be able to produce long-range alignments of matrix fibers and that these fibers could serve to guide nerve fibers to their proper points of connection. By orienting matrix fibers radially around themselves, masses of growing cells (limb buds, for example) could steer nerve fibers toward themselves. These ideas were very influential at one time.

Simulated by a conversation with Michael Abercrombie, I attempted to show that the alignment effects were actually mechanical in nature, being caused by the same forces that the cells of the body use for their locomotion (Harris, 1973). The initial evidence was that the distortions of plasma clots by individual cells were longitudinal, along the axis of cell spreading as opposed to being radially symmetrical about the main cell body, and also that this distortion was reversed every time the cell underwent mitotic division, or if it were poisoned with cytochalasin (which inhibits actin assembly), or if the cell died. Proponents of induced shrinkage stoutly maintained, however, that all these effects could be accounted for by various combinations of localized water uptake and release. They did not give up until we developed the silicone rubber substratum technique (Harris, Wild and Stopak; 1980), showed that cultured cells produce identical patterns of distortion on sheets of inert silicone rubber, that glial cells are much more strongly contractile than nerve cells (recalling Weiss' interpretation of the differing gel-distortion capabilities of these two cell types), and that localized water uptake (using a strongly hygroscopic salt) will actually not produce the localized protein gel shrinkages that had been postulated (Stopak and Harris, 1982). Everyone now seems to accept that the one and two center effects result from active cell contractility.

The traction forces
by which tissue cells crawl
are made visible by wrinkling
of thin silicone rubber substrata.

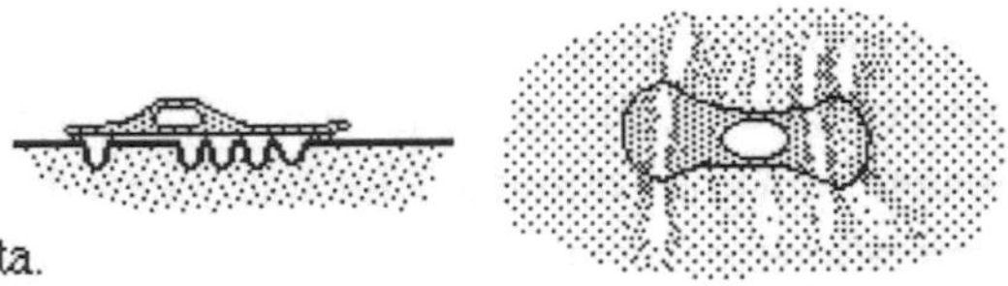

On the other hand, our results lent increased support to Weiss' ideas about the embryological significance of these effects. Stopak and I showed that collagen gels are even more sensitive than plasma clots to the two center effect. Rectilinear alignments are produced over distances at least an order of magnitude greater (centimeters rather than millimeters) within only a day or so. When fragments of embryonic heart (or other tissues containing either numerous fibroblasts or glial cells) were placed in gels made of collagen (dissolved from rat tail tendons), we found that the cell's pulling forces were sufficient to distort the collagen radially around each mass of cells, and to align the collagen into linear tracts running from one mass of cells to another. These tracts frequently extended all the way across petri dishes and were clearly visible to the naked eye. When observed between crossed polarizing filters, the birefringence of the aligned collagen fibers produced effects reminiscent of searchlight beams.

Two center effects occur
even more strongly
and over longer distances
in collagen gels than in
plasma clots.

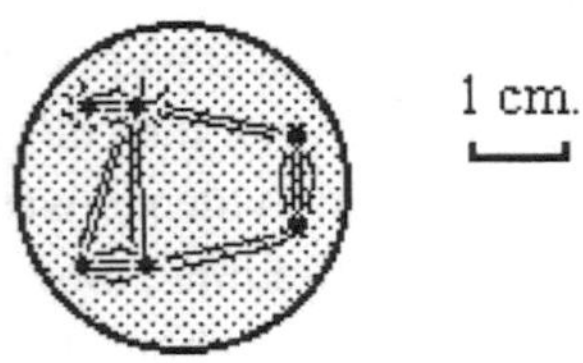

Stopak and I first thought of these tracts in terms of the sorts of nerve fiber guidance pathways that Weiss had hypothesized. When we did experiments to test this, however, we found that nerve explants placed next to the collagen tracts did not result in as much steering of the outgrowing nerve fibers as we had hoped. Gradually, we concluded that the collagen tracts themselves correspond to anatomical structures, with ligaments, tendons and even skeletal muscles being the specific structures we had in mind (Harris, Stopak and Wild, 1981). Our basic idea was that collagen, elastin or other extracellular proteins would become aligned along axes of mechanical tension produced within developing embryonic tissues by pulling activities of cells at specific locations, for example at the so-called "origin" and "insertion" sites on the surfaces of developing bones and other skeletal elements. When myoblast cells are present within the area of this tensile alignment, then these would likewise be aligned and concentrated together to form typical skeletal muscles.

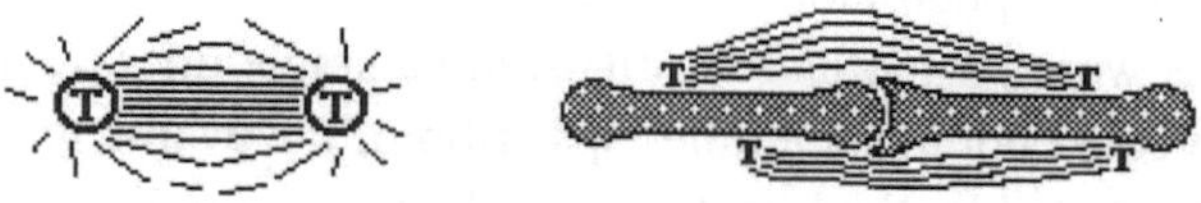

The Stopak-Harris Theory proposes that ligament, tendons and skeletal muscles are formed as mechanical effects of traction forces exerted by cells at the sites (so-called "origins" and "insertion" sites) of attachment to the skeleton.

This proposal differs greatly from the long-accepted view that the main bodies (or "bellies") of skeletal muscles first organize themselves (by unspecified mechanisms) and then actively migrate later to seek out their proper sites on the skeleton and form attachments to them. For example, they might be attracted chemotactically to these sites. Our proposal is that the muscles are stretched into shape by tensile stresses exerted by cells at what will become the attachment sites. Use of the names "origin" and "insertion" to refer (respectively) to the more proximal and the more distal of these attachment sites seems to reflect this traditional class of theory.

Despite personal discussions with some very tenacious advocates of such views, I have been unable to learn what actual evidence there is supporting them, other than that such views have been believed for so long that most histological studies of developing muscles describe their static sections of fixed material on the assumption that the muscles were somehow actively seeking out and finding their connections (for example, see Cihák, 1972). But if anything in these sections should have been different had cellular traction forces been responsible for pulling muscle tissues toward their attachment sites, then it would help to know what these differences might be. Our hypothesis is at least equally consistent with the very important observations of Chevallier, Keiny and Mauger (1977), who deprived developing chick limb buds of muscle cells and showed that tendons then develop without muscles. Those result are exactly what our theory would predict.

Our experimental evidence in favor of what we called our "tractional structuring" mechanism was of two main kinds. First, there were organ culture experiments in which Stopak explanted pairs of embryonic chicken cartilages into collagen gels. These cartilages were positioned end to end, as they would have been inside a developing limb bud, and a dozen or so fragments of embryonic back muscle tissue were scattered near them. These preparations were then maintained in a tissue culture incubator for a week, with the positions and rearrangements of cells and collagen matrix being recorded periodically in photomontages. (Unfortunately, the dimensions of these preparations were far too large to fit within a single microscope field, which is why we could not make a time-lapse film of the events.) The results were exactly as predicted by the hypothesis: collagen was pulled in toward the cartilage surfaces by the locomotory activities (traction) of perichondrial cells, with a progressive concentration and alignment of collagen parallel to the axes of the cartilages. The muscle fragments were likewise pulled in toward the cartilages, and were stretched along their axes. The masses of muscle cells became attached to the different spots on the cartilage surfaces by dense strands of aligned collagen fibers. The aligned masses of muscle cells, with the strands of collagen at their ends, closely mimicked the normal morphology of leg muscles. When they began twitching spontaneously, their contractions flexed the cartilages relative to one another, exactly as if they had been real, anatomical muscles.

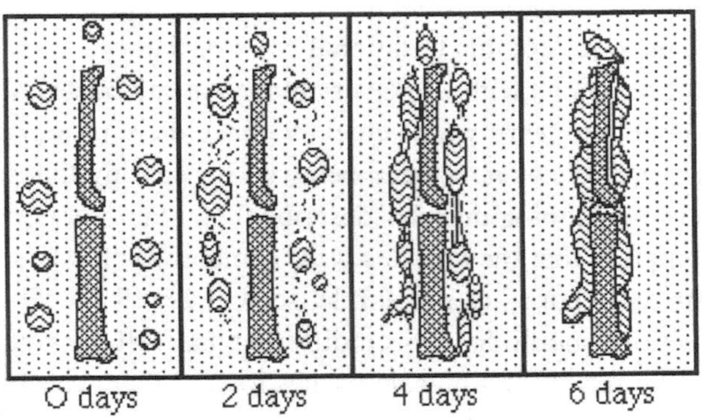

Our second line of experimentation was based on the injection of fluorescently labeled collagen directly into developing chicken limb buds (Stopak, Wessells and Harris; 1985). After injection, the collagen quickly gelled, and the embryos were allowed to continue development for a few days. They were then fixed and (paraffin) sectioned for histological examination. The idea was to see whether the injected collagen would be mechanically rearranged to become parts of whatever anatomical structures happened to be forming at the sites of injection. This was exactly what occurred, with labeled rat collagen having become reorganized into parts of tendons in some sections, into parts of perichondria in others, and sometimes into parts of the dermis, as well as parts of the walls of arteries.

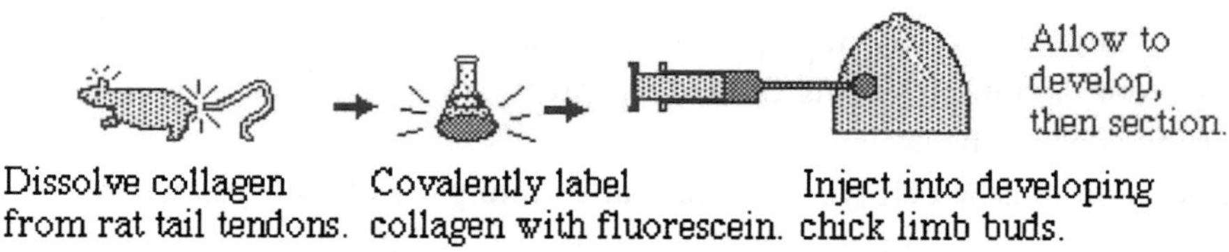

These results are what should occur if imposed mechanical forces are the cause of collagen's geometrical arrangement in the body. In contrast, there would be no reason to expect any such rearrangement if, as the main opposing hypothesis supposes, collagen is simply secreted into place. Likwise, if the causation of collagen patterns had been a matter of molecular self assembly, then collagen derived from rat tails should not have been able to become part of all different sorts of anatomical structures in chicken limbs.

It has long been known that when myoblasts are grafted from one part of the embryonic body to another, they become arranged according to the morphological pattern of muscles and skeletal attachments that is normal for the site to which they have been grafted. This is as opposed to developing whatever geometrical patterns would have been appropriate for the site <u>from which</u> the graft had been derived. For example, if you graft muscle precursor cells from the back to the leg, then the muscles into which they will develop will have the shapes and skeletal attachments appropriate to leg muscles, not those of back muscles. The converse is also true. This implies that muscle geometry is somehow controlled or created by some influence from the skeleton (or possibly other tissues, although the skeleton seems most likely).

Not that this influence would have to be mechanical, of course. But if development of the musculature were really a matter of muscles first forming and then seeking out appropriate tendons and skeletal attachments, then one would expect grafted precursors either not to connect at all, or to form shapes appropriate to the souce of the grafted tissue. Even without grafting, one would also expect that muscles would sometimes develop without skeletal attachments at one or both ends, as a result of failure to form connections; but this seems not to occur. Abnormalities of other kinds are surprisingly common in the musculature (Grant, 1962). Two cadavers seldom if ever have quite the same combinations of muscular attachments, nor match the idealized drawings in the textbook. The usual nature of these abnormalities, however, is for muscles either to have extra points of skeletal attachment, to attach where some other muscle should have, or to be missing altogether. It is not unusual for extra muscles to form so as to connect between the origin and insertion sites for two entirely different and unrelated muscles (Barlow, 1935). These are exactly the types of abnormality that would be expected to result from the muscles being formed by forces exterted at these origin and insertion sites on the skeletal surface.

Even non-muscle cell types, when grafted into parts of embryos where skeletal muscles are forming, can become rearranged into the correct geometric patterns and alignments of the particular muscles formed at that site. This was discovered by Grim (1991), who took cells from the lateral epiblast region (not muscular in nature) of quail embryos and grafted these into premuscular masses of chicken embryo wing buds. In one experiment, histological sections through the wing that developed showed that considerable portions of one of the wing muscles (the extensor medius longus) consisted of fibroblast-like quail cells. As a believer in the theory of "Positional Information", Grim himself interpreted these results as indicating that the grafted cells were receiving chemical signals that had been meant for myoblasts and (despite not being muscle cells) were able to respond to these signals by rearranging and aligning themselves into the geometry appropriate for cells at that location. He was not receptive to the idea that the "signals" could have been physical forces. Perhaps it will be neceassary to show that some entirely inert material, such as silicone rubber, will also be remolded into muscle-like shapes and inserted into the right place at the critical time.

Toward a Concept of Geometrical Homeostasis:

A major step toward understanding morphogenesis would be the extension of the concept of homeostasis to make it apply to structural arrangements of cells, instead of just to the usual physiological quantities like body temperature and ion concentrations (Harris, 1987). Without the homeostasis concept, the science of physiology would have stagnated at the stage of mere description. Much of development will likewise remain incomprehensible until we learn how to think about anatomy in terms of balances between sets of opposed forces that form negative feedback cycles. Whenever anatomical patterns arise spontaneously, or resist change, or restore themselves after external disturbances, we need to look for balances between opposed forces. This is especially true of cases where cells rearrange themselves spontaneously, jumping from one arrangement to another, as in gastrulation and neurulation. By regarding such events as consequences of one homeostatic state being supplanted by another, we can make sense of cases where embryonic cells arrive at the same eventual geometry by way of two or more different alternative pathways. Instead of just paying attention to the physical forces that drive a given rearrangement, we must also ask why cells first "prefer" one spatial arrangement, but then begin to prefer a different one. For most of us, such ways of thinking will be a big change.

There are two main obstacles to thinking of embryonic development in terms of geometrical homeostasis. First, there is the common misinterpretation of thermodynamics that regards any tendency to gravitate to one state in preference to others as evidence that the preferred state minimizes free energy. The second obstacle is the lack of good metaphors for self creating structures. A few technological examples were mentioned in the second section of this paper, and D'Arcy Thompson's classic book "On Growth and Form" catalogued many parallels between inorganic patterns and anatomical ones. But such concepts as "surface tension" have been greatly over-extended and taken much too literally as if Thompson's point had been that the actual physical forces creating biological structures were the same as those that cause water droplets to round up, rather than that the mathematical rules governing the biological process were sufficiently similar to provide openings for deeper analysis. This second obstacle can best be overcome by computer simulation.

The fallacy of energy minimization

I am not aware of anyone claiming that homeostatic maintenance of body temperature proves that 37 degrees Celsius must therefore be some kind of thermodynamic minimum free energy state. But when the variables being held constant are geometrical, many people begin to suppose that energy minimization must be responsible. Whole auditoriums full of biologists have been intimidated into accepting this fallacy, on pain of being thought not to understand thermodynamics. It deserves detailed rebuttal.

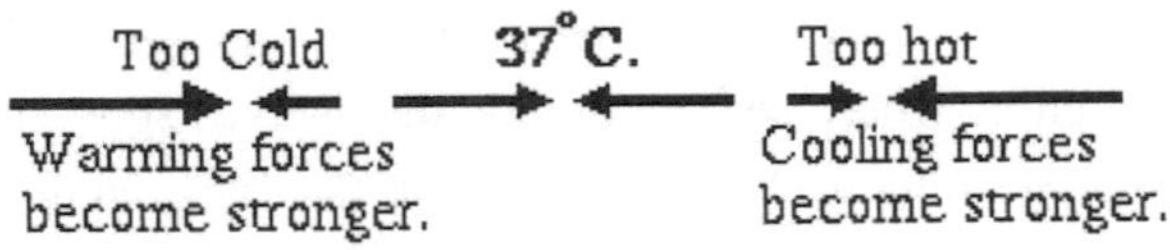

To begin, recall how body temperature is kept constant. The rates of heat production and heat loss both vary as functions of current body temperature. Temperatures above 37 stimulate increased sweating, as well as other mechanisms for dispersing heat, such as dilation of peripheral arteries. Conversely, cooler body temperatures stimulate shivering and other means for producing more heat, and for preventing its loss. Only at 37 are the heat-producing and heat-dispersing mechanisms exactly balanced against one another. Equally essential, but usually taken for granted, is that these imbalances between heating and cooling are always in whichever direction will pull body temperature back to normal, at least at all nearby temperatures. This last point is very important. It is not enough that the opposing "forces" be balanced at one point; they also have to be imbalanced at all nearby points, and the direction of all these imbalances must be toward the balance point. Otherwise, you get things like heatstroke and hypothermia, where the heating and cooling forces become imbalanced, all right, but the directions of these imbalances are away from 37 degrees, rather than toward it!

These same principles apply to the homeostatic control of geometrical variables. Consider the following examples: first, a suspended weight pulling against the elasticity of a spring; and second, a motor boat moving upstream against a current, with this opposing current becoming stronger the further the boat moves upstream.

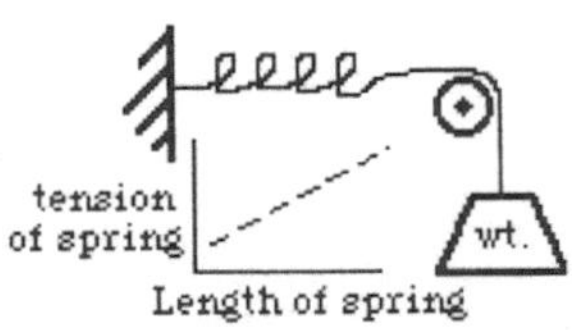

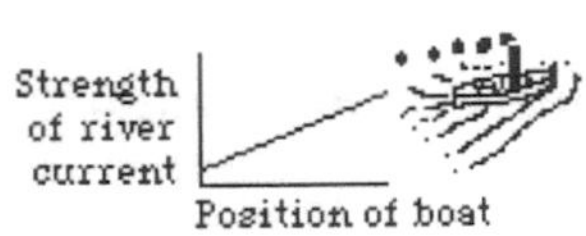

In both cases, the systems gravitate to a certain geometrical configuration in preference to all others. This is the configuration at which the opposing forces are stably balanced against each other. In the first example, stability occurs when the spring has been stretched enough so that its pulling force is exactly equal to the gravitational force on the weight. If you perturb this system, for example by pulling the weight downward and then letting go, then the tension in the spring will become temporarily stronger than the weight's gravitational force; so the weight will be pulled back up just enough to bring the forces back into balance.

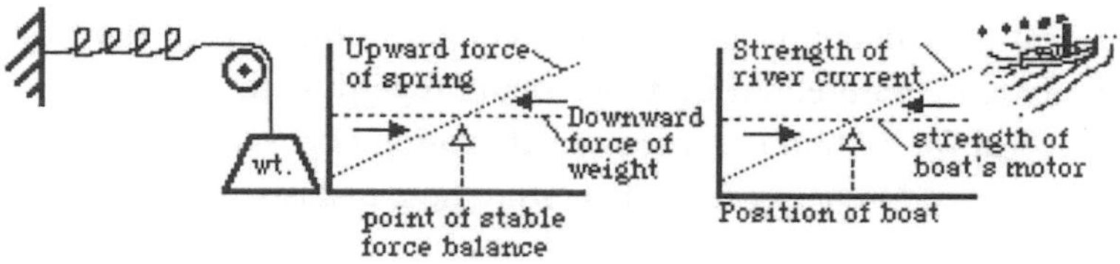

Similarly, the boat will move upstream until the force of its motor is exactly equalled by the opposing force of the current (because the current gets stronger the further upstream the boat is located). If you push the boat further upstream, then the leftward force of the current becomes stronger than the rightward force of the motor, so the boat will be carried downstream just enough to restore the balance.

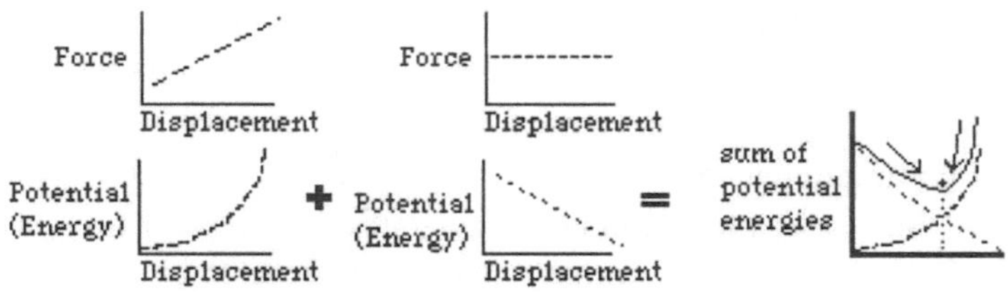

These examples also illustrate the relationship of stability to energy minimization. In the case of the spring and the weight, the potential energy of the weight itself increases linearly with its height, and thus in proportion to leftward movement of the point where the string is attached to the spring. The potential energy of the spring, on the other hand, increases as this point of attachment moves to the right (because the spring is more and more stretched); and it increases more than linearly (approximately as the square of the amount of stretching). In both cases, the potential energies are the integrals of the forces, integrated relative to the displacements. If you sum these integrals, then this sum passes through a minimum at the point where the forces are exactly equal and opposite.

What about the potential energies of the boat? Of course, there is nothing to prevent you from trying to take integrals of the forces acting on the boat moving against the current. Indeed, the sum of these integrals would also be a minimum at the point where the two forces are exactly balanced against one another. Even in the case of the homeostatic control of body temperature, you could calculate some kind of integral of the various "forces" acting, such as the amounts of sweating and shivering at different temperatures; and the sums of these integrals would likewise have their minima where these opposing sets of "forces" are stably balanced. The trouble is, the nature of these forces are not such that their integrals are energies, neither in the case of the boat nor that of body temperature. At best, one could regard them as pseudo-energies, possibly useful for certain purposes of calculation or visualization. For example, the pseudo-energy for temperature homeostasis would look something like this.

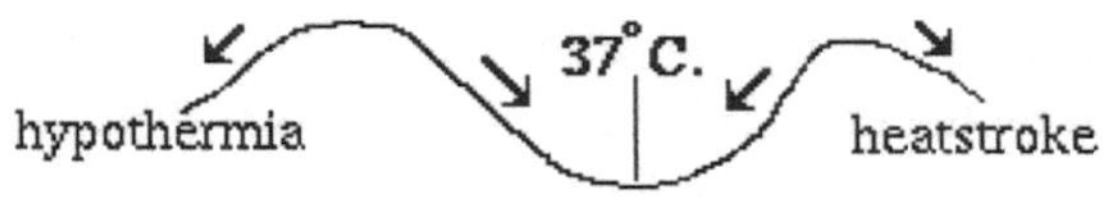

By taking integrals of the relative strengths of all the different heating and cooling "forces", one could add these integrals together to produce a curve whose slope at each point equals the net amount of heating or cooling as a function of temperature, with the aid of which we could visualize the cooling and warming effects as if a little ball were rolling down hills into a valley (such a valley being called an "attractor basin"). This also dramatizes the fatal effects of getting either too hot or too cold, so that the little ball escapes from the valley. Such metaphors can be very useful, so long they aren't taken too literally. Only when all the "forces" involved happen to be of certain kinds, like springs or weights (that, among other things, don't continue to expend energy in a stable state), will the integrals of their strengths over their displacements really be potential energies. Conversely, it can help to regard such forces as derivatives of their potential energies. This strategy is the basis of many applications of mathematics to physical problems, which allows vectors (e.g. forces) to be dealt with in terms of variables that are scalars (e.g. energies). Engineers sometimes use this method to predict how materials will behave under stress. The same principles apply in the case of thermodynamic free energies, which are in these respects equivalent to potential energies.

The forces that create shape must themselves change in strength as functions of the current shape of the material whose shape they serve to change

If a system does not need to minimize energy for it to gravitate to one geometrical arrangement in preference to others, what then are the actual requirements?

a) There must be two or more mutually opposed forces, that collectively are strong enough to change the geometry of the system.

b) The relative strengths of at least some of these forces must change as functions of the existing geometry.

c) These changes in strength must be such that the forces are exactly counterbalanced when (and only when) the correct geometrical shape or arrangement exists.

d) For all (nearby) shapes or arrangements, the directions of the imbalanced forces must be such as to pull the system back toward this correct geometry.

Notice that the essential controlling elements are the rules of variation of the different forces. The forces themselves can be almost anything; so can the reasons for the variations in the relative strengths of these forces. But just as when you want to make a system gravitate to a certain temperature, then you have to change the relative strengths of the heating and cooling "forces" according to whatever temperature currently exists, the regulation of geometrical properties requires changing the relative strength of physical forces. If you want a system to gravitate to a certain size, then those "forces" that tend to make it grow have to become weaker as its size increases, and whatever "forces" tend to make it shrink need to become stronger. The same is true for length, curvature, or any variable or combination of variables whatsoever. The essential principle is that those forces that increase the variable have to get weaker when the variable gets too large, and they have to get stronger when it gets too small.

This is really just common sense, once you start to look at shapes in terms of forces. Any system that gravitates to a certain preferred geometry is plainly subject to forces whose effect is to change that geometry. Likewise, the relative strengths of these forces obviously must be changing as some function of the existing geometry, else they couldn't be balanced in one state but not in others. So it follows that if you could control these rules (by which the forces vary as functions of the existing geometry) then you could make the system gravitate to whatever shape or arrangement you wish. What I therefore propose is that (above the size level of molecular self-assembly) genes control larger-scale anatomical geometry in just this way. This approach has the virtue of explaining how cells "home in" on a given shape, and reach the same geometry by more than one series of intermediate stages.

There are, of course, many different kinds of reasons why the strengths of morphogenetic forces can vary in strength as functions of the geometric properties that they are serving to modify. Some of these reasons amount to little more than leverage effects. A good example is that the lateral force exerted by a stretched sheet or cord will vary in proportion to the sharpness of its curvature. The sharper the bend, the stronger the lateral component of force (for a given amount of tension). On the other hand, in the case of an actively contracting element such as acto-myosin, one can imagine situations in which properties like curvature or size actually feed back to control the force being generated. Any changes in the effective strengths of the forces can accomplish the desired effect — for forces to be balanced only when the one desired geometry exists.

If we regard the capacity for self-repair as evidence for some kind of geometrical homeostasis, then what sense can we make of something like gastrulation, where the cells previously had been gravitating to one geometrical arrangement, that of the blastula, and that includes tending to restore that arrangement following disturbances, but now jump spontaneously to new spatial arrangements? The answer is that the rules must have changed. It could be that new forces have suddenly come into existence. Our greater familiarity with outside-in manufacturing methods tends to make us assume that new movements mean new forces. But if we think of morphogenesis in terms of internal balances of forces, then it should become clear that the true cause might be some change in the rules obeyed by some of the forces, specifically the rules by which their strengths vary as functions of the existing geometry. A fever is not due to new sources of heat generation, but rather to some change in the negative feedback that has shifted the temperature at which the heating and cooling mechanisms become counterbalanced. Morphogenesis can likewise be due to changes in the rules controlling the relative strengths of already-existing forces. In that case, it would fall in the category of what the French mathematician René Thom has called "Catastrophes", and the whole subject of "Catastrophe Theory" may need to be restudied by someone who understands both the mathematics and the biology.

A reconsideration of histotypic cell sorting in terms of contractile forces, combined with selective cell-cell adhesion

The energy minimization fallacy discussed above has been most fully developed in relation to the specific phenomenon of cell sorting. I refer, of course, to Malcolm Steinberg's differential adhesion hypothesis (Steinberg,1962a, 1962b, 1963, 1970). The essence of this theory was to regard cell sorting as logically equivalent to the behavior of combinations of immiscible liquids (oil and water for example) because of surface tension, except with cell-cell adhesion forces playing the role of molecule-molecule attraction forces. Just as a liquid with a higher surface tension will form droplets enclosed by drops of another liquid having a lower surface tension, cells of one differentiated cell type will sort out to an interior position relative to another. This hypothesis proposed that cells sort out internally because their cell-cell adhesion forces are stronger. It is well known that the surface tension of liquids is caused by imbalances in their intermolecular attractive forces, with those molecules that happen to be located at surfaces (as well as those at interfaces between one liquid and another) tending to be pulled into the interior; the net result is almost exactly as if each surface and interface were contractile, with the strength of its contractile force being proportional to the strength of the intermolecular attractive forces. In fact, early physicists believed that surface tension must be caused by the formation of some kind of contractile film at liquid surfaces; this is still widely believed, and the macroscopic behavior of liquids would be essentially the same if it were true.

Three main pieces of evidence were put forth by Steinberg and his students in support of this interpretation:

1. A transitive hierarchy predicts the relative positions to which different combinations of cell types will sort out.

for example:

Heart sorts internally to liver. But heart sorts externally to pigmented retina. Pigmented retina sorts internally to liver.

2. If one cell type sorts out internally to another, then the more internally sorting cell type is the one that will be engulfed by the other.

3. Those cell types that sort out more internally also resist flattening of their aggregates more.

These observations are extremely important, whether or not one accepts the hypothesis that motivated them. Alternative explanations are possible for each of them. In the first case, the same transitivity would be predicted by any and <u>all theories based on quantitative differences</u>. If A is more strongly contractile than B, and B is more strongly contractile than C, then it follows that A will also be more strongly contractile than C. No matter what quantitative variable anyone had proposed, it would have made this same prediction; and so the laborious experimental demonstration that the hierarchy really is transitive, for all combinations of 8 different cell types that were tried (Steinberg, 1970), does not support one quantitative theory any more than another. Likewise, B merely indicates that the same forces are responsible for engulfment as for sorting out, while C indicates that these same forces are also responsible for the tendency of cell aggregates to round up. All three sets of observations are equally consistent with the possibility that the quantitative variable responsible is the cells' cortical contractility, rather than the cells' adhesiveness to one another (Harris, 1976). This would be an ironic contrast to the surface tension of liquids, where mutual attraction of subunits simulates the effects of a contractile surface layer. In the case of cells and cell aggregates, surface contractility really is the cause of the various surface tension-like effects, including rounding, engulfment and sorting.

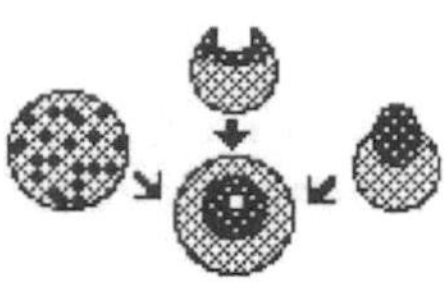

"Whereas…neural material moved inward either by way of infolding of a plate of cells, by immigration of scattered cells (or) as a solid lump of cells [or by sorting out]….One might wish to find a common mechanism for these 3 modifications of inward movement."

Townes and Holtfreter (1955) J. Exp. Zool. 128: 53-120

These observations on engulfment recall earlier observations by Townes and Holtfreter (1955). They observed that combinations of neural plate cells with certain other cell types would consistently arrive at the same end result, namely a ball of neural plate cells surrounded by a mass of the other cells. When both cell types were dissociated and mixed, then the neural cells would sort out to the interior position; when the neural cells were left as a sheet and laid on the top of the other cells, then they would fold up and sink into the interior position; and if aggregates of neural plate cells were put in contact with aggregates of the other cell type, then the neural cells would be engulfed by them. In all three cases, the same geometrical arrangement would be created, with this arrangement being remarkably similar to the normal anatomical positioning and structure of the neural tube (even to the point of forming a cavity within the mass of neural cells). As quoted below, Townes and Holtfreter surmised that some single basic mechanism must be able to produce all three types of rearrangement, not just sorting but also invagination and engulfment.

Subsequently, the infolding of the neural plate has been shown to be caused by active contraction of acto-myosin in the exposed surface of the neural plate (Karfunkel, 1974). The obvious logical implication would be that increased cortical contraction should also be the cause of the engulfment of one aggregate by another, as well as of sorting out to an internal position. This would mean that Phillips and Steinberg's (1969) measurements of what they thought were the reversible works of cell-cell adhesion were actually measurements of the force of active contraction in their different cell types.

Active cortical contraction can exactly mimic surface tension.

The essential factors are that each cell's cortex should contract actively, that this contraction should be stronger where they are touching either cells of other differentiated types or the outside medium, and that this cortical contraction should be stronger in some cell types than others. In any combination of cell types, the one whose cortical contractions are stronger should sort out to the more internal position. It should likewise be engulfed by the other.

Of course, none of these cellular behaviors could occur if the cells involved did not adhere to one another. But there are two different questions here: first, what causes like cell types to get back together and stay together; and second, what controls the relative positioning of cells, inside versus outside. Except for Holtfreter, almost everyone has assumed that both these questions have the same answer, namely differences in cell-cell adhesion, with Steinberg advocating quantitative differences and most others qualitative differences in the cell adhesion molecules. Some of the strongest evidence for the latter view was provided by Nose, Nagafuchi and Takeichi (1988). They transfected genes for two different cell adhesion molecules into two groups of fibroblasts, and showed that these fibroblasts would sort out from one another according to which kind of adhesion molecule they expressed. Since the cells were otherwise identical, this seemed to clinch the argument in favor of qualitative differences. But when you examine their illustrations carefully, it becomes apparent that the geometric patterns of sorting differed greatly from what occurs when one differentiated cell type sorts out from another; the patterns lack those specific features that Steinberg's theory was designed to account for. Neither "cell type" adopts a preferred position inside versus outside; neither do the boundaries between cell types tend to minima.

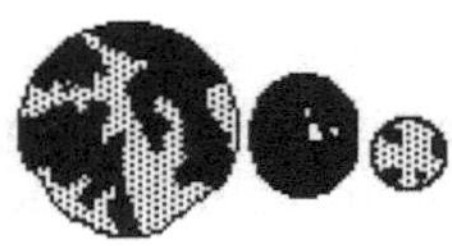

Cells transfected with P-cadherin genes
Cells transfected with E-cadherin genes
Cells differing only in adhesion proteins will sort out, but do not adopt any consistent relative position (inside versus outside)

I interpret this to mean that there are two elements to cell sorting. One element is that of getting like cell types together with one another, and this is caused by the specificity of the cell adhesion molecules, such as N-CAM and the different cadherins. But this is not what determines the relative positioning of the different cell types. Positioning is controlled by difference in the strengths of the cells' cortical contractility, with those having the strongest contractility sorting out to the more internal position. Steinberg and his collaborators mistook these differences in contractility for differences in adhesiveness, but they were right to attribute the relative positioning of cells to a quantitative variable; and their measurements of this variable probably are also essentially correct, even if they are really forces per length rather than energies per area.

How can mechanicians, engineers and applied mathematicians help to solve these and other embryological problems?

Systems of interacting forces and stimuli don't have to be very complicated before the unaided human intuition can no longer predict accurately what the net result should be. At this point computer simulations, or other mathematical models, become necessary. Without the aid of mechanicians, and others skilled in simulation and modeling, developmental biology will remain a prisoner of our inadequate and conflicting physical intuitions and metaphors. How far would physicists have progressed if restricted to their imaginations in predicting what net result ought to follow from a given combination of forces?

Physicists, of course, make their predictions with mathematics, much of which was invented for their use. Many think that if biology could only be made more quantitative, this same mathematics could be applied to it. I think they are naive, and that we will almost always have to use simulations instead (as do physicists, when confronted with systems anywhere near as complex as biological ones).

As we have seen in the preceding several sections, the more usual situation for developmental biologists is that of already knowing the net result (the spontaneous formation of a certain anatomical pattern, for example), and needing to figure out what set of interactions produced this result. Given the solution, what is the equation? Sometimes, the biologist is unable to think of even one possibility. Unfortunately, there is something even worse than that — namely when the biologist can think of one possibility (but only one)! In fact, you almost always have to pass through one, on the way to two, three and four.

Nothing has been more common than for erroneous theories to dominate thinking for long periods, not because any alternatives to them had been tested and disproven, but merely because no alternatives had come to mind. This is where the modeler is needed: to provide as many alternatives as possible, and to direct attention to experimental criteria by which we can eventually decide which alternative is correct. Unfortunately, modeling too often plays an opposite role: that of shoring up dominant hypotheses. How many published computer simulations even consider more than one hypothesis, let alone proposing any new alternatives? Instead of seeking out the weakest points in each theory, the aspects where each would (unless correct) be most vulnerable to disproof, most modelers try their best to make their chosen theory work convincingly. As a result, the best they can hope for is to prove their theory is not self-contradictory, or at least not irretrievably so. The main effect is that the prestige and weight of mathematics and computers seem to have been set onto the evidentiary scales on the side favoring whatever hypotheses the model happened to be based on. Simulations are taken as evidence that a theory is true, and often are presented as such.

Most theories about morphogenesis started out as metaphors, sometimes to natural phenomena (such as the surface tension analogies discussed above), and occasionally to human construction methods. (If the genes contain the blueprints to anatomy, do the gas laws contain the blueprints to weather?) Modelers can be of most service in broadening and generalizing each metaphor as much as possible. In fact, opportunities to do this constantly present themselves whenever you try to reduce a given physical phenomenon to its most essential abstractions. You just have to ask yourself, "Aren't there some other physical phenomena that have abstract properties similar enough that, if you had been modeling them instead, you would have gotten very similar equations, etc. to the ones you did get?" The point is that any two combination of forces that would have been reducible to the same equations would, therefore, also have produced the same net results. That means it should be added to the list of alternative hypotheses. For example, in modeling surface tension, it can be very helpful to regard exposed surfaces as contracting with forces proportional in strength to those forces with which the subunit elements tend to maximize their contacts with one another. The net effect is almost indistinguishable. This should alert us to the possibility that any behavior resembling surface tension might really be caused by active contraction. What seem to be side effects could really be causes.

Working together, each lending his special skills to the other, the engineer and the experimental biologist may expect to discover mechanisms that are both important and ironic.

Literature Cited:

Ankel WE Eigenbrodt H (1950) Über der Wuchsform von *Spongilla* in sehr flachen Raumenn. Zool Anz 145: 195-204

Bard, J (1990) Morphogenesis: The Cellular and Molecular Processes of Developmental Anatomy. Cambridge U. Press, Cambridge

Barlow RN (1935) The sternalis muscle in American whites and Negroes. Anat Rec 61: 413-426

Bjorklund NK Gordon R (1993) Nuclear state splitting: a working model for the mechanochemical coupling of differentiation "waves" with the controlling genes (master genes) Metastabil'noe sostoianie iadra: rabochaia model' mekhanokhimicheskoi sviazi "voln" differentsirovki s upravliaiushchimi genami (master genes). Ontogenez 24: 5-23

Bond C (1992) Continuous cell movements rearrange anatomical structures in intact sponges. J Exp Zool 263: 284-302

Bond C Harris AK (1988) Locomotion of sponges and its physical mechanism. J Exp Zool 246: 271-284

Campbell RD (1967a) Tissue dynamics of steady state growth in Hydra littoralis. I. Behavior of specific cell types during tissue movements. J Exp Zool 164: 379-392

Campbell RD (1967b) Tissue dynamics of steady state growth in Hydra littoralis. II. Patterns of tissue movement. J Morph 121: 19-28

Chevallier A, Kieny M, Mauger A (1977) Limb-somite relationship: Origin of the limb musculature J Emb Exp Morph 41: 245-58

Cihák R (1972) Ontogenesis of the skeleton and intrinsic muscles of the human hand and foot. Ergbn d Anat u Entw Gesch Bd 46, Heft 1, 1-194

Elsdale TR (1968) Parallel orientation of fibroblasts in vitro. Exp Cell Res 51:439-450

Elsdale TR (1973). The generation and maintenance of parallel arrays in cultures of diploid fibroblasts. In: Biology of Fibroblasts, ed. E. Kulonen and J. Pikkarainen, Academic Press, pp. 41-58

Fung YC (1977) A first course in continuum mechanics. second edition Prentice Hall Englewood Cliffs, New Jersey

Galtsoff P S (1925) Regeneration after dissociation, an experimental study on sponges. J Exp Zool 112: 465-483

Gordon JE (1978) Structures, or why things don't fall down. Plenum Press, New York.

Gordon R Brodland GW (1987) The cytoskeletal mechanics of brain morphogenesis. Cell state splitters cause primary neural induction. Cell Biophys 11:177-238

Grant BCJ (1962) An Atlas of Anatomy. fifth edition,Williams and Wilkins, Baltimore

Grim M (1991) Control of muscle morphogenesis and endplate pattern in limb muscles of avian chimeras. pp. 293-297 in Developmental Patterning of the Vertebrate Limb. edit J R Hinchliffe et al. Plenum Press New York

Harris, AK (1973) Cell surface movements related to cell locomotion. pp. 3-26 in "Locomotion of Tissue Cells" Ciba Foundation Symposium, Elsevier, Amsterdam

Harris AK (1976) Is cell sorting caused by differences on the work of intracellular adhesion? A critique of the Steinberg hypothesis. J Theoretical Biol 61: 267-285

Harris AK (1987) Cell motility and the problem of anatomical homeostasis. J Cell Sci Supp 8: 121-140

Harris AK , Wild P, Stopak D (1980) Silicone rubber substrata; a new wrinkle in the study of cell locomotion. Science 208: 177-179

Harris AK, Stopak D ,Wild P (1981) Fibroblast traction as a mechanism for collagen morphogenesis. Nature 290: 249-251

Holtfreter J (1939) Gewebaffinität, ein Mittel der embryonalen Formbildung. Arch. Exp. Zellf. 23: 169-209

Houwink R (1958) Elasticity, Plasticity and the Structure of Matter. Cambridge U. Press. Cambridge

Huxley JS (1911) Some phenomena of regeneration in *Sycon* : with a note on the structure of its collar cells. Phil Trans Roy Soc Lond Ser B 202: 165-189

Huxley JS (1921) Further studies on reconstitution bodies and free tissue culture in *Sycon*. Quart J Micros Sci 65: 293-322

Karfunkel P (1974) The mechanism of neural tube formation. Int Rev Cytol 38:245-271

Katzberg, AA (1951) Distance as a factor in the development of attraction fields between growing tissues in culture. Science 114:431-432

Lawrence PA (1992) The Making of a Fly. Blackwell, London.

Lenhoff SG, Lenhoff HM (1986) Hydra and the Birth of Experimental Biology: Abraham Trembly's memoires concerning the polyps. Boxwood Press, Pacific Grove CA.

Moscona AA (1957) The development in vitro of chimeric aggregates of dissociated embryonic chick and mouse cells. Proc Natl Acad Sci USA 49: 184-194.

Nose A, Nagafuchi A, Takeichi M (1988) Expressed recombinant cadherins mediate cell sorting in model systems. Cell 54:993-1001

Otto JJ (1977) Orientation and behavior of epithelial cell muscle processes during Hydra budding. J Exp Zool 202: 307-322

Otto JJ, Campbell RD (1977) Budding in Hydra attenuata: bud stages and fate map. J Exp Zool 200: 417-428

Phillips HM, Steinberg MS (1969) Equilibrium measurements of embryonic chick cell adhesiveness I. Shape equilibrium in centrifugal fields. Proc Natl Acad Sci USA 64:121-127

Phillips HM, Steinberg MS (1978) Embryonic tissues as elasticoviscous liquids I. Rapid and slow shape changes in centrifuged cell aggregates. J Cell Sci 30:1-20

Phillips HM, Steinberg MS, Lipton BH (1977) Embryonic tissues as elasticoviscous liquids II. Direct evidence for cell slippage in centrifuged aggregates. Dev Biol 59:124-134

Ruoff AL (1973) Materials Science. Prentice Hall Englewood Cliffs New Jersey.

Shames IH (1964) Mechanics of Deformable Solids Prentice Hall Englewood Cliffs N.J.

Steinberg MS (1962a) On the mechanism of tissue reconstruction by dissociated cells I. Population kinetics, differential adhesiveness and the absence of directed migration. Proc Natl Acad Sci USA 48: 1577-1582

Steinberg MS (1962b) On the mechanism of tissue reconstruction by dissociated cells III. Free energy relations and the reorganization of fused heteronomic tissue fragments. Proc Natl Acad Sci USA 48: 1769-1776

Steinberg MS (1963) Reconstruction of tissues by dissociated cells. Science141: 401-408

Steinberg MS (1970) Does differential adhesion govern self-assembly procsses in histogenesis? Equilibrium configurations and the emergence of a hierarchy among populations of embryonic cells. J Exp Zool 173: 395-434

Steinberg MS (1975) Adhesion-guided multicellular assembly; A commentary upon the postulates, real and imagined, of the differential adhesion hypothesis, with special attention to computer simulations of cell sorting. J Theor Biol 55: 431-143

Stopak D, Harris AK (1982) Connective tissue morphogenesis by fibroblast traction I: Tissue culture observations. Dev Biol 90: 383-398

Stopak D, Wessells NK, Harris AK (1985) Morphogenetic rearrangement of injected collagen in developing limb buds. Proc Natl Acad Sci USA 82: 2804-2808

Townes PL, Holtfreter J (1955) Directed movements and selective adhesion of embryonic amphibian cells. J Exp Zool 128: 53-120

Trinkaus J P (1984) Cells Into Organs, 2nd edition. Prentice-Hall, New Jersey

Weiss P (1952) Comments and Communications: "Attraction fields" between growing tissue cultures. Science 115: 293-295.

Weiss P (1955) Nervous system. in "Analysis of Development" (B H Willier, P Weiss and V Hamburger eds.) pp 346-401 Saunders, Philadelphia.

Weissenfels N (1981) Bau und Funktion des Susswasserschwams Ephydatia fluviatalis L. VIII. Die Entstehung und Entwicklung der Kragengeisselkammern und ihre Verbindung mit dem auffuhrunden Kanalsystems. Zoomorph 98: 35-45

Wilson HV (1907) On some phenomena of coalescence and regeneration in sponges. J Exp Zool 5: 245-258

Wilson HV (1911) On the behavior of the dissociated cells in hydroids, Alcyonaria and Asterias. J Exp Zool 11: 281-338

Wilson HV, Penny JT (1930) The regeneration of sponges (Microciona) from dissociated cells. J Exp Zool 56: 73-147

Wolpert L (1969) Positional information and the spatial control of cellular differentiation. J Theor Biol 25: 1-47

The Interplay of Active Forces and Passive Mechanical Stresses in Animal Morphogenesis

L.V.Beloussov
Department of Embryology
Faculty of Biology
Moscow State University
119899 Moscow
Russia

1. A biomechanical view on morphogenesis: scratching over the surface or penetrating inside?

Both for an experienced and for a naive observer the development of a living sample, be it plant or animal, looks, first of all, as a regular succession of complicated changes in the shapes and mutual arrangement of its parts; such a succession is usually defined as a morphogenesis while its components as morphogenetic processes. Invaginations, evaginations and the bending of epithelial layers, condensations of freely moving mesenchymal cells, as well as the changes in shapes and overall proportions of the large masses of almost immobile plant cells may serve as the examples of morphogenetic processes. As was shown by the molecular biology within several last decades, all of these processes are based upon a highly regulated motile activity of the molecular and supramolecular components of the living cells. In the first approximation, all of these processes may be considered as *mechanical*, what means that they are associated with the production of mechanical forces and changes in space positions of the material constituents.

This conclusion may look as almost trivial, until we do not make another step and put the following question: are the mechanical forces and the changes in space positions no more than the end results of the large chains of essentially

NATO ASI Series, Vol. H 84
Biomechanics of Active Movement
and Division of Cells
Edited by N. Akkaş
© Springer-Verlag Berlin Heidelberg 1994

non-mechanical processes or they are themselves deeply involved into the *regulation* of the developmental events? To express this question in a more precise manner, we should remind a widely accepted conclusion that the organic development shares the main properties of so-called self-organizational processes, obeying both dynamical and parametrical regulation and requiring the effective non-linear feedbacks between its dynamical components (Nicolis & Prigogine, 1977; Haken, 1978; Beloussov, 1988; Murray, 1989). Thus, the question may be reformulated in the following manner: do the mechanical factors play any important role in the morphogenetical feedbacks? Even prior to any experimentation we may specify this question in the following way: it is almost obvious (and have been numerously tested - see Bereiter-Hahn, 1987) that the developing organisms have the properties of visco-elastic bodies. In this kind of bodies, the "active" mechanical forces would inevitably create more or less long-termed and long-ranged stresses of tension and/or compression. Any change in the shape or in the particles arrangement within these bodies should be immediately (with a mechanical deformation wave rate) and adequately reflected by the stress pattern. Isn't it possible, under these circumstances, that the morphogenesis is regulated just by the feedbacks between "active" forces and the "passive" stresses of tension and/or compression, which these forces generate? This is the main problem to be discussed in this paper.

The views of the developmental biologists upon this problem are controversial. On one hand, as will be shown later, within at least two latter decades some constructive ideas and models of this kind have been elaborated. On the other hand, still now the dominating positions belong to the view, that "morphogenesis is essentially a slave process which is determined once the chemical pattern has been established" (Murray, 1989). The opposition between the "mechanical" and the "chemical" (or the "reactionary-diffusional", see Meinhardt, 1982) approaches in the modern developmental biology remains to be rather harsh.

By my view, to a large extent such an irreconcibility is a result of misunderstanding. Within the molecular-supramolecular area at least, the "chemical", "mechanical" and also electrical events are really indivisible, creating a common "electro-mechanochemical field" (Goodwin, 1985). The importance of the slowly relaxed mechanically stressed states of the protein molecules and supramolecular structures for their specifical activities have been pointed out recently by several authors, and in particular by Bluemenfeld (1983). This paper deals however mostly with the upper levels processes (cellular and supracellular) where the distinctions between the chemical and mechanical factors seems to be at the first glance better substantiated.

Even here, meanwhile, some bridges between the both concepts can be established. It have been shown, for example (Symes & Smith, 1987), that certain so-called inductive substances in amphibian embryos, which are known to generate the development of the axial rudiments (neural tube, notochord, somites and so on) from the competent embryonic tissue very soon after their application (and prior to other visible effects) stimulate a particular kind of cooperative cell migrations, so-called convergence-extension movements, inevitably associated with the formation of a rather complicated mechanical stress pattern. It seems highly probable therefore, that in some important cases at least there are just mechanical effects of the chemical factors which are crucial for their specifical action. In the other words, the effects of the chemical substances well may be mechanically mediated, and vice versa. In any case it would be strange if the living nature, being confronted with the mechanical forces from the very beginning of its evolution, could not make a use from them for regulating its own developmental processes. In brief, the following advantages of the mechanical factors as morphogenetic regulators over the purely chemical ones may be stressed:

(1) As mentioned previously, the mechanical impulses are spread very rapidly (by biological time scale) and, in the premise of elasticity, are not damped within large distances,

up to entire embryonic body. Hence, they may serve as an effective tool for the long-range interactions.

(2) They produce automatically quite regular geometrical patterns, with a marked orthogonality (e.g. Poisson deformation: longitudinal stretching - transversal compression). This may be responsible for a wide spread of the precisely orthogonal patterns and their combinations in the developing organisms.

(3) They are to a large extent universal, requiring no highly specifical intercellular machinery which is necessary for any kinds of chemical regulations.

If at least some of these assumptions are true, the study of the biomechanics of the developing organisms is not just a scrathing over the surface of developmental processes, as some people believes, but may contribute to our understanding of their fundamental principles.

2. Force-generating, deformation-stabilizing and stress-providing intercellular devices

A set of these devices is highly restricted and conservative. Let us consider them one after another.

2.1. Molecular-supramolecular force-generating devices.

2.1.1. Devices for producing osmotically driven turgor pressure in the cytoplasm, vacuoles and intercellular cavities.

It is well known that the osmotical pressure in cytoplasm, intracellular vacuolar compartments and embryonic cavities (blastocoel, subgerminal cavities in Amniota embryos, gastrocoel, eye cavity) considerably exceeds that of external environment. For cell cytoplasm, this is caused by the excess in intercellular potassium and various anions, partly trapped by actine matrix (Oster et al., 1985). The motile role of cytoplasmic osmotical pressure, often distributed quite unevenly throughout a cell body, have been extensively discussed (e.g. Bereiter-Hahn & Strohmeimer, 1987) and will not be commented here in more details. As to the turgor pressure in embryonic cavities, it was shown to be a necessary prerequisite for normal morphogenesis in the budding fresh-water hydra

(Wanek et al., 1980), vertebrate eye rudiment (Coulombre, 1956) and early amphibian embryos (our own observations, see section 4). An excessive turgor pressure in the above mentioned embryonic cavities is supported mostly by the active transepithelial sodium transport (Chain, 1980; Slack & Warner, 1983; Stern, 1984), accompanied, at least in the case of amphibian embryos blastocoel, by chloride ions. By very rough estimations, based upon the chemical content of so-called Danilchek solution, which is claimed to be isotonical to blastocoel fluid (Wilson et al., 1989) the osmotical pressure within Xenopus laevis blastocoel is about 300 mOsm, exceeding thus the osmotical pressure of a tape water in about 30 times. When comparing with the plant organisms such a value is not so high.

The main morphogenetical function of the turgor pressure seems to be in supporting the stretched state of the epithelial envelopes. Assuming that the blastocoel shell (so-called blastocoel roof) in amphibian embryos is of a hemispherical shape we can calculate its tensile stress by a standart formula $S = rP/2t$ where r is a radius of a hemisphere, P is osmotical pressure value and t is the thickness of a shell. For Xenopus laevis embryos, with $r = 600$ mkm and $t = 100$ mkm we obtain $S = 200\ N/sm^2$. While being dissected, the blastocoel roof, by our observations, immediately contracts in an elastic manner, up to 10-50% of its initial length. That permits to estimate its Young's modulus being within a range from 16 to 80 MN/m^2. This is typical for biological samples.

Let us consider now in more details the morphogenetical role of osmotically driven turgor pressure in cell vacuoles. This was mostly studied in the developing buds of hydroid polypes (Beloussov et al., 1984, 1989, 1992; Labas et al., 1992). These species (as well as sponges, vertebrate embryos at some early stages and also some algae - Labas et al., 1980) are characterized by a so-called pulsatorial growth, being consisted of a series of the regular several minutes periods oscillations in the length and shape of the developing

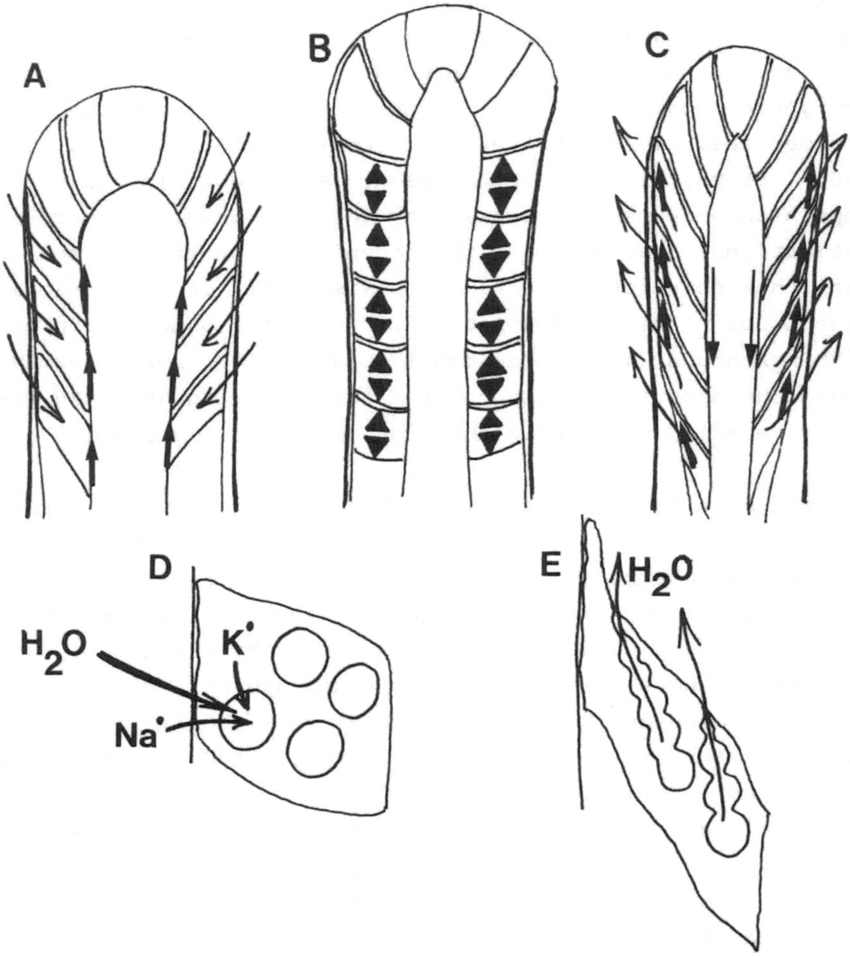

Fig. 1. A hydraulic mechanisms of the linear growth and shaping: growth pulsations in the buds of hydroid polypes. A-C: schematical representation of a growing tip (cells unproportionally enlarged) at the retraction, extension and the next retraction phases correspondingly. Thin curved arrows indicate water inflow (A) and outflow (C), small vertical arrows in A and C show the displacement of cell walls, black triangles in (B) denote the mutual tangential pressure of swelled cells. D illustrates ionic transport within cell vacuoles and the resulted stretching of cell walls, E shows water pumping out of the fused vacuoles.

rudiments. Each next extension phase is associated with the water pumping inside cell vacuoles while the retraction phase by these vacuoles fusing together and the water pumping out (Fig. 1 A,B). The water pumping-in phase is associated with the increase of intervacuolar potassium and sodium concentrations, accompanied probably by bicarbonate anions (Fig. 1 C,D).

These oscillatorial processes (growth pulsations - GP) are the examples of a markedly non-linear phenomenae, self-regulated by a number of positive and negative feedbacks. A coherent picture of these regulatorial patterns is still lacking, but it is obvious that it includes both chemical mediators and the mechanical stresses, interacting with each other (see for discussion Stern, 1984, Beloussov et al., 1989). In any case, the vacuolar pumping dynamics, creating itself the substantial mechanical forces, is itself highly mechanosensitive (Foissner, 1985, as concerning Nitella; our own observations on hydroid polypes).

The morphogenetical effect of these peculiar hydraulic phenomenae is in generating the tangential pressure stresses within epithelial layers of growing buds. The temporal patterns of these stresses are strictly regulated both on the genetical and epigenetical levels. The morphogenetical consequence of these stress patterns will be discussed in 5.1-5.2.

2.1.2. Devices for producing contraction.

As it is well known, those are the antiparallel mutually sliding actin filaments, connected by myosin bridges (microfilaments). Two main kinds of microfilamentous structures involved in morphogenetic movements can be distinguished: the apical bundles of linearly arranged microfilaments (Burnside, 1971; Spooner & Wessells, 1972) and the randomly oriented filaments creating what is often called actin gel (Oster et al., 1985). The both kinds of structures are localized predominantly in the submembraneous (usually called subcortical) cell areas. The apical bundles, while acting cooperatively in a number of adjacent cells, create tangential contractions over the relatively large distances along cell sheets. The actin gels, on the other hand, can in principle generate isotropical cell contractions, diminishing cell surface areas. The both kinds of structures act as a rule in a coordinated manner (see 2.2).

2.1.3. Devices for elongating cell and producing cell processes.

These shape changes are produced by growing and mutually sliding microtubuli. The best known event of this kind is the anaphase elongation of cell body during cell division. In the animal embryonic tissues, contrary to the plant meristems, such an elongation is only rarely involved in the morphogenesis and production of mechanical stressess: the animal embryonic cells in most cases are divided at the periphery of cell layers and only later on are inserted within. On the other hand, the microtubuli of animal cells participate in the formation of directed processes (fillopodia) which are regularly produced by morphogenetically active cells and bear cell tensions (see 2.3. (c), 2.5.1).

2.2. Space-temporal coordination of biomechanical events.

The tangential and isotropical contractions and microtubuli-mediated extensions are as a rule perfectly coordinated with each other. The best example is a bending of an epithelial cell sheet during gastrulation, neurulation, formation of sensory placodes, differentiation of the neural tube, intestine wall etc. All of these processes start from the formation of a continuous group of the radially (i.e. perpendicular to a sheet surface) elongated and often apically converged cells. This phenomenologically simple process, known as a contact cell polarization (CCP) involves actually a whole set of coordinated cell activities, including apical contraction, radial extension, redistribution of cell organelles and so on (Beloussov & Lutchinskaia, 1983; Belintzev et al., 1987). By this process, the cells often take an oblique shape, being converged towards a certain point. or towards a line in a 3- dimensional space (Fig. 2 B). As was noted long ago (Gurwitsch, 1914), such a convergence, so to say, predicts further deformation of a cell sheet, since the latter is directed towards the surface oriented perpendicularly to the long axes of converging cells (Fig. 2B, surface aa). This may be explained if assuming that the next phase of the morphogenesis consists in *isotropical subcortical contraction* of the previously extended-converged cells (Fig. 2C). The transition from the first phase to the second one, which takes normally from several dozens minutes to several hours may be extensively accelerated if separating a part of a sheet consisting of the already polarized cells, from the underlying tissues: in this case it becomes deformed practically immediately closely imitating the normal, "slow" morphogenesis of the same region. That permits us to make the following important biomechanical generalization:

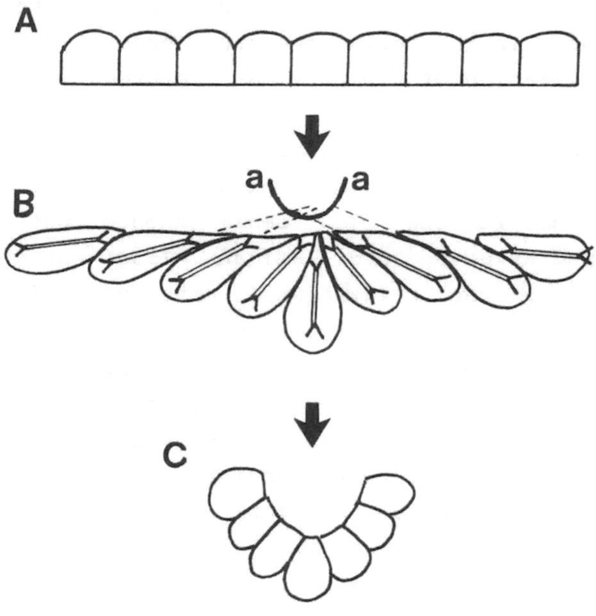

Fig. 2. Two steps of an epithelial layer folding. A: initial configuration. B: formarion of a group of converged cells. aa is a "predicted" configuration arranged perpendicularly to the long axes of converged cells. Double converged arrows indicate the directions of further cell contractions. C: final configuration arised as a result of these contractions.

All of the embryonic epithelial sheets in the process of their shape changes pass two successive biomechanical phases. The first one (Fig. 2 A-B) is characterized by the *increase* in the cells surface-volume (S/V) ratio and by the storage of mechanical (quasi -elastic) energy in the cortical-subcortical cell parts. This phase is also associated with the reduction of the symmetry order for each individual cell and, in some sense, with generation of the "morphogenetical information" about the forthcoming bending (since it is possible to predict the latter knowing the achieved configuration of the converged cells). The second phase (Fig. 2 B,C) is of a quasirelaxational character: it is associated, on the contrary, with a *decrease* of the cells S/V ratio, increase in the cells symmetry orders and with the physical realization of the previously stored "morphogenetical information". By defining this phase as just "*quasi*"-relaxational, we imply that under the normal conditions it should consume indeed the mechanical energy (the energy of the neural plate folding is small enough but measurable, being estimated as 8×10^{-3} erg - Selman, 1958) but goes, at the same time, along the *relaxational* pathway. This may be illustrated by the movement of a body down along the slope with a substantial friction: the energy is required for overcoming friction, but the direction of the body movement is determined by the slope itself, rather than by the energy source. We shall return later on to this important point.

2.3. Supramolecular devices for fixing and stabilizing the achieved deformations.

The microfilament contraction cannot last itself for more than several minutes or, in the best case, several dozens of minutes; same is true for the microtubuli-generated cell extensions. This is due to instability of the both kinds of cytoskeletal structures. On the other hand, the local contractions and extensions of the individual cells and entire cell sheets are much more stable. How can this be achieved? The main fixing and stabilizing mechanisms are obviously associated with cell membrane. Both the active microfilament contraction and the passive relaxation of a cell surface is followed, as a rule, by the rapid *resorption* of the corresponding part of a cell membrane (due to endocytosis and, in some cases, probably also by shedding), while its active extension may be followed by the *insertion* of additional membraneous and extracellular matrix subunits due to exocytosis. These statements are supported by the following findings:

(a) The apical membrane endocytosis following the contraction of a given area have been shown unambiguously in the fish (Fundulus) blastoderm (Betchaku & Trinkaus, 1986) and is also highly probable, according to electron-microscopy pictures and some label experiments (Radeva et al., 1986) for the cells of the amphibian embryos blastopore and neural plate.

(b) The resorption of the relaxed cells membrane have been demonstrated in fish zygotes (Ivanenkov et al., 1990) and in the ectodermal cells of amphibian embryos at the gastrula stage (personal observations). Whereas in the first case the resorption is obviously due to endocytosis, in the second case the shedding (throwing outside) of the membrane subunits is propable (our unpublished observations).

(c) The evidences for membrane insertion in the extended cell parts are less direct but nevertheless convincing. First of all, we know definitely that immediately after (or during) the anaphase extension of a dividing cell the additional portions of cell membrane are produced, either in the polar (White, 1990) or in the furrow regions. Secondly, the artificial stretching of the suprablastoporal zone in Xenopus laevis early gastrula embryos is followed by the rapid (taking few minutes) formation of the cellular processes elongated in the stretching direction and requiring new membrane (Beloussov et al., 1988). Similar processes are produced by the stretched epiectodermal cells of the normal embryos in the immediate vicinity of the blastopore. Assuming that the membrane extension is as a rule microtubuli-mediated (2.1.3) one of the most probable mechanisms for membrane insertion may be the vesicular transport along the microtubuli towards their distal (+) ends.

2.4. The cellular-supracellular events imitating and amplifying the activities of the above described molecular-supramolecular devices.

The fundamental property of any complicated and, in particular, self-organizing system is its hierarchical structure or, in the other words, multileveleness. In these systems the activity of the energy-supplying and energy-transforming lower levels devices (molecular mashines) is reinforced, rather than being damped, on the upper levels. This general statement is particularly true for the biomechanical morphogenetical events. Almost each out of the above described subcellular activities finds striking parallels on the cellular-supracellular level.

In particular, the events associated with the increase of internal pressure and extension of a cell body or its parts are closely imitated on the supracellular level by a widely spread process known as *cell intercalation* (Keller, 1987; Keller et al., 1989). Two main kinds of cell intercalation can be distinguished: a less cooperative, so-called radial cell intercalation (Fig.3A), and a highly cooperative convergent cell intercalation directed towards an embryo midline (Fig. 3B, C). The first one results in more or less isotropical, while the second one in the markedly anisotropical (antero-posterior) extension of the embryonic surface. To the similar kind of processes the *cell emigration* may be attributed, described by Steinberg (1978) for the artificially stretched aggregates of chick embryonic cells. The radial cell intercalation and cell emigration are associated with the increase of S/V ratio while the convergent cell intercalation with the increase of L/S ratio, where L is the length of the elongating axis.

Another group of cell activities is associated, on the contrary, with *the decrease* of S/V ratio. These are exemplified

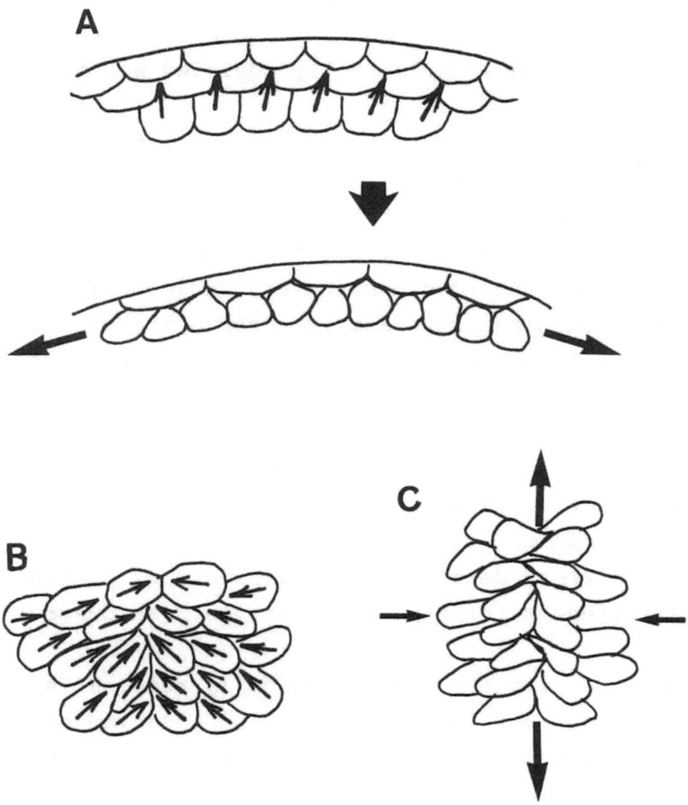

Fig. 3. The radial (A) and convergent (B, C) cell intercalation. The first one consists in cell insertion perpendicularly to the layer's surface and results in the isotropical extension of a cell sheet, while during the second one the cells are migrating in convergent directions along the sheet's plane leading to its anisotropical extension-contraction.

by the immigration of cells from the surface of a tissue piece and by the rolling of the pieces. These events are very common both for normal and experimentally modified morphogenesis (see 3.4.1).

2.5. Mechanical stresses: methods of detection and space-temporal patterns.

For developing our initial idea on the important role of the active forces - passive stresses feedbacks in regulating morphogenesis one should demonstrate unambiguously the presence of the real mechanical stresses in embryonic tissues, spread over the large enough distances and lasting for some not too short time periods. Not so much the absolute stress values as the existence of their regular patterns and their relations to the fundamental morphogenetical processes should be revealed. In the next section we describe the results of our research group in this field.

2.5.1. Stress detection by dissections tests; maps of mechanical stresses for amphibian embryos (Beloussov et al., 1975).

One of the simplest and biologically reasonable ways for a rough detection of the presence and location of mechanical stresses consists in dissecting or incising embryonic tissues and tracing their deformations immediately after dissections under the conditions which do not permit an active contractile response of a dissected tissue (moderate cooling, low concentractions of cytoskeletal or energetical inhibitors). In such a way, the "passive", quasielastic deformations, characterized by the temperature and inhibitors insensitivity, the lack of latent period and, in addition, by a pronounced stage- and regional specificity could be distinguished from the less specifical "active" responses sharing the opposite characterics. Most of the passive deformations consisted in immediate and considerable contractions and/or outwards bendings of dissected tissue pieces which could be qualified as relaxations of preexisted stresses. By making a number of

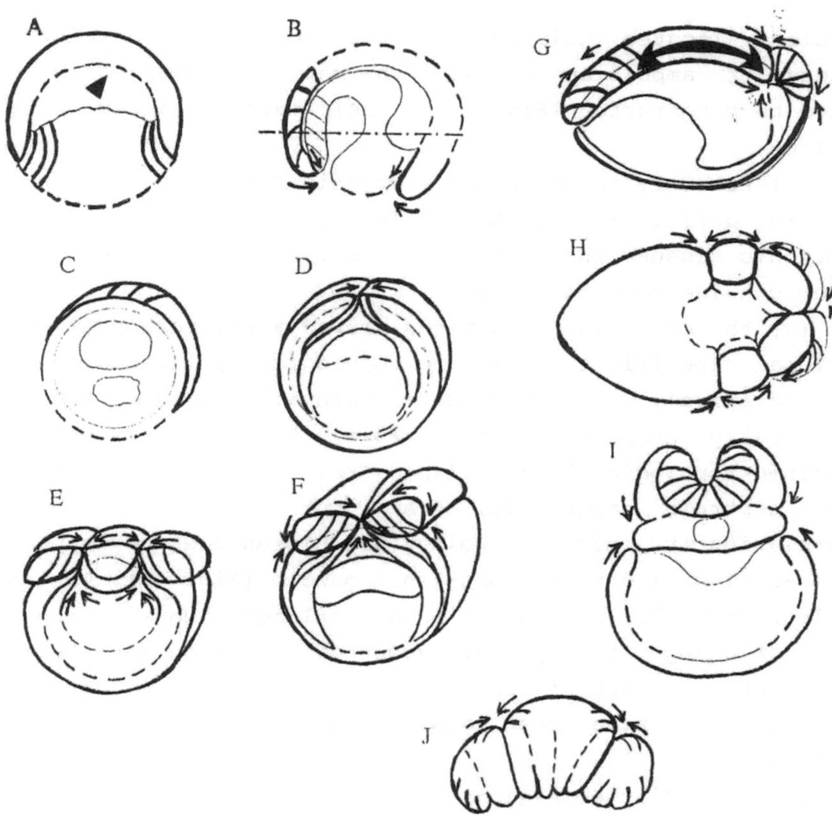

Fig. 4. Maps of mechanical stresses for several successive stages of *Rana temporaria* development. A: late blastula, B: mid -gastrula, saggital section, C: same stage, transversal section along the line indicated on B, D: transition from gastrula to neurula, posterior region, E: anterior region of early neurula, F: posterior region, same stage (D-F - transverse sections), G: early-middle neurula, saggital section, H: middle-late neurula, frontal section, I: similar stage, transverse section, J: an ectomesodermal fragment from neurula, 10-15 min after its isolation. Heavy contours represent distinct stress lines, dotted contours - dispersed stress-lines, fine lines - non-tensed surfaces, dense double-headed arrow in G shows longitudinal pressure stress in the dorsal embryonic area, filled triangle in A - turgor pressure in blastocoel.

precisely located incisions at the successive developmental stages of amphibian embryos the maps of mechanical stresses could be constructed (Fig. 4). Their main features were as follows:

(1) not only the external embryonic surfaces but as well certain definitely located files of internal cells crossing embryonic tissues throughout ("cross-lines") were found to be under considerable tensions;

(2) the cross-lines patterns showed a remarkable stage- and regional specificity remaining topologically invariable within the relatively prolonged developmental periods and then drastically changing. For example, the cross-lines pattern remained topologically invariable from the late blastula to late gastrula stage (Fig. 4 A-C) but before the beginning of neurulation its initial topology was broken due to appearance of so-called middorsal cross-lines bundle (Fig. 4D). During the whole neurulation period the bundle enlarged considerably but did not change its topology except of splitting longitudinally in the anerior part of embryo (Fig. 4F). Further topological complications have been associated with the tail-bud stage (Fig. 4 G-I).

(3) the tissue regions characterized by the pronounced cross - lines system showed after explantations a remarkable structural stability of developmental pathways whereas those lacking pronounced cross-lines developed in quite variable and unstable way (Beloussov, 1979).

Beginning from early neurula stage, not only the tensile but as well compression stresses could be revealed in the dorsal embryo wall, being directed longitudinally and checked by the immediate overlapping of dissected tissue edges (Fig. 4B, doble-headed solid arrow).

Worth mentioning, no more than 20-30 min was enough for a detailed tensile pattern to appear de novo in an extirpated and hence relaxed fragment of embryonic tissue (Fig. 4J).

2.5.2. Detection of stresses by measuring surface angles.

This classical method consisted in measuring angles formed by cell walls in their joint points (Thompson, 1942; Steinberg, 1978). It is based upon two general presumptions:

(1) all the mechanical forces which determine the shape of a cell are equilibrated by tensile forces oriented parallel to cell membranes and localized in their immediate vicinity. It is a cell membrane together with a submembraneous (cortical) cell layer which we consider as a "cell wall";

(2) in each time moment a net of cell walls tensions within the whole rudiment or entire embryo is mechanically equilibrated;

The latter means that in each cell apex (cell walls intersection point) $\Sigma\, \mathbf{F}=0$ where $\mathbf{F}$ is a force of tension. The principle of forces equilibrium (lack of inertial effects) is widely accepted in biomechanics (Odell et al., 1981; Goodwin & Trainor, 1985) saving us for discussing it in more details.

In a print of a cell net drawn from a histological section (Fig.5) we begin from an arbitrary chosen apex point O_1 ascribe to the vertical cell wall directed from it a certain tension force **a** and by measuring the angles α and β calculate with the use of above assumptions the tension forces **b** and **c** as related to **a**. Then we pass to the next apex point O_2 measure the corresponding angles γ and σ and calculate the tensile forces **e** and **f** in relation to the force **d** = **c** (see Beloussov & Lakirev, 1988 for more details). In such a way we can pass through all the cells apices in a given area and even return to the initial apex from the opposite side, measuring some tensions for the second time. If the secondly made measurments of the same walls tensions did not differ significantly from the first ones we may conclude that the measured area have been more or less mechanically closed

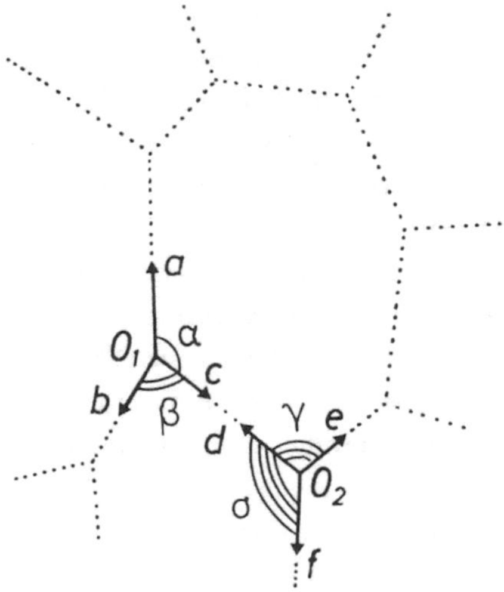

Fig. 5. A geometrical method for measuring the relative tensions values in cell walls. For explanations see text.

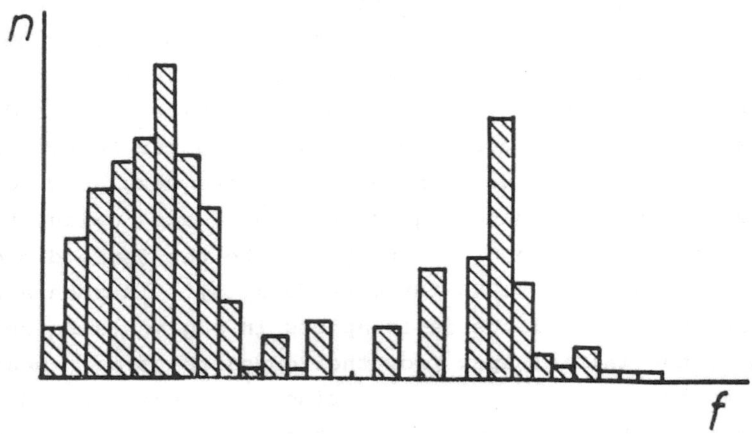

Fig. 6. A markedly bimodal histogram of tensile forces in cell walls of ealy neural plate. Abscissa: tensile forces, ordinate: number of cell walls. Those comprising the right-hand peak are strongly tensed (From Beloussov & Lakirev, 1989).

(equilibrated within itself) and our measurments were correct (in the opposite case, we have to accept that some considerable tensions have been localized out of a studied area or of a print plane).

The corresponding measurments performed on the saggital section of early gastrula suprablastoporal region and on the transversal section of neural plate not only satisfied these demands but also coincided generally with the above described stress maps based upon dissections tests, supporting thus their validity. Another conclusion was the pronounced bimodal distribution of tensile forces values in embryonic tissues (Fig. 6). That confirms the possibility to distinguish, in the stress maps, just two classes of tensions (strong and weak) and also points to the bistability (non-linearity) of the tension-generating mechanisms.

3. Models of the feed-back loops linking active forces and passive stresses.

3.1. Odell et al. (1981) model.

By my knowledge, Odell et al. (1981) were the first who postulated the existence of a regulatorial feed-back loop between the active mechanical forces and the stresses which they produce in the neighbouring tissues. By their view, the active tangential contraction of the vegetal plate cells in sea -urchin gastrula, while leading to the tangential stretching of more animal cells of a gastrula wall, induce the latter for producing the similar tangential contraction forces. Therefore, a positive feed-back loop "passive tangential extension - active tangential contraction" have been postulated. The authors believe, that in such a way the new cells can be successively recruited into the gastrulation movements. The model does not explain, however, why such an involvement at a certain moment (and as concerning Echinodermata embryos very soon after the beginning of gastrulation) will be arrested sharply delimiting the embryo body into the columnarized involuting (endodermal) and tangentially stretched non-involuting (ectodermal) cells.

On the other hand, it would be unreasonable to reject completely the idea of such a feed-back loop. In any case, it seems to work in some simple systems, like the stretched cytoplasmic strands of Physarum (Wohlfarth-Bottermann, 1987) and probably in the multicellular systems characterized by a relatively high viscosity and low elasticity (see 3.4.1.9a). For most of embryonic cell sheets meanwhile, the next model looks as more suitable.

3.2. Belintzev et al. (1987) model.

The idea of this model is almost opposite to the previous one. It postulates, that (1) embryonic cells possess two alternative dynamically stable shapes, an isotropical and highly polarized (columnarized) one, with instability in between; (2) the impulse for cell polarization is spread from one contacting cell to another as a diffusion-like wave; this is just above mentioned (2.2) contact cell polarization (CCP) which really propagates within amphibian embryonic tissues with a rate of about 5 mkm/min. (3) the tangential tissue stretching *inhibits*, rather than stimulates cell polarization (and, consequently, tangential cell contraction). Hence, in Belintzev et al model a *negative*, rather than positive feed-back loop is postulated between the passive tangential stretching and the active tangential contraction. This is supported by the experiments showing that the relaxation of tensions stimulates, rather than inhibits (as would be expected according to Odell et al. model) CCP (Beloussov et al., 1990). In general, Belintzev et al model exploits, within a biomechanical realm, a widely accepted principle (Murray, 1989) of a short-range stimulation (exemplified now by CCP) and a long-range inhibition (caused, in the model terms, by stretching).

This model have an advantage to interpret the self-limitation of cells involvement into CCP, since the CCP automatically increases tangential tensions. Correspondingly, it perfectly imitates the universal phenomenon of the epithelial sheets segregation into the continuous domains of highly polarized and tangentially stretched cells. In addition, the length relations of the polarized and stretched cell domains appeared to be, within the model framework, scale invariant; in such a way the model reproduces, without any additional assumptions, the fundamental event of embryonic regulations (retaining of structural proportionality under the different absolute dimensions), which is not an easy task for the self-organizational models. Worth mentioning, according to the model, a cell should not exactly "recognize" its position within a sheet - all what it "knows" is the delocalized tangential tension value and the polarization state of the adjacent cell. In the other words, the model closely imitates an important step of morphogenesis without implying a detailed "positional information". Whatever being its advantages, meanwhile, it reflects but a single, even if important step of morphogenesis without giving an idea of how an entire developmental succession can be realized. The attempt to trace the way for resolving this fundamental task will be made in the next sections.

3.3. Hyperrestoration of the mechanical stresses as a possible driving tendency of morphogenesis.

Any living entity, starting from the primitive cells, is confronted with the osmotically generated internal pressure stresses tending to disrupt its external envelopes. The nature invented different ways for overcoming this inevitable physico-chemical challenge. The most direct one, namely the reinforcement of the envelopes, is not the best one because it leads to excessive isolation of a cell from its environment. Another, much more suitable possibility to reduce the dangerous tension of an outer shell is to increase its square, that is to produce new portions of a cell envelope. If all of the internal turgor pressure P is used for stretching the envelope, we get

$$PdV = TdA$$

where dV is a volume increase, $T = \sigma r$ (where σ is a stretching stress, r is a thickness of the surface) and dA is an increase of the surface square (see Harold, 1990). A condition for non-exceeding a certain "safety threshold" of T value should be in preserving $dA \geqslant dV$ or, in the border case, $dA = dV$. Is it compatible with the shape constancy? One can easily see that this is so only when applied to cylindrical (filamentous) shapes. All of the convexed shapes, meanwhile, for non-exceeding during growth their critical stretching stress value should increase their initial S/V ratio by making folds, ingressions and so on. It is but natural to assume now that, in a general case (and also for ensuring itself) the living sample will try to diminish a bit T value, that is not just to restore, but to *hyper*restore the initial stress value. That will lead to the folding of even initially cylindrical shapes. We come to the plausible general conclusion, that the morphogenesis might arise, within the course of evolution, as an inevitable side-result of a struggle against excessive stretching stresses , or, in the other words, as a result of the repeated (hyper)restorations of the initial stress values. If this were true, it would be quite natural to assume that the same principle will work out also in the contemporary morphogeneses. The corresponding suggestion have been formulated as a hyperrestoration, or HR rule (Beloussov & Mittenthal, 1992):

A cell or a tissue piece, after being shifted by any internal or external force out of its initial mechanical equilibrium generates an active mechanical response directed towards the restoration of the initial stress value but as a rule overshooting it to the opposite side.

(NB: what we mean by this formulation is the (hyper)restoration of just stress value, but in no way of the initial shape of a cell or a tissue piece; in the overwhelming majority of cases the shapes should be drastically changed in the course of stress (hyper)restoration. This is an important and delicate point which should be made absolutely clear).

Now we would like to show, first of all, that this reaction is widely spread and, next, that the morphogenesis of different animal species may be considered as a chain of the hyperrestorative (HR-) reactions.

3.4. The main HR reactions

3.4.1. After tissue stretching.

Suggest that a cell or a tissue piece is stretched and tends to (hyper)restore its initial stress value. This can be done in two general ways:

(a) if a stretching force is not too great and a sample is capable of active contraction it contracts indeed up to its initial dimensions or overshooting them by shortening even more;

(b) if a stretching force is too large and/or prolonged to be overruled in such a way, and/or if a sample is incapable to the active contraction, it (hyper)relaxes the stretched surface by rearranging its material in such a way to increase either its S/V, or L/S of a sample, assuming in the latter case that the sample is stretched just along L direction.

The (a) possibility have been already mentioned (3.1), as related to Physarum cytoplasmic strands. We suggest that it may take place also in the highly viscous Metazoan tissues, with the rapidly damped elastic tensions. Assume for example that an actively contracting cell stretches effectively no more than its next neighbour. That may induce the contractile response in the latter, stretching the next cell and so on. In such a way a relay of a contractile activity may spread throughout a tissue bulk. This is meanwhile in the contrast with the situation typical for most of embryonic epithelia where tangential cell contraction immediately produces a long-range elastic response (Belintzev et al. situation). In the latter case a HR response will go along (b) way, which is hence more universal.

First of all, it corresponds to the above mentioned universal processes of cell intercalation (2.4.), either radial (isotropically increasing S/V), or convergent (anisotropically increasing L/S). Our experiments on amphibian embryos showed that the both kinds of intercalation are suppressed by relaxation of tangential tensions (Beloussov et al., 1990 and in preparation) and that the convergent intercalation may be effectively reoriented by stretching artificially the fragment of a suprablastoporal tissue of Xenopus laevis embryo perpendicularly to its normal (antero-posterior) elongation direction (Beloussov et al., 1988). In each case the convergent cell movements appeared to be oriented perpendicularly to the stretching direction (Fig. 7 A,B). The initial external stretching could be rather small and not so prolonged (30 min is enough). As a consequence, meanwhile, a long-termed effect have been developed, producing a large longitudinal active *pressure* stress (Fig. 7C), which actively elongated the rudiment more than 3-fold . The arising of a pressure force could be tested by the overlapping of tissue edges after its dissection. The entire reaction may be considered as a vivid example of the overshooting response, directed towards negative tension (pressure) stresses (Fig. 7E).

A similar phenomenon of a stretch-initiated cell repacking leading to S/V increase have been described by Steinberg (1978) - see 2.4. In this paper unfortunately no data have been given permitting to conclude whether a real overshooting of the stress values took place.

Another very important modus of HR of tangential tensions up to production of the compression stresses is in making the beforehand columnarized cells to contract radially (Fig. 7F). This usually leads to the epithelial folding (see section 5).

By Kolega (1986) and our (Beloussov et al., 1988) electron-microscopy observations, the microfilament bundles are rapidly formed in the stretched tissues, being oriented just along the stretching direction. One may speculate that it is the subsequent contraction of these bundles which flattens cells perpendicularly to this direction promoting both intercalation reactions and folding movements. In this aspect, the both categories of movements do not differ so much, on the cytological level, from the purely contraction response (a):

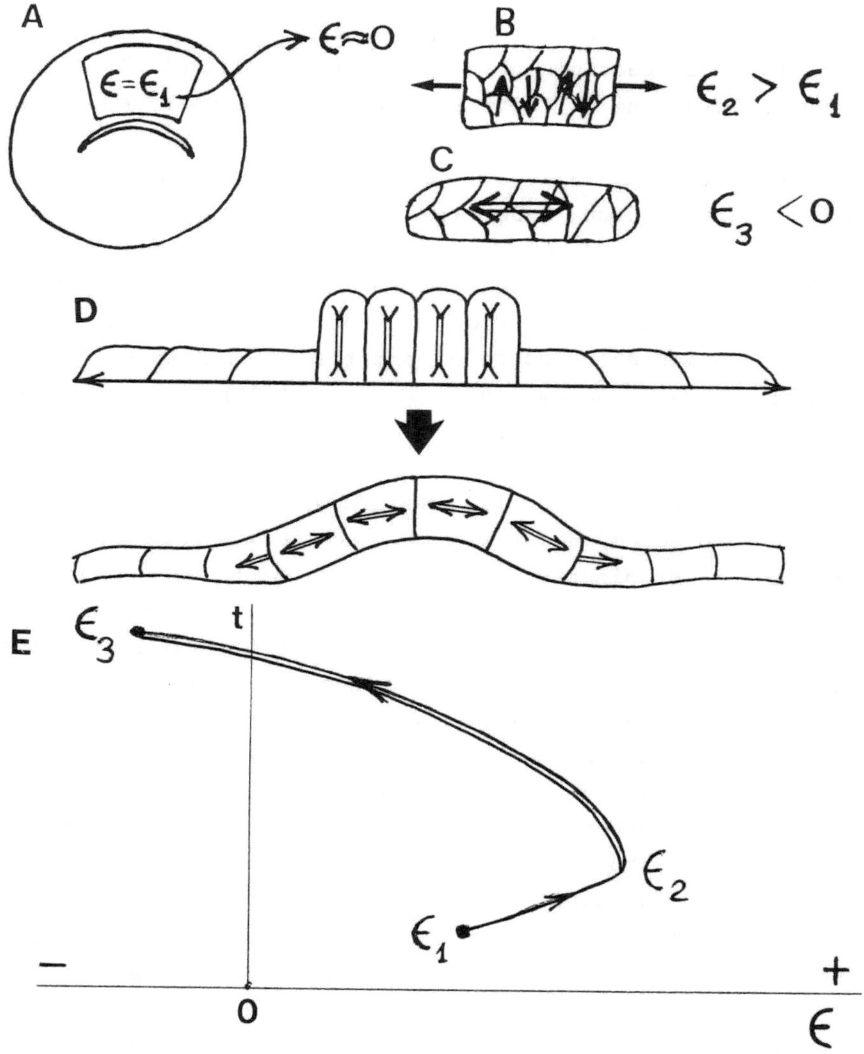

Fig. 7. Hyperrestoration of stress values after the artificial stretching of a piece of embryonic tissue. A-D: for explanation see text. E: a biomechanical interpretation of the experiment. Horizontal axis: stress values (stretching to +, compression to -); vertical axis: time. Up from here the passive (caused by an external force) stresses are shown by single lines while the active stress responses by double lines.

their main differences are related to the supracellular level.

3.4.2. After tissue relaxation.

The typical reactions of the embryonic epithelial tissues to the relaxations of initial tensions are summarized in Fig. 8.

Suggest that initially a layer is slightly stretched by the turgor pressure from underlaid cavity (Fig. 8A, black triangle), its tangential stretching ε_1 proportional to the surface angle α being shown to the right. Then the layer is relaxed by dissections (Fig. 8 A,B), the surface angles immediately diminishing up to β and the tension up to ε_2. (for numerical data see Beloussov et al., 1990). The correspon ding stress may be increased up to certain ε_3 in different ways depending largely upon the boundary conditions:

If a relaxed tissue is immobilized on the adhesive substrate (a piece of a latex film), it is dissociated, within several dozens of minutes, into single cells with numerous lobopodia (Fig. 8C). If no adhesive substrate is provided and a relaxed tissue piece have no free edges the typical reaction is a columnarization of a certain continuous domain of its cells, in a close relation to Belintzev et al. (1987) model (Fig. 8D). Same kind of reactions could be also observed in Echinodermata blastulae relaxed by external hypertonicity (Beloussov & Zhadan, 1993): after 10 min incubation in 0.05 M sucrose solutions the blastocoel volume decreased in 2 times, while cell polarization index (length /width ratio) increased in 2.5 times. The latter led to a rapid disappearance of the initially arised folds so that the perfect spherical shape have been restored.

In the relaxed pieces which have free edges a complicated reaction is developed, including immigration of some amount of cells, formation of swelled cavities (these two reactions

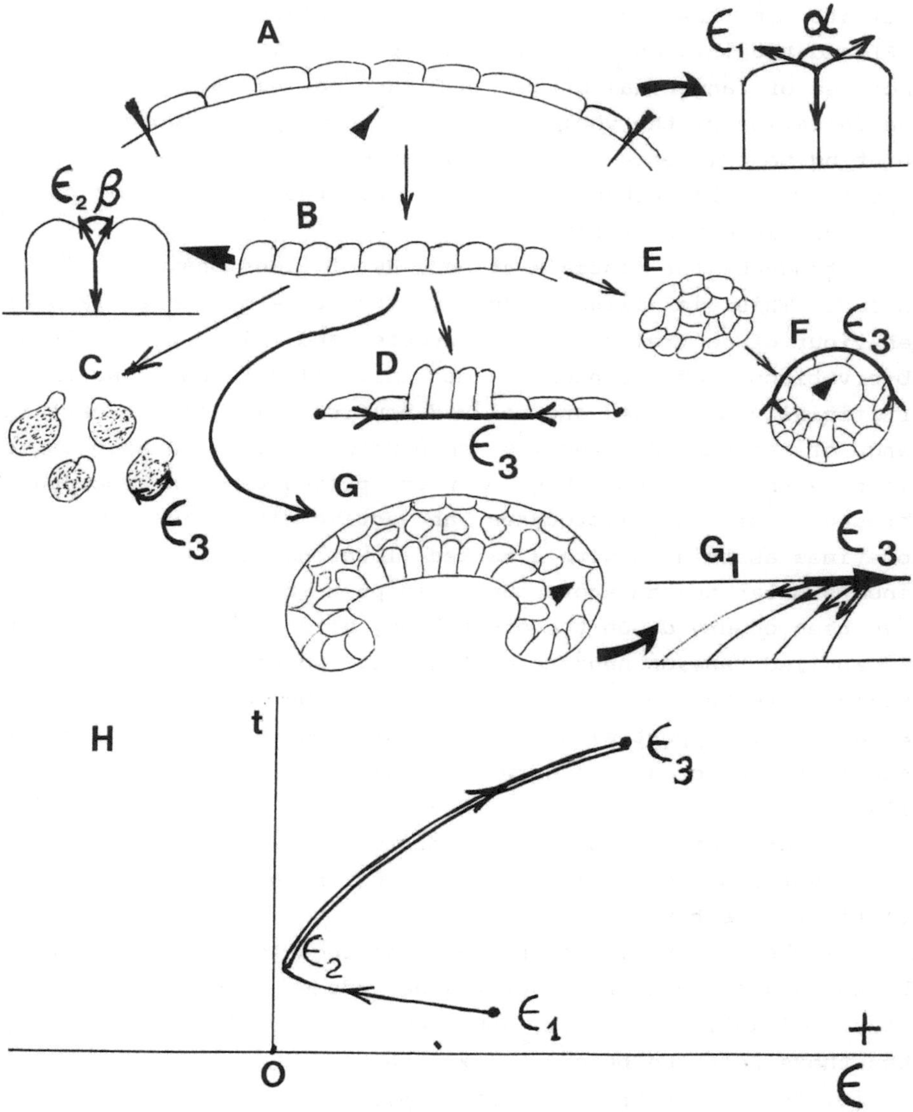

Fig. 8. Hyperrestoration of stress values after relaxation: schemes of experiments (A-G) and biomechanical interpretation (H). For further explanations see text.

dominate in the isolated pieces of a blastocoel roof - Fig. 8,F) and the inwards rolling of the piece edges (Fig. 8G). The increase of tangential stress in G case may be easily tested by the formation of the whole series of highly obliqued cells, pointing to existence of a considerable tangential stress (Fig. $8G_1,\varepsilon_3$). The biomechanical scheme of the postrelaxational reactions is given in Fig. 8H.

A tendency for postrelaxational HR of tensions is widely spread. This is illustrated by the above mentioned (2.3.b) behaviour of a relaxed fish zygote and also by exciting observations of Dennerli et al. (1989) as concerning artificially stretched and relaxed neurits. Within a certain range of stretching neurits responded to tension as passive visco -elastic bodies. However when applied tension exceeded a threshold of 100 microdynes an active elongation response, sometimes associated with a marked lowering of tension (and hence similar to the above mentioned post-stretching reactions) have been observed. On the other hand, after a sudden reduction of applied tension neurits contracted, sometimes doubling the steady-state tension that existed before the experiment; this is a typical postrelaxational HR of tensions. For a similar interpretation of axonal growth see also Heidemann & Buxbaum (1990).

Here are some of our semi-quantitative data indicating the real overshooting of initial tensions in the postrelaxational reactions of embryonic tissues.

(a) The mostly direct ones are the local dissections tests: the immediate moving apart of tissue edges observed in 1-2h after relaxations always appeared to be considerably greater than those detected before relaxation.

(b) In the postrelaxed tissue pieces a number of obliquely shaped cells with the largely increased S/V ratio was much greater than in the same sample of a pre-relaxed (intact) tissue.

(c) Sea-urchin blastulae incubated in hypertonical solutions (see above) are developed into gastrulae with abnormally widened blastopores; opposite to blastopores the domains of extensively columnarized cells are situated. Due to mechanical integrity of the embryo wall the width of the blastopore is directly determined by these cells tangential contraction. Consequently, the abnormal widening of the blastopore points to the excessive postrelaxational tangential contraction, that is to the overshooting of the initial (pre-relaxational) tangential tension of a cell layer.

(d) Isolated pieces of amphibian embryonic tissue within few dozens of minutes are rolled into smooth spherical balls completely covered by ectoderm. Very simple geometrical considerations show that:

(1) Since a square of a hemisphere is in about 25% smaller than the square of the same volume sphere, the solid hemisphere - sphere transformation (usually observed in the isolated vegetal halves of amphibians early gastrula embryos) should be inevitably accompanied by a considerable surface stretching. The more will be the stretching if the shape of a rolling piece will be deviated from the perfectly spherical one, what is just taking place. This stretching may be definitely considered as a post-relaxational one since the isolation of a vegetal half is associated with the reduction of its initial tangential tensions.

(2) If a flat piece of tissue, only one of its surfaces being covered by ectoderm, is rolled into a sphere, the ratio between the spherical surface completely covered by ectoderm (S ect.sph) to the ectodermal surface of a flat layer (S ect.l.) depends upon the relation r/h where r is the tangential diameter of a cell sheet and h its height (radial dimension). Under about $r/h<10$ S ect.sph>S ect.l. while under $r/h>10$ the reverse is taking place. Consequently, for getting a smooth overstretched spherical surface the small tissue pieces do not require migration of some surface cells to the inside, while in the larger ones this should take place. This is just what is really observed: small tissue pieces rapidly form solid tissue balls, while in the largest ones one or several invagination or immigration areas appear.

In no way the experimentally induced relaxations, followed by the overshooting responses should be considered as a kind of artifacts alien to normal morphogenesis. On the contrary, as have been already mentioned (2.2), all of the investigated epithelial morphogeneses are going along the quasi-relaxational pathways, increasing, at the same time, tensions on the adjacent embryonic surfaces. In the other words, relaxations and stretchings within the normal development are going hand-by-hand. That gives us a hope to consider the main processes of normal morphogenesis as a succession of the stress HR responses.

4. A presentation of the main morphogenetic processes in amphibian embryos from blastula to neurula stage as a chain of the HR responses

For the sake of simplicity, let us roughly divide the corresponding period of morphogenesis to several successive steps and try to interpret each one of them within the framework of HR rule.

Step 1. Pregastrular animo-vegetal epibolic movements. These movements consist mostly in the vegetal spreading of the animal (ectodermal) cap due to radial cell intercalation. Our main suggestion is that the latter process may be considered as a HR response of the blastocoel roof cells to their stretching caused by turgor pressure within blastocoel (Fig. 9A). This is supported by the following experiments:

If relaxing the stretching of the blastocoel roof by inserting a sector of endodermal tissue (Fig. 10A), not only the radial cell intercalation immediately stops but, instead, within few hours the number of cell layers in the blastocoel roof considerably increases. As a result, quite abnormal embryos are developed (Fig. 10A, arrow 1). Their development could be however to a great extent recovered, and the radial cell intercalation reinitiated if the operated embryos were

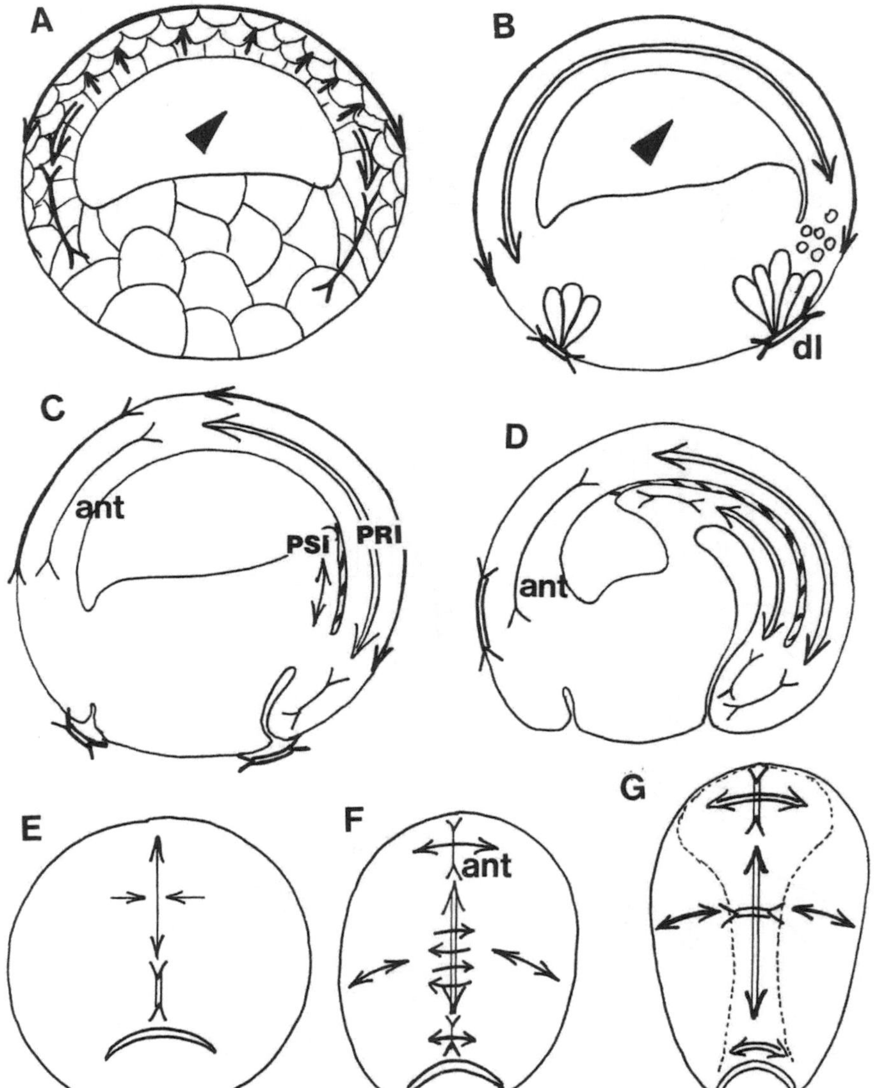

Fig. 9. The biomechanical interpretation of the gastrulation and neurulation in amphibian embryo. dl - dorsal lip of the blastopore. For other explanations see text.

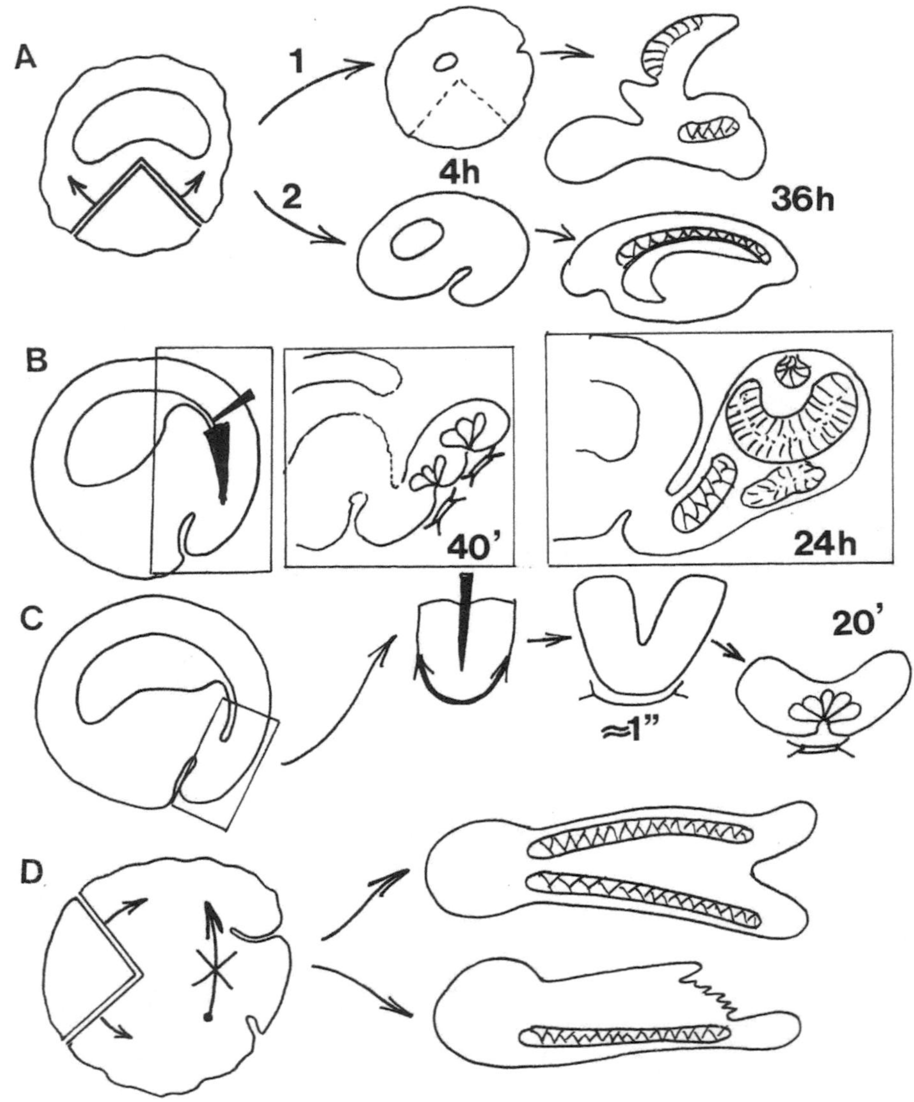

Fig. 10. The experiments testing the biomechanical interpretation of amphibians gastrulation. In A,D the relaxation is caused by inserting an additional tissue sector in the vegetal or ventral positions correspondingly. Black edges in B and C show the direction of dissections.

placed soon after operation in a highly hypotonical solution (25-fold diluted normal medium) - Fig. 10A, arrow 2. Under these conditions the relaxed tensions have been restored within no more than few hours (detailed description in preparation).

Step 2. Initiation of gastrulation: formation of tangentially contracted and contact-free cell groups within the marginal zone.

We consider these events as HR responses to the relaxation (and probably slight compression) of the presumptive marginal zone by the actively stretched (due to radial cell intercalation) blastocoel roof (Fig. 9A, double-headed single convergent arrows). As shown by Fig. 8 C,D a typical response to the relaxation is just the loss of cell contacts in the presence of an adhesive substrate (inner part of the marginal zone) and the tangential contraction (cell columnarization) on the surfaces lacking such a substrate (outer surfaces) - Fig. 9 B. The experimental evidences favouring this point of view are as follows:

(1) If relaxing (by dissection) a preinvoluted part of a blastocoel roof the numerous areas of cell columnarization and tangential contraction are produced within few dozens of minutes - Fig. 10B, 40′ (Beloussov, 1988).

(2) If isolating and then dissecting the dorsal blastoporal lip (Fig. 10C), this is followed firstly by the rapid relaxation of preexisted tensions around the lip (Fig.10C, 1′′) and then by the formation of a bunch of highly columnarized and tangentially contracted bottle-shaped cells increasing again these tensions, - Fig. 10C , 20′ (Beloussov et al., 1975). The data presented in (1) and (2) show, that all the cells of the intact preinvoluted zone are potentially capable to extensive polarization which is triggered by the relaxation.

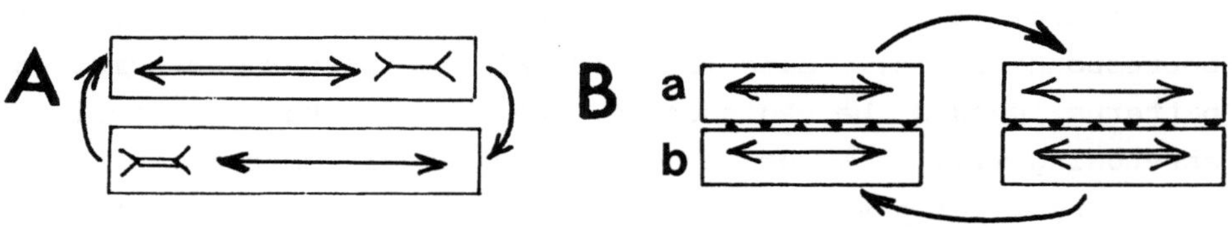

Fig. 11. Schemes of the uniplanar (A) and biplanar (radial) (B) positive feed-back loops causing the extension of the interacting biomechanical entities

The expanding blastocoel roof together with the contracting blastoporal area create what we may define as a *uniplanar positive biomechanical feed-back loop* (Fig. 11A): the active (intercalation -mediated) extension of the roof (left part of the scheme) relaxes /compresses the blastoporal part (right), causing thus its active contraction. The latter, in its turn, by stretching the left part stimulates further cell intercalation in this area, etc. One can also show that such a process while going on within a radially symmetrical surface will be highly unstable, tending to transform itself into a bilaterally symmetrical one (with a single stretching-contraction axis to become dominated). We suggest that at the beginning of gastrulation just this kind of symmetry breaking is taking place, using the earlierly established dissymmetry (associated with the rotation of fertilization (Gerhart et al., 1989)) as an initial (border) condition. (Note that the rotation of fertilization is itself also a biomechanical reaction!)

Step 2a. Invagination (inwards buckling) of columnarized cells of the blastoporal region may be easily interpreted as a local hyperrelaxation of the newly established tangential stress in this region, according to Fig. 7D .

Step 3. Convergent (latero-medial) cell intercalation. I agree with Keller (1987) that this is the main engine of gastrulation. What I suggest in addition is that its stability is provided by just described uniplanar positive feedback loop (Fig. 11 A). The following should be also added: (1) the intercalation-mediated stretching creates the relaxation /compression zone not only to the posterior, but to the anterior as well (Fig. 9 C,D,F, ant). The latter will be even more expressed than the posterior one, since the posterior relaxation is permanently diminished due to the involution of corresponding cell material. (2) all the planar deformation patterns should be Poissonian, that is the longitudinal stretching will be accompanied by transversal compression, and vice versa. The passive stresses will be transformed into the active ones and back again due to the same uniplanar feed-back loop (Fig. 10A). The most important morphogenetic consequences will be the transversal extension of the head region (Fig.9F,G) as well as the formation of a series of transversal folds and the placodes (columnarized cells areas) in this region, due to its longitudinal relaxation.

However, the planar interactions are not enough for providing the stability of a stress pattern. By our suggestion, they are supported by the biplanar ones, occuring between the adjacent parts of the pre- and postinvoluted cell sheets (Fig. 9C, pri and psi). A scheme of this *biplanar positive feedback loop* is shown in Fig. 11 B. Suggest one of the mechanically binded cell sheets (A) to be actively extended due to cell intercalation. Then B sheet will be passively stretched by A sheet. According to HR rule, that induces cell intercalation in B sheet; becoming active, it stretches A sheet, stimulating it for further intercalation, and so on. As a result, the both layers will be extended cooperatively, until the possibilities for cell intercalation will not be exhausted. Even if one of the layers (say A) is itself incapable to cell intercalation, a local cell intercalation in B sheet will create an overall passive stretching of A sheet, the same stretching being spread, via mechanical binding, to the entire B sheet back. In such a way, further intercalation within B will be stimulated again. In amphibian gastrulation the psi is assumed to be firstly stretched passively by pri (Fig. 9C), thus bringing into action the entire biplanar feed-back.

Our evidences for the reality of the uniplanar feedback-loop within amphibian gastrulation are as follows :

Kinematical evidences.

As shown by Keller (1978) and also by Petrov & Beloussov (1986) the convergent cell intercalation within suprablastoporal zone (SBZ) of embryo is preceeded by a short-termed longitudinal contraction of the blastoporal zone, causing a slight SBZ stretching. This may be enough for switching-on cell intercalation within SBZ.

Experimental evidences.

(a) relaxation of tangential tensions at the early gastrula stage inhibits the extensive ventro-dorsal convergent cell movements (Fig. 10D, crossed vertical arrow), required for the normal cell intercalation towards dorso-medial line (Beloussov et al., 1990 and in preparation). That results in the reduplication of axial organs or in a complete reversal of dorsoventral symmetry - Fig. 10D.

(b) as mentioned above (3.4.1), the reorientation of external tensions reorients correspondingly the direction of cell intercalation (Beloussov et al., 1988).

The importance of the biplanar feedback loop (Fig. 11B) is confirmed by following:

(a) cell intercalation within either pre- or postinvoluted sheet appeared to be impossible without this sheet to be overlaid by another epithelium, of either ectodermal or endodermal origin (Keller & Danilchik, 1988, Wilson et al., 1989).

d) If incubating an embryo operated as shown in Fig. 10B for no less than a day without permitting the separated layers to fuse back again cell intercalation becomes largely inhibited in the both layers. The postinvoluted layer remains in this situation completely non-differentiated while the preinvoluted one gives rise to fore-brain structures instead of trunk ones, as normally (Fig. 10B, 24 h).

Step 4. Segregation and rolling of neural plate. We consider the segregation of neural plate out of an initially homogeneous embryonic ectodermal epithelium as a kind of a process imitated by Belintzev et al. (1987) model, that is as a formation of a continuous domain of extensively columnarized cells, its dimensions being self-restricted by the tangential stretching of an entire layer. The normal location of the domain on the dorsal embryo side may be explained, from this viewpoint, by the initial Poissonian transversal relaxation/compression of this very side. Such a suggestion is supported by the following findings:

(a) Relaxation of transversal tensions on the embryonic surface by inserting an additional tissue sector in the ventral location of early gastrula embryos leads to abnormal

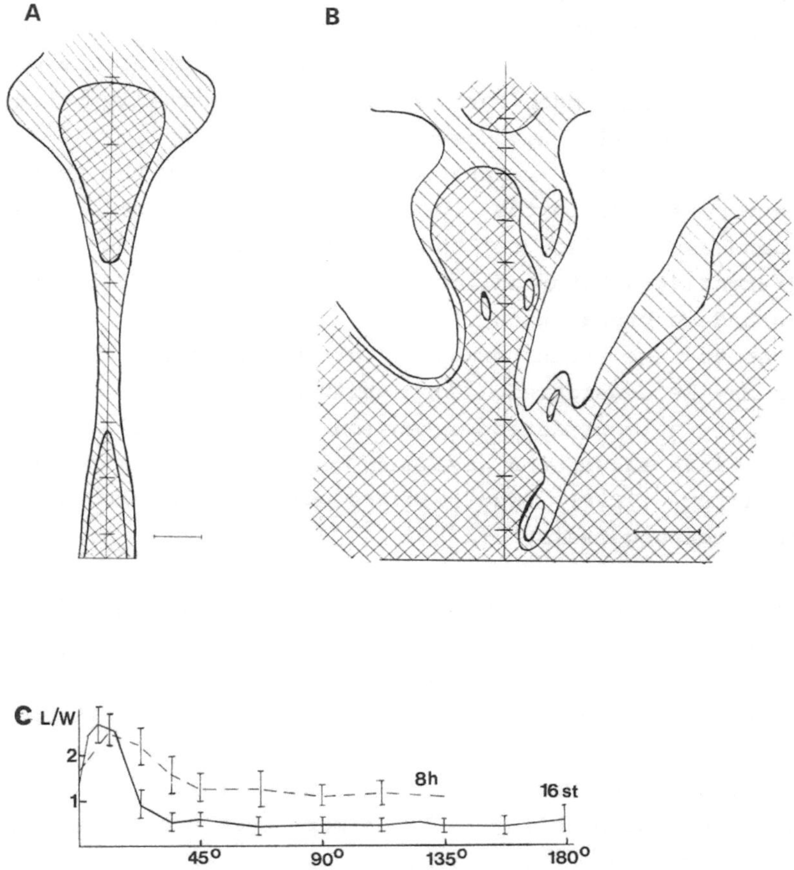

Fig. 12. Changes in the polarization patterns of ectodermal cells on the dorsal side of amphibian embryos in 8 h after relaxation of circumferential tangential stresses. A,B: planar maps of the dorsal sides of an intact (A) and operated (B) embryos. Cross-hatched are highly polarized cells (length/width (L /W) ratio >2). Hatched areas correspond to 2>L/W>1.5 and empty areas to L/W<1.5. C: a profile of cell polarization in the intact (solid line) and relaxed (dotted line) embryos. Horizontal axis: angular distance from the dorsomedial axis. Vertical axis: L/W ratio (from Beloussov et al., 1990).

transversal extension of the columnarized cells domain and to the increase of the amount of neural tissue. Both the outlines of the columnarized cell domain and the structure of neural rudiments become much more irregular than normally - Fig. 12 A-C (see for details Beloussov et al., 1990).

(b) a similar relaxation, while leaving the ventral area intact, results in the formation of additional domains of columnarized cells on the ventral embryo side which are differentiated later on into neural structures (Beloussov & al. i990 and in preparation).

As to the transformation of a neural plate into a tube, this may be considered, for amphibian embryos at least, as a kind of a contraction-extension cell-cell relay (see 3.4.1. (a)). This assumption have been tested by modelling (Beloussov & Lakirev, 1988). The results obtained imitated closely the normal morphogenesis.

5. Biomechanical approach to the buckling of epithelal layers.

5.1. General suggestions.

Buckling of epithelial layers is one of the most ubiquitous morphogenetical processes and the first one to be treated biomechanically (His, 1874; Gurwitsch, 1914). While His still considered the buckling to be no more than a passive mechanical reaction to the outer lateral pressure, Gurwitsch gave already the quantitative arguments for the active character of the neuroepithelial folding, its driving forces being located within the buckled region itself. This view is supported by the modern findings revealing the internal tangential pressure within buckling epithelia, as detected by their edges overlapping immediately after dissections (Beloussov, 1973; Beloussov et al., 1993). The internal pressure may be of the different origin. As mentioned above, in hydroid polypes it is osmotically driven and fluctuates within several minutes periodicity, creating growth pulsations (GP). In the large number of other rudiments, including embryonic tissues of vertebrates, it is more constant and caused either by cell insertion (intercalation) within a layer or by the radial cell contraction. Since practically all of the presumptively buckled zones consist of columnarized cell domains, being integrated into the tangentially stretched epithelia, any buckling process may be considered as a kind of HR reaction illustrated by Fig. 7F (3.4.1). That permits us to give a general biomechanical scheme of the buckling process which is independent of any details.

Let us consider a part of a layer with its edges fixed which starts to buckle under the influence of internal pressure. If a layer is completely flat, its buckling direction is unpredictable (this is a classic Eulerian instability). From the instant of its initial buckling, however, it will continue to buckle to the same side, inevitably creating bendings of the opposite sign on the flanks. Those, as pressure continues, will also increase, creating new adjacent bendings and so on, generating, as a result, a wave-like pattern (Fig. 13). A more

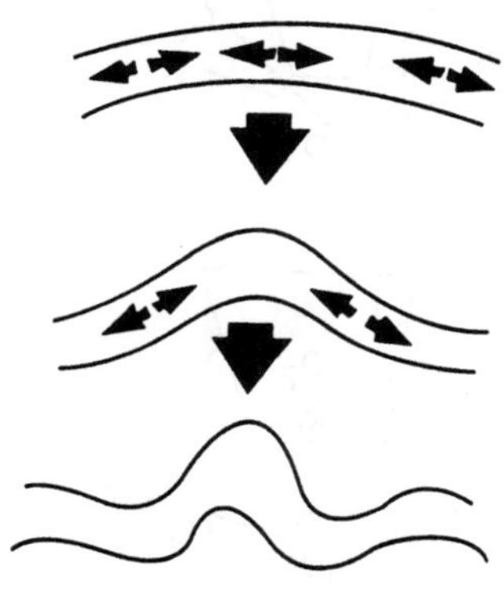

Fig. 13. A scheme of a wave-like folding of an epithelial layer under the action of tangential pressure forces.

detailed approximation is given by the following model.

5.2. The Internal Pressure Model (IPM). (Beloussov & Lakirev, 1991).

The main model algorithm is very simple (Fig. 14a). In any area of a nonzero curvature the pressure forces exerted by cells j - 1 and j + 1 upon cell j have a radial centrifugal component F_j, tending to shift cell j upwards. We suggest that the distance of its real active cell shift D_j is proportional to F_j. Now it is easy to demonstrate that the active cell shift is approximately inversely proportional to the radius of local curvature R:

$$D_j = K/R$$

Here K is a coefficient which just reflects the activity of a j cell's response to the external tangential pressure. In our

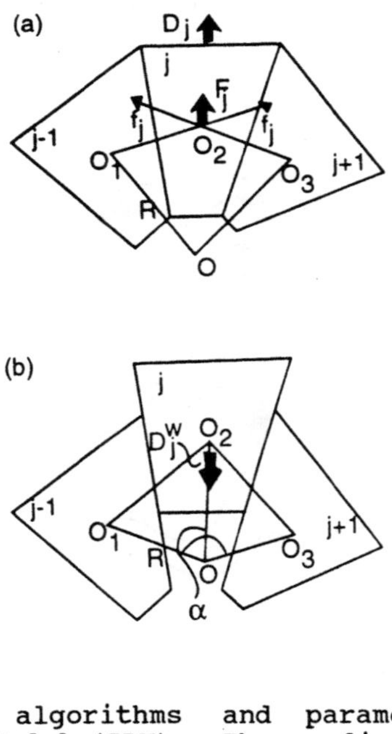

Fig. 14. The algorithms and parameters employed in the Internal Pressure Model (IPM). Three adjacent cells j-1, j and j+1 are shown, with their centers O_1, O_2 and O_3. O is a curvature center. For other designations see text.

constructions, we used either constant, or periodically fluctuated K values, in the latter case these values showing various oscillatorial patterns (so called narrow, intermediate and wide pulsations), which corresponded to the different kinds of species-specifical GP in hydroid polypes (see for details Beloussov et al., 1993).

The model includes also the parameter W, which expresses the viscoelastic linkage between cells, resisting their centrifugal shifts and smoothing the deformed surface. We assume that W varies from 0 to 1. Its mode of action is illustrated by Fig. 14b by the symbol D_w.

A simplest version of the model implies the synchronity of the radial impulses applied to all the cells. By this version, while starting from a simple smooth shape shown in Fig. 15A, we get a set of quite realistical complicated buckled configurations being the functions of W parameter and of the duration of the periodical radial impulses (Fig. 15 B-I). If

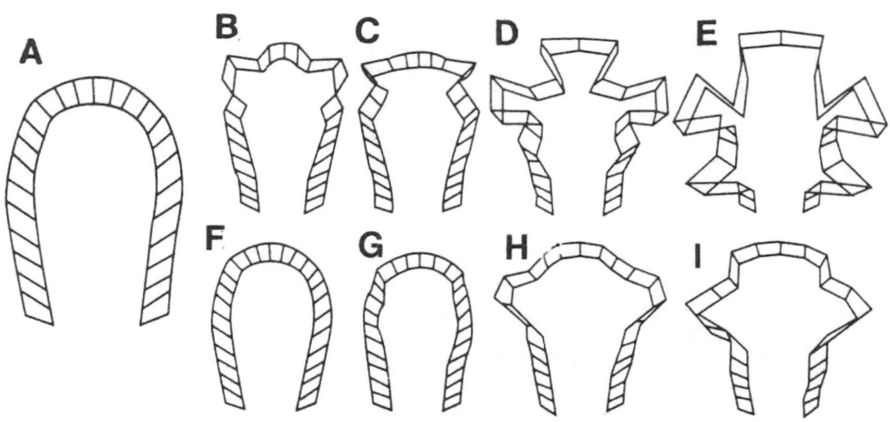

Fig. 15. Shape changes of an initial configulation A as imitated By IPM under the assumptions of "narrow" (B,F), "intermediate" (C,G), "wide" pulsations and the constant action (E,J) of a tangential pressure force. The upper row B-E differs from the lower one (F-I) by the reduced values of elasto-viscous resistance parameter W. In this version of the model the radial cell shifts are assumed to occur synchronously within the entire perimeter of the rudiment.

assuming, in addition, that the impulse for the local curvature increase is asynchronous, as spreading from a certain point in a wave-like manner around the rudiment perimeter some other patterns may be generated, even if starting from the perfectly circular shapes (Fig. 16). It is worth mentioning, that the

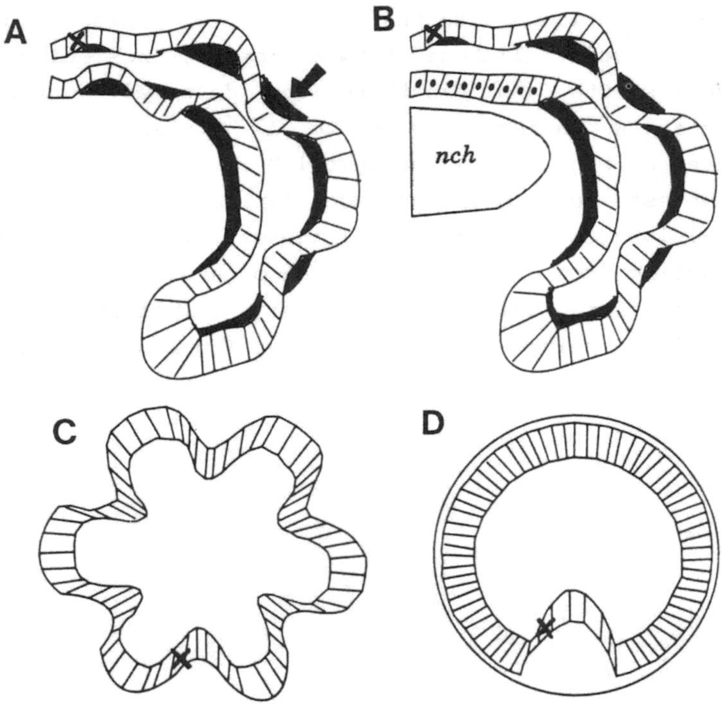

Fig. 16. Shape changes imitated by a wave-like version of IPM. In all the frames a wave of the radial cell shifts moves in a clock-wise direction out of a cross-marked point. A: the modelling of shape changes of the neural tube in vertebrate embryos; B: same under the presence of the anterior portion of a notochord (dotted) assumed to anchoring the overlaid neural cells. In the both frames the initial contours are shown by black. C,D: modelling of the shape changes of the initially circular configurations. C: under certain values of parameters and in assumption of a free extensibility a circle is transformed to a 6 - fold star. D: if introducing a non-extensible external membrane removed at a half cell height distance from the outer surface a gastrula-like shape is generated under the same parameters values (see for details Beloussov & Lakirev, 1992).

patterns are highly sensitive to the boundary conditions, for example switching rapidly from the star-like to the gastrula-like shapes after introducing a non-extensible envelope around the rudiments surface (Fig. 16, cf C and D).

6. Some generalizations.

6.1. Extrapolation of a uniplanar feed-back loop to the closed biomechanical entities surrounded by quasi-circular shells.

Suggest that a small area of a closed planar or spherical shell is contracted (Fig. 17A). As usually, this will initiate

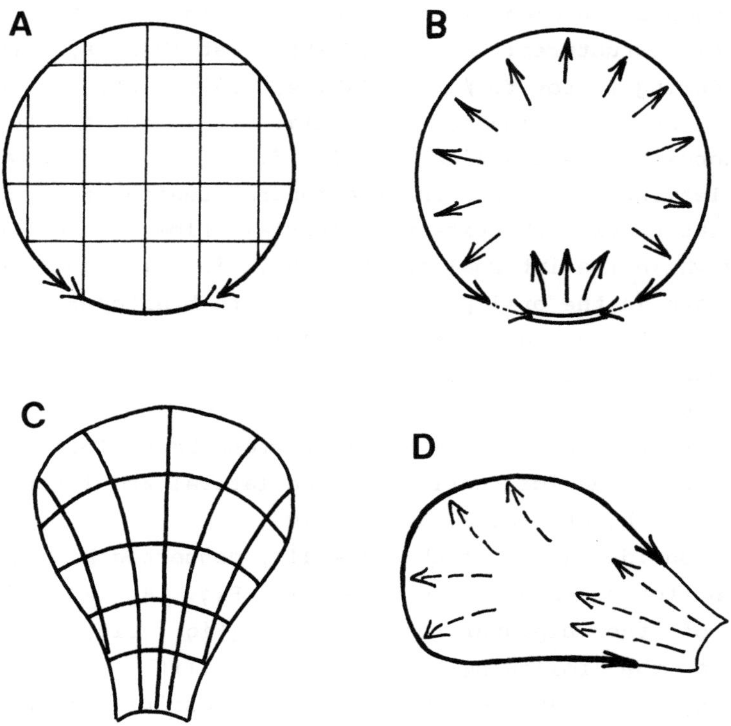

Fig. 17 A-D. A self-perpetuating biomechanical structure. A -C: surface view, D: side view, optical section. For explanations see text.

the active HR reaction throughout the entire shell, consisting in resorption (internalization) of a material within the contracted zone (to increase the locally relaxed tension) and in exteriorization (insertion) of a material within the stretched zone (to decrease the locally increased tension) - Fig. 17B. Due to this, a grid of the tension lines assumed initially to be isotropical (Fig. 17A), will be deformed as shown in Fig. 17C, converging now towards the contracted zone. Such a configuration shows a considerable dynamical stability and, in the case of a material flow between the contracted and expanded areas is self-perpetuating. This flow should be directed from the expanded to contracted area on a sample's surface and in the opposite direction in its interior (Fig. 17D). Such a dynamical scheme seems to imitate the various dynamical structures: a moving fibroblast (Harris, 1990), a zygote during furrow formation (White, 1990) and, to a certain extent, an invaginated gastrula. The contracted area corresponds to the trailing end of a fibroblast, furrow region of a cleaving zygote and the blastoporal region of a gastrula. In the latter case the external material flow corresponds to epiboly while the internal one to involution. The joining point of the both is the oral pole, opposite to blastopore. We can see that at the different organizational levels the living nature exploits the same structurally stable biomechanical principles.

6.2. Extrapolation of a biplanar feed-back loop to more than 2 biomechanical entities: the lateral and latero-radial spreading of biomechanical activities.

(1) Consider a set of the laterally connected biomechanical entities (for example, single cells) - Fig. 18. Exploiting the scheme of the biplanar feedback loop (Fig. 11B) we can come easily to the following conclusions: (1)if the cells are contacting with each other along the entire length of the lateral surfaces (Fig. 18A, 1) an active elongation (=polarization) of a left cell will induce each next one to do the same, firstly passively and then actively. Such a polarization wave, identical to the previously described CCP,

will be self-restricted due to the gradual rotation of the maximal stretching axis from vertical to horizontal, by obvious geometrical reasons (Fig. 18A, 2.3). In such a way the Belintzev et al. (1987) situation may be at least qualitatively reformulated in the framework of HR concept.

(2) Now consider an actively elongating cell *a*, fixed by its *f* surface and only partly contacting with *b* cell (Fig. 18B, 1). By the similar reasons *ab* contact zone will be increased at the expense of *bc* contact zone (cell *c* assumed to be non-elongated and /or non-adhesive to *b* cell). That explains a number of facts known from cell biology: (a) a tendency of homologous cells to increase contacts with each other; (b) segregation of a cells mixture according to cells adhesive potentials; (c) movement of cells up along either adhesive (haptotaxis) or tensions gradients. From the biomechanical point of view, the both kinds of gradients are intimately coupled.

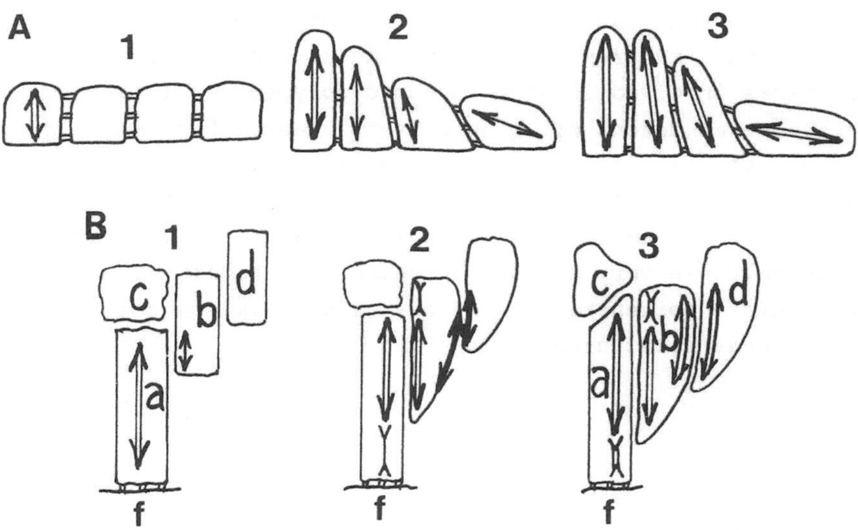

Fig. 18. Examples of the lateral spreading of biomechanical activities. A, 1-3: a self-restricted spreading of contact cell polarization to the right of an actively elongated cell. B, 1-3: the establishment of a cell flow up along the tension and/ or adhesion gradient.

Such a cooperative spreading of the biomechanical activities of individual cells looks as a main way for transforming the microevents into space-temporal macrostructures.

6.3. From biomechanics to the behavioural reactions of embryonic tissues.

Being it wrong or true, the HR concept means that the embryonic tissues are endowed by some elements of behaviour, including "epigenetical memory". In the other words, an embryonic cell or a tissue piece not only percepts the instantaneous stress value, but is also able to compare it with that occuring some time earlier. Paradoxical as it may look at the first glance, the development is moving forth by each embryonic part striving to return back into its "biomechanical past". Due to resistance of the neighbouring parts, meanwhile, such an attempt should necessarily produce some structural novelties, rather than being a simple repetition of what have been already happened.

The characteristical times of both epigenetical memory storage and HR responses may range from several minutes (feedbacks involved in GP of hydroid polypes, 2.1.1. and 6.2) up to several dozens of minutes at least (the duration of topological invariability of tensile fields in amphibian embryos, 2.5.1). One of the main functions of the entire intercellular regulatorial machinery may be in modulating and adjusting these very times. It seems also highly probable that the main epigenetical memory store is associated with cytoskeletal structures.

REFERENCES

Belintzev B N Beloussov L V Zaraisky A G (1987) Model of pattern formation in epithelial morphogenesis. J theor Biol 129: 369-394.

Beloussov L V (1973) Growth and morphogenesis of some marine Hydrozoa according to histological data and time-lapse studies. Publ Seto Mar Biol Lab 20: 315-366.

Beloussov L V (1979) Experiments on changing tensile fields of the axial rudiments in amphibian embryos. Ontogenez (Sov J Devel Biol) 10: 120-129.

Beloussov L V (1988) Ontogenesis and self-organization theory. Rivista di Biologia (Biology Forum) 81 (2): 159-183.

Beloussov L V Dorfman J G Cherdantzev V G (1975) Mechanical stresses and morphological patterns in amphibian embryos. J Embr exp Morphol 34: 559-574.

Beloussov L V Labas J A Kazakova N I Zaraisky A G (1989), Cytophysiology of growth pulsations in hydroid polypes. J Exp Zool 249: 258-270.

Beloussov L V Labas J A Kazakova N I. Growth pulsations in hydroid polypes: kinematics, biological role and cytophysiology. In Oscillations and Morphogenesis (L Rensing ed) Marcel Dekker New York Basel Hong Kong 1993 pp 183-193.

Beloussov L V Lakirev A V. Self-organization of biological morphogenesis: general approaches and topo-geometrical models. In Thermodynamics and Pattern Formation in Biology (Eds I Lamprecht and A I Zotin). W de Gruyter Berlin New York 1988 pp 321-336.

Beloussov L V Lakirev A V (1991) Generative rules for the morphogenesis of epithelial tubes. J theor Biol 152: 455-468.

Beloussov L V Lakirev A V (1992) Why five? Biology Forum 85 (1): 93-101.

Beloussov L V Lakirev A V Naumidi I I (1988) The role of external tensions in differentiation of Xenopus laevis embryonic tissues. Cell Diff & Devel 25: 165-176.

Beloussov L V Lakirev A V Naumidi I I Novoselov V V (1990) Effects of relaxation of mechanical tensions upon the early morphogenesis of Xenopus laevis embryos. Int J Dev Biol 34: 409 -419.

Beloussov L V Luthinskaia N N (1983) A study of relay cell interactions in the explants of amphibian embryonic tissues. Cytologia (Russ) 25: 939-944.

Beloussov L V Mittenthal J (1992) Hyperrestoration of mechanical stresses as a possible driving mechanisms of morphogenesis. Zhurn Obsch Biol (Russ) 53: 797-807.

Beloussov L V Zhadan A A (1993) Morphological reactions of echinodermata embryos upon the lowering of osmotical gradient between blastocoel and the external environment. Ontogenez (Russ J Devel Biol) 24: 32-36.

Bereiter-Hahn J. (1987) Mechanical principles of architecture of Eukaryotic cells. In Cytomechanics. The Mechanical Basis of Cell Form and Structure. (Bereiter-Hahn, J, & Anderson, O.R., eds). Springer-Verlag. p 3-30.

Bereiter-Hahn J Strohmeier R Hydrostatic pressure in Metazoan cells in culture: its involvement in locomotion and shape generation. In Cytomechanics. The Mechanical Basis of Cell Form and Structure. (Bereiter-Hahn, J, & Anderson, O.R., eds). Springer-Verlag. p 261-272.

Betchaku T Trinkaus J P (1986) Programmed endocytosis during epiboly of Fundulus heteroclitus. Am Zool 26: 193-199.

Bluemenfeld L A (1983) Physics of Bioenergetic Processes Springer Verlag Berlin etc

Burnside B (1971) Microtubules and microfilaments in Newt neurulation. Dev Biol 26: 416-441.

Chain B M (1980) The transepithelial potential and osmotic regulation in the green hydra J Exp Biol 88: 161-173.

Coulombre A J (1956) The role of intraocular pressure in the development of the chick eye I. Control of eye size. J Exp Zoology 33: 211-226.

Dennerli T J Lamoureux P Buxbaum R E Heidemann S R (1989) The cytomechanics of axonal elongation and retraction. J Cell Biol 109: 3073-3083.

Foissner I (1985) Reversible dilatation of endoplasmic reticulum cisternae during pressure and plasmolysis. Phyton 25: 225-231.

Gerhart JC Danilchik M Doniach T Roberts S Rowning B Stewart R (1989) Cortical rotation of the Xenopus egg: consequences for the anteroposterior pattern of embryonic dorsal development. Development 107 Supplement: 37-51

Goodwin B C (1985) What are the causes of morphogenesis? BioEssays 3: 32-36.

Goodwin B C Trainor L E H (1985) Tip and whorl morphogenesis in Acetabularia by calcium-regulated strain fields. J theor Biol 117: 79-106.

Gurwitsch A G (1914) Die Vererbungsmechanismus der Form. Arch Entw-mech 39: 516-577.

Haken H (1978) Synergetics. Springer Verlag Berlin Heidelberg New York.

Harold F M (1990) To shape a cell: an inquiry into the causes of morphogenesis of microorganisms. Microbiol Rev 545: 381 -431.

Harris A K (1990) Protrusive activity of the cell surface and the movements of tissue cells. In: Biomechanics of Active Movement and Deformation of Cells. (N. Akkas ed) Springer-Verlag Berlin-Heidelberg etc p. 249-294.

Heidemann S Buxbaum R E (1990) Tension as a regulator and integrator of axonal growth. Cell Motility and Cytoskeleton 17: 6- 10.

His W (1974) Unsere Korpersform. Springer Verlag, Leipzig.

Ivanenkov V V Meshcheryakov V N Martynova L E (1990) Surface polarization in loach eggs and two-cell embryos: correlation between surface relief, endocytosis and cortex contractility. Int J Dev Biol 34: 337-349.

Keller R E (1978) Time-lapse cinematographical analysis of superficial cell behavior during and prior to gastrulation in Xenopus laevis. J Morphol 157: 223-248.

Keller R E (1987) Cell rearrangement in morphogenesis. Zool Sci 4: 763-779.

Keller R Cooper M S Danilchik M Tibbetts P Wilson P A (1989) Cell intercalation during notochord development in Xenopus laevis. Journ Exp Zool 251: 134-154.

Keller R Danilchik M (1988) Regional expression, pattern and timing of convergence and extension during gastrulation of Xenopus laevis. Development 103: 193-209.

Kolega J (1986) Effects of mechanical tension on protrusive activity and microfilament and intermediate filament organization in an epidermal epithelium moving in culture. J Cell Biol 102: 1400-1411.

Labas Ju A Badenko L A Beloussov L V Letunov V N (1980) On the pulsatorial growth in multicellular organisms. Dokl Akad Nauk SSSR (Russ) 257: 1247-1250.

Meinhardt H (1982) Models of Biological Pattern Formation Acad Press New York.

Mittenthal J. & Beloussov L.V. (1993). A biomechanical model for gastrulation of Xenopus laevis. In Lectures on Theoretical Biology Vol 2 (Ed K Kull T Tiivel) Valgus Tallin (Estonia) (in press).

Murray J D (1989) Mathematical Biology Berlin etc Springer Verlag.

Nicolis G Prigogine I (1977) Self-organization in non-equilibrial systems. J Wiley & Sons New York London etc.

Odell G M Oster G Alberch P Burnside B (1981) The mechanical basis of morphogenesis. I.Epithelial folding and invagination. Devel Biol 85: 446-462.

Oster G F Murray J D Odell G M (1985) The formation of microvilli. In Molecular Determinants of Animal Form. Alan R Liss Inc (UCLA Symp Mol Cell Biol New Ser Vol 31) pp 365 -384.

Petrov K V Beloussov L V (1986) Movements of cell material of a dorsal wall of Xenopus laevis embryos during gastrulation and neurulation. Ontogenez (Sov J Devel Biol) 17: 78-83.

Radeva V Ichev K Ovtscharoff W Dolapchieva S Roussev G (1986) Some aspects of endocytosis of neuroepithelial cells in early newt's neurula. Z mikr-anat Forsch 100: 861- 866.

Selman G (1958) The forces producing neural closure in Amphibia. J Embryol exp Morphol 6.

Slack C Warner A E (1983) Intracellular and intercellular potentials in the early Amphibian embryo. J Physiol (London) 232: 313-330.

Spooner B S Wessells N K (1972) An analysis of salivary gland morphogenesis: role of cytoplasmic microfilaments and microtubules. Devel Biol 27: 38-52.

Steinberg M S (1978) Cell-cell recognition in multicellular assembly: levels of specificity In Cell-cell recognition (A S G Curtis ed) Cambridge Univ Press p 25-49.

Stern C D (1984) A smple model for early morphogenesis. J theor Biol 107: 229-242.

Symes K Smith J C (1987) Gastrulation movements provide an early marker of mesoderm induction in Xenopus laevis Development 101: 339-349.

Thompson D Arcy (1942) On Growth and Form Cambridge University Press Cambridge.

Wanek N Marcum B A Lee H T Campbell R D (1980) Effect of hydrostatic pressure on morphogenesis in nerve-free hydra. J Exp Zool 211: 275-280.

White J G (1990) Laterally mobile, cortical tension elements can self-assemble into a contractile ring. In Cytokinesis. Mechanisms of furrow formation during cell division. G W Conrad T E Schroeder eds N Y Acad Sci p 50-59.

Wilson P A Oster G Keller R (1989) Cell rearrangement and segmentation in Xenopus: direct observation of cultured explants. Development 105: 155-166.

Wohlfarth-Bottermann K-E (1987) Dynamic organization and force production in cytoplasmic strands. In Cytomechanics. The Mechanical Basis of Cell Form and Structure. (Bereiter-Hahn, J, & Anderson, O.R., eds). Springer-Verlag. Berlin-Heidelberg etc p. 154 -166.

THE ROLE OF ELASTICITY IN THE MOTILE BEHAVIOUR OF CELLS.

Jürgen Bereiter-Hahn, Holger Lüers
Cinematic Cell Research Group, Dept. of Zoology, Johann Wolfgang Goethe Universität Frankfurt/M., Marie Curiestr. 9, D 60439 Frankfurt am Main.

I. Introduction

In Metazoans cells are either mobile or they are tied up in a multicellular arrangement. Bound and free states of a cell may change with time depending on differentiation and external requirements. Thus, a mobile cell may become integrated into a tissue and immobilized, or a tissue cell may leave its environment and become motile. Such events are of importance for example during embryonic development, wound healing, inflammation or metastasis. Motile and static forms are characterized by distinct organizations of the cytoskeleton, the expression of receptor molecules on their surface and therefore their responsiveness to external stimuli. As a consequence the competence of cells to interact with the extracellular matrix (ECM) or to mechanical stimulations is changed. This principle has been termed dynamic reciprocity by Bissell et al. (Bissell and Aggeler, 1987). Similar thoughts have been forwarded recently by several groups (Bereiter-Hahn, 1988; Clegg, 1992; Carver et al., 1991; Ingber and Folkman, 1987; Lapiere and Nusgens, 1989).

In the following, all motile phenomena based on cilia or flagellae have been excluded. We are dealing exclusively with the type of movement which is generally referred to as *ameboid movement*.

The question "*how do cells crawl*" still bears many unsolved problems which can be approached on different organizational levels:

- By biochemical and molecular biological methods characterizing the types of molecules involved and their mutual interactions.
- The constructional principles underlying the arrangement of the macromolecular constituents. These are closely related to
- the level of cell mechanics locating force generation and force distribution at different states of motility.
- The control of cell motility by external factors as are chemical, mechanical or electrical stimuli.

Understanding locomotion of cells includes their interactions with the ECM or any other solid providing a substrate for adhesion, the plasma membrane bound receptors responsible for contact to the ECM receiving external stimuli and transducing them to the cell interior. In the cytoplasm ion concentrations may change, kinases being activated or de-activated altering the interaction of cytoskeletal elements with membrane bound molecules and with each other, thus altering force generation and propagation. In addition, these events are supposed to

NATO ASI Series, Vol. H 84
Biomechanics of Active Movement and Division of Cells
Edited by N. Akkaş
© Springer-Verlag Berlin Heidelberg 1994

influence proliferative activity and gene expression in the nucleus. All these parameters finally have to be integrated into a comprehensive model of cell locomotion which must explain the generation of a given cell shape and its macroscopic continuity throughout cycles of motion as well.

Locomotion studies have been focussed on a few model systems in the past.
The first cell type which has been intensively investigated were amoebae (see Grebecki, this Vol.). Eminent investigations have been performed by for example by Mast (1926) and Goldacre (1961). Lateron the discussion on ameboid movement was dominated by two conflicting hypotheses on motive force generation, the fountain zone contraction hypothesis propagated by R.D. Allen (1961, 1970; Allen et al., 1972), and the hydrostatic pressure hypothesis favorized by the cytologists in Bonn (Wohlfahrt-Bottermann, 1964 a,b; 1979; Komnick. et al. 1973).
Amoebae of the slime mold *Dictyostelium* became widely investigated models for cell locomotion because of the availability of a great variety of mutants of cytoskeletal elements (for review see Gerisch et al. 1991).
Among vertebrate cells, fibroblasts, leukocytes, spreading blood platelets and epithelial cells are the favorites for *in vitro* locomotion studies. Fibroblasts and leukocytes (i.e. granulocytes) are easily to obtain. They represent different types of motion, as fibroblasts exhibit extensive contact to ECM or glass or plastic surfaces and granulocytes develop their pseudopodia in almost all directions and to move they dependend much less on strong adhesion to a solid substratum. Highlights in the studies of fibroblast motility were the investigations of Abercrombie and his group (Abercrombie, 1961; Abercrombie et al. 1970 a,b,c; 1971; Heaysman et al., 1982), and of Vasiliev (1982). Due to the ease of short time culture and the significance of their motility in inflammatory and immune responses granulocyte and macrophage motility has been studied in many details (e.g. Keller, 1983; Stossel, 1982; 1983; Stossel et al., 1987). More recently the potential of epidermal cells of aquatic vertebrates (teleost fish and amphibian tadpoles), in particular, had been recognized (Bereiter-Hahn et al. 1981; Cooper and Schliwa, 1985; Korohoda, 1992; Theriot and Mitchison, 1991). The main advantages of these cells is their large size, relatively high speed and the broad, organelle-free and very flat lamella clearly distinct from the cell body (Bereiter-Hahn, 1967).

Engagement of different elements of the cytoskeleton, influence of protein phosphorylation and the effects of external stimuli have been described for these cell types (and for others). Less is known on the mechanical properties of cells on the subcellular level. The report on the last NATO Advanced Study Institute held in Istanbul in 1989 includes some very significant contributions to cytomechanics

(Evans and Dembo, 1990; Harris, 1990; Skalak and Zhu, 1990; Simon and Schmid - Schoenbein, 1990a). Detailed knowledge has been obtained for erythrocytes (e.g. Waugh and Hochmuth, 1987) and for the rheological properties of isolated cytogels (e.g. Janmey, 1991; Müller et al., 1991; Pollard, 1982).

The results and their interpretation derived from experiments with these different model systems quite often is difficult because the principles according to which a generalization can be made are still under debate. Nevertheless mathematical modelling (Alt , 1990; DiMilla et al. 1991; Evans and Dembo, 1990) may help to design the right experiments. Most of the experimental results presented in this contribution have been obtained by studying tadpole epidermis cells, and the discussion is focussed to animal cells in culture. With this presentation we intend to contribute to an understanding of the mechanical basis of cell form and locomotion.

II. What determines the mechanical properties of cells ?

From a mechanical point of view an animal cell can be described as a visco-elastic gel enclosed in a semi-permeable membrane. The gel is in intimate contact with the membrane and it is composed of several types of fibrillar elements. Many proteins bind specifically to these fibrils and may achieve homologous or heterologous crosslinks. The fibrils are polymers arising from the aggregation of mono- or oligomers. The group of fibrils which resist extraction of the cells with anionic detergents (like TritonX100) is called **cytoskeleton** (Osborn and Weber 1977).
Regarding the vast literature in this field and the many articles reviewing recent advancements in discovery of new proteins interacting with elements of the cytoskeleton, the biochemistry of the main elements of the cytoskeleton and their associated proteins is not discussed here. The interested reader is referred to the contributions by Alt (1990), Evans and Dembo (1990), Skalak and Zhu (1990), Harris (1990) in the former ASI volume on "Biomechanics of active movement and deformation of cells", to the comprehensive reviews by M. Schliwa (1986), Bershadsky and Vasiliev (1988) and a very recent review on the cortical actin layer scholarly written by T. Stossel (1993).

Using the machinery outlined above, a cell can develop *mechanical forces* by

- ion transport through the membrane giving rise to osmotic differences (Na+/K+-ATPase),
- active contractions of the actomyosin fibrillar network
- swelling pressure inside the cytogel due to charge density and the development of Donnan equilibria,

- extension of bundles of fibrils by polymerisation (e.g. acrosome reaction for actin or formation of bundles of microtubules in the axopodia of Heliozoans). In this context it has to be pointed out that all these fibrils are primarily tensile elements. The only stiff rods are microtubules. Actin fibrils gain some stiffness by parallel assembly (Gittes et al., 1993, Tilney, 1980, 1983, Tilney et al. 1973; 1988)

Therefore *mechanical strength (*i.e. stiffness) of a cell may result from

- the hydrostatic pressure generated by ion transport, contraction of a cortical actomyosin layer (Ambrose et al. 1970, Bray et al. 1986), swelling pressure of the cytogel,
- bundling of fibrillar elements (microspikes, microvilli, acrosomal process, axopodia, cortical microtubule layer in some protozoa)
- contraction of fibril bundles (e.g. stress fibers) against either an incompressible core cytoplasm or between adhesion zones to an external solid.
- elastic coil of intermediary filaments acting between attachment zones (e.g. subcortical layer of "tonofilaments" spanning between desmosomes in the surface layer of fish and tadpole epidermis (Anniko and Arnold, 1990; Schliwa, 1975; Kunzenbacher et al. 1982)
- the viscous resistance of cytoplasm against deformation. All the elements of the cytosekeleton may contribute to this viscosity, actin (Cortese and Frieden, 1990; Grazi et al., 1990; Janmey, 1991; Janmey et al. 1991; Müller et al., 1991)) and vimentin (Janmey et al. 1991) in particular.

The *shape* of a cell results from mechanical interactions of the constituing cellular and extracellular elements. The cytoskeleton interacting with the plasmamembrane comprises the main cellular structure responsible for the generation of a given shape and for the interaction with the extracellular matrix via specific adhesion molecules. Also multicellular arrangements are shaped by these interactions: In epithelia, for instance, specifically differentiated contact sites connect the cells with each other. Fibrils terminating in these adhesion zones propagate forces from one cell to the other, thus distributing tension throughout the epithelial layer as can easily be demonstrated for skin (Fig. 5) (Kunzenbacher et al. 1982), intestine (Mooseker et al., 1978) or capillary endothelium (Franke et al. 1987 ; Levesque et al. 1989).

A deeper understanding of cell shape can be achieved from the intracellular distribution of the cytoskeletal elements and from force distribution throughout a cell. Both these approaches are complementary to each other because structures (and fibrils in particular) can be comprehended as force guiding elements. The term "*structure*" should be considered as a highly dynamic entity which results from continuous and ordered turnover of its chemical constituents. As a consequence shape (morphology) of living organisms appears static only on a macroscopic scale. On a micro-scale the elements arranged to a certain shape are in steady turnover and movement, however, with very different time constants for different structures.

A few methods have been developed which allow measurements of cell elasticity (stiffness) on a subcellular scale. Cell deformability is probed by sucking of some cytoplasm into the tip of a capillary (for overview see Hiramoto, 1987) or by poking with a small ball on the cell surface (Duszyk et al., 1989). Atomic force microscopy also uses this principle (Butt et al., 1990; Henderson et al., 1992; Hörber et al. 1992). Another recent method is scanning acoustic microscopy (SAM) which uses the relationship of the propagation velocity of longitudinal waves to the elasticity of the medium and provides a resolution in the micrometer or submicrometer range (for details see appendix I). Using this method areas of cytoplasm with different stiffness can be determined. Experiments influencing the organization and activity of cytoskeletal elements reveal their contribution to the mechanical properties determined with this method.

Living XTH2-cells (a line derived from *Xenopus laevis* tadpole heart endothelia - Schlage et al., 1981; Schlage, 1986) were observed with SAM (Lüers et al., 1991, 1992) After control recordings these cells were treated with the Ca^{2+}-specific ionophore ionomycin in the presence of 1.2 mM calcium in the culture medium. Most XTH2 cells exhibit a typical pattern of stiffness distribution. In the thin peripheral parts of the cells, the stiffness is rather high and decreases toward the cell center. At the transition region to the dome-shaped center, stiffness often reaches a minimum and increases again along the margins of the cell body (Fig. 1). In most cells a sudden decrease in sound velocity is obvious at the site where the slope of the cell body starts (Fig. 2). An almost linear relationship exists between the angle of the surface slopes of the peripheral cytoplasm and that at the margin of the cell body and with the local decrease of the longitudinal sound velocity in this area. These observations indicate that cell body and lamella region mechanically may be almost separated, a finding which is of importance for understanding the mutual interactions between lamella and cell body regions (see below), predisposes this place for gathering surface attached particles and for endocytotic ingestion.

The reactions of cells to extremely increased Ca^{2+}-levels are manifold. Microtubules are destroyed (e.g. Schliwa, 1980; Stolz and Bereiter-Hahn, 1987), the actomyosin network may be induced to contract, and lateron actin organization is destroyed, most of the F-actin becomes disassembled. Due to these actions, ionomycin treatment should reveal the contrubution of actin fibrils and of actomyosin based contractions to the mechanical properties of cells. The time-course of the ionomycin effects on the acoustic properties is biphasic: A short transient increase of stiffness of the cortical cytoplasm is followed by a permanent decrease (Fig. 1c). After 30s of incubation with ionomycin, the increase in sound velocity, and thus in stiffness, is most prominent in the cell body and the more central parts of the lamella. Thickness and width measured along a scanning line traversing the cell are also reduced. The subsequent decrease of stiffness starting around 4 min after addition of ionomycin affects primarily the same regions. In these experiments a contraction phase was indicated by morphological changes and a concomittant increase in stiffness (elasticity), followed by a phase with pronounced loss of stiffness accompanying the disassembly of cytoplasmic fibrils.

The relative high stiffness at the edge of the cells corresponds well to distinct F-actin bundles in these regions before and after ionomycin treatment. Likewise a correspondence between low stiffness at the edge and peripheral parts and a reduced content of F-actin can be observed. From these observations we conclude that the mechanical properties of these thin peripheral cell regions are determined by the distrubution and contractile state of actomyosin fibrils composing a stiffened basal plate (Bereiter-Hahn et al. 1990a). In the thicker central regions of the cells the cortical actomyosin layer is difficult to visualize by normal fluorescence microscopy and the relative influence of the basal layer is reduced by the overlaying organelles, therefore a correlation between stiffness and actin distribution is no longer detectable.

Fig. 1. Changes of stiffness distribution and cell shape induced by ionomycin.
a) Acoustic image of a living XTH-2 cell on glass immediately before addition of ionomycin (defocussing +12 µm, 1 GHz). The upper curve represents the amplitude of the acoustic signal recorded along the scan line traversing the cell (lower line).
b), **c)**, and **d)**: Acoustic parameters of the cell calculated before **(b)**, after 30 s **(c)** and after 4 min **(d)** incubation with 1 µM ionomycin in the presence of calcium. Thickness (bold line) and distance were obtained from the interference pattern in the acoustic image. The ratio between y axis and x axis is 1:15, i.e., the cells are very flat. In the very center no reliable measurements of thickness and sound velocity could be made. Sound velocity (bars) is calculated for each pair of interference fringes. After 30 s **(c)** of incubation with ionomycin, an increase of stiffness (revealed by sound velocity measurement) occurs that is most prominent in the cell periphery. The thickness is reduced, due to the contraction of the cell. After 4 min of incubation **(d)** the sound velocity in the cell body was decreased, whereas in the very periphery the values changed only slightly.

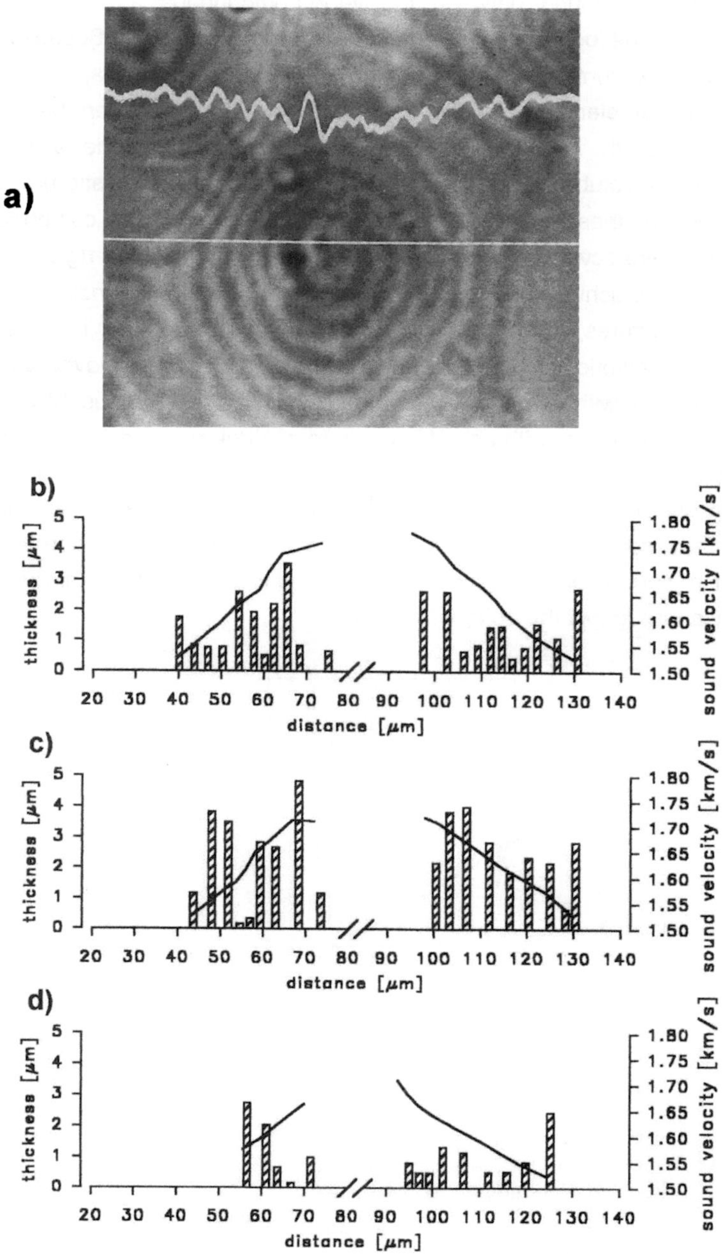
a)
b)
thickness [µm]
sound velocity [km/s]
distance [µm]
c)
thickness [µm]
sound velocity [km/s]
distance [µm]
d)
thickness [µm]
sound velocity [km/s]
distance [µm]

No correlation was found between the very homogneous arrangement of the microtubules and the local stiffness variations measured with SAM. Because of the codistribution of vimentin and microtubules in these cells, no major influence of these cytoskeletal elements on the reflectivity of ultrasound can be expected, although not directly measured til now. This does not exclude a substantial contribution of microtubules and intermediate filaments to the elastic properties in those cells where these elements are the dominating structural components (i.e nerve cells, keratocytes). Microtubules represent relatively rigid structures responsible for the achievement and maintenance of cell polarity (Inoue, 1982). The rigidity of structures formed by crosslinked microtubules is qualitatively demonstrated by mitotic spindles (Hirokawa, 1985), which can be moved around in a cell when pushed with a needle. Microtubules linked by various MAP's to the cytoplasmic membrane or with actin filaments (e.g. Hirokawa and Tilney, 1982) may play a role as stiffening elements in shape generation. Intermediate filaments form highly visco-elastic gels (Janmey et al. 1991). Considering their arrangement in cells in culture they can be assumed to determine the mechanical properties of the perinuclear region. In epithelia they form a subcortical layer distributing forces via desmosomes throughout the epithelial layers.

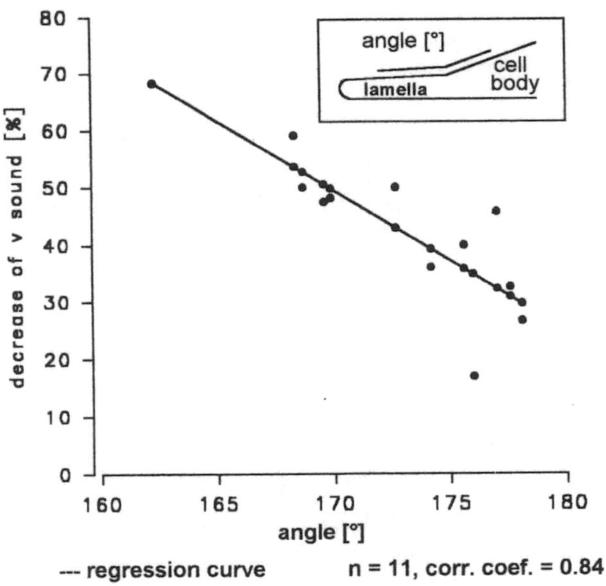

Fig 2. Relationship between the decrease of sound velocity at the cell body / lamella transition region (in % of the value in the lamella immediately adjacent to the increasing slope of the cell body) and the angle formed between lamella and cell body (see insert).

III. Structural basis for cell locomotion.

The expression and relative amount of the main groups of cytoskeletal fibrils, actin, microtubules and intermediate filaments varies among different cell types. Their contribution to cell locomotion may be elucidated by comparing the distribution of these elements in cells moving very actively and those which are almost static. In the following such a comparison is presented for the three main types of fibrils.

Cell motility and actin distribution.

F-actin has been fluorochromed with TRITC-phalloidin (Faulstich et al. 1983) in a co-culture of the mesenchyme derived 10T1/2 cells and mouse leukemia BWO-Li cells (Verschueren et al. 1991) (Fig. 3). These cells represent two extremes in motile activity, 10T1/2 no longer move having reached cell density of confluence, whilst the BWO cells are highly motile. The mobile cells are totally devoid of stress fibers which are prominent in the 10T1/2 cells. This difference seems to be typical. All fast moving cells lack stress fibers and thus focal contacts (Couchman and Rees, 1979). Focal contacts are assumed to represent zones of strong contact to the substratum which gives support to the formation of taut actin bundles (stress fibers) most probably arising by contraction from a previously isotropic actin network (Bereiter-Hahn, 1987; Burridge, 1981; Guiliano and Taylor, 1990; Kreis and Birchmeier, 1980; Korohoda and Kajstura, 1982; Opas, 1987). On the other hand the highly motile cells exhibit a distinct cortical actin (and myosin) layer. Therefore we conclude that stress fibers inhibit locomotion rather than participating in motive force generation. This does not exclude that contraction of stress fibres after release of focal adhesions adds to the process of movement (see for instance Chen, 1979, 1981). They primarily represent stabilizing elements, their destruction only favors cell motility.

In most migrating cells F-actin is more concentrated at the leading edge than elsewhere in the cell (Fig. 4a,b). The relative concentrations of F-actin can be determined by micro-fluorometry of cells stained with TRITC-phalloidin under certain precautions (Kajstura and Bereiter-Hahn, 1989). The ratio of the fluorometer readings and thickness of the cell at the same site (as determined by reflection interference microscopy - RIM) provides a measure of the relative actin concentrations. In the case of tadpole epidermis cells like the one shown in Fig. 4a, F-actin concentration at the leading edge is about 25 times as high as in the cell body (Bereiter-Hahn et al. 1990). In most of these cells the central part of the lamella contains only a small amount of actin (Bereiter-Hahn et al. 1981) (Fig. 4 a,b), but this distribution varies considerably, therefore the almost homogenous actin distribution described for the very similar teleost epidermal cells by Theriot and Mitchison (1991)

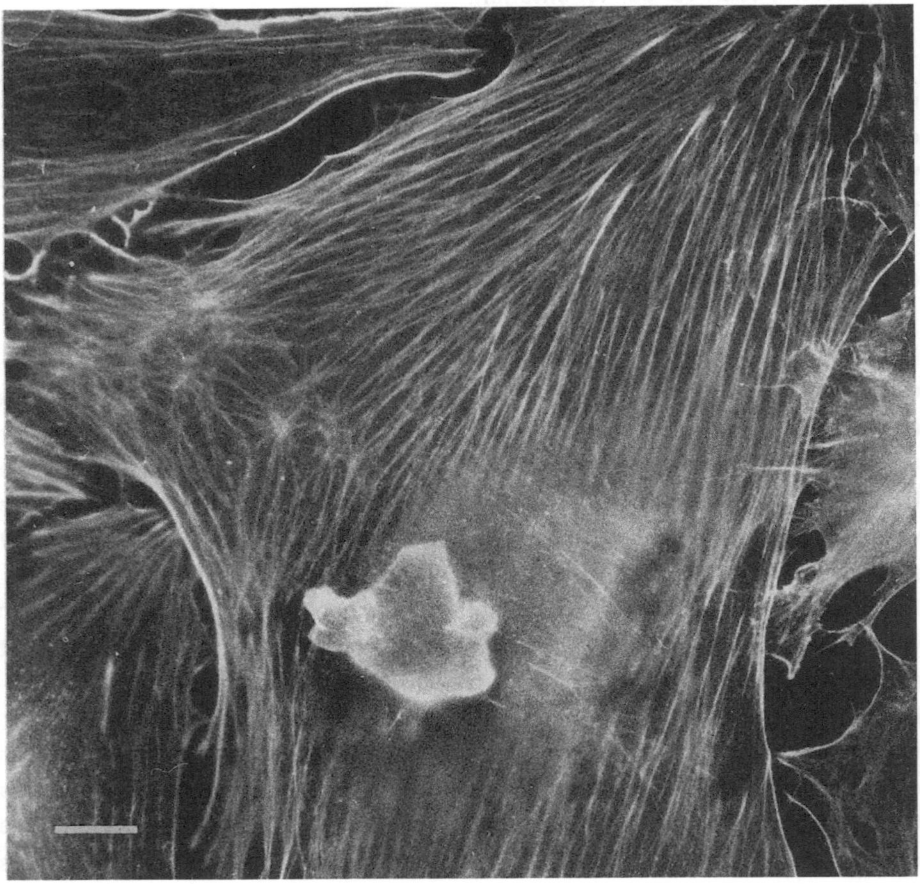

Fig. 3. F-actin pattern (as revealed by TRITC-phalloidin fluorescence) of mouse 10T½ cells (exhibiting a distinct pattern of thick fibrils / stress fibers) and a BWO-Li cell, a mouse leukemia line (Verschueren et al., 1991): diffuse staining, brightest fluorescence beyond the cell membrane and in the pseudopodia.

Fig. 4. Xenopus tadpole epidermis cells (XTE-cells), distribution of F-actin (**a, b, e,f**), myosin (**c,d**) in single migrating cells (**a-f**), after collision with another cell (**e**) and after treatment with La^{3+} to stop locomotion. The graphs show the fluorescence intensity along the lines marked in **a** and **c**, respectively. The high actin concentration at the leading front and the lack of myosin in the periphery of the lamella are obvious. After collision, the most intense actin staining is observed in the area of contact (**e**).

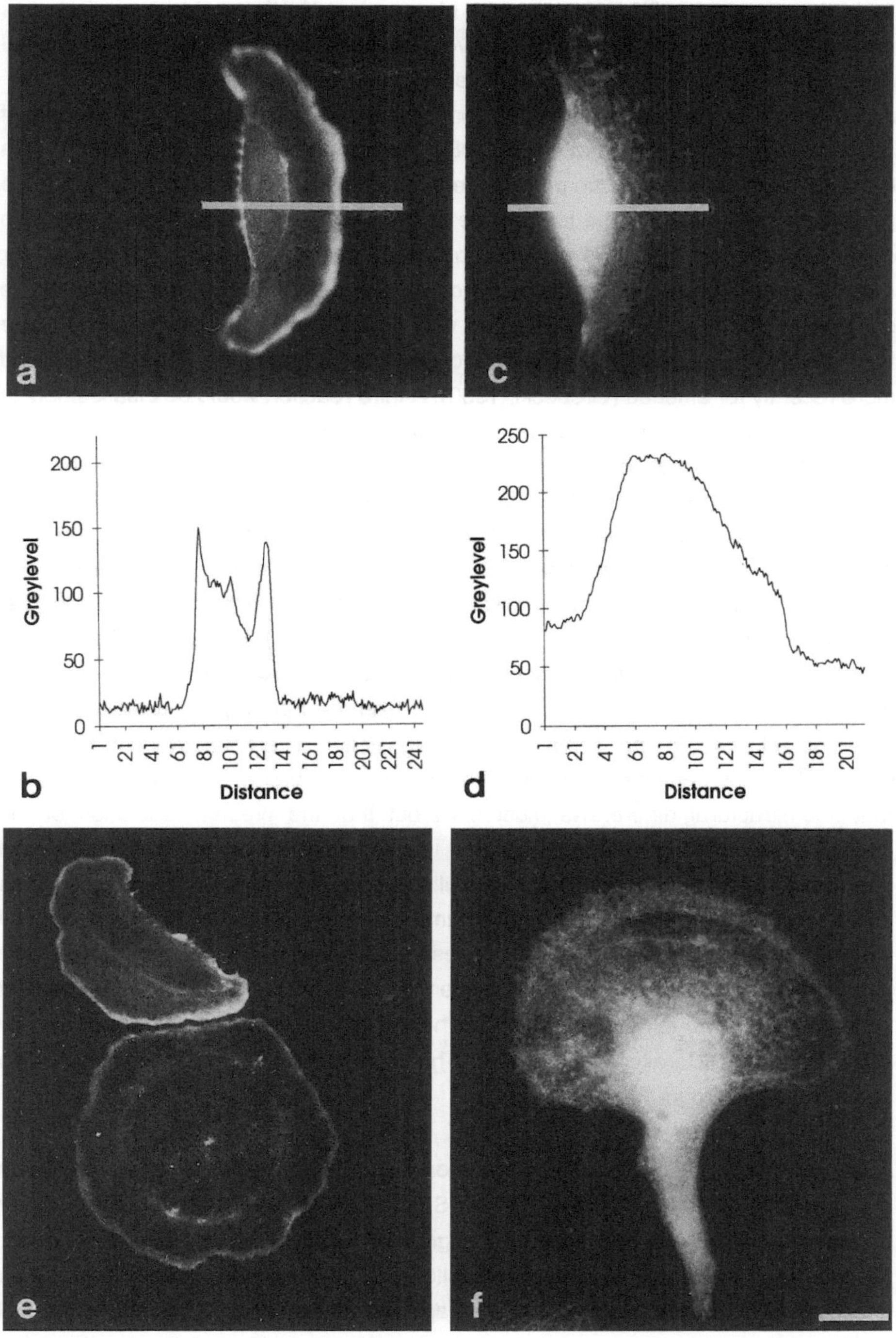
a
c
200
150
100
50
0
Greylevel
1
21
41
61
81
101
121
141
161
181
201
221
241
b
Distance
250
200
150
100
50
0
Greylevel
1
21
41
61
81
101
121
141
161
181
201
d
Distance
e
f

Fig. 4.

is in the range of individual variability. More recent immuno-cytochemical studies proved that ß-actin is located exclusively at the leading edge. The functional consequences of this observation are totally unclear at the moment.
Ample evidence has been presented for the leading edge being the main site of actin polymerization (Hall et al. 1988, 1989; Harvath, 1990; Small et al. 1982). This led to the conclusion that actin polymerization itself might be involved in driving force generation (see below). The further fate of the membrane bound F-actin seems to depend on the cell type and a variety of mostly unknown factors. The actin layer inserted in the membrane may move retrograd with the integral membrane molecule complexes (Harris and Dunn, 1972) or it may detach from the membrane and move backwards forming arcs as has been described first for fibroblasts (Heath, 1983) and more recently for amebae (Grebecki, 1990). A third reaction would be disassembly of the newly formed F-actin (e.g. Cassimeris et al., 1990).
Any further interpretation of the role of actin in a cytoplasmic lamella requires detailed knowledge of its fine-structural distribution. The most convincing studies on this topic have been presented by the groups of J.V. Small in Austria studying macrophage lamellae (Rinnerthaler et al., 1991; Small et al., 1980; 1982) and Höglund in Sweden working on fibroblasts and glia cells (Höglund 1985; Höglund et al., 1980; Lindberg et al. 1981). Rinnerthaler et al. (1991) fixed macrophages with lamellipodia showing a net advance over the terminal 30 s in either a phase of withdrawal, protrusion or rest at the time of fixation. For regions exhibiting these different activities in the same cell, no significant difference in actin filament-organization has been detected under the leading membrane. The authors presume that fine structural differences must exist but that the present resolution of the method does not allow to recognize them. Chemoattractant or cytochalasin induced depolymerization of F-actin in the lamellipodia of polymorphonuclear leukocytes argue for a change in G/F-actin equilibrium being involved in the control of lamella extension (Cassimeris et al. 1990). These observations correspond to our own experiences with Xenopus tadpole epidermis cells (XTE-cells). Only the contact zones between two lamella of cells after having collided a thin row of high density actin was found (Bereiter-Hahn, 1986). This increase in F-actin deposition is also seen in fluoromicrographs (Fig. 4e).

A wealth of observations has been made on distribution of actin binding proteins on a subcellular level (e.g. Grebecki, 1994; Stossel, 1993; Weeds et al., 1991). Their engagement in driving and contolling migration, however, is not fully understood. Troponin and alpha-actinin distribution follow that of F-actin in diffuse networks, in stress fibers, however, banding patterns appear. Myosin-2 is abundant in the cell body region and the cortical fibrillar layer but almost lacking at the leading front

(Couchman and Rees, 1979; Warrick and Spudich, 1987). Considering actomyosin based contractions as a source for driving forces for cell locomotion, the significance of myosin deserves special attention. In XTE-cells most of the myosin (myosin-2) is located at the cell body / lamella transition region (Fig. 4c, d) and in the cortical layer. As soon as motile activities of the cell are blocked (which also induce smoothening of the cell body / lamella transition region) myosin and actin become homogenously distributed (Fig. 4f). New light was thrown on the involvement of myosin in driving cell locomotion by the identification of mutant Dictyostelium ameba which were obviously devoid of myosin and still were moving (Fukui et al., 1990; Gerisch et al. 1991). Injection of anti-myosin antibodies in PtK2 cells destroyed stress fibres, the cells developed extensive motile lamellae but did not locomote (Höner et al. 1988; Höner and Jockusch, 1988). All these observations concern the role of myosin-2, myosin-1 still may be active. Myosin-1 is related to membrane and vesicle transport and it may also drive contraction of an actin network.

Cell motility and microtubule distribution.

The involvement of microtubules in control and maintenance of cell locomotion is very cell type specific. In XTE and teleost epidermis cells disassembly of microtubules by cold treatment or nocodazole may be without influence on the speed of locomotion (Euteneuer and Schliwa, 1984). However, the situation seems to be somewhat more complicated. Measuring the influence of colchicin on the locomotion of XTE cells we found a mean speed reduction of about 30% in many cells. However, some cells did not react at all. In the lamella of XTE cells only few microtubules are found, in some of the lamellae no microtubules are present at all. In moving cells microtubules never reach the very front of the lamella. Microtubules around the main direction of locomotion are short, they are longer in the lateral areas of the lamella where they tend to lay parallel to the long axis of the cell (which is perpendicular to the direction of movement) (Bereiter-Hahn et al., 1981). We presume that this alignment indicates the flux of cytoplasm inside the lamella. This view is supported by observations on the arrangement of microtubules in cells immobilized by La^{3+}: Abundant and almost straight microtubules emanate from the cell body into the lamella.

In fibroblasts disassembly of microtubules by colcemid reduces the motile activities of the lamellae and in leukocytes and endothelial cells the position of the centrosome changes on transition of the cells from a non-motile to the migratory form (Euteneuer and Schliwa, 1985). The hypothesis that microtubules propagate the transport of cytoskeletal elements and membrane vesicles and thereby

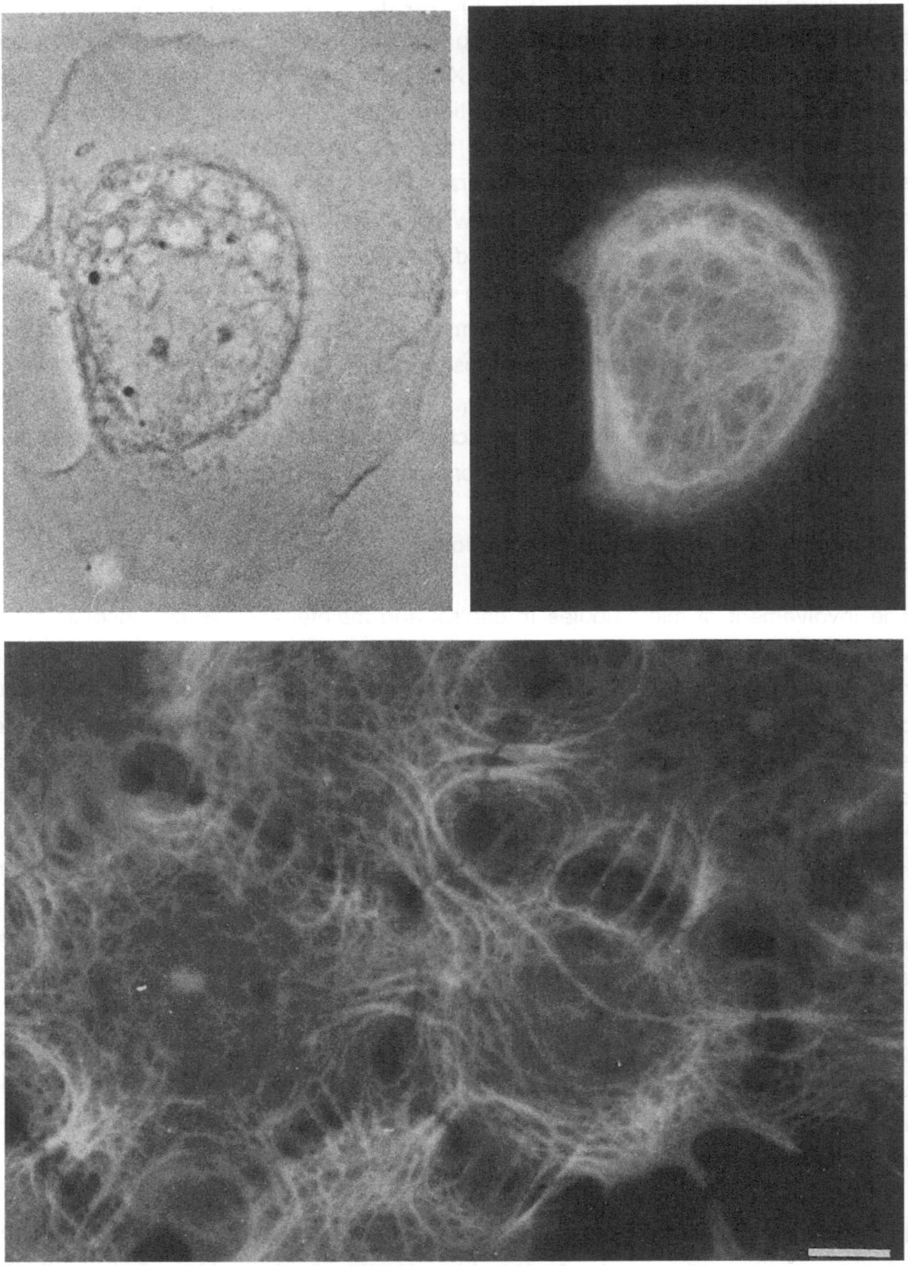

Fig. 5. Distribution of cytokeratin in XTE-cells in primary culture. **a, b**. single migrating cell. **a**: phase contrast. **b**. The same cell stained for cytokeratin. No fibrils extend into the lamella. **c**: Cytokeratin fibers in an epithelial layer of XTE-cells. The cytokeratin filaments form a network propagating forces via desmosomes from cell to cell (U. Weich phot.).

contribute to locomotion (Dunn, 1980) seems very reasonable. Recent observations on colchicine-induced stimulation of polymorphonuclear leukocyte lomotion support this view (Keller and Niggli, 1993). In addition, the stiff microtubules may stabilize the lamella and by this the directionality of the movement (Domnina et al. 1985). Extension of neuronal processes and locomotion of cnidarian nematocytes (Agosti et al., 1992) fully depend on the presence of intact microtubules. Although their role in these cases is not yet clear, the need of microtubules could reside in transport phenomena.

Cell motility and intermediate filament distribution.

Although intermediate filaments seem not to be as stable as has been assumed for many years (Skalli et al., 1992), active participation of these structures in cell locomotion is improbable. The slow turnover and the intracellular arrangement of intermediate filaments are more in favor of a passively stabilizing role. Regarding this situation, only one example is discussed in more detail. In single, moving XTE cells cytokeratin fibrils surround the nucleus and form a "basket" enclosing the membraneous organelles. Only a few fibers extend into the proximal part of the lamella (Fig. 5b). In larger cell aggregates where the cells have no free edge the distribution is totally different. Strands of cytokeratin extend in between desmosomal spots and from the desmosomes to a ring of fibrils surrounding the nucleus (Fig. 5 c). In this case the cells may be deformed but they do not slide relative to each other.

IV Motive force generation for cell locomotion.

Location and type of the motive force driving cell locomotion are still under debate. The minimum requirement for a hypothesis to be acceptable must be to explain the continuity (or change) of cell shape during migration and the extension of a pseudopodium or a lamella. Some topics of cell locomotion are generally accepted (Fig.6), most of them have been discussed in the previous chapter:

- at the lamella front actin polymerization takes place
- dissassembly of actin is achieved somewhere in the cell body or close to the cell body / lamella transition region (Symons and Mitchison, 1991).
- myosin penetrates only into the proximal part of the lamella
- mechanical tension is developed between the front part of the lamella and the cell body

- membrane bound proteins move retrograde in the lamellar region but anterograd from the trailing edges of locomoting cells (Harris and Dunn, 1972; Holifield and Jacobson, 1991; Kucik et al., 1989).

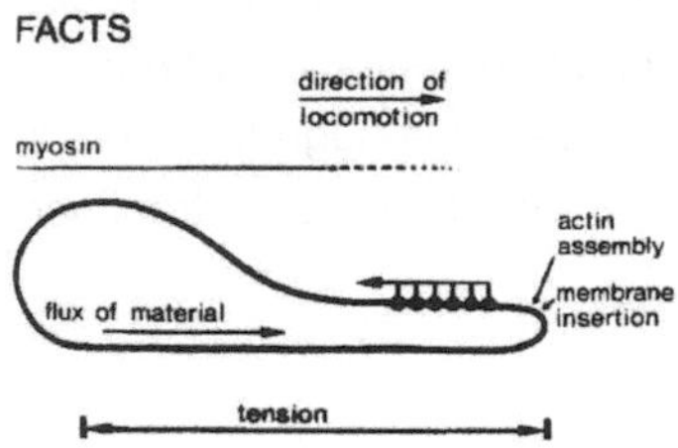

Fig. 6. "Facts" = observations which are in general accepted to be significant for cell locomotion.

These features have to be kept in mind discussing how cells crawl. The distribution of elasticity and of forces according to some hypotheses is summarized in Fig. 7. The six main hypotheses for driving force generation are:

1. Active sliding of actin fibrils in the lamella.

Fibrils in the lamella are supposed to slide against each other and thus providing a force for lamella extension (Huxley, 1976). This hypothesis may be extended by distinguishing fibrils which are supposed to be fixed for instance beyond the membrane and others (in the more central endoplasm) sliding against the fixed layer. This principle may well account for intracellular plasma streaming in plant cells or nerve cells. To explain cell locomotion, this mechanism would require the presence of myosin all along the lamella. A prerequisite which is not proven for myosin-1 at the moment and can be excluded for myosin-2. Following the dynamics of fluorochromed actin in the lamella of a moving cell (Theriot and Mitchison, 1991) or a cell which is impeded in its motility by strong adhesion (Wang et al. 1982) does not support that sliding plays a major role in edge propagation.

2. Frontal contraction of cytoplasm anchored to the cell membrane in the advancing tip, thus pulling the cytoplasm forward.

This hypothesis has been worked out for the fountain zone streaming in Amoeba proteus (Allen 1961, 1970; Allen et al., 1972) and a very similar hypothesis for leukocytes (Stossel, 1982). The main point is, the flow of cytoplasm from less ordered regions to regions with higher order and stiffness. A good test would be to show the pressure or tension gradient from the lamella to the cell body which must be generated if the leading front would pull the rest of the cell. Such determinations have been performed on motile active neutrophils following the intracellular dislocation of particles during crawling (Simon and Schmid-Schönbein, 1990). However the method reveals strain and does not allow decisions on the origin of the stresses responsible for these strains.

A definitive proof or disproof of this hypothesis (and most of the others) is rather difficult.

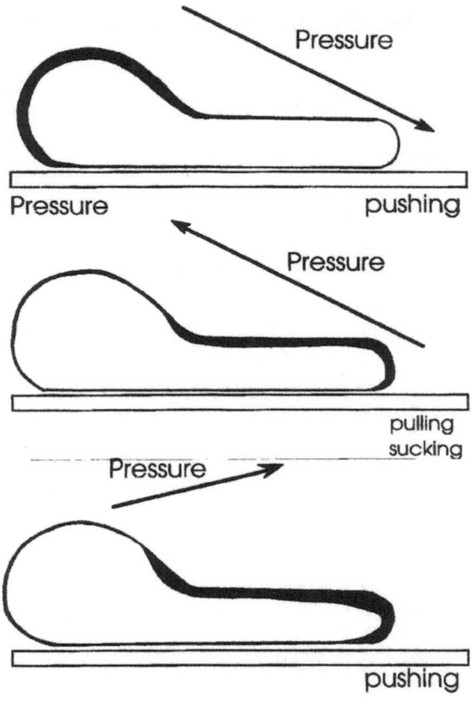

Fig. 7 Schematic view of forces and local elasticity which have to be assumed according to different hypotheses of motive force generation. Line thickness represents elasticity (stiffness), not a certain structural detail. Uppermost graph: If hydrostatic pressure provides the motive force for the flow of cytoplasm and extension of a lamella, extension should occur at weak zones (thin lines). Central graph: According to the frontal zone contraction hypothesis, the hypothesis of osmotic swelling in the front zone and to the "actin-polymerization - hypothesis" the front of the extending lamella should be stiff, and the rest of the cell is pulled or sucked towards the leading front (reverse pressure gradient) Bottom graph: The "tug of war"-principle, introduced by Stossel (1982). The lamella is extended by pushing.

3. Hydraulic pressure produced by contraction of the cortical actomyosin layer.

Contraction of the cortical actomyosin layer would induce a volume decrease of the cell by pressing water out of the cell until the resulting osmotic pressure difference equals the mechanical pressure exerted by the cytoplasm (Layrand et al. 1972). So far only an internal pressure has been generated. Motility requires a pressure difference[1], i.e. extension of the cytoplasm at a site where the cortical layer is weakened (less stiff) and as a consequence it is bulged out. If the intimate contact between plasma membrane and fibrillar layer is loosened, blebs are formed (see below). Blebbing (Fujinami, 1976; Harris, 1990) as well as speed of locomotion (Strohmeier and Bereiter-Hahn, 1987) can be abolished by increasing external osmolality (Fig. 8).

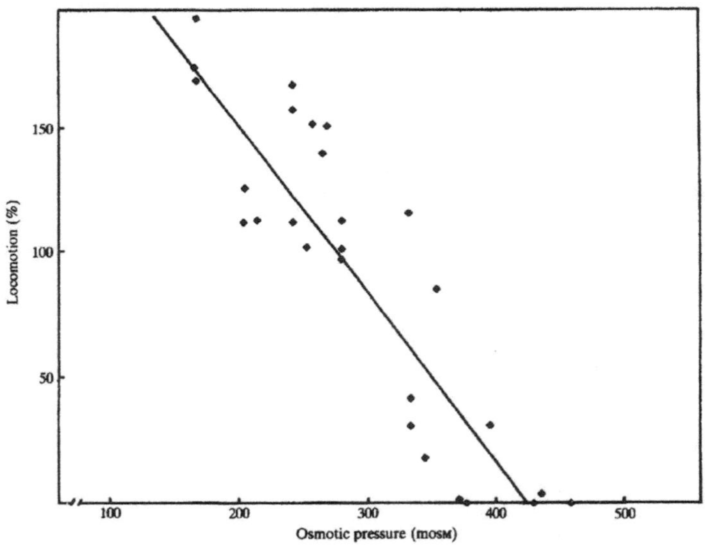

Fig. 8. Dependence of the speed of locomotion of primary XTE cells on the osmolality of the culture medium. Osmolality was controlled by sucrose content. (From Strohmeier and Bereiter-Hahn, 1987)

[1]The existence of a pressure difference in a hydraulic system refers to the dynamic character of the whole system: Local weakening of cytoplasm causes bulging at this site, during this process of bulging a (probably very small) lowering of the hydrostatic pressure must exist in this region which is balanced as soon as the process of bulging is stopped. During crawling continuous lamella extension (a special case of bulging) requires a continuous pressure gradient. Dissipation of this gradient is achieved by material flux (cytoplasm, water). The material flux rate is limited by viscosity and may lead to phase separation as seen amoeba where a hyaline, fluid-filled cap is formed which is almost devoid of polymerized components.

According to this hypothesis (Bereiter-Hahn et al. 1981, DiPasquale 1975, Harris 1973; Komnick et al. 1973) extension of a lamella would occur at mechanically weak sites by a flow of cytoplasm percolating the fibrillar network traversing the lamella. If the lamella is very thin, the material flow needed for lamella extension can be small, therefore no measurable volume or pressure decrease is going to occur. Due to their large size oligomeric myosin aggregates are facing more resistivity during their percolating flux than for instance mono- or oligomeric actin. This results in size dependent separation of macromolecules under steady state conditions of cytoplasmic flow and locomotion. Concommitantly with the extension of the lamella at its front, it is shortened by actin and myosin based contractions at the cell body / lamella transition region. These contractions pull the cell body "into" the lamella and thus reducing it. This model includes all the features present in moving cells which have been listed above.

A few more observations need to be discussed in the light of this hypothesis, the appearance of ruffles and of stress fibers. Both these structures are related to adhesion and speed of locomotion. Ruffles are presumed to result from excessive incorporation of membrane material and polymerization of actin at the leading edge. "Excessive" in this context means more than is required for the extension of the lamella. No ruffles are displayed in fast moving cells, all the membrane material which may be inserted at the front can be used for lamella extension. Stress fibers originate from focal adhesion sites. If the cell body is fixed by focal contacts at the rear end of the cell it cannot follow lamella extension, and if the state of integral membrane proteins giving rise to focal contacts is in favor of spot-like assembly and binding of actin, the retrograd movement of the fibrillar network is inhibited: tension develops between the contact zone and the cell center (as a consequence of contraction), a stress fiber emerges from the previously isotropic actin network, while on the dorsal cell surface the retrograd movement may still proceed (e.g. De Brabander et al., 1990, 1991).

The existence of hydraulic pressure has been proven experimentally for a variety of cells including amoebae (Kamyia, 1964), cytoplasmic strands of the acellular slime mold Physarum (Wohlfahrt-Bottermann 1977, 1979), XTE cells (Strohmeier and Bereiter-Hahn 1987). This does not exclude that other cells, as are erythrocytes or adult muscle cells (Matsubara and Elliot, 1972; Rome, 1968) do not develop a hydraulic pressure. These cells lack the cortical actomyosin layer and do not move actively.

4. Local osmotic pressure changes.

By changing the physicochemical properties of the cytogel beyond the tip of a lamella osmotic swelling should result in extension of the lamella and flux of water driving cytoplasmic flow towards the extending front (Oster 1984; Oster and Perelson, 1987). For the explanation of cell crawling additional reactions are required as are F-actin-disassembly and membrane flux which may be caused for instance by contraction in the lamella / cell body transition region.

5. Cytogel expansion by actin polymerization.

The idea that pressure exerted by polymerizing actin at the leading edge of a lamella could provide the motive force for crawling has found its way into a leading text book of cell biology (Alberts et al. 1989). Indeed, actin becomes polymerized at the leading front of moving cells (Symons and Mitchison, 1991). However, this does not necessarily mean that this process provides the motive force for lamella protrusion. The main support for this hypothesis comes from the acrosome reaction of *Thione* spermatozoa during fertilization (Tilney, 1980; Tilney et al. 1973), similar structures are filopodia which may form very fast and can be induced by application of strong electric fields (Popov et al., 1991). Those, however, are no organelles of cell crawling . The hypothesis of lamella (or pseudopodia) extension by actin polymerization can be tested against the hydraulic pressure model by examining whether lamella extension occurs at specifically weak or stiff zones of the advancing edge.

6. Surface extension of the plasma membrane

The contribution of the plasma membrane to cell movements is still unclear. The leading front of polarized cells seems to be the preferred location for exocytotic events as is the release of viruses (Singer and Kupffer, 1986), and in ameba continuity of locomotion depends at least to a small amount on cytotic activity (Stockem, 1972). Endocytosis in moving cells is primarily found in the cell body / lamella transition region. So far the morphological basis for motion driven by coupled exo- and endocytosis is given. Quantitative approaches to determine whether the membrane turnover would be sufficient for providing the source for the movements (i.e. total turnover of plasmamembrane when a cell moved about twice its length) which have been carried out for amebae by Stockem (1972) and by us for tadpole epidermis cells (unpublished) do not support this model.

V. Cell shape and locomotion

Locomotion of cells always is connected to the maintenance of a certain shape.The main characteristics are the polar development of pseudopodia or of a lamella into the direction of movement. Therefore locomotion can only be understood by comprehending cell shape, that of pseudopodia and lamellae in particular.

1. What keeps a lamella thin ?

Crawling cells exhibit a more or less distinct lamella. If several lamellae are present the largest one determines the direction of locomotion. This might be a consequence of the specific properties of lamellae. To understand this function the development and fine structure of a lamella (Fig. 9) has to be understood, because migration can take place without an overall shape change of the cell, thus the continuity of cell shape has to be preserved while all constituents are in steady turnover. A certain region of cytoplasm, as a lamella is, may be kept thin by three different mechanisms which are not mutually exclusive:

- Crosslinking the "dorsal" and the "ventral " plasma membrane, the ventral membrane adhering to the surface of a solid.
- Pulling the dorsal cortex, while the ventral membrane adheres to the surface of a solid. Both parts being loosely connected.
- Squeezing through narrow slits: The restrictions imposed from the environment flatten the cells.

The third possibility seems to be realized for instance for leukemic cells or activated lymphocytes penetrating a cell layer and crawling through the space between cover glass and the cell layer (Verschueren et al. 1991). It will not be discussed further.
In the case a hydrostatic pressure is present in a cell, this pressure tends to thicken the lamella. Thus the thickness of a lamella would result from an equilibrium of forces (Fig. 11): The hydrostatic pressure extending the lamella, the tension pulling at the surface the lamella towards the cell body and the adhesion of the ventral cell membrane. The forces from the dorsal cortex are submitted to the adhering ventral lamella via some cross-connecting fibrils. The presence of these forces should be demonstrated and their organizational effect as well.

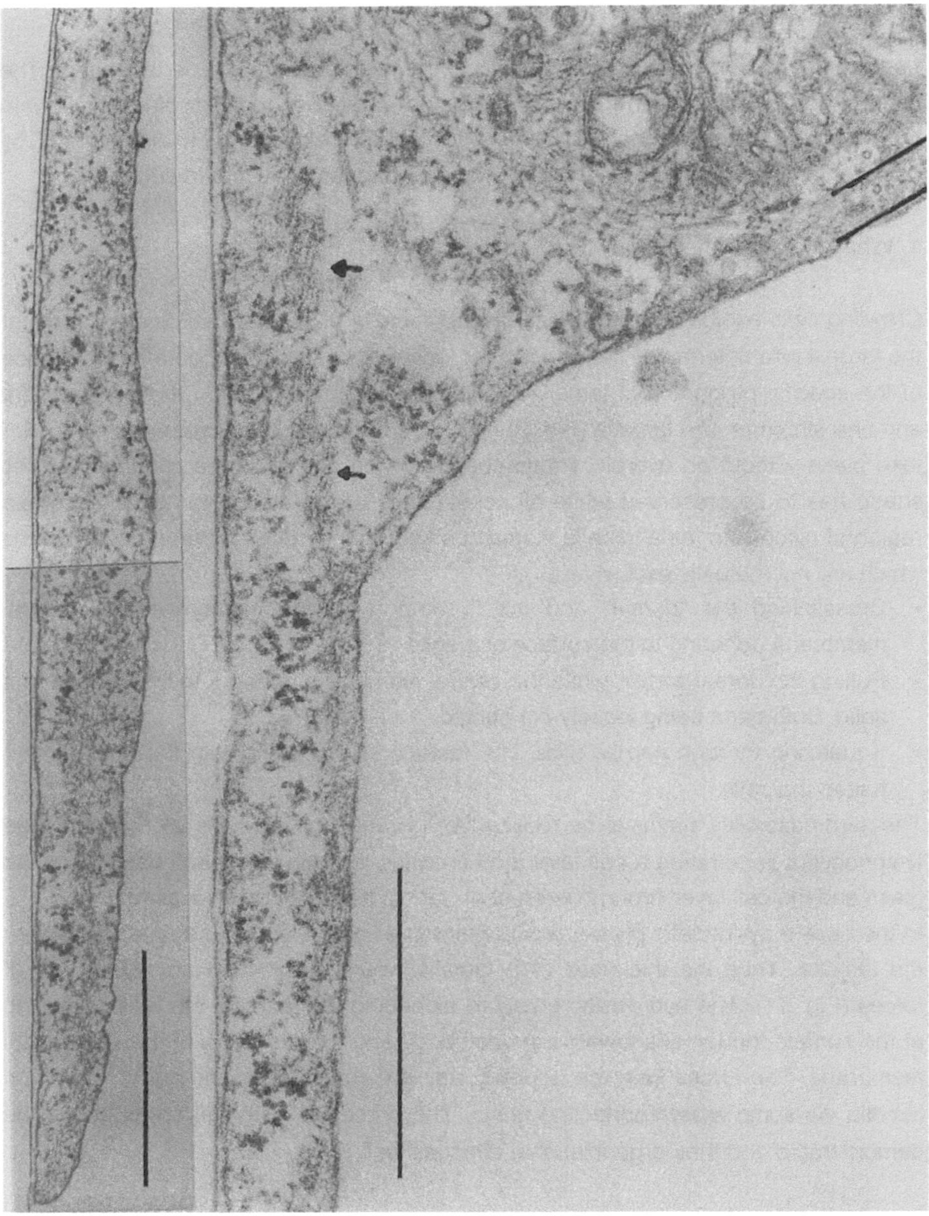

Fig. 9. Electron micrographs of the front part of a lamella (lower image) and of the cell body / lamella transition region (upper image). The cortical actomyosin layer is visible at the cell body (two streaks) and penetrates from the dorsal plasmamembrane of the lamella to the cell center (arrows). In the frontal part of the lamella, the actin network fills the whole space of the lamella and develops a cortical layer towards the cell center (left part of the picture).

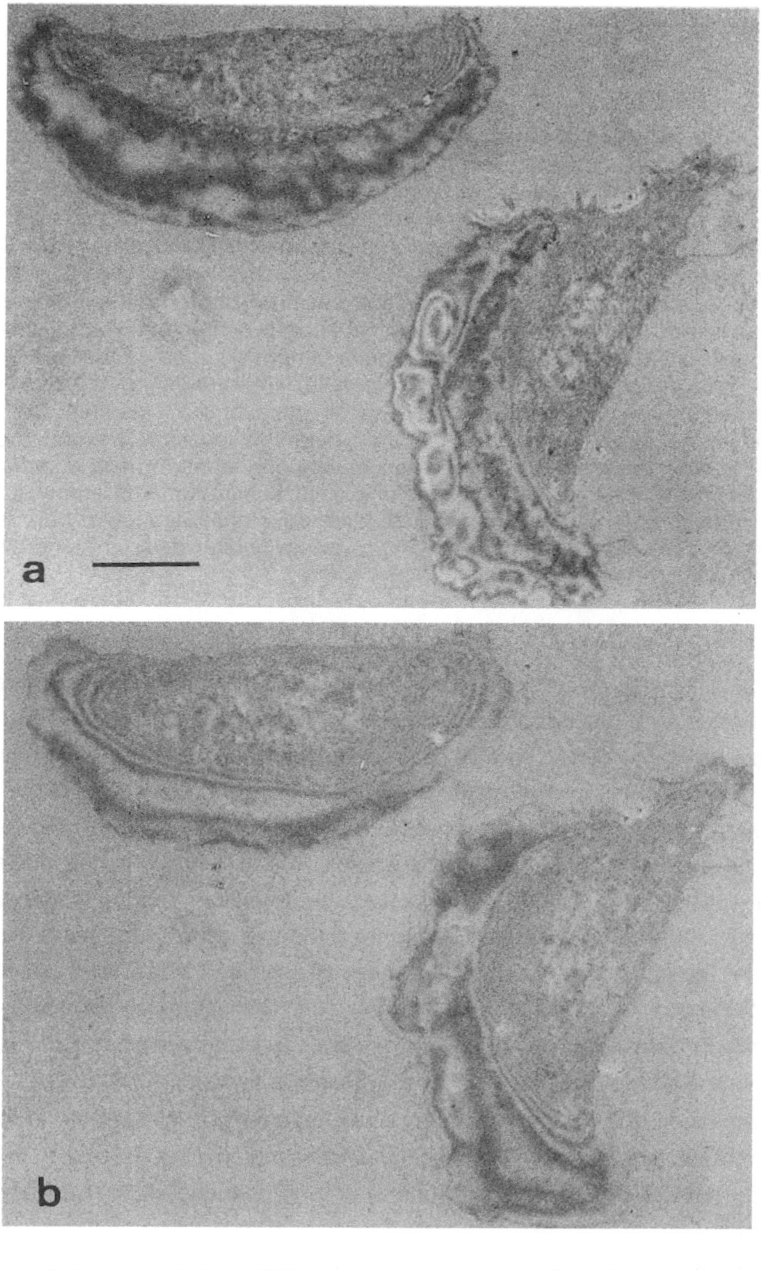

Fig. 10. RIK images of two XTE-cells. a. control condition immediately before addition of La^{3+} (1mM). b. About 10 min in presence of La^{3+}. The lamella of both cells became thinner, in the cell body / lamella transition region appear interference lines indicating a flattening of the surface slope in this area. (For help in interpretation of this type of micrographs see appendix 1)

Fig. 11. Forces which are involved in maintenance of shape of a lamella and in the development of the submembranous
actomyosin cortex. The basal cell membrane is supposed to be fixed to the solid substratum by adhesion. The dorsal cell region is losely connected to the basal membrane via cross-linking fibril; they stabilize the thickness of the lamella which tends to increase by hydrostatic pressure, the force difference between that exerted by the fibrils connecting the dorsal and the basal parts and that extending lamella thickness is represented by the small vertical arrow. Contraction of actomyosin fibrils (myosin extends into about 2/3 of the lamella periphery) pulls on the isometric actin network (organized at the tip of the lamella) thus orienting it. This force is represented by the horizontal arrow. The direction of the resultant force vector determines the slope of the lamella. Movement is possible beyond the dorsal membrane only, therefore actin fibrils accumulate only there forming a cortical layer. In pseudopodia which do not adhere to a solid substratum (no immobilization of parts of the membrane) either no cortical layer develops because of bulk retraction of the three-dimensional fibrillar network, or a thin fibrillar layer is found all around the plasma membrane.

The action of the hydrostatic pressure thickening the lamella can easily be demonstrated by inhibition of Ca^{2+}-influx using Ca-antagonists or La^{3+} (Strohmeier and Bereiter-Hahn, 1984). In this case locomotion ceases after a few minutes of exposure and first the microcolliculi (the small swellings at the margin of the lamella) disappear, the lamella collapses to about half its previous thickness (Fig. 10). After prolonged exposure to La^{3+} (2-3 mM $LaNO_3$) the difference in the slope of the lamella surface and the cell body / lamella transition region disappears, the thickness of the cell increases steadily from the margin of the lamella to the cell center. At the same time the typical pattern of actin and of myosin (see above) is lost, both these cytoskeletal elements become homogenously distributed. The contractile forces caused by actomyosin are supposed to induce a flux of F-actin and its membrane bound links from the tip of the lamella toward the cell center. Because of the adhesion of the ventral membrane, the material flux is restricted to the dorsal cell membrane where it develops a subplasmalemmal layer of F-actin. In those cells which do not adhere strongly with one side of a pseudopodium, no clear cortical layer arises.

According to our interpretation the primary collapse of the lamella is a consequence of a loss of hydrostatic pressure, allowing elastic recoil of the elements connecting the dorsal and the ventral lamella membrane. Subsequently the contractility of the actomyosin fibrils ceases which extend about 2/3 the lamella from the cell body , therefore the tension between the lamella and the cell body is lost, the lamella can thicken. The process of thickening proceeds from the cell body / lamella transition region towards the periphery. This thickening can be explained by a decline of the contractile force in the area with the highest myosin density (see above, Fig. 4c).

Thus the polarized distribution of cellular constituents must develop in a kind of morphogenetic process. The spreading of trypsinized cells represents an experimental model to follow the transformation of a spherical cell to a flattened cell with peripheral lamella areas. The model which has been derived from observations of XTH-2 cell spreading, including the action of hydrostatic pressure, is outlined in Fig. 12.

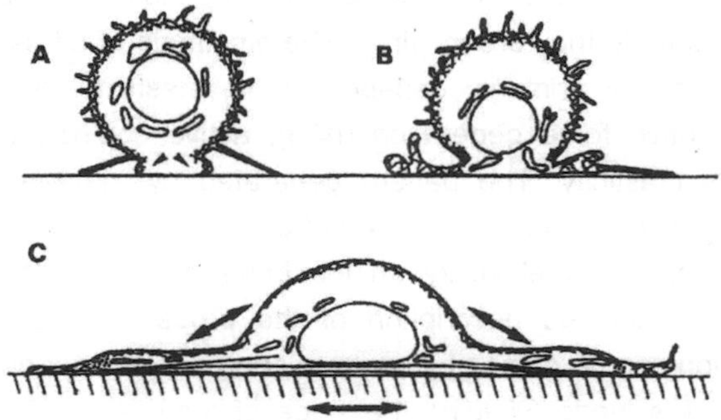

Fig. 12 Scheme of spreading of trypsinized cells.
A. Spherical cell, the cortical fibrillar layer (crosses) is just broken close to the substratum and a bulk flow (arrowheads) of cytoplasm is causing a bleb to protrude and to adhere to the glass surface. Some filopodia may support adhesion. The nucleus and the other organelles are located almost in the cell center.
B. Cytoplasm is extruded due to the pressure exerted by the contraction of the cortical actomyosin layer. The orifice of outflow is widened, thus the lining firillar layer becomes more compact, the nucleus and other organelles are dislocated towards the bottom. The peripheral cytoplasm starts flattening and thus forming a primordial lamella by cross linking of newly polymerized F-actin.
C. After starting flattening the surface smoothens, the bulk flow of cytoplasm compresses the fibrillar meshwork to the periphery giving rise to a peripheral actin ring (dots), which does not allow further progressio of larger organelles and of microtubules. Distal to this ring lamellae have been formed. General contraction of the actomyosin fibrils finally results in isometric contraction between adhesion points giving rise to stress fibers forming a stiff basal plate. Tension in the upper cortical layer (double arrows) pulls the central part of the cell body flat causing a further flow of cytoplasm towands the periphery until a new equilibrium of forces is reached. The direction s of tension are marked by arrows outside the cell. (From Bereiter-Hahn et al. 1990b).

2. Control of the direction of movement

Locomotion is based on the presence of certain structures, these by themselves, however, are subjected to steady turnover and change. The processes underlying cellular locomotion and its control should become more comprehensible if cells are induced to change the direction of their locomotion. Two approaches seem reasonable, studying subcellular elasticity and its relation to the direction of movement and studying the role of potential messenger molecules.

2.1 Subcellular elasticity related to the direction of locomotion.

By means of **S**canning **A**coustic **M**icroscopy (SAM) elasticity changes can be followed in living cells while they are moving. The amplitude of ultrasound reflected from the surface of a cell primarily depends on its elasticity. According to the considerations on motive force generation (IV.3), active contractions in the cell cortex determine its elasticity. The pattern generated by interferences between soundwaves reflected from the cell surface and from the solid substratum was used to analyse the distribution of elasticity on the basis of the contrast among the interference lines (for detailed description of the procedure, see appendix II). Qualitative information can be gained by contrast analysis: the first derivation of the grey level distribution is made (yielding the slope of grey level changes and thus being a measure of contrast regardless of the absolute grey values). Inversion of the image resulting from the derivation transforms all values indicating high contrast to dark areas while those zones with low contrast appear bright (this procedure is done for psychological reasons: the association of dark values with high contrast seems to be easier than the association with bright level values). This procedure is exemplified for a tadpole epidermis cell which has collided with the adjacent cells

---

Fig. 13. SAM images of XTE-cells.
Upper triade: Original SAM image (left) and the inverse of the first derivative of this image (representing the contrast of the original). The lower one is a repeat of the contrast image but with red overlay marking areas of specifically high contrast revealing zones of high elasticity. The theoretical basis for this procedure is outlined in appendix II.
Lower triade: Two subsequent images of the same migrating XTE-cell processed in the way outlined in the upper triade. Stiffness increases from green to red. Extension of the lamella occurs at front zones of relative weakness, no extension is found in the small "red" area in the most advanced part of the lamella. The relative migration can be seen in the bottom picture which is an overlay of the original SAM micrograph of the 1:25 min state and an outline (violet) of the 1:00 min state.

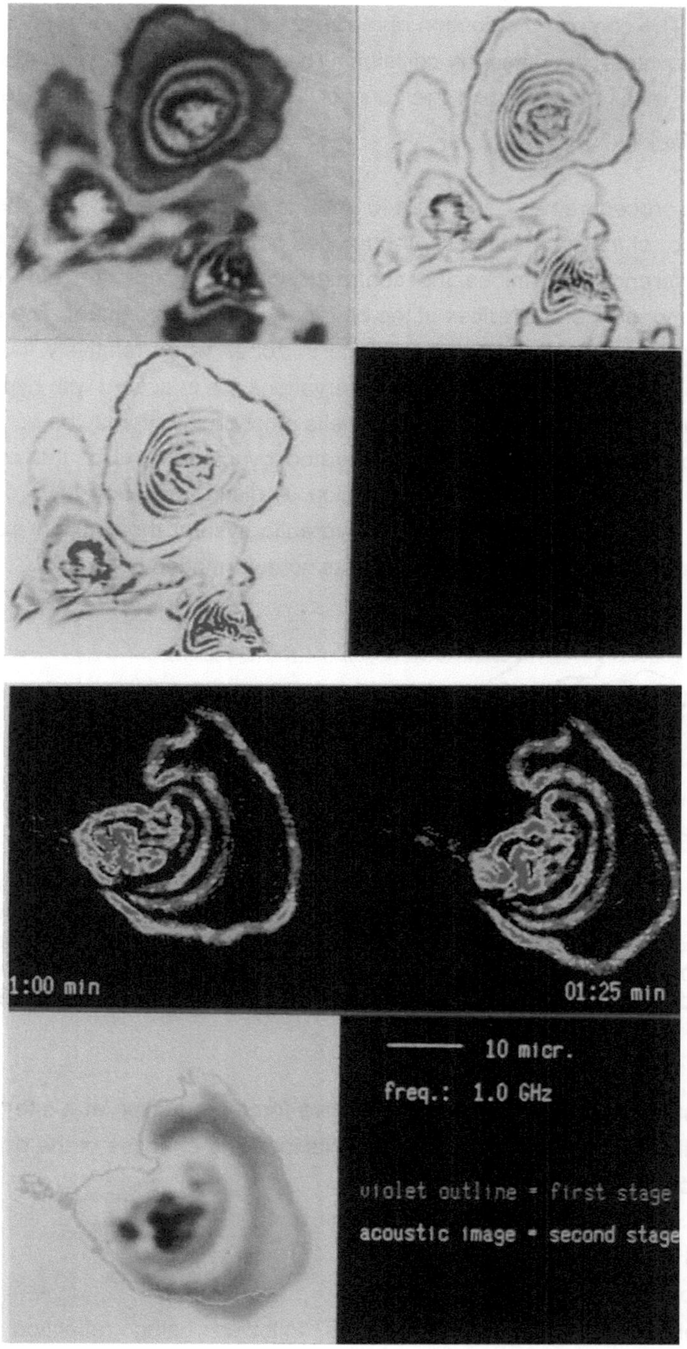
1:00 min
01:25 min
10 micr.
freq.: 1.0 GHz
violet outline = first stage
acoustic image = second stage

Fig. 13

and now is going to change the direction of movement away from the contact zones (Fig. 13). The contrast distribution reveals zones of higher stiffness in the cell areas directed towards the sites of collision: The moving cell is retracting itself, the movement takes place toward the zone of lower elasiticity and thus tension in the cortex (Fig. 13).

The same procedure has been used to analyse whether the lamella extends where it is of high or of low local stiffness. In the case of the lamella margin diffraction rather than interference determines the contrast. However, diffraction also depends on acoustic impedance differences at the boundary cell / environment. The results show that extension of the lamella takes place in zones whith relatively low impedance (and thus elasticity) (Fig. 13). These observations are in accord with hypotheses IV 3 and 6. listed above. Extension of the lamella most probably may be achieved by the release of a hydrostatic pressure generated by contraction of the cell cortex, in particular at the lamella / cell body region. A distinct location of force production does not make sense, because in a hydraulic system the tension exerted at the surface is the same everywhere as long as no deformation occurs.

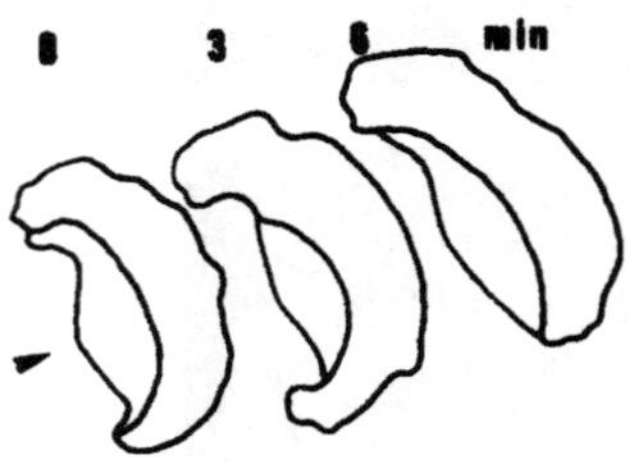

Fig. 14. Local release of calcium at the rear margin of the cell body (arrowhead) of Xenopus tadpole epidermis cells permeabilized with the Ca-ionophore A23187 causes contraction of the cell body (smaller projection area at 3 min) and activates locomotion away from the tip of the capillary. Cessation of calcium-release allows re-spreading of the cell body (6 min). (From Mittal and Bereiter-Hahn, 1985).

Further support for the hypothesis of motive force production at a site different from the lamella front is provided by experiments inducing changes of the direction of cell locomotion by local application of

- Ca^{2+} to cells permeabilized for calcium
- cytochalasin D
- phalloidin

The reaction to these substances is followed using reflection interference microscopy (RIC) which allows to monitor changes in the surface topography of the very thin lamella (0.1-1.0 µm) in the range of a some nanometers to several micrometers (see Appendix I).

Local application of Ca^{2+} by a micro-capillary to cells permeabilized with the calcium-ionophore A23187 evokes dual response: If applied to the cell body at the rear end of the cell, the projection area of the cell body diminishes and the cell moves away from the capillary (Fig. 14).

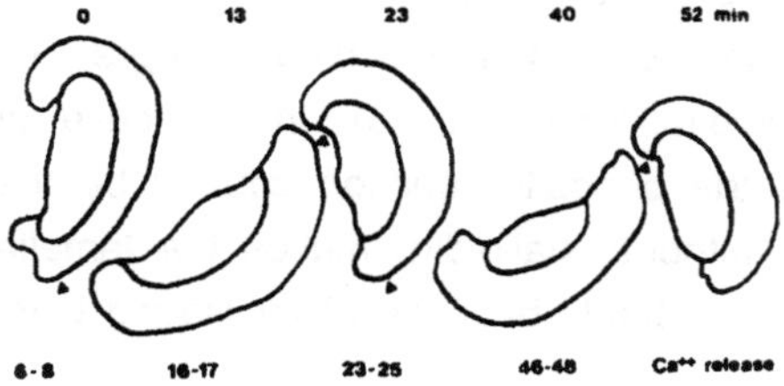

Fig. 15. Local release of calcium at the rear margin of the cell body (arrowhead) of Xenopus tadpole epidermis cells permeabilized with the Ca-ionophore A23187 causes contraction of the cell body (smaller projection area at 3 min) and activates locomotion away from the tip of the capillary. Cessation of calcium-release allows re-spreading of the cell body (6 min). (From Mittal and Bereiter-Hahn, 1985).

If Ca^{2+} is released at the margin of the lamella, the cell follows the Ca^{2+} - gradient, the lamella thickens locally and is extended towards the tip of the micro-capillary (Fig. 15). Because the whole experiment has been performed in the presence of EGTA a steep gradient of Ca^{2+} can be assumed, its action being restricted to a very small (however, undefined) space. Such a dual response is not unique (Kohama, 1992; Taylor et al., 1976; Taylor and Fechheimer, 1982). Contraction of the cortical layer of the cell body is assumed to be induced by Ca^{2+} release at the cell body, whilst losening of actin crosslinking (e.g. by α-actinin) is supposed to allow extension of the lamella towards the calcium gradient. Variations in calcium concentrations may well be part of the physiological control of crawling (i.e. receptor mediated stimulus which activates proteinkinases, and finally stimulate for instance the calcium/calmodulin - dependent phosphorylation of the myosin light chain [Cande and Ezzel, 1986; Shirinsky et al., 1992; Wilson et al., 1991]. Details for this type of control have been investigated for leukocytes only [Keller, 1990; Keller et al. 1990, Keller and Niggli, 1993b]).

The interpretation regarding the mechanical changes induced by Ca^{2+} needs some support by a more direct approach. A set of two experiments seems to be conclusive: The local release of two substances at the margin of the lamella of a moving cell: cytochalasin D and phalloidin. Cytochalasin D shortens the length of the actin fibrils, thus decreasing their capability to form crosslinks which should cause local weakening of the cytoplasm. The opposite effect is to be expected on the local release of phalloidin which stabilizes F-actin inhibiting its depolymerization. In addition this experiment should allow to distinguish whether lamella extension is due to actin polymerization - then phalloidin should evoke lamella progression toward the capillary tip - or whether inhibition of F-actin decomposition stabilizes the leading front. Phalloidin does not permeate the membranes of living cells (with the exception of liver cells). Therefore the phalloidin had been dissolved in digitonin-saline rendering the membrane permeable for phalloidin. The direction of locomotion can be changed, the cells extend their lamella towards the cytochalasin D gradient and stop lamella extension towards phalloidin (Figs. 14, 15). In the case of phalloidin, this substance has to overrule the opposite reaction caused by the digitonin used for permeabilization which (without phalloidin) makes the lamella extend to the tip of the micro-capillary, most probably because of facilitating Ca^{2+} influx.

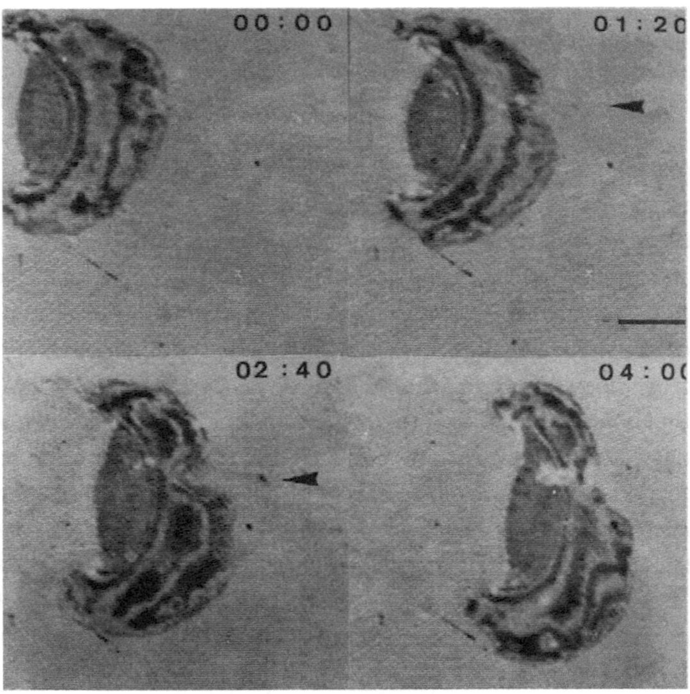

Fig. 16. Local release of phalloidin (and digitonin to permeabilize the cell locally) towards the progressing lamella of a locomoting Xenopus keratinocyte (1 min 20 s to 2 min 40 s). The cell is turning away from the microcapillary (arrowhead), no further extension of the lamella takes place. Reflection interference micrgraphs. Scale bar = 10 µm.

a)

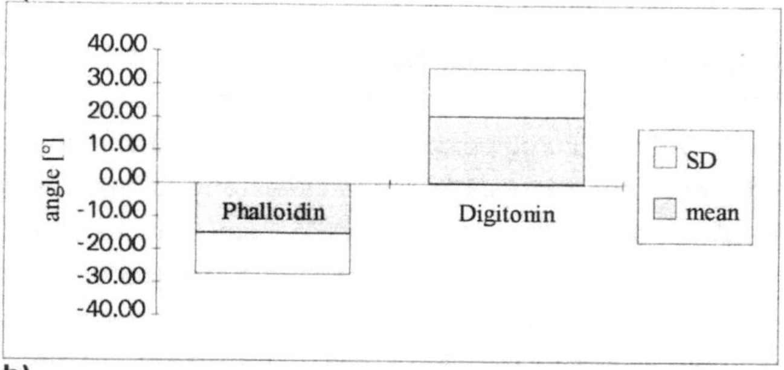

b)

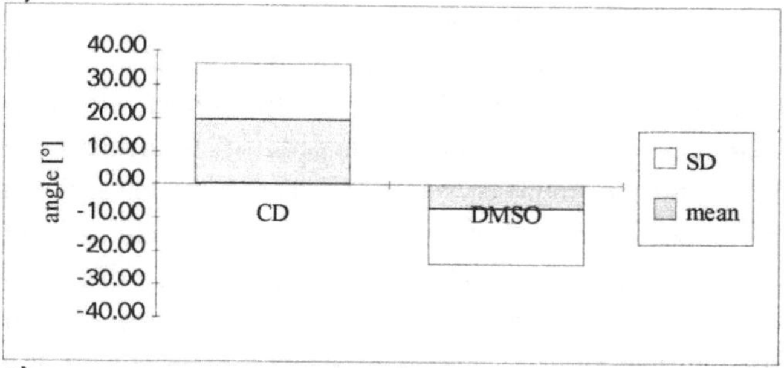

c)

	Phalloidin	Digitonin
mean	-15.08	20.39
SD	12.40	14.38
N	17	17
t-test	7.70	
p<	0.001	

	CD	DMSO
mean	19.71	-7.05
SD	16.35	16.73
N	26	25
t-test	5.77	
p<	0.001	

Fig. 17.Change in the direction of locomoting primary keratinocytes (Xenopus laevis tadpoles) after release of

a) 1 µM phalloidin (with 0.05% digitonin in ACM). Control: release of 0.05% digitonin in ACM. The bars represent the change of the direction of locomotion following the release relative to the direction of locomotion before the release. If the cell is turning towards the microcapillary the angle was set positive and turning away from the capillary is indicated by a negative angle.

b) 4µg cytochalasin D / ml in amphibian culture medium (with a final concentration of dimethylsulfoxide [DMSO] of 0.002%). Control: 0.002% DMSO in ACM.

c) The results of the experiments and control experiments differ significantly ($p < 0.001$, paired Student T-test).

The substances were released by pressure from a microcapillary manoeuvered close to the lamella of the cells, the behavior of the cells was followed in the reflection interference contrast mode and has been analyzed using a Quantimet 520 image analyzing system.

2.2. Motile activities following collision between cells.

Collision of an advancing cell with a solid object or another is cell is a situation which induces a change of the direction of locomotion in a chemically undisturbed environment. For this purpose we co-cultured epidermis cells (XTE) and endothelial cells (XTH-2) cells, both derived from Xenopus laevis tadpoles and followed their behavior with SAM. This method allows the visualization of local changes of cell thickness and of their elastic properties (Fig. 18). Subtraction of subsequent images (Fig. 18, left column) reveals all motility events which happened in an appropriate time interval (Vesely et al. 1994). Curved movement in the XTE cells requires local thickenings in proximity to the lamella margin in the area which is going to extend whilst margins which are determined to become retracted are almost stable (Fig. 18 left column 5s, 10s). In some cells the total thickness of the lamella increases towards the direction of extension and is very thin in the area which is withdrawn (Fig. 19). Also the cell body / lamella transition region remains static towards this direction (Fig. 19 lower part of the keratocyte, left column). Thus the whole cell area is acting as a static unit where the tension forces between the central part of the cell and its periphery are relatively high keeping the lamella thin. They prevent further flux of cytoplasmic material into this direction and are balanced by adhesion. It is obvious that the local thickenings (which have been termed microcolliculi [Bereiter-

---

Fig. 18 Series of SAM images of a keratocyte from Xenopus tadpole epidermis (about semicircular cell on the right) colliding with an endothelial cell (large well spread cell on the right).
Right column: Original images. The cell bodies of both the colliding cells are clearly distinguished from the lamella area, in some parts interference fringes delineating surface topography are obvious. The brighter zones at the periphery of the keratocyte lamella are local thickenings (microcolliculi). On collision the keratocyte changes its direction of migration by extending the upper part of its lamella backwards (comp. 5 s with 30 s).
Left column: Difference of two subsequent pictures with a time interval of 5s (SUBSAM-images). The absolute differences between the corresponding picture in the right column and the following one (not shown for the 20s and 30s states) are shown in dark. Bright areas indicate that no changes took place. The dark zones at the rear end and at the upper part of the cell body / lamella transition region of the keratocyte reveal the dislocation of the margins. At the front of the lamella motility domains are visible: dark (motile) zones which are separated from each other by narrow zones of quiescence. Regarding their probable function as areas of specific actin organization these have been termed "nucleating domains" (Vesely et al. 1994). For further explanations see text.
Ultrasound frequency: 1GHz. Width of the single image is 100µm.

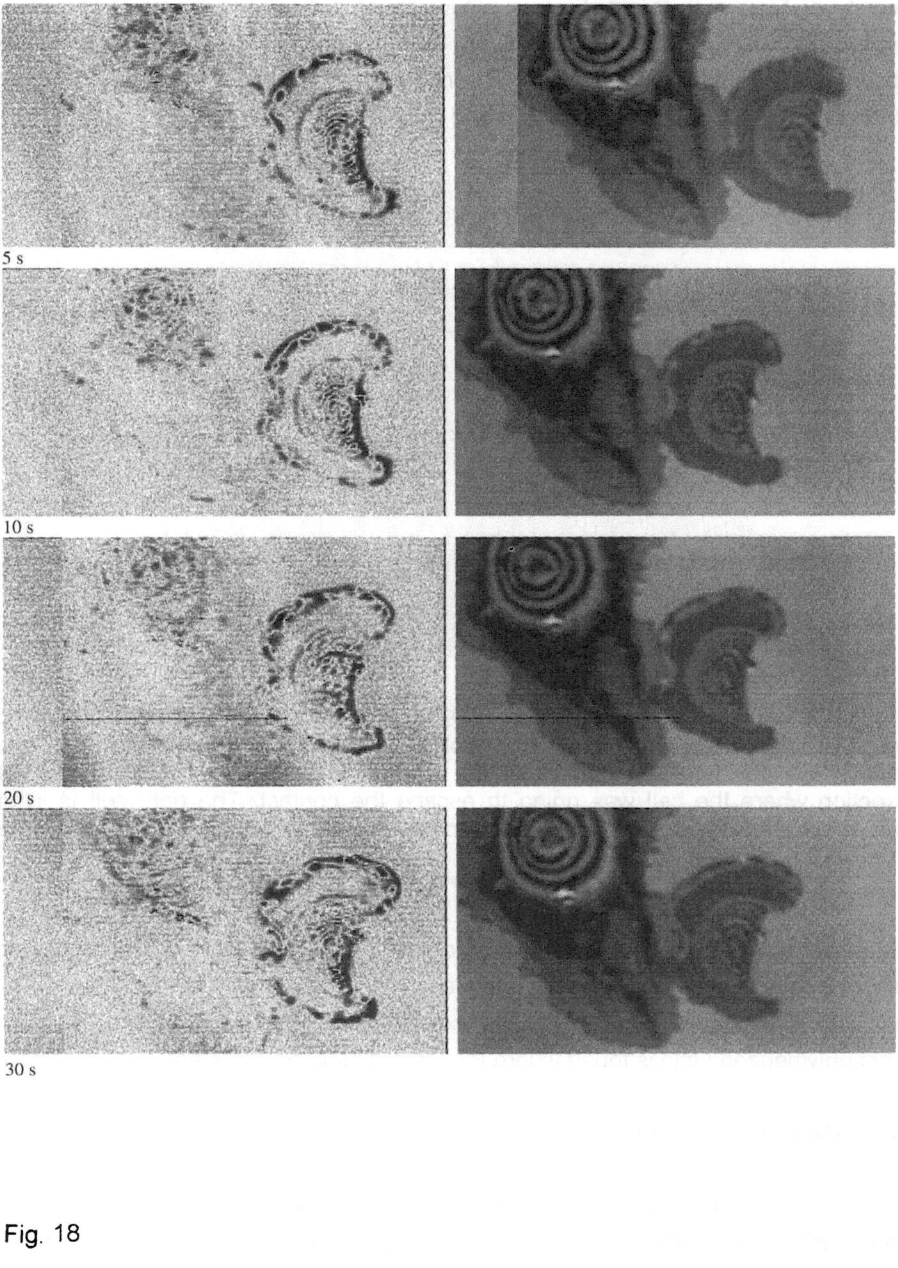

Fig. 18

Hahn et al. 1981]) do not appear at the very margin of the lamella rather than one or two microns apart from it. We presume that these microcolliculi arise from the flux of cytoplasm towards the tip of the lamella which is impeded at the front where high density of F-actin is found. Further extension of the lamella therefore would be caused by material flux into relatively weak areas (comp. Fig. 13). These observations on material flux are in accord with interference microscopic analyses made for fibroblast locomotion (Brown and Dunn, 1989).

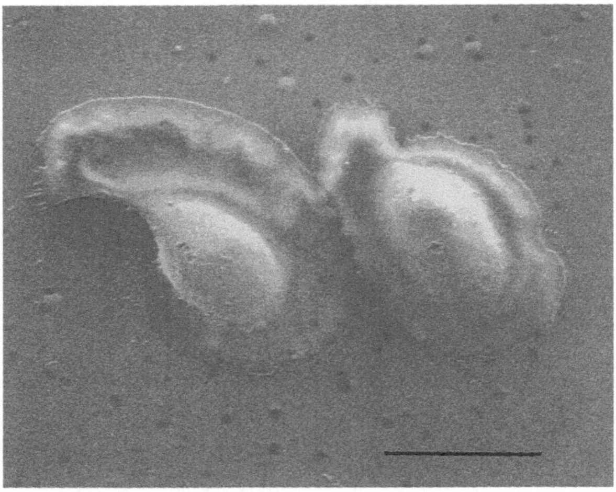

Fig. 19. Scanning electron micrograph (0.9 KV, field emission SEM, Hitachi 4000) of two XTE-cells fixed immediately after collision. Differences in the thickness of a lamella are obvious. The lamella of the left cell is thickened into the upper left direction where the cell was going to escape the contact. The right cell is in an earlier stage of response, it just started to realize the contact, and tends to spread towards the left. Bar: 20 µm.

Fluorometry reveals a further interesting reaction of cells to contact: At the contact sites the relative amount of F-actin is increased (Fig. 4b, Bereiter-Hahn, 1986). This is a strong arguments against lamella extension by actin polymerization because most polymerized actin is found where the lamella stopped extension.

VI. Model of cell crawling

Taken together, the distribution of actin, myosin, the arrangement of microtubules and intermediate filaments, the existence of hydrostatic pressure in the cells, the SAM observations, and the experiments on local application of substances, a model for cell locomotion arises: Crawling is based on passive extension of the lamella by release of hydrostatic pressure, followed by enhanced actin polymerisation and crosslinking at the advancing sites. Actomyosin driven contractions organize a cortical layer and orientate the actin fibrils towards the cell body. This contraction in the cell body / lamella transition region pulls the cell body into the lamella space, and thus shortens the lamella. F-actin depolymerization is presumed to occur in the cell body region. The direction of cell locomotion results from a subtle interplay of forces, the contraction moving the cell body toward the lamella front, the resistence of the lamella against expansion by the hydrostatic pressure and thus against the flux of cytosol.

VI. Generalization of the model.

Are the principles outlined for the locomotion of epidermis cells valid for other cell types too? The differences in the motile behavior of different cell types depend on their construction and interaction with the environment. Let us compare the general descriptors of locomotion (*forward transport of cytoplasmic material, adhesion to the external, organization of cytoskeletal elements and speed of locomotion*) for four cell types in culture: fibroblasts, leukocytes, epidermis cells and for the advancing growth cone of a nerve cell (Fig. 20).

Fibroblasts in culture adhere strongly to the solid substratum via focal contacts and they form some broad adhesion zones. Stress fibres emanate from the focal contacts. The elongated triangular shape results from the resistence of the rear part against the advancing front, considerable tension is developed along these cells (Harris, 1986; Harris et al. 1980). The internal pressure of these cells is relatively high (Bereiter-Hahn, 1987), speed of locomotion is slow because of the strong adherence. As soon as the adhesion is loosened experimentally, the cells develop blebs in the direction of the leading lamella and increase speed for a short time (Chen, 1981). The flux of cytoplasmic components (cytosol and membrane material) exceeds its need for lamella extension because of the strong contractile forces pulling the lamella material back to the cell body. These are the conditions to develop ruffles, which can be presumed to represent a surplus of membrane and newly polymerized F-actin which is pulled backwards.

Leukocytes, like PMNLs (polymorphonuclear leukocytes) represent an almost opposite organization. They adhere weakly to the glass or plastic, therefore they cannot develop thin lamellae rather than pseudopodia of approximately cylindrical shape, actin fibrils form an isotropic network. They move relatively fast, their speed may be limited by their ability to adhere and thus transmitting motive force to the environment.

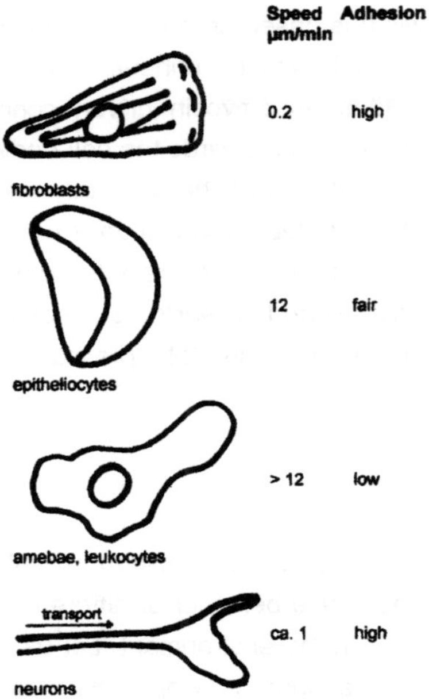

Fig. 20. Summary of the structural principles involved in the migration of different cell types. In all the cells except neurons, hydrostatic pressure can be supposed to provide the driving force for the cytoplasmic flow extending the pseudopodia or lamellae. In neurons the anterograd transport of material may fulfill this demand. Cell shape primarily depends on adhesion to the environment. For further explanation see text.

Epidermis cells are fast moving cells with broad areas of close contacts. Many of them are devoid of focal contacts (and only those are moving fast). Their internal hydrostatic pressure is in the same range as in the fast moving Amoeba proteus (Kamiya, 1964; Bereiter-Hahn, 1987). The broad lamella allows an intensive, pressure driven flux of material from the cell body to the leading front.

Extending neuronal processes differ from all the other examples by the very long and thin extension, which restricts material flow by hydrostatic pressure. In this case the flux of cytoplasmic material can be assumed to be limiting for the advance of the front (called growth cone). Flux of cytoplasm in nerve cells depends almost exclusively on microtubule and dynein or kinesin mediated transport. The growth cone is well adhering to environmental structures. In this case the driving force for the extension of the lamella may be produced by the anterograd cytoplasmic flux along the microtubules.

Thus the differences in the motile behavior of the different cell types are based on different constructional properties. If those are taken into account, locomotion of cells can be described using the principle determinants flux of cytoplasmic constitutuents ("hyaloplasm", ground plasm), adhesion, actin polymerization and depolymerization, and structural restrictions guiding the forces.

Appendix I

Reflection Interference microscopy

Reflection interference microscopy (RIM) is used to view cells in culture, adhering to a plane glass surface. Assuming the cytoplasm being optically homogenous, three optical interfaces may occur: glass/culture-medium, medium/cell, cell/medium (two thin layers on top of the glass). In areas where the cells are intimately attached to the glass, the glass/medium interface is replaced by a glass/cytoplasm interface (one thin layer on the glass). Light beams reflected from the various interfaces may interfere with each other. The resulting intensities depend on the differences in optical path lengths and the **RI** at the boundaries. As far as interference fringes appear, an estimation of the distances can be made because the distance between adjacent constructive (bright line) and destructive (dark lines) interferences corresponds to a quarter wavelength (the phase difference between adjacent maxima and minima corresponds to $\lambda/2$, but the light must pass the layer twice before interfering). Therefore, the geometrical difference of layer thickness (d) between adjacent maxima and minima is given by:

$$d = \lambda \cos \beta / 4 n \qquad (1)$$

d: thickness of layer between maxima and minima of adjacent interference fringes, λ: wavelength of light in vacuum, n: refractive index of the layer, ß: angle of incidence of the illuminating beam (normal incidence: $\cos \beta = 1$).

Zones of *close contact* of the cells (Izzard and Lochner, 1976) to the glass appear dark grey, *focal contacts* almost black. Their brightness does not change considerably by changing either the wavelength or the angle of illumination.

Epithelial cells normally fulfill the condition of smooth adherence to the glass.

The upper cell surface is best revealed using an angle of incidence $0° \leq 22°$. The surface profile is delineated by the course of interference rings (Fig. 16). Under the conditions used for Fig. 16, a change from maximum to minimum brightness of an area corresponds to a thickness change of about 100 nm.

Appendix II

Scanning acoustic microscopy

SAM images are determined by the interaction of sound waves with cellular structures, the substratum to which the cells adhere and with the culture medium.

Due to the mechanical nature of these interactions the acoustic image of a cell allows to deduce its mechanical properties and its topography (thickness) (Johnston et al., 1979; Hildebrand et al., 1981; Hoppe and Bereiter-Hahn, 1985).

Principles of the interpretation of SAM images of living cells.

The contrast of SAM images of cells is mainly produced by interferences of reflections from the cell surface and from the substratum. Contrast differences reflect variations in the acoustic impedance (which is a product of density and sound velocity). Sound velocity is related to the appropriate elastic modulus of the the specimen. In cases with unknown Poisson ratio (as is the case for cytoplasm) the qualitative description of elasticity using the term "stiffness" is more appropriate (for this relation of elasticity to sound velocity see Kundu and Bereiter-Hahn 1991). Variations in acoustic impedance of the cortical cytoplasm are revealed by the reflectivity of the cells thus showing variations in stiffness of the cytoplasm:

Like in RIM, the boundaries between the cells and the solid substratum and with the culture medium determine the reflections of sound waves which may interfere with each other. The amplitude of the reflected sound waves depends on the differences in the acoustic impedances of the two media forming a boundary. Model calculations on the amplitude of sound waves resulting from the interference of sound reflected at the upper surface of a wedge shaped thin layer and its lower surface being attached to glass clearly show that the resulting intensity variations in the interferences resulting from thickness changes along the wedge increase with increasing differences of acoustic impedance (Litniewski and Bereiter-Hahn, 1990). This principle is used for a qualitative display of cell stiffness in Fig. 13.

SAM investigations were carried out with an ELSAM (Leitz, Wetzlar). A frequency of 1.0 Ghz was selected, corresponding to a wavelength of approximately 1.5 µm in water. The radius of curvature of the acoustic lens was 40 µm, and the opening angle 100°. Acoustic coupling of the lens to the cells was provided by the saline which was assumed to be acoustically identical to water (longitudinal wave velocity: 1510 m/s at 28°C, attenuation coefficient at 28°C: 0.015 μm^{-1}). The cells were attached to glass slides with known acoustic parameters (longitudinal wave velocity: 5870 m/s, density: 2472 kg/m^3).

The method for the evaluation of the acoustic parameters of cells in culture followed the procedure developed by Litniewski and Bereiter-Hahn (1990). For this purpose the images were taken with the focus level about 12 µm above the glass surface. This out of focus imaging eliminates sound waves other than of normal incidence. Defocussing does not considerably reduce the resolving power of the SAM.

Acknowledgements

We are indebted to Mrs. Monika Vöth and Silke Engel for excellent technical assistence. The experimental part of our own work reported in this review has been supported by the Gesellschaft der Freunde und Förderer der Johann Wolfgang Goethe Universität Frankfurt am Main and by the Deutsche Forschungsgemeinschaft (Be 423/13 -16).

List of References

Abercrombie M. (1961) The bases of the locomotory behaviour of fibroblasts. Exptl Cell Res. 8: 188-198

Abercrombie M., Heaysman J.M., Pegrum S.M. (1970a) The locomotion of fibroblasts in culture. I. Movements of the leading edge. Exptl. Cell. Res. 59: 393-398

Abercrombie M., Heaysman J.E.M., Pegrum S.M. (1970b) The locomotion of fibroblasts in culture II: "Ruffling" Exptl. Cell Res. 6O: 437-444

Abercrombie M., Heaysman J.E.M., Pegrum S.M. (1970c) The locomotion of fibroblasts in culture III Movements of particles on the dorsal surface of the leading lamella. Exptl. Cell Res. 62: 389-398

Abercrombie M., Heaysman J.E.M., Pegrum S.M. (1971) The locomotion of fibroblasts in culture IV electron microscopy of the leading lamella. Exptl Cell Res 67: 359-367

Agosti C.G., Stidwill R.P. (1992) The contributions of microtubules and F-Actin to the in vitro migratory mechanisms of Hydra nematocytes as determined by drug interference experiments. Exp Cell Res 200: 196-204

Alberts B., Bray D., Lewis J., Raff M., Roberts K., Watson J.D. (1989) Molecular biology of the cell. 2nd ed. 1219 pp. Garland Publ. Inc. New York and London.

Allen R.D. (1961) A new theory of ameboid movement and protoplasmic streaming. Exptl. Cell Res. 8: 17-31

Allen R.D. (1970) Comparative aspects of amoeboid movement. Acta Protozoologica VII: 291-299

Allen R.D., Zeh R., Condelis J. (1972) Capillary suction test of the pressure gradient theory of amoeboid motion. Science 177: 636-638

Alt W. (1990) Dynamic mechanical properties of physarum cytoplasm. NATO ASI Series H42: 403-422

Ambrose E.J., Batzdorf U., Osborn J.S., Stuart P.R. (1970) Sub-surface structures in normal and malignant cells. Nature 227: 397-398

Anniko, M.; Arnold, W. (1990) Cytoskeletal Network of Intermediate Filament Proteins in the Adult Human Vestibular Labyrinth Acta Oto-Laryngol : 40-50

Bellomo, G.; Mirabelli, F. (1992) Oxidative stress and cytoskeletal alterations Aging and Cellular Defense Mechanisms 663: 97-109

Bereiter-Hahn J. (1967) Dissoziation und Reaggregation von Epidermiszellen der Larven von Xenopus laevis (Daudin) in vitro. Z. Zellf. 79: 118-156

Bereiter-Hahn J. (1986) Epidermal cell migration and wound repair. In: J. Bereiter-Hahn, A.G. Matoltsy, S. Richards (eds) Biology of the Integument. II. Vertebrates pp.443-471, Springer Vlg. Heidelberg, New York..

Bereiter-Hahn J. (1987) Mechanical principles of architecture of eukaryotic cell. In: J. Bereiter-Hahn, O.R. Anderson, W.E. Reif (ed.) Cytomechanics pp.1-30, Springer Vlg., Heidelberg, New York.

Bereiter-Hahn J. (1988) Involvement of microcompartmentation in the regulation of cell proliferation. In: D.P. Jones (ed.) Microcompartmentation. pp.55-70, CRC Press, Boca Raton.

Bereiter-Hahn, J., Kajstura, J. (1988) Scanning microfluorimetric measurement of TRITC-phalloidin labelled F-actin in cultured cells: Dependence of F-actin content on density of normal and transformed cells. Histochem. 90: 271-276

Bereiter-Hahn J., Strohmeier R., Kunzenbacher I., Beck K., Vöth M. (1981) Locomotion of Xenopus epidermis cells in primary culture. J. Cell Sci. 52: 289-311

Bereiter-Hahn J., Braun N., Vöth M. (1990a) Continuity of movement and preservation of architecture during cell locomotion. In: W. Alt (ed.) Biological Motion, Lecture Notes in biomathematics pp., Springer Vlg., Heidelberg, New York.

Bereiter-Hahn J., Lück M., Miebach T., Stelzer H.K., Vöth M. (1990b) Spreading of trypsinized cells: cytoskeletal dynamics and energy requirements. J. Cell Sci. 96: 171-188.

Bershadsky A.D., Vasiliev J.M. (1988) Cytoskeleton. 298pp. Plenum Press, New York, London.

Bissell M.J., Aggeler J. (1987) Dynamic reciprocity: How do extracellular matrix and hormones direct gene expression? In: Cabot M.C., McKeehan W.L. (eds.) Mechanisms of signal transduction by hormones and growth factors. pp.251-262, Alan R. Liss, Inc. New York

Bray D., Heath J., Moss D. (1986) The membrane-associated "cortex" of animal cells: its structure and mechanical properties. J. Cell Sci. Suppl. 4: 71-88

Brown A.F., Dunn G.A. (1989) Microinterferometry of the movement of dry matter in fibroblasts. J. Cell Sci. :

Burridge K. (1981) Are stress fibres contractile? Nature 294: 691-692

Butt, H.J., Wolff E.K., Gould S.A., Northern B.D., Peterson C.M:, Hansma P.K. (1990) Imaging cells with the atomic force microscope. J. Struct. Biol. 105: 54-61

Cande W.Z., Ezzell R.M. (1986) Evidence for regulation of Lamellipodial and tail contraction of glycerinated chicken embryonic fibroblasts by myosin light chain kinase. Cell Mot. Cytoskel. 6: 640-648

Carver, W.; Nagpal, M.L.; Nachtigal, M.; Borg, T.K.; Terracio, L. (1991) Collagen Expression in Mechanically Stimulated Cardiac Fibroblasts Circ Res 69: 116-122

Cassimeris, L., Mcneill, H., Zigmond, S.H. (1990) Chemoattractant-stimulated polymorphonuclear leukocytes contain 2 populations of actin filaments that differ in their spatial distributions and relative stabilities. J. Cell Biol. 110: 1067-1075

Chen W.-T. (1981) Mechanism of retraction of the trailing edge during fibroblast movement. J.Cell Biol. 90: 187-200

Chen Wen-Tien (1979) Induction of spreading during fibroblast movement. J.Cell Biol. 81: 684-691

Chen Wen-Tien (1981) Mechanism of retraction of the trailing edge during fibroblast movement. J. Cell Biol. 90:187-200

Clegg J.S. (1992) Cellular infrastructure and metabolic organization. Cell. Regulation 33: 3-14

Cooper M.S., Schliwa M. (1985) Electrical and ionic controls of tissue cell locomotion in DC electric fields. J. Neuroscience Res. 13: 223-244

Cortese, J.D.; Frieden, C. (1990) Effect of filamin and controlled linear shear on the microheterogeneity of F-actin/gelsolin gels. Cell Motil Cytoskeleton 17: 236-249

Couchman J.R., Rees D.A. (1979) Actomyosin organisation for adhesion, spreading, growth and movement in chick fibroblasts. Cell Biol. Intern.Reports 3: 431-440

De Brabander M., Nuydens R., Geerts H. (1990) Dynamic cytomatrix-membrane interactions in cell shape and organization. In: N. Akkas (ed.) Biomechanics of active movement and deformation of cells. pp.215-234, Springer Vlg.Berlin.

De Brabander, M.; Nuydens, R.; Ishihara, A.; Holifield, B.; Jacobson, K.; Geerts, H. (1991) Lateral Diffusion and Retrograde Movements of Individual Cell Surface Components on Single Motile Cells Observed with Nanovid Microscopy J Cell Biol 112: 111-124

DiMilla, P.A.; Barbee, K.; Lauffenburger, D.A. (1991) Mathematical Model for the Effects of Adhesion and Mechanics on Cell Migration Speed Biophys J 60: 15-37

DiPasquale A. (1975) Locomotory activity of ephithelial cells in culture. Exp.Cell Res. 94: 191-215

Domnina L.V., Rovensky J.A., Vasiliev J.M., Gelfand I.M. (1985) Effect of microtubule-destroying drugs on the spreading and shape of cultured epithelial cells. J.Cells Sci. 74: 267-282

Dunn G.A. (1980) Mechanismus of fibroblasts locomotion. In: A.S.G.Curtis, J.D.Pitts (ed.) Cell adhesion and motility. pp.409-423, Cambridge Univ.Press.

Duszyk M., Schwab III B., Zahalak G.I., Qian H., Elson E.L. (1989) Cell poking: Quantitative analysis of indentation of thick viscoelastic layers. Biophys. J. 55: 683-690

Euteneuer U., Schliwa M. (1984) Persistent, directional motility of cells and cytoplasmic fragments in the absence of microtubules. Nature 310: 58-61

Euteneuer U., Schliwa M. (1985) Evidence for an involvement of actin in the positioning and motilityof centrosomes. J.Cell Biol. 101: 96-103

Evans E., Dembo M. (1989) Dynamic cytomatrix - membrane interactions in cell shape and organization. NATO ASI Series H42: 185-214

Faulstich H., Trischmann H., Mayer D. (1983) Preparation of tetramethylrhodaminyl-phalloidin and uptake of the toxin into short-term cultured hepatocytes by endocytosis. Exp. Cell Res. 144: 73-82

Franke R.P., Gräfe M., Dauer u., Schnittler H., Mittermayer C. (1987) Stress fibres (SF) in human endothelial cells (HEC) under shear stress. In: H. Kiesewetter, F. Jung (ed.) Blood fluidity and endothelial influences on microcirculation. pp.989-992, Spr

Fujinami N. (1976) Studies on the mechanism of circus movement in the dissociated embryonic. J.Cell Sci. 22: 133-147

Fukui Y., de Lozanne A., Spudich J.A. (1990) Structure and function of the cytoskeleton of a dictyostelium myosin-defective mutant. J. Cell Biol. 110: 367-378

Gerisch, G.; Noegel, A.A.; Schleicher, M. (1991) Genetic Alteration of Proteins in Actin-Based Motility Systems Annu Rev Physiol 53: 607-628

Gittes, F.; Mickey, B.; Nettleton, J.; Howard, J. (1993) Flexural rigidity of microtubules and actin filaments measured from thermal fluctuations in shape J Cell Biol 120: 923-934

Giuliano, K.A.//Taylor, D.L. (1990) Formation, Transport, Contraction, and Disassembly of Stress Fibers in Fibroblasts Cell Motility and the Cytoskeleton 16: 14-21

Goldacre R.J. (1961) The role of the cell membrane in the locomotion of amoebae and the source of the motive force and its control by feedback. Exp.Cell Res. 8: 1-16

Grazi, E.; Trombetta, G.; Magri, E.; Cuneo, P. (1990) The actin gelling activity of chicken gizzard alpha-actinin at physiological temperature is triggered by water sequestration. FEBS Lett 272: 149-151

Grebecki, A. (1990) Dynamics of the Contractile System in the Pseudopodial Tips of Normally Locomoting Amoebae, Demonstrated Invivo by Video-Enhancement Protoplasma 154: 98-111

Grebecki, A. (1994) Membrane and cytoskeleton flow in motile cells with emphasis on the contribution of free-living amoebae. Int. Rev. Cytol. (in press)

Hall A.L., Schlein A., Condeelis J. (1988) Relationship of pseudopod extension to chemotactic hormone-induced actin polymerization in amoeboid cells. J. Cell. Biochem. 37,: 285-299

Hall A.L., Warren V., Condeelis J. (1989) Transduction of the chemotactic signal to the actin cytoskeleton of Dictyostelium discoideum. Developm. Biol. 136: 517-525

Harris A., Dunn G. (1972) Centripetal transport of attached particles on both surface of moving fibroblasts. Exp.Cell Res. 73: 519-523

Harris A.K. (1973) Cell surface movements related to cell locomotion. In: Ciba Found.Symp. (ed.) Locomotion of Tissue Cells. pp.3-26, Elsevier North-Holland.

Harris A.K. (1986) Cell Traction in relationship to morphogenesis and malignancy. In: M.S. Steinberg, (ed.) Developmental Biology pp.339-357, Plenum Publ. Corp. New York

Harris A.K. (1990) Protrusive activity of the cell surface and the movements of tissue cells. Nato ASI Series H: Cell Biology 42: 249-294

Harris A.K., Wild P., Stopak D. (1980) Silicone rubber-substrata: a new wrinkle in the study of cell locomotion. Science 208: 177-179

Harvath, L. (1990) Regulation of neutrophil chemotaxis - correlations with actin polymerization Cancer Invest 8: 651-654

Heath J.P. (1983) Behaviour and structure of the leading lamella in moving fibroblasts.I. Occurence and centripetal movement of arc-shaped microfilament bundles beneath the dorsal cell surface. J.Cell Sci. 60: 331-354

Heaysman J.E., Pegrum S.M., Preston T.M. (1982) Spreading chick heart fibroblasts. A correlated study using phase contrast microcopy, RIM, TEM, and SEM. Exp.Cell Res. 140: 85-93

Henderson E., Haydon P.G., Sakaguchi D.S. (1992) Actin filament dynamics in living glial cells imaged by atomic force microscopy. Science 257: 1944-1946

Hildebrand J.A., Rugar D., Johnston R.N., Quate C.F. (1981) Acoustic microscopy of living cells. Proc. Natl. Acad. Sci. USA 78: 1656-1660

Hiramoto Y. (1987) Evaluation of cytomechanical properties. In: J. Bereiter-Hahn, O.R. Anderson, W.-E. Reif (eds.) Cytomechanics pp.31-46, Springer Vlg., Berlin, Heidelberg, New York.

Hirokawa N., Takemura R., Hisanaga S.I. (1985) Cytoskeletal architecture of isolated mitotic spindle with special reference to microtubule-associated proteins and cytoplasmic dynein. J.Cell Biol. 101: 1858-1870

Hirokawa N., Tilney L.G. (1982) Interactions between actin filaments and between actin filaments and membranes in quick-frozen and deeply etched hair cells of the chick ear. J.Cell Biol. 95: 249-261

Hoath, S.B. (1990) Considerations on the Role of Surface Tension and Epidermal Growth Factor in the Mechanism of Integumental Pattern Formation J. Theoret. Biol. 143: 1-14

Holifield B.F., Jacobson K. (1991) Mapping trajectories of Pgp-1 membrane protein patches on surfaces of motile fibroblasts reveals a distinct boundary separating capping on the lamella and forward transport on the retracting tail. J. Cell Sci. 98: 191-203

Hoppe M., Bereiter-Hahn J. (1985) Applications of scanning acoustic microscopy - Survey and new aspects. Transactions on Sonics Ultrasonics SU-32: 289-301

Huxley H.E. (1976) Introductory remarks: The relevance of studies on muscle to problems of cell motility. In: R.Goldman, T.Pollard, J.Rosenbaum (ed.) Cell Motility Book A. pp.115-126, .

Höglund A.S. (1985) The arrangement of microfilaments and microtubules in the periphery of spreading fibroblasts and glial cells. Tissue & Cell 17: 649-666

Höglund A.S., Karlsson R., Arro E., Fredriksson B.A., Lindberg U. (1980) Visualization of the peripheral weave of microfilaments in glia cells. J. Muscle Res. Cell Mot. 1: 127-146

Höner B., Citi S., Kendrick-Jones J., Jockusch B.M. (1988) Modulation of cellular morphology and locomotory activity by antibodies against myosin. J. Cell Biol. 107: 2181-2189

Höner B., Jockusch B.M. (1988) Stress fiber dynamics as probed by antibodies against myosin. Europ. J. Cell Biol. 47: 14-21

Hörber J.K., Häberle W., Ohnesorge F., Binnig G., Liebich H.G., Czerny C.P., Mahnel H., Mayr A. (1992) Investigation of living cells in the nanometer regime with the scanning force microscope. Scanning Micr. 6: 919-930

Ingber D.E., Folkman J. (1987) Regulation of endothelial growth factor action: solid state control by extracellular matrix. In: Ingber D.E. (ed.) Mechanism of signal transduction by hormones and growth factors. pp.273-282, Alan R. Liss, Inc. New York.

Inoué S. (1982) The role of self-assembly in the generation of biological form. In: Alan R.Liss, New York (ed.) Developmental order: Its origin and regulation. pp., .

Izzard C.S. (1976) Cell-to-substrate contacts in living fibroblasts: an interference reflexion study with an evaluation of the technique. J. Cell Sci. 21: 129-159

Janmey, P.A. (1991) A Torsion Pendulum for Measurement of the Viscoelasticity of Biopolymers and Its Application to Actin Networks J Biochem Biophys Methods 22: 41-53

Janmey, P.A.; Euteneuer, U.; Traub, P.; Schliwa, M. (1991) Viscoelastic Properties of Vimentin Compared with Other Filamentous Biopolymer Networks J Cell Biol 113: 155-160

Johnston R.N., Jipson V, Atalar A., Heiserman J., Quate C.F. (1979) Acoustic microscopy: resolution of subcellular detail. Proc. Natl. Acad. Sci. U.S.A. 76: 3325-3329

Kamiya N., The motive force of endoplasmic streaming in the ameba (1964) In: Allen R.D. and Kamiya N. (eds.), Primitive motile systems in cell biology. 257-277, Academic Press, New York, London.

Kohama K. (ed.), Calcium inhibition: A new mode for Ca^{++} regulation (1992) CRC Press, Boca Raton .

Keller H.U. (1983) Motility, cell shape, and locomotion of neutrophil granulocytes. Cell Motility 3: 47-60

Keller H.U. (1990) Diacylglycerols and PMA are particularly effective stimulators of fluid pinocytosis in human neutrophils. J. Cell. Physiol. 145: 465-471

Keller H.U., Niggli V., Zimmermann A. (1990) The protein kinase C inhibitor H-7 activates human neutrophils: effect on shape, actin polymerization, fluid pinocytosis and locomotion. J. Cell Sci. 96: 99-106

Keller H.U., Niggli V. (1993) Colchicine-induced stimulation of PMN motility related to cytoskeletal changes in Actin, a-Actinin, and Myosin. Cell Motility & Cytoskeleton 25: 10-18

Keller H.U., Niggli V. (1993) The PKC-inhibitor RO 31-8220 selectively suppresses PMA- and diacylglycerol-induced fluid pinocytosis and actin polymerization in PMNS. Biochem & Biophys Res,Commun. 194: 1111-1116

Komnick H., Stockem W., Wohlfarth-Bottermann K.E. (1973) Cell motility: mechanisms in protoplasmic streaming and ameboid movement. Intern.Rev.Cytol. 34: 169-249

Korohoda W. (1992) Migration of epithelial cells. Kosmos 41: 39-50

Korohoda W., Kajstura J. (1982) Demonstration of the significance of isometric cintraction for the formation of stress fibres in chick embryo fibroblasts. Folia Histochem.Cytochem. 20: 153-156

Kunzenbacher, I., Bereiter-Hahn, J., Osborn, M., Weber, K. (1982) Dynamics of the cytoskeleton of epidermal cells in situ and in culture. Cell Tiss. Res. 222: 445-457.

Kreis T.E., Birchmeier W. (1980) Stress fiber sarcomeres of fibroblasts are contractile. Cell 22: 555-561

Kucik D.F., Elson E.L., Sheetz M.P. (1989) Forward transport of glycoproteins on leading lamellipodia in locomoting cells. Nature 340: 315-317

Kundu T., Bereiter-Hahn J., Hillmann K. (1991) Measuring elastic properties of cells by evaluation of scanning acoustic microscopy V (Z) values using simplex algorithm. Biophys. J. 59: 1194-1207

Kunzenbacher, I., Bereiter-Hahn, J., Osborn, M., Weber, K. (1982) Dynamics of the cytoskeleton of epidermal cells in situ and in culture. Cell Tiss. Res. 222: 445-457.

Lapiere, C.M., Nusgens, B. (1989) Mechanical powers are responsible for differentiation of fibroblasts cultured within a 3 dimensional collagen support In: Hybrid artificial organs: concepts and development. pp.77-90, .

Layrand D.B., Matveeva N.B., Teplov V.A., Beylina S.I. (1972) The role of elastoosmotic parameters in locomotion of myxomycete plasmodia. Acta Protozoologica 11: 339-354

Levesque M.J., Sprague E.A., Schwartz C.J., Nerem R.M. (1989) The influence of shear stress on cultured vascular endothelial cells: The stress response of an anchorage-dependent mammalian cell. Biotechn. Progress 5: 1-8

Lindberg U., Höglund A.S., Karlsson R. (1981) On the ultrastructural organization of the microfilament system and the possible role of profilactin. Biochimie 63: 307-323

Litniewski J., Bereiter-Hahn J. (1990) Measurements of cells in culture by scanning acoustic microscopy. J. Microsc. 158: 95-107

Lüers H., Bereiter-Hahn J., Litniewski J. (1991) SAM investigations: The structural basis of cell surface stiffness of cultured cells. Acoustical Imaging 19: 511-516

Lüers H., Hillmann K., Litniewski J.,Bereiter-Hahn J. (1992) Acoustic microscopy of cultured cells: Distribution of forces and cytoskeletal elements. Cell Biophysics 18: 279-293

Mast S.O. (1926) Structure, movement, locomotion and stimulation in amoeba. J.Morph. Physiol. 41/2: 347-423

Matsubara I., Elliot G.F. (1972) X-ray diffraction studies on skinned single fibres of fro skeletal muscle. J. Mol. Biol. 72:657-669.

Mittal A.K., Bereiter-Hahn J. (1985) Ionic control of locomotion and shape of epithelial cells: I. role of calcium influx. Cell Motility 5:123-136.

Mooseker M.S., Pollard Th.D., Fujiwara K. (1978) Characterization and localization of myosin in the brush border. J.Cell Biol. 79: 444-452

Müller O., Gaub H.E., Bärmann M., Sackmann E. (1991) Viscoelastic moduli of sterically and chemically cross-linked actin networks in the dilute to semidilute regime: Measurements by an oscillating disk rheometer. Macromol. 24: 3111-3120

Opas M. (1987) The transmission of forces between cells and their enviroment. In: J. Bereiter-Hahn, O.R. Anderson, W.-R. Reif (ed.) Cytomechanics pp.273-285, Springer-Vlg., Heidelberg, New York.

Osborn M., Weber K. (1977) The detergent-resistent cytoskeleton of tissue culture cells includes the nucleus and the microfilament bundles. Exp.Cell Res. 106: 339-349

Oster G.F. (1984) On the crawling of cells. J. Embryol. Exp. Morphol. Suppl. 83: 329-364

Oster G.F., Perelson A.S. (1987) The physics of cell motility. J. Cell Sci. Suppl. 8: 35-54

Pollard T.D. (1982) Methods to characterize actin filament networks. In: L. Wilson (ed.) Methods in Enzymology pp.211-233, Acad. Press, New York.

Popov, S.V.; Svitkina, T.M.; Margolis, L.B.; Tsong, T.Y. (1991) Mechanism of Cell Protrusion Formation in Electrical Field - The Role of Actin Biochim Biophys Acta 1066: 151-158

Rinnenthaler G., Herzog M., Klappacher M., Kunka H., Small J.V. (1991) Leading edge movement and ultrastructure in mouse macrophages J. Struct. Biol. 106: 1-16

Rome E., (1968) X-ray diffraction studies of the filament lattice of striated muscle in various bathing media. J. Mol. Biol. 37:331-334

Schlage W., Kuhn, C., Bereiter-Hahn J. (1981) Established Xenopus tadpole heart endothelium (XTH) cells exhibiting selected properties of primary cells. Europ. J. Cell Biol. 24: 342

Schlage W.K. (1986) Determination of human factor VIII (AHG-associated protein)-antigen in endothelial cells from Xenopus laevis (XTH cells). Evaluation of a sensitive ELISA technique. Cell Tissue Res. 245: 445-451

Schliwa M. (1975) Cytoarchitecture of surface layer cells of the teleost epidermis. J. Ultrastruct. Res. 377-386:

Schliwa M. (1980) Pharmacological evidence for an involvement of calmodulin in calcium-induced microtubule disassembly in lysed tissue culture cells. In: M. de Brabander, J. de Mey (eds.) Microtubules and microtubule inhibitors. pp., Elsevier N.H..

Schliwa M. (1986) The cytoskeleton. An introductory survey. In: (ed.) Cell Biology monographs, Vol 13 pp.322 pp, Springer Vlg., Wien.

Shirinsky, V.P.; Biryukov, K.G.; Hettasch, J.M.; Sellers, J.R. (1992) Inhibition of the Relative Movement of Actin and Myosin by Caldesmon and Calponin J Biol Chem 267: 15886-1589

Simon S.I., Schmid-Schoenbein G. (1990) Three position controller for tensile regulation of axon length. NATO ASI Series H42: 429-436

Simon S.I., Schmid-Schönbein G.W. (1990) Cytoplasmic strains and strain rates in motile polymorphonuclear leukocytes. Biophys. J. 58: 319-332

Singer S.J., Kupfer A. (1986) The directed migration of eukaryotic cells. Ann. Rev. Cell Biol. 2: 337-365

Skalak R., Zhu C. (1990) Physical model for phagocyte motility: Local growth of a contractile network from a passive body. NATO ASI Series H42: 155-184

Skalli O., Chou Y.-H., Goldman R.D. (1992) Intermediate filaments: not so tough after all. Trends in Cell Biol 2: 308-312

Small J.V., Celis J.E., Isenberg G. (1980) Aspects of cell architecture and locomotion. In: J.E.Celis, A.Graessmann, A.Loyter (eds.) Biophysics and Molecular Biology pp.75-111, Plenum Annonc. Publ. Corp..

Small J.V., Rinnerthaler G., Hinssen H. (1982) Organization of actin meshworks in cultured cells: the leading edge. Cold Spring Harbor Symp. 46: 599-611

Stockem W. (1972) Membrane-turnover during locomotion of Amoeba proteus. Acta Protozool. 11:83-94.

Stolz B., Bereiter-Hahn J. (1987) Calcium sensibility of cytoplasmic microtubules is correlated with the cell cycle. Europ. J. Cell Biol. 43,Sup.17: 57

Stossel T.P. (1982) The structure of cortical cytoplasm. Phil. Trans. Soc. Lond. B 299: 275-289

Stossel T.P. (1983) The spatial organization of cortical cytoplasm in macrophages. In: J.R. Intosh (ed.) Modern Cell Biology pp.203-223, Alan R. Liss Inc. N.Y..

Stossel T.P. (1993) On the crawling of animal cells. Science 260: 1086-1093

Stossel T.P., Janmey P.A., Zaner K.S. (1987) The cortical cytoplasmic actin gel. In: Bereiter-Hahn J., Anderson O.R., Reif W.E. (ed.) Cytomechanics. The mechanical basis of cell form and structure. pp.131-153, Springer Verl. Berlin, Heidelberg, New York.

Strohmeier R., Bereiter-Hahn J. (1984) Control of cell shape and locomotion by external calcium. Exp. Cell Res. 154: 412-420

Strohmeier R., Bereiter-Hahn, J. (1987) Hydrostatic pressure in epidermal cells is dependent on Ca-mediated contractions. J. Cell Sci. 88: 631-640

Symons, M.H.; Mitchison, T.J. (1991) Control of Actin Polymerization in Live and Permeabilized Fibroblasts J Cell Biol 114: 503-513

Taylor D.L., Fechheimer M. (1982) Cytoplasmic structure and contractility: the solation-contraction coupling hypothesis. Phil. Trans. R. Soc. Lond. B 299: 145-327

Taylor D.L., Moore P.L., Condeelis J.S., Allen R.D. (1976) The mechanochemical basis of amoeboid movement.I.Ionic Requirments for maintaining viscoeleasticity and contractility of amoeba cytoplasm. Exp. Cell Res. 101: 127-133

Theriot J.A., Mitchison T.J. (1991) Actin microfilament dynamics in locomoting cells. Nature 352: 126-131

Tilney L.G. (1980) Polymerization of actin.V.A new organelle,the actomere,that initiates the assembly of actin filamnets in thyone sperm. J. Cell Biol. 77: 551-564

Tilney L.G. (1983) Interactions between actin filaments and membranes give spatial organization to cells. In: B.H. Satir (ed.) Modern Cell Biology pp.163-199, A.R.Liss Inc., N.Y..

Tilney L.G., Hatano S., Ishikawa H., Mooseker M.S. (1973) The polymerization of actin:its role in the generation of the acrosomal process of certain echinoderm sperm. J. Cell Biol. 59: 109-126

Tilney L.G., Tilney M.S., Cotanche D.A. (1988) Actin filaments, stereocilia, and hair cells of the bird cochlea. V.How the staircase pattern of stereociliary lengths is generated. J. Cell Biol. 106: 355-365

Vasiliev J.M. (1982) Spreading and locomotion of tissue cells: Factors controlling the distribution of pseudopodia. Phil. Trans. R. Soc. Lond. B299: 159-167

Verschueren H., De Baetselier P., Bereiter-Hahn J. (1991) Dynamic morphology of metastatic mouse T-lymphoma cells invading through monolayers of 10T 1/2 cells. Cell Motility & Cytosk. 20: 203-214

Vesely P., Lüers H., Riehle M., Bereiter-Hahn J. (1994) Subtraction scanning acoustic microscopy reveals motility domains in cells in vitro. Nature (submitted)

Wang Y.L., Lanni F., McNeil P.L., Ware B.R., Taylor D.L. (1982) Mobility of cytoplasmic and membrane associated actin in living cells. Proc. Natl. Acad. Sci. 79: 4660-4664

Warrick H.M., Spudich J.A. (1987) Myosin structure and function in cell motility. Am.Rev. Cell Biol. 3: 379-421

Waugh R.E., Hochmuth R.M. (1987) Forces shaping an erythrocyte. In: Bereiter-Hahn J., Anderson O.R., Reif W.E. (ed.) Cytomechanics. The mechanical basis of cell form and structure. pp.249-260, Springer Verlag, Berlin, Heidelberg, New York.

Weeds, A.G.; Gooch, J.; Hawkins, M.; Pope, B.; Way, M. (1991) Role of Actin-Binding Proteins in Cytoskeletal Dynamics Biochem Soc Trans 19: 1016-1020

Wilson, A.K.; Gorgas, G.; Claypool, W.D.; Delanerolle, P. (1991) An Increase or a Decrease in Myosin-II Phosphorylation Inhibits Macrophage Motility J Cell Biol 114: 277-283

Wohlfarth-Bottermann K.E. (1964a) Differentiations of the ground cytoplasm and their significance for the generation of the motive force of ameboid movement. In: (ed.) Primitive motile systems in cell biology. pp.79-107, Acad. Press, Inc., N.Y..

Wohlfarth-Bottermann K.E. (1964b) Cell structures and their significance for ameboid movement. Int. Rev. Cytol. 16: 61-132

Wohlfarth-Bottermann K.E. (1977) Oscillating contractions in protoplasmic strands of Physarum: simultaneous tensiometry of longitudinal and radial rhythms, periodicity analysis and temperature dependence. J. exp. Biol. 67: 49-59

Wohlfarth-Bottermann K.E. (1979) Contraction phenomena in Physarum: New results. Acta Protozool. 18: 59-73

ON FREE BOUNDARY PROBLEMS AND AMOEBOID MOTION

Micah Dembo
Theoretical Division
Theoretical Biology and Biophysics Group (T-10)
Los Alamos National Laboratory
Los Alamos, New Mexico 87545
U.S.A.

Abstract

The actin cytoskeleton of amoeboid cells can be regarded as a composite material composed of an aqueous solvent and a network of filamentous polymers. Accordingly, a family of models of the cytoskeletal motility of amoeboid cells can be formulated using continuum mixture theory as a basic language. Our aim is to elucidate the structure of such models and to give some examples of applications to representative problems of cytoskeletal dynamics and amoeboid motion.

Historical Background

We currently possess no lack of imaginative hypotheses and qualitative concepts about the basic physical interactions and chemical couplings that conspire to produce amoeboid motility. The development of these views can be traced to the theories and observations of Aristotle. This author was attempting to make some sense of the remarkable fact that unlike the stars and the planets, nothing of the earthly world moves in a simple and predictable fashion. Nevertheless, consideration of the basic experiential data with regard to the subluminal chaos led him to note the subclassification of earthly objects as either animate

NATO ASI Series, Vol. H 84
Biomechanics of Active Movement
and Division of Cells
Edited by N. Akkaş
© Springer-Verlag Berlin Heidelberg 1994

(capable of some degree of autonomous or self-directed motility and sensation) or inanimate. To explain animation, Aristotle endorsed the doctrine of the "soul" or vital spirit. In the sense of Aristotle, the soul of a living being was simply a kind of prime mover that could enter and temporarily inhabit its inanimate corpse (Aristotle: *de Anima* 403b26, 413a21).

Although Galileo demonstrated that celestial matter was not of a special character, his investigations failed to shed much light on the distinction between the animate and inanimate states. Nevertheless, the discovery of "animalicules" by Leeuwenhock in the seventeenth century served to demonstrate that even very small amounts of matter could have a soul. Thereafter, efforts to study the basis of animation in its simplest setting fueled the development of microscopy and the early investigations of cell motion. By the 1830's there was a vigorous controversy as to the theoretical basis for animation. This pitted the structuralist views of Ehrenberg (1830, 1832) against the idealist concepts of Dujardin (1835, 1838).

Ehrenberg believed that microorganisms contained internal organs and muscles just like higher organisms. In the case of amoeba he believed that pseudopodia where caused by muscular contractions in the body leading to pressure gradients and outward herniations. In contrast, Dujardin proposed that the amoeba consisted of a pure and homogeneous globule of an essential plasma. He called this elementary substance the "sarcode." To explain animation Dujardin felt that the sarcode must possess at least two opposing tendencies: contraction and expansion. These two tendencies could then interact in a Hegelian fashion to produce motility.

The demise of the concept of living matter as an elementary substance came from improvements in the light microscope that allowed the definitive observation of internal structures; variations in the refractile index and flows of the cytoplasm in amoeba (Wallich, 1863a,b); and myxomycete plasmodia (de Bary, 1864). These authors both observed characteristic differences in the consistency and dynamics of cortical versus interior cytoplasm. They explained their observations in terms of theories in which structures in the cell cortex (ectoplasm) acted on a passive fluid material in the cell interior (endoplasm). Wallich felt that active contraction of the cortex was the sole mode of force generation; de Bary agreed that contraction was important but felt that active expansion might also be occurring, particularly at the leading edge.

An important contribution to this sort of theory was the attempt by Heitzmann (1873) to provide some structural bases for cortical contractility. To do this he proposed that the cell contained a reticulum of elastic filaments that could expand and contract. Variations in the density of this reticulum could account for the relative passivity of the interior cytoplasm. To our knowledge this is the first suggestion that cell motility arises from a structure that we today would vaguely recognize as a cytoskeleton. Porter (1984) has given an excellent historical account of the later development of the concept of the cytoskeleton.

The first careful observations of the full pattern of fountain streaming in moving amoeba by Schulze (1875) presented a major challenge to the simple models based on contractility of fixed cytoplasmic structures (see Figure 1). In analyzing his observations Schulze noted that, as in muscle, contraction of the cell cortex at any given point must eventually be followed by relaxation or expansion in order to complete the working cycle. In contrast to the resulting expectation, the data are that the entire cytoplasm of amoeba undergoes a continuous steady flow without any periodic reversals. Since it is very difficult to see how an arrangement or action of fixed structures could account for such motion, Schulze concluded that there must be continuous transfer of contracted material from the posterior part of the ectoplasm to the passive interior fluid or endoplasm and an opposite conversion must occur at the anterior.

Schulze's theory was met with skepticism in some quarters for several reasons. First the basic concept of contractility was very vague and lacked a secure physical-chemical basis. It was particularly troubling that no one was able to directly visualize or otherwise demonstrate the contractile filaments of Heitzmann in the cell cortex. Furthermore,

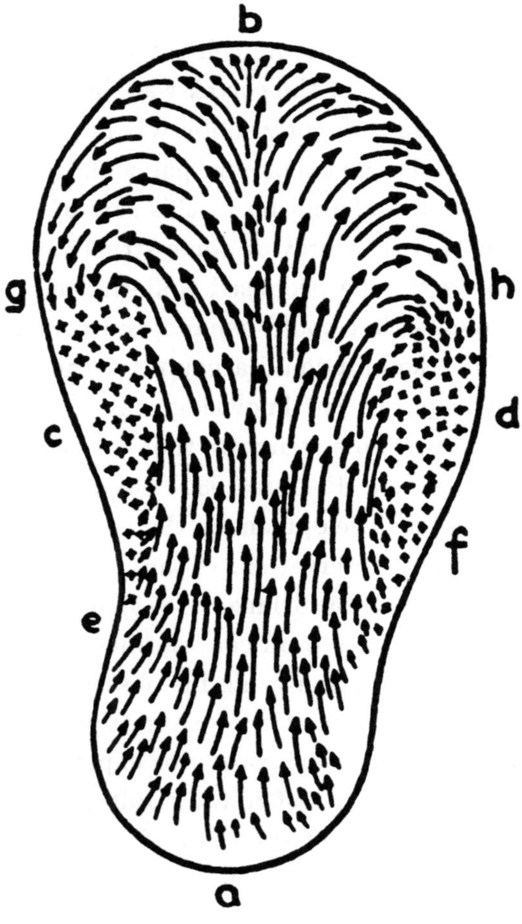

Figure 1. Schulze's (1875) Observations of Fountain Streaming in *Pelomyxa*. Crosses indicate regions where cytoplasm is at rest with respect to the substrate; arrows indicate the direction and speed of flows. The cortical layer of the cytoplasm (i.e., the ectoplasm) appears to be of greater density and viscosity when compared to the interior cytoplasm (the endoplasm). Contraction of the cortical cytoplasm along the arc e-a-f appears to drive the endoplasm forward. The cortical cytoplasm is relaxed or weakened along the arc h-b-g so as to accommodate the forward motion. In order to account for the continuous cycling of material there must be conversion from endoplasm to ectoplasm in the anterior region d-h-b-g-c. In the posterior region c-e-a-f-d it is necessary to suppose that the opposite reaction predominates.

knowledge of polymer chemistry was insufficient to provide an understanding of how Heitzmann's reticulum of filaments could have sufficient connectivity and structure to generate contractile force and still be sufficiently fluid and dynamic so as to undergo the required flow or transition from endoplasm to ectoplasm.

As a result of these issues there was a period from 1880 until about 1920 during which the predominant view was that cytoplasmic contractility was negligible or irrelevant. It was proposed that variations in surface tension were primarily responsible for driving cytoplasmic streaming and amoeboid movement. This hypothesis was attractive because it invoked only simple chemistry that could be demonstrated in nonliving systems. Representative of this school are the views of Rhumbler (1898). Surface tension theories were ultimately discredited by careful observations of particles on the surface of moving amoeba (Jennings, 1904), by observations of the discrete adhesions and walking movements of amoeba when viewed from the side (Dellinger, 1906), and by demonstration of the plasmalemma by microdisection (Seifriz, 1921).

The views of subsequent authors like Pantin (1923, 1924) and Mast (1926) were in basic agreement with the cortical contraction model proposed by Schulze although various refinements and embellishments are offered. For example concepts of colloid chemistry (i.e., gellation and solation) are introduced to lend some legitimacy to the conversion of endoplasm and ectoplasm at the anterior and posterior of the moving amoeba. Nevertheless, for both Pantin and Mast the physical basis of contractility was still mysterious.

Mast seems to be the first author to clearly incorporate the plasma membrane into his model of amoeboid movement. In Mast's model the plasma lemma provided a tough elastic barrier between the cortical cytoplasm and the exterior. He realized that this barrier was rapidly repaired after puncture, and he felt that it had a role in controlling and organizing the transitions of gelation and solation. Mast also discussed the anchorage of the gellated cytoplasm to the plasma membrane, and he fully realized that this anchorage could be disrupted particularly at the tip of pseudopodia (see Figure 2). It is interesting that these developments preceded the electron microscope.

In recent years the major development has been the explosion of detailed information concerning the molecular composition and organization of the cytoskeleton and the introduction of new biophysical methods for studying its dynamics. For example, we now recognize that the animation of living matter can be categorized according to the underlying molecular machinery (i.e., microtubule system or actin filament system) upon which they are based. The actin filament system is thought to be predominantly responsible for the so called "amoeboid" motions: phagocytosis, cytokinesis, pseudopodal extension/retraction, cytoplasmic streaming, cell spreading, and crawling locomotion on surfaces (Taylor and Condeelis, 1979; Hartwig et al., 1985; Stossel, 1993). Although there are important nuances and variations of the actin filament system from one sort of amoeboid cell to the next, its basic structural and regulatory elements are of great phylogenetic antiquity (Korn, 1982; Pollard and Cooper, 1986; Matsudaria and Janmey, 1988).

Unfortunately, although these breakthroughs have changed the language of the debate, they have really not led to a definitive resolution of the controversies of the past. For example, there are still a large number of competing and seemingly contradictory proposals concerning elementary issues such as the role of contractility and various other forces in pseudopodal extension. The traditional contractility school, dating back to Schulze (1875) and Mast (1926), is still well represented by authors such as Grebecki (1984), Beireiter-Hahn et al. (1990), Taylor and Condeelis (1979), Allen (1961). Except for details these authors regard contractility, powered by the various isozymes of myosin, as sufficient to account for pseudopodal extension.

A variation on the Mast type model substitutes osmotic pressure for the contractile pressure as a means of generating the protrusive force of pseudopodal extension (Oster and Perelson, 1987; Stossel 1990). In this approach a local increase of osmotic pressure adjacent to the cell boundary is generated by the depolymerization of the actin cytoskeleton and the consequent release of osmotically active fragments. As with the

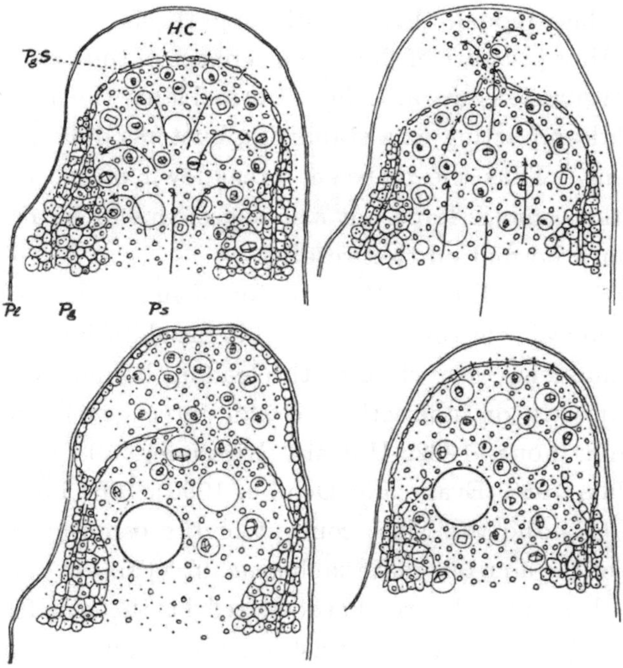

Figure 2. Mast's (1926) Concept of Events at the Tip of an Advancing Pseudopod.

Pl denotes the plasma lemma: a thin semipermeable membrane that surrounds the cytoplasm.

Ps denotes the plasma sol or endoplasm: This is the passive fluid form of the cytoplasm. It is rich in granular material and predominates in the interior of the cell.

Pg denotes the plasma gel: This is the gellated and contractile form of the cytoplasm (also called ectoplasm). Plasma sol can be converted to plasma gel by unknown chemical conditions characteristic of the tip of a pseudopod. The reverse reaction in which plasma gel is converted to plasma sol is signalled by conditions prevailing at the posterior of the cell. The control of conditions leading to sol-gel transitions could be a function of the plasma membrane. Note that Mast has rendered the plasma gel with a foam-like ultrastructure. This bit of artistic license is an attempt to account for his observations of the restricted Browian motion of granules trapped in regions of plasma gel.

HC denotes the Hyaline Cap: a zone of agranular plasmasol frequently observed at the tip of the pseudopod.

PgS denotes the plasma gel sheet: a thin layer of plasma gel formed as the result of a disruption of plasma gel from the plasma membrane at the leading edge of the pseudopod. This structure was not directly visualized by Mast but was inferred because of the exclusion of granules from the hyaline cap.

classical Mast model, this osmotic hypothesis is distinguished by the assertion that pressure of the aqueous phase of the cytoplasm pushes the nascent protrusion outward. In its normal state the network is viewed as being solid and acting as a stabilizing buttress during periods between protrusions. Thus the transient breakdown of the network not only provide the osmotic force but also the means of localizing or focusing the outward herniation that results.

An echo of de Bary's (1864) notion of an expansive force is provided by the proposal that the free energy liberated during the polymerization of actin filaments very close to the leading boundary of the psudopod can be directly or indirectly used for force generation (Dunn, 1980; Tilney and Inoue, 1982; Hill and Kirschner, 1982; Wang, 1985; Zhu and Skalak, 1988; Evans and Dembo, 1990). The important difference between this and the Mast concept or the osmotic model is that the pressure of the aqueous material has no direct role in causing protrusion. Instead, these models are characterized by the belief that filiments of the cytoskeleton are able to push on the membrane during the act of polymerization. In most cases these models leave open the underlying question of molecular mechanism for how this is accomplished. Such ambiguity is probably prudent; but to address the minimal feasability of a "polymerization force" hypothesis, the notion of a Brownian ratchet is sometimes invoked (reviewed by Peskin et al., 1993).

Continuum Representation of the Cytoplasm

To hasten the day when we can bring some final closure to the discussion of amoeboid motion, it is clearly desirable to introduce a language wherein theories can be formulated and analyzed with exactness and rigor. The reactive interpenetrating flow (RIF) model is the most successful formalism of this type that we have yet devised (Dembo and Harlow, 1986; Dembo et al., 1984). The basic structure of the RIF model follows that of classical continuum mixture theory (Trusdell and Toupin, 1960; Drew and Segel, 1971; Bowen, 1980). The idea is to view the cytoplasm as a finely divided composite material composed of two (or more) distinct but interacting material phases (incompressible aqueous solution and contractile actin/myosin network). As we shall see, there is no need to be extremely precise about the exact chemical definition of the two phases. The motion and spatial distribution and rheology of the cytoskeleton and aqueous solution are then given in terms of continuous fields. In the case of the aqueous solution, these fields correspond to the pressure, the velocity field, and to various "compositional descriptors" (the latter are discussed in the next section). In like manner, the cytoskeleton is characterized by a density, a velocity field, and further compositional descriptors.

In formulating the the RIF model, one of the most useful simplifying postulates is to neglect any elasticity of the network phase. There is ample justification and experimental evidence for this assumption, at least in actively motile amoeboid cells under most conditions (see subsequent discussion of cytoplasmic fluidity). In addition to the fluidity of the cytoskeleton the version of the RIF theory presented below represents a compromise that leaves out some rather unimportant details in the interests of simplicity. Thus we assume that the viscosity of the solvent is negligible compared to that of the polymeric fluid. Furthermore, we neglect the effects of inertia, of body forces (e.g., gravity), and of network anisotropy.

Under these conditions, the basic balance laws governing deformation of the cytoplasm are:

incompressibility

$$0 = \partial_{x_j}\left[\theta_s V_j^{(s)}\right] + \partial_{x_j}\left[\theta_n V_j^{(n)}\right] \ , \tag{2a}$$

network mass balance

$$\partial_t \theta_n = -\partial_{x_j}\left[\theta_n V_j^{(n)}\right] + \mathcal{J} \ , \tag{2b}$$

solvent linear momentum

$$0 = \theta_n \theta_s \mathcal{H}\left[V_i^{(n)} - V_i^{(s)}\right] - \theta_s \partial_{x_i} P \ , \tag{2c}$$

and network linear momentum

$$0 = \partial_{x_j}\left[\theta_n \mathcal{M} \mathbf{e}_{ij}^{(n)}\right] - (1/3)\partial_{x_i}\left[\theta_n \mathcal{M} \mathbf{e}_{kk}^{(n)}\right] + \theta_n \theta_s \mathcal{H}\left[V_i^{(s)} - V_i^{(n)}\right] + P\partial_{x_i}\theta_n + \partial_{x_i}\left[\theta_n(\mathcal{C} - P)\right] \ . \tag{2d}$$

In Eqs. (2a)-(2d), x_j are Cartesian coordinates, θ_n, and $V_j^{(n)}$ are the fields representing the volume fraction and velocity components of the network phase. $V_j^{(s)}$ and $\theta_s = 1. - \theta_n$ are the velocity and volume fraction of the solution phase; P is the so-called pore pressure, and

$$\mathbf{e}_{ij}^{(n)} \equiv \partial_{x_i} V_j^{(n)} + \partial_{x_j} V_i^{(n)} \tag{2e}$$

is the rate of strain tensor for the network. The pore pressure is actually the complete mechochemical potential of the solvent phase and therefore includes so called osmotic stresses.

Macroscopic field equations necessarily involve a level of abstraction wherein certain details of a fundamental molecular description are lost. This is neither good nor bad; it is simply unavoidable, and one must learn to deal accordingly. The nature of the abstractions inherent in the present theory are apparent if we consider the physical meaning of the scalar coefficients $\mathcal{J}$, $\mathcal{H}$, $\mathcal{M}$, and $\mathcal{C}$. These are, respectively, the mass supply of network due to polymerization/depolymerization, the specific hydraulic drag (inverse of the permeability) for flow of the solvent relative to the network, the specific viscosity of the network, and finally the specific contractility of the network. The physical processes that govern these distinct attributes of the cytoplasm occur at the molecular level, and yet there is a curious disconnection between the molecular details and the macroscopic consequences.

For example, if $\mathcal{C}$ is negative, it simply means that the balance of forces acting between elements of the network phase tends to favor swelling rather than contraction. On a molecular level the swelling stress could be due to electrostatic repulsion between fixed charges on the network polymers or because of osmotic effects that reduce the chemical potential of the solvent in regions of high network density. Additional data, beyond that inherent in the continuum level, would be needed to decide between these alternatives. Fortunately, it is possible to consider the macroscopic consequence of negative contractility in the absence of any molecular information whatsoever. In a similar fashion, an increase in the value of $\mathcal{C}$ could be due to a change in the contractile activity of myosin (see subsequent discussion of compositional descriptors) or due to a decrease in the repulsive interactions. In summary, "contractility" has some kind of molecular explanation, but it is also an abstract collective concept that can be studied independently from any immediate knowledge of molecular details.

Proceeding in a similar vein, $\mathcal{M}$ is the specific viscosity of the network. This coefficient can vary in space and time due to differences in the density and the orientational anisotropy of the cytoskeletal filaments and also due to regulatory factors that influence the average length of filaments and the number and strength of crosslinks between filaments (Taylor and Condeelis, 1979; Stossel et al., 1985; Pollard and Cooper, 1986; Hartwig and Yin, 1988; Sato et al., 1987; Janmey, 1991). Reactions that change $\mathcal{M}$ without any change in network mass are said to cause "gelation" (thickening, decreased compliance) or "solation" (thinning, increased compliance). Once again, these are abstract concepts that can be examined without the need to invoke any particular molecular model.

In addition to changes of consistency, the actin cytoskeleton also undergoes changes in total mass; this means that the various constituents of the cytoskeletal polymers exist as pools of monomeric or oligomeric precursors that undergo dynamic processes of assembly/disassembly. Specialized proteins (e.g., gelsolin, profilin) exist for the apparent purpose of regulating these reactions (Korn, 1982; Stossel et al., 1985; Pollard and Cooper, 1986; Zigmond, 1993). The important point is that in going to the macroscopic level, all these complex molecular processes have become lumped into a single scalar coefficient: $\mathcal{J}$.

The final important constitutive law required for closure of the RIF theory is that for $\mathcal{H}$. This gives the specific hydraulic drag between the solvent and network phases. The drag is primarily a function of the surface area per unit mass of network. It is also proportional to the microscopic or true viscosity of the solvent. Thus even if the mass per unit volume of network material is held constant, $\mathcal{H}$ can be decreased by bundling the network filaments into thick cables. Alternatively, $\mathcal{H}$ is increased by spreading the network filaments apart so that they form a diaphanous arrangement in which interaction with the solvent is maximized.

Compositional Descriptors

Equations (2a-2d) are complete only if density, velocity, and pressure are the only dynamical variables required to completely specify the state of the cytoplasm. Usually, however, this is not the case. The additional dynamical variables needed for the so-called "closure" of a model of the cytoplasm are called compositional descriptors. The RIF formalism allows considerable flexibility with regard to the choice of compositional descriptors, depending on what is considered physically or biologically important for a particular problem. In the case of the solution phase the compositional descriptors could be such things as the concentrations of various small ions or enzymes, monomeric or "disassembled" components of the network, or second messengers produced as a result of hormone-receptor interactions. In the case of the network, the compositional descriptors could include such things as the ratio of myosin to actin, the number of free barbed of pointed ends per unit actin density, the density of crosslinks, etc.

A special class of compositional descriptors, called "cybernetic factors" is comprised of small molecules that function to encode spatial information for purposes of control. For example, Taylor and Condeelis (1979) discuss the possibility of a cybernetic factor (i.e., Ca^{++}) that can simultaneously increase contractility and decrease viscosity in a concerted fashion (solation-contraction coupling). Whether or not such cybernetic factors really exist, we will use the basic concept to provide us with a simple example of the more general way that compositional descriptors are incorporated into the RIF model. Thus, if a cybernetic factor is important, then Eqs. (2a-2d) must be supplemented by a transport equation for this factor:

$$\partial_t q = -\partial_{x_j} \left[q V_j^{(s)} \right] + \partial_{x_j} \left[\delta_q \partial_{x_j} q \right] + \mathcal{S}. \tag{2f}$$

In this equation, q is the concentration of factor, δ_q is the diffusion constant of q and $\mathcal{S}$ is a new constitutive law giving the rate of chemical synthesis or loss of q.

Some Illustrative Constitutive Laws

As mentioned above, the constitutive laws for $\mathcal{J}$, $\mathcal{H}$, $\mathcal{M}$, $\mathcal{C}$, (and also $\mathcal{S}$) represent the collective result of a large number of complex molecular processes. The best or correct choice of these laws cannot be derived *ab initio*, so we are forced to make empirical approximations. The latter can be informed by what is known biochemically and also by physical intuition about what is needed to produce the particular behavior observed in various kinds of amoeboid motion. As an illustrative example, we now consider the constitutive laws that have been used (Dembo, 1989), to model the cytoplasm of *Amoeba proteus* during the simple process of steady fountain flow:

$$\mathcal{J} = \left[\theta_{eq} - \theta_n \right] / \tau_n \ , \tag{3a}$$

$$\mathcal{H} = \mu_s / \kappa \ , \tag{3b}$$

$$\mathcal{M} = \mu_n \exp\left[(\theta_n / \theta_{gel})^2 \right] \ , \tag{3c}$$

$$\mathcal{C} = \psi[q - \xi] \ , \tag{3d}$$

and

$$\mathcal{S} = -q / \tau_q \ . \tag{3e}$$

The constitutive law for $\mathcal{J}$ implies that the chemical synthesis and breakdown of the network is governed by simple linear mass action. The constant parameters θ_{eq} and τ_n correspond to the volume fraction of network at chemical equilibrium and the relaxation time of the reaction, respectively. Implicit in this constitutive law for $\mathcal{J}$ is the notion that homeostatic processes maintain the monomeric or oligomeric precursors of the network in constant supply. Furthermore, the chemical reaction is homogeneous throughout the interior of the cell; i.e., the model does not invoke any special processes or signals that cause rapid network synthesis at the front of the cell or rapid depolymerization at the tail.

The constitutive law for $\mathcal{H}$ means that the hydraulic drag between the network and solution phases follows Darcy's law for flow through porous media (μ_s is the solvent viscosity and κ is the permeability coefficient). A discussion of hydraulic drag between aqueous fluids and filament networks can be found in Happel and Brenner (1973). These authors show that κ is in general a measure of the cross sectional area of the typical filament of the network and that κ is only very weakly dependent on network geometry and density.

The constitutive law for $\mathcal{M}$ is meant to empirically describe the effects of the increased opportunities for formation of crosslinks between network filaments as a result of changes in filament density. If θ_n is below θ_{gel} (the "gel point"), then the specific viscosity approaches a constant, μ_n. As θ_n gets much larger than the gel point, however, the specific viscosity increases explosively. In this model the processes of gelation and solation of the network are not subject to control by any special factors or signals, and like the chemical reaction they are homogeneous in space.

The constitutive law for $\mathcal{C}$ embodies the assumption that the contractile tendency of the network filaments depends on the concentration of a cybernetic factor. In other models we might choose to let $\mathcal{M}$ or $\mathcal{J}$ be controlled by such a compositional descriptor. The present law for $\mathcal{C}$ is expressed in terms of a constant of proportionality, ψ, as well as a parameter, ξ (the contraction threshold). The latter gives the amount of cybernetic factor at which contraction exactly balances the swelling of the network. It is usually most convenient to express the concentration of cybernetic factor in nondimensional units. If this is the case, then ξ is also nondimensional; and the parameter ψ, which has units $dynes/cm^2$, is called the characteristic contractile stress.

The final constitutive law used in the *proteus* model, that for $\mathcal{S}$, prescribes a simple linear decay of cybernetic factor due to sequestration or enzymatic degradation within the bulk of the cytoplasm. The parameter τ_q is the characteristic lifetime of the cybernetic factor. Since no interior sources of q are included in $\mathcal{S}$, use of this simple constitutive law implies that there must be a supply of q through the boundaries if one is to obtain nontrivial solutions (see subsequent discussion of boundary conditions).

The physical idea behind the choice of constitutive law for $\mathcal{S}$ is to force the development of a sharp concentration gradient of the cybernetic factor; q will be high in the cortex of the cell and low in the interior. Since q controls contractility of the network, its cotico-medullary gradient will cause network to swell in the cell interior (giving rise to the endolpasm) and contract in the cortex (giving rise to the ectoplasm).

Evidence of the Fluidity of the Actin Cytoskeleton

An important assumption of the field equations we have presented is the notion that the actin cytoskeleton of amoeboid cells behaves as a viscous fluid. This proposition is an extreme limit, but one may nevertheless cite any number of commonplace and well known observations in support of its validity. For example, consider the formation of hybridoma cells in which two separate cytoskeletons or pieces of cytoskeletons gradually intermix and thereafter act as one. Fluids frequently possess this ability to "anneal" when brought into contact; solids are not known to anneal. Interestingly, recent work has shown that fusion of the membranes and cytoplasm will proceed in the presence of agents that disassemble microtubules although the fusion of the cell nuclei and the reorganization of intermediate filaments is inhibited

(Zheng and Chang, 1991). This beautiful result indicates that the distinct roles played by the two filament systems during cell division is reiterated during cell fusion.

The fluidity of the cytoskeleton is also seen dramatically in studies of cell wounding and healing (McNeil 1991; McNiven et al., 1984; Gelfand et al., 1985; Albrecht-Buehler, 1980; Jeon and Jeon, 1975). For example, it has long been known that an amoeba that is completely impaled on a microneedle is able to continue to crawl and to eventually "flow" around and past the needle without damage (McClendon, 1909). The ability of cells to divide and undergo other kinds of topological transitions (excytosis, phagocytosis) is also indicative of fluid rheology.

One should also consider the implications of the various forms of continuous motility. For example, cytoplasm of suspended specimens of *proteus* or *physarium* (Nowakowska, 1978; Kamiya, 1986) is capable of undergoing an indefinite "twisting" motion in which neighboring layers of cytoplasm continuously rotate past each other throughout the organism. Obviously, one has no hope of explaining such kinematics using a model in which the cytoskeleton has solid structure. As discussed previously, similar conclusions can also be derived from a consideration of the kinematics of fountain streaming (see Figure 1). From such general arguments, we find it fruitless to maintain that the actin cytoskeleton of motile amoeboid cells is able to retain a significant "memory" of a putative "unstressed" configuration. At least for such cells, one can thus eliminate any macroscopic rheological model with a true parallel elastic component (e.g., standard viscoelastic solid). There remain many possibilities for complex fluid rheology (e.g., Maxwell fluid), but in our view the model of two interpenetrating viscous fluids is equally tenable and simpler.

The concept of a composite fluid creates confusion partly because it is not widely understood that fluids are perfectly able to transmit tensile stresses such as those created by actin-myosin interactions, and partly because the network is capable of processes of gelation that can cause its effective viscosity, $\mathcal{M}$, to become extremely high (more than 10^5 poise). Similar enormous viscosities are encountered in studies of

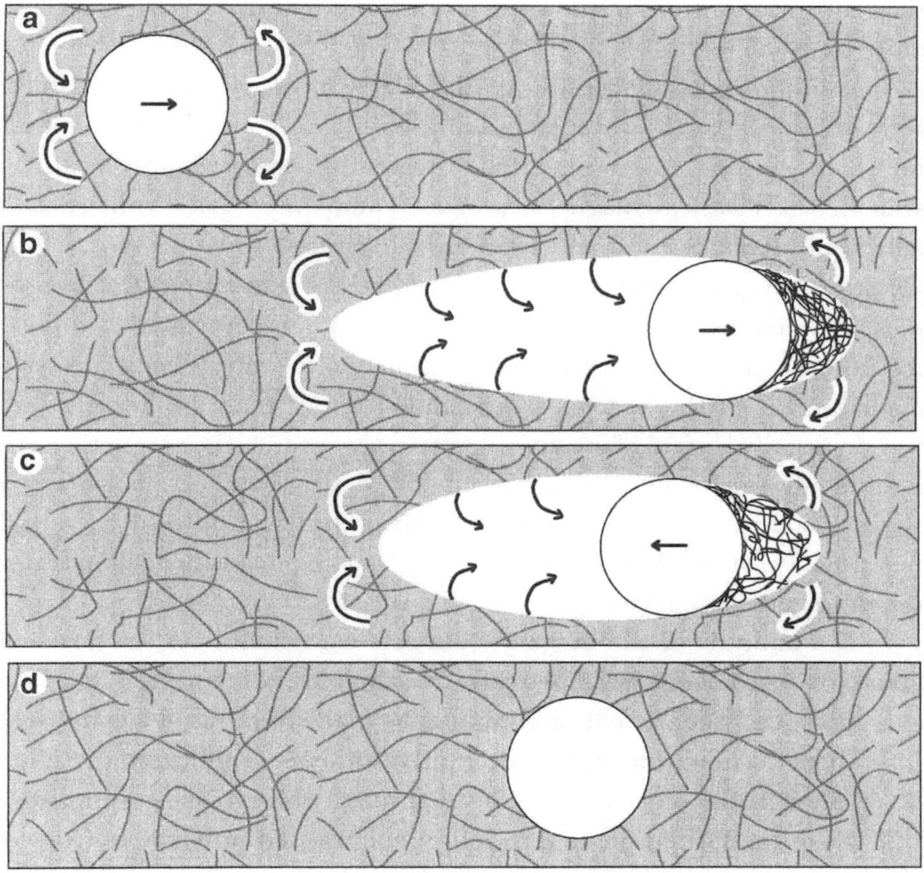

Figure 3. Interpretation of a Compliance Test.

A) Initial state of the magnetic bead just after turning on the field. The network filaments are uniformly distributed around the bead.

B) After the bead has been displaced for some distance a steady state is reached wherein the bead moves with constant velocity; the filaments are compressed in front and rarefied behind.

C) When the magnetic field is turned off, the swelling stresses between the network filaments, in conjunction with the gradient of network density, cause a recoil of the particle toward its initial location.

D) Finally, when the energy stored in the form of gradients of network density has been dissipated, the motion of the bead ceases.

the geological motions of the earth's crust and mantle but are otherwise quite unusual.

Another cause of confusion that leads investigators to attribute solid structure to the cytoplasm is its ability to store energy, at least for short periods. Experimental evidence for energy storage can be seen in the classical magnetic particle compliance test (Valberg and Albertini, 1985; Sato et al., 1983, 1984; Hiramoto, 1969; Yagi, 1961; Crick and Hughes, 1950). In this sort of study a small bead of iron or nickel is introduced into the cytoplasm and a constant force is applied to the bead by turning on an external magnetic field. The progressive displacement of the bead is accompanied by a period of deceleration, but eventually a state of steady velocity is reached. If the magnetic force is suddenly turned off during this steady phase, then there is usually a slight recoil of the bead so that it returns part way toward its initial position. It is this phenomenon of recoil that conclusively demonstrates the storage of some energy.

The analysis of the magnetic particle compliance test in terms of standard viscoelasticity is perfectly reasonable (Janmey, 1991). However, an equally consistent analysis can be made in terms of a composite fluid theory. An outline of this alternative analysis is indicated in Figure 3. As shown, the motion of the bead through the cytoskeletal network will cause the formation of a region of compressed polymer in front of the bead and a region of rarefaction behind. If $\mathcal{C} < 0$, then this compression and rarefaction are opposed by the swelling stresses that act between the network filaments leading to a flow of the filaments around the obstruction and the eventual development of a steady state. Finally, when the field is turned off, the net swelling force due to the gradient of network density acting over the particle surface will cause a recoil just as observed in the experiments.

Finally, there is persuasive data for the fluidity of actin networks from direct micropipet measurements on whole cells (Evans and Yeung, 1989; Dong et al., 1988; Evans and Kukan, 1984). A detailed analysis of these experiments in terms of the RIF model is presented in the final section of this paper.

Boundary Conditions for Confined Motion

In most cases of amoeboid motion (e.g., cytokinesis, phagocytosis, pseudopodal extension) the cytoplasm is contained in a spatial domain that undergoes large variations of shape. These shape changes are determined not only by the bulk cytoplasm but also by the internal rheology of the surface layer of the cell and by interactions between the surface layer and external objects (i.e., adhesion). To temporarily avoid the added problems evoked by such interactions, it is expedient to focus on systems that involve only bulk cytoplasm contained in fixed spatial domains of relatively simple geometry (fortunately such systems exist).

Although one can restrict the cytoplasm to a fixed domain, there are still many possible choices of the boundary conditions at the edges of this domain (Dembo and Harlow, 1986). To illustrate the role of boundary conditions in defining a model of the cytoplasm, let us once again consider the case of *Amoeba proteus* (Dembo, 1989). In this study, the shape of the organism was approximated by a right circular cylinder, Ω, of length ℓ_Ω and radius r_Ω. The boundary of the domain was divided into the three natural segments: i.e., anterior, posterior, and lateral. We let L_j and N_j denote components of a unit tangent and outward-normal vector on a boundary segment.

From direct observations it is clear that the plasma membrane of *proteus* is able to slip tangentially with respect to the cytoskeleton. As a result, the *proteus* model assumed "slip" boundary conditions for tangential velocity components:

$$\mathbf{e}_{ij}^{(n)} N_i L_j = 0 \ , \qquad (4a)$$

on all three boundary segments.

There is also very good evidence that the plasma membrane of *Amoeba proteus* is virtually impermeable to flux of the solution phase (Lovtrup and Pigon, 1951). In light of this, the normal condition on solution velocity was taken to be:

$$N_j V_j^{(s)} = 0 \ , \qquad (4b)$$

on all three boundary segments.

The normal boundary condition on the network velocity was taken to be "stick" on the posterior and lateral surfaces

$$N_j V_j^{(n)} = 0 \ . \tag{4c}$$

Such a "stick" condition implies that there are adhesive sites on the inner surface of the plasma membrane that are able to bind to the cytoskeleton and anchor it. In contrast, we found it necessary to invoke a "no-stick" boundary condition on the anterior boundary segment of the amoeba. This boundary condition simply means that the membrane acts as a sieve or screen that prevents outward movement of the network but that offers no resistance to inward movement. Such a boundary condition is quite important since disruption of the network from the membrane (see Figure 2) seems to be the basis for the hyaline cap which in turn underlies the entire differentiation of the amoeba along the anterior/posterior axis. The mathematical statement of such a semi-stress condition takes the form of two simultaneous inequalities:

$$\begin{aligned} &0 \geq N_j V_j^{(n)} \quad \text{and,} \\ &0 \geq \mathcal{M}\mathbf{e}_{ij}^{(n)} N_j N_i - \frac{1}{3}\mathcal{M}\mathbf{e}_{kk}^{(n)} + \mathcal{C} \ . \end{aligned} \tag{4d}$$

In the case of the cybernetic factor, very simple Dirichlet boundary conditions,

$$q = const. = 1 \ , \tag{4e}$$

were used on all boundary segments. The choice $const. = 1$ in this Dirichlet condition involves no loss of generality since the boundary value would in any case be used as the scale for nondimensionalization of q. Notice that because the cybernetic factor was assumed to spontaneously decay upon entering the bulk cytoplasm [(see Eq.(3e)], Eq. (4e) implies that at steady state there will be a cortico-medullary gradient of q.

Because we regarded the boundaries of the domain Ω as rigid and impenetrable to volume fluxes, the pressure field within Ω is only determined up to an arbitrary additive constant [i.e., only gradients of pressure appear in Eqs. (2a)-(2e)]. To eliminate such ambiguity, we took the additive constant such that the lowest value of P at any point in the computational domain was equal to zero. Thus P, by definition, is a relative pressure. The absolute pressure is:

$$P_{abs} = P + P_{ext} + P_0 \quad , \tag{4d}$$

where P_{ext} is the pressure in the external medium and P_0 is the pressure jump between the point of minimum pressure inside the cell and the exterior. P_0 is a measurable quantity corresponding to what is usually called the "turgor" pressure.

In Vitro Studies of Unreactive Contractile Networks

The first serious application of the RIF theory was to the so-called superprecipitation reaction (Dembo et al., 1986). In the generic form of this reaction, a curvette or other reaction vessel is uniformly filled with an artificial mixture or cell extract that supplies the basic ingredients of the contractile network. In the first phase of the reaction, the network is allowed to spontaneously assemble from the supplied ingredients. Once the network is formed, there is a lag of variable duration; then it is seen to undergo a simple irreversible contraction forming a clump (or clumps) in the middle of the reaction vessel. To obtain contraction the starting mixture must contain actin, myosin, and ATP; but of course many other factors (e.g., crosslinking proteins) can also be present.

The actual contractile event in the superprecipitation reaction takes two qualitative forms. Form #1 involves a simple self-similar shrinkage of the network. In the ideal limit, this means that the shape of the contracting clump of network material preserves the proportions, but not the size, of the original reaction vessel. In order to do this, the flat edges of the contracting material must move as very sharp fronts and must remain flat; angles between edges, as at the corners of the original reaction vessel, must be at least approximately preserved. Form #2 involves a rupturing or tearing of the network during contraction. In the mild case a transient fissure or channel is opened in the contracting clump so as to allow solution material to escape more easily; this channel may close again as the contraction proceeds. In the extreme case the network literally tears itself into innumerable small pieces.

Dembo et al. (1986) has studied the simplest possible RIF model ($\mathcal{J} = 0$, $\mathcal{C} = const.$, $\mathcal{M} = const.$, $\mathcal{H} = const.$ and no cybernetic factor). This regime is sufficient to provide an account of both limiting forms of the superprecipitation reaction and of the transition between them. According to this simple model, the bifurcation between the two modes of the reaction is governed by a nondimensional quantity that describes the relative importance of hydraulic drag against the solution versus the internal viscosity of the network (the rending number). Although we easily obtain agreement with the qualitative observations of superprecipitation, it is unfortunate that really detailed and carefully controlled studies to this reaction have never been attempted. Thus much work of a quantitative nature remains to be done.

Motility of Isolated Cytoplasm and the Basic "Working Cycles" of the Cytoskeleton

Another elementary application of the RIF theory is to the phenomena observed in isolated cytoplasm of giant *amoeba* and *physarium*. These phenomena include the chaotic or turbulent streaming patterns observed in large volumes of pooled cytoplasmic extracts (Thompson and Wolpert, 1963; Pollard and Ito, 1970) and also more coherent motions such as flair-streaming (Taylor et al., 1973; Kuroda, 1979).

The most remarkable feature of these experimental systems is the ability of the isolated cytoplasm to maintain a state of continuous motility. This may be contrasted with the simple irreversible and transient contraction of the superprecipitation reaction. As realized since Schulze

(1875) such continuous motility implies the existence of dynamic processes that effectively create a closed working cycle. This cycle must allow for the continuous injection of energy and the regeneration of all necessary material components of the cytoplasm in their appropriate locations. Moreover, since we are dealing here with isolated cytoplasm, working cycles that involve external sources or the participation of membranes or surfaces can be discarded. In terms of the RIF theory this means that we are left with only two possible forms for the basic "working cycle" of the cytoskeleton.

The contraction-depolymerization cycle: (a) A spatial gradient of total contractility ($\theta_n \mathcal{C}$), causes cytoskeletal material to flow into a localized region(s) (the contraction zone). This naturally removes material from some surrounding region(s) (the dilation zone); (b) The chemical reaction of network polymerization/depolymerization ($\mathcal{J}$) counteracts the accumulation of mass in the contraction zone by causing cytoskeleton to undergo local disassembly. (c) The resulting depolymerized material diffuses and convects away from the contraction zone and finds its way into the dilation zone where it is assembled into new cytoskeletal network. This replaces the material removed in step (a) and completes the cycle.

The polymerization-swelling cycle: (a) A spatial gradient of chemical reaction ($\mathcal{J}$) causes depolymerized material to assemble into new cytoskeleton in some localized region(s) (the polymerization zone). Simultaneously there is disassembly and generation of depolymerized material in some other region(s) (the depolymerization zone); (b) Swelling stresses ($\theta_n \mathcal{C}$) must be less than zero to counteract the effect of $\mathcal{J}$ by causing the cytoskeletal material to flow away from the polymerization zone and toward the depolymerization zone; (c) To complete the cycle of all components, diffusion and convection must create a flux of depolymerized material in the opposite sense from that of the polymerized material (i.e., out of the depolymerization zone and into the assembly zone).

A general aspect of the behavior of isolated cytoplasm in all the experimental systems reported to date is the spontaneous organization of the network phase into easily delineated regions of high and low density. Furthermore in all cases the motion of the polymerized cytoskeleton is directed toward regions of high density and away from regions of low density. This means that the polymerization-swelling cycle must be eliminated as an explanation of the observed *in vitro* motility. At least within the context of the RIF theory this leaves the contraction-depolymerization cycle as the only viable alternative.

Note that the polymerization-swelling cycle presupposes the existence and maintenance of a spatial gradient of $\mathcal{J}$. One can imagine any number of mechanisms by which this could be accomplished under *in vivo* conditions (e.g., by controlling activity of an enzyme with a cybernetic factor). However, all such mechanisms depend on some preexisting spatial structure. Thus it is much more difficult to establish and maintain the requisite spatial gradient in isolated cytoplasm. We conclude that the inability (up till now) to experimentally demonstrate an artificial system that works by the polymerization-swelling cycle is not surprising. Hopefully work on devising such a system will one day be successful since such cycles are important in many theories and there are indications that they can and do occur in intact cells (e.g., in the leading lamella of fibroblasts).

The contraction-depolymerization cycle presupposes the existence and maintenance of gradient of total contractility ($\theta_n \mathcal{C}$). Obviously such a gradient could be maintained by extrinsic control, but it turns out that even in the absence of such control (i.e., even if $\mathcal{C} = const.$), the equilibrium solution of zero motility becomes unstable if $\mathcal{C}$ is positive and sufficiently large (see Dembo et al., 1984, for proof). This means that states of continuing motility based on the contraction-depolymerization cycle can be generated in a very minimal system by spontaneous bifurcation from the uniform equilibrium solution without any extrinsic direction or preexisting information. This inherent robustness is probably the underlying reason for the prevalence of this sort of cycle in the available experiments.

Dembo (1986) numerically examined the predictions of a simple RIF model of isolated cytoplasm. In particular this study focused on the spontaneous bifurcations predicted for the contraction/depolymerization cycle and the qualitative form of the spatial patterns of flow and density obtained under different conditions. The constitutive laws used to model the isolated cytoplasm in this study were:

$$\mathcal{J} = [\theta_{eq} - \theta_n]/\tau_n \ , \tag{5a}$$

$$\mathcal{H} = const. \ , \tag{5b}$$

$$\mathcal{M} = const. \tag{5c}$$

and

$$\mathcal{C} = \psi(1. - \sigma\theta_n) > 0 \ , \tag{5d}$$

Note that the constitutive laws for contractility and viscosity differ from the laws used in the model of fountain streaming [see Eq. (3a-e)]. Also note that no cybernetic factor is needed for closure of the isolated cytoplasm model. According to Eq. (5d), a balance point between contraction and swelling occurs when $\theta_n = \sigma^{-1}$. This feature was incorporated to guarantee the occurrence of a stable high density phase of the network. At the time of these calculations such a phase seemed like a reasonable hypothesis for the nature of ectoplasm, but subsequent studies have not supported this (cf. discussion of the proteus model).

Our calculations confirmed the existence of the spontaneous states of continuous motility. The details of these emergent modes depend in a complicated way on the relative rates of contraction versus chemical reaction, on the relative amounts of endoplasm and ectoplasm, and also on the amount of hydraulic drag. Our computations demonstrate that, for different choices of these nondimensional parameters, the emergent streaming modes can be simple steady states, chaotic states, or (for intermediate parameter values) coherent states that resemble flair streaming (Taylor et al., 1973).

In a related study of isolated cytoplasm, Alt (1987) has applied the RIF theory to the experiment of Ezzel et al. (1983). In the Ezzell experiment a rather crude extract of the cytoplasm of frog eggs is prepared and supplemented with ATP and ATPγS. After warming to room temperature, the initial stages of gelation and contraction proceed in a manner similar to the superprecipitation process. After the initial contractile event, however, subsequent waves of network formation and contraction appear and continue every few minutes for several hours.

According to the analysis of Alt, the trick for obtaining periodic waves from the RIF model is to choose the constitutive law for contraction in such a way that $\mathcal{C}$ is negative – when the network density is small but becomes positive at higher network densities. If this is so, then as soon as the process of network synthesis causes the average network density to rise above a critical level, a contractile wave starts at the boundaries of the reaction vessel and moves inward. The passage of the wave then causes the average network density to drop below the threshold. Thus a refractile period occurs, and more chemical synthesis is needed before the next wave can start. The analysis is replete with many detailed and easily tested predictions, and it will be interesting to see if these are born out by further experiments.

The Steady Fountain Flow in *Amoeba proteus*

In attempting to extend the model of isolated cytoplasm to the intact amoeba (Dembo, 1989), we found that our "phase separation" model for the physics of the endoplasm/ ectoplasm dichotomy [i.e., Eq.(5d)] was unable to account for the high rate of contraction of the ectoplasmic tube, nor was it able to account for the very high viscosity of the ectoplasm relative to the endoplasm. We were also unable to devise a system of constitutive laws and boundary conditions that would lead to self assembly and hydrodynamical stability of the endoplasmic channel. We finally solved these problems and devised a family of workable models by postulating the gelation mechanism, Eq. (3c), as opposed to phase separation, Eq. (5), to explain the endoplasm/ectoplasm dichotomy. The hydrodynamic stability of the ectoplasmic tube was insured by introducing a cybernetic factor to control the network contractility and by arranging the dynamics of the cybernetic factor in such a way as to obtain a cortico-medullary gradient of contractility [see discussion of Eqs. (3e) and (4e)]. The organization of the anterior/posterior axis of the amoeba was accomplished by and on control of the adhesive interaction between the cytoskeleton and the inner surface of the plasma membrane. In other words, this difference depended on putting a special boundary condition at the leading edge of the cell [see discussion of Eq. (4d)]. Finally, the success of the *proteus* model, depends on a very vigorous "working cycle" of network contraction and depolymerization.

With appropriate parameter values, the *proteus* model quantitatively accounts for the self assembly of the cytoskeleton of monopodial form of the organism; i.e., for the formation and detailed morphology of the endoplasmic channel, the ectoplasmic tube, the uropod, the plasma gel sheet, and the hyaline cap. The model accounts for the kinematics of the cytoskeleton: the detailed velocity field of the forward flow of the endoplasm, the contraction of the ectoplasmic tube, and the inversion of the flow in the fountain zone. The model also gives a satisfactory account of measurements of pressure gradients, measurements of heat dissipation, and measurements of the output of useful work by amoeba. By balancing normal forces on the plasma membrane as closely as possible, the model also predicts the turgor pressure and surface tension of *proteus*.

If one counts the length and radius of the amoeba, then the *proteus* model in its minimal form involves 12 parameters. Some of these parameters are known independently; the rest can be determined by fitting the model to some of the available data on *proteus* as described elsewhere (Dembo, 1989). In any event, the best available values for a typical specimen are summarized in Table I. It will be seen from the table that the characteristic time for complete turnover of the cytoskeleton in *proteus* is only 12 seconds! Initially this seemed like a very rapid reaction, but it has since been reported that the turnover time

TABLE I

Parameters of the *proteus* Model

	Parameter	Symbol	Value	cgs units
1.	Length of Ω	ℓ_Ω	5.0×10^{-2}	cm
2.	Radius of Ω	r_Ω	5.0×10^{-3}	cm
3.	Solution viscosity	μ_s	3.0×10^{-2}	poise
4.	Volume fraction of network at chemical equlibrium	θ_{eq}	7.2×10^{-3}	–
5.	Characteristic lifetime of network elements	τ_n	1.2×10^{1}	sec
6.	Viscosity of network (at $f = 0$)	μ_n	1.2×10^{3}	poise
7.	Volume fraction of network at gel point	θ_{gel}	5.4×10^{-3}	–
8.	Characteristic contractile stress of network	ψ	1.8×10^{4}	dynes/cm^2
9.	Network permeability	κ	2.5×10^{-11}	cm^2
10.	Diffusion constant of cybernetic factor	δ_q	3.7×10^{-6}	cm^2/sec
11.	Characteristic lifetime of cybernetic factor	τ_q	8.4×10^{-1}	sec
12.	Contraction threshold	ξ	1.4×10^{-1}	–

in leukocytes is just as fast (Zigmond, 1993). Thus the rapid polymerization/depolymerization of the cytoskeleton may well turn out to be a general phenomenon.

Another interesting physical quantity in Table I is the characteristic contractile stress of the network phase (i.e., ψ). To convert this quantity into more familiar form, it is useful to multiply it by the cross sectional area of an actin filament. We thereby obtain the characteristic force per actin filament during steady fountain flow of 4×10^{-9} *dynes*. Similar quantities are available in only two other cases. The force per filament during isometric contraction of skeletal muscle is determined by Oosawa et al. (1973) to be about 1×10^{-5} *dynes*. Interestingly, the latter is also very close to the force required to snap an actin filament in two. The force per actin filament of the constriction ring of the sea urchin egg at mid mitosis is determined by Schroder (1975) to be about 5×10^{-7} *dynes*. Ishijima et al. (1991) report the average force generated by a single myosin to be in the range $3.0 - 0.4 \times 10^{-7}$ *dynes*.

Continuum Formulation of A Free Boundary Problem

In order to make definitive contact between continuum theories and experimental observations of important biological motions such as cell division, phagocytosis, and crawling locomotion, it is necessary to move beyond problems involving simple confined flow and address the issue of free boundary dynamics in its full complexity. Although we cannot claim to have completely solved all the relevant problems, we can at least give some indications of the current state of the art. In what remains of this discussion we will therefore address ourselves to the analysis of the dynamics of the cytoskeleton of a polymorphonuclear leukocyte (neutrophil) as it enters a pipet under the influence of a pressure difference.

The diameter of a typical human neutrophil in its inactivated or resting state is about twice that of a typical capillary. As a result, neutrophils repeatedly undergo cycles of large deformation and recovery as they are forced through the capillary vessels of the circulation. Experiments to characterize the ability of these cells to withstand such repeated mechanical insult have focused on the aspiration of cells into micropipets and on the recovery of cells following aspiration.

The main conclusion of these studies is that cells are very similar to droplets of incompressible Newtonian fluid under surface tension (Evans and Kukan, 1984; Evans and Yeung, 1989; Yeung and Evans, 1989; Needham and Hochmuth 1990; Tran-Son-Tay et al., 1991; Needham and Hochmuth, 1992; Hochmuth et al., 1993). Other groups have taken the view that the Newtonian model is deceptively simple and that it does not seem to explain certain aspects of the aspiration process. In particular, the very early phase of the aspiration wherein the leading edge of the cell seems to instantaneously "jump" into the pipet for distances of 1-2 capillary diameters within the space of a single video frame has been puzzling. Also the effective viscosity seems to change somewhat depending on the extent and rate of aspiration. To account for such phenomena it has been suggested that the cell interior is a Maxwell fluid inside a pre-stressed elastic shell (Dong et al., 1988) or a laminate structure consisting of three concentric shells: two inner shells of Maxwell fluid together with the outer elastic cortex (Dong et al., 1991). The RIF theory offers a distinct alternative that seems to have some advantages over these classical rheological models.

As previously, we will approximate the cytoplasm by a composite consisting of an incompressible aqueous fluid with negligible viscosity interpenetrated by a polymeric fluid of very high viscosity. We let this material occupy a region Ω of volume $v_\Omega \equiv 4\pi r_c^3/3$. This region is embedded in a Euclidian space with Cartesian coordinates x_i. Thus the cell surface $\partial\Omega$ at time t is given parametrically by functions

$$x_i(t) = f_i(s_1, s_2, t) \ . \tag{6}$$

We usually think of the s_α as intrinsic Lagrangian coordinates. For example, the coordinates s_α could be the longitude and latitude of a point on the cell surface, established at the initial spherical state and thereafter moving with the cell surface as it deforms. We require that the surface of the cell be smooth at all points except possibly at the nozzle. This means that first derivatives of the f_i with respect to s_α exist and are continuous. Also note that we follow the convention of using Greek subscripts and superscripts for surface fields and Roman indices for three-dimensional fields.

In order to simplify our discussion without sacrificing essential principals we will regard the pipet as a cylindrical tube bored into the half space $z > 0$. We take the interior caliber of the pipet to be a constant r_p. In the initial state $t = 0$, the cell is round and contacts the pipet only along the ring formed by the sharp corner of the nozzle. At all times $t > 0$ we assume the existence of a discontinuous or "stepfunction" drop in ambient pressure between the outside and the inside of the pipet. Physically such a pressure jump is stable only if the contact between the cell and the pipet at the nozzle constitutes a prefect gasket.

Within the time dependent domain Ω, we assert that the cytoplasm is governed by Eqs. (2a)-(2d). At high pressures, it is reasonable to neglect the chemical assembly and dissembly of the cytoskeleton during the brief interval required for complete aspiration. We therefore assume that $\mathcal{J} = \prime$. All the other constitutive laws of the RIF model will be of the simplest possible type: $\mathcal{M} = const.$, $\mathcal{H} = const.$, and $\mathcal{C} = const.$.

A complexity inherent in applying the RIF model in the setting of a free boundary problem arises because the surface area of the leukocyte is approximately twice the amount needed to enclose a sphere of equal volume (Evans and Yeung, 1989; Simon and Schmidt-Schoenbein, 1988). It is therefore necessary to introduce an additional dynamical variable in order to follow the wrinkling and unwrinkling of the plasma membrane. This is accomplished by averaging over those irregularities of the plasma membrane falling below some cutoff in size. Such averaging leads to a special surface field, $\rho(s_1, s_2, t)$: defined as the cm^2 of actual membrane bilayer per cm^2 of "projected" surface area. Values of ρ much greater than one describe a highly wrinkled or ruffled membrane; ρ exactly equal to one implies a completely smooth membrane.

Since lipid bilayers must maintain approximately constant ratio of surface area to mass, values of ρ less than one can occur only following failure or rupture of the plasma membrane.

As regards the transport of the surface layer of the cell we simplify matters by assuming that the normal component of membrane velocity W and the two tangential components U^α are simply corresponding projections of the $V_i^{(n)}$. The defining relationship is:

$$V_i^{(n)} \equiv U^\alpha f_{i,\alpha} + W N_i \ , \tag{7a}$$

where the N_i are the Cartesian components of the outward unit normal. Note that the velocity components U^α are contravariant with respect to the intrinsic coordinates; whereas the vectors $f_{i,\alpha}$ are covariant. The $f_{i,\alpha}$ are also vectors with respect to the embedding space; they are parallel to the intrinsic coordinate axes and define the tangent plane, but they are not necessarily unit vectors. At the nozzle, Eq. (7a) leads to some ambiguity since the $f_{i,\alpha}$ may be discontinuous. In this event we define the tangent direction by averaging the limits obtained by approaching the nozzle from the interior and exterior of the pipet.

With these understandings the time evolution of ρ is governed by a continuity equation:

$$D_t\rho = -0.5\rho(\mathbf{e}_{kk}^{(s)} - \mathbf{e}_{ij}^{(s)}\eta_i\eta_j) = -\rho(D_\alpha U^\alpha + 2WC) \ , \tag{7b}$$

where D_t is the substantial derivative for an observer moving with the membrane, D_α is the covariant derivative with respect to one of the intrinsic coordinates, and C is the mean curvature. It is important to note that Eq. (7b) neglects the possibility of production or removal of membrane (e.g., by endocytosis or exocytosis of vesicles). It also implies negligible tangential "slip" between the membrane and the cytoplasm and negligible permeation of membrane by aqueous material.

The surface structures of the cell are assumed to exert only an isotropic tension. This leads to a stress tensor of the form

$$S^\alpha_\beta = \gamma \delta^\alpha_\beta \ . \tag{8a}$$

where δ^α_β is the Kronecker delta and γ is a scalar called the surface tension. In general, γ is not a constant but is related to ρ – and other descriptors of deformation history – by means of a constitutive function. We will take this law to have the functional form:

$$\gamma(\rho) \approx \gamma_0 + \omega(\rho_0 - 1)/(\rho - 1) \ , \tag{8b}$$

where $\gamma_0 + \omega$ is the surface tension of the resting rounded cell, ρ_0 is the membrane area density of the resting rounded cell, and $\omega\rho_0/(\rho_0 - 1)$, is the small strain area modulus. This equation is a slight modification of the constitutive law proposed by Needham and Hochmuth (1992). We regard our formulation as slightly preferable because in order to enforce the constraint on total membrane area the constitutive law needs a singularity as $\rho \rightarrow 1$.

Since the pipet exerts no tangential tractions on any segment of the cell periphery, it is possible to require the continuity of the tangential components of stress on all boundary segments:

$$\theta_n \mathcal{M} \mathbf{e}^{(n)}_{ij} N_j f_{i,\beta} = D_\alpha S^\alpha_\beta = \partial_\beta \gamma \ . \tag{9a}$$

Note that in the case of $\omega = 0$ the surface stress is constant and the right of Eq. (9a) vanishes.

To characterize the normal boundary conditions on the aspirating cell we need to classify each infinitesimal interval of the cell periphery with respect to whether it is "free" or "constrained" by contact with the pipet. If a segment is part of the free boundary inside the pipet, then it is said to belong to the set, $\partial\Omega_{fi}$. In like manner the union of constrained segment(s) in contact with the interior of the pipet constitutes the set $\partial\Omega_{ci}$; constrained segments in contact with the outside of the pipet constitutes $\partial\Omega_{co}$; and the free segments outside the pipet comprise the set $\partial\Omega_{fo}$.

The boundary condition on $\partial\Omega_{fi}$ and $\partial\Omega_{fo}$ is the usual "free boundary" or Neumann condition on the normal stress:

$$\theta_n \mathcal{M}[\,\mathbf{e}_{ij}^{(n)} N_j N_i - (1/3)\mathbf{e}_{kk}^{(n)}\,] - (P - P_{ex}) + \theta_n \mathcal{C} + 2\gamma C = 0 \ . \tag{9b}$$

On $\partial\Omega_{ci}$ and $\partial\Omega_{co}$ we enforce a semi-Derichlet constraint on the normal component of velocity. Just as in the case of Eq. (4d), this can be expressed in terms of two simultaneous inequalities:

$$\begin{aligned} &0 \geq V_j N_j \quad \text{and,} \\ &0 \geq \theta_n \mathcal{M}[\,\mathbf{e}_{ij}^{(n)} N_j N_i - (1/3)\mathbf{e}_{kk}^{(n)}\,] \\ &\quad - (P - P_{ex}) + \theta_n \mathcal{C} + 2\gamma C \ . \end{aligned} \tag{9c}$$

This boundary condition has a trivial physical interpretation, but it is highly nonlinear and requires some special numerical procedures (see the next section).

In Eqs. (9b) and (9c) P_{ex} refers to the ambient external pressure, and in general it is a function of position. Since pressure is relative, without loss of generality we can take $P_{ex} = 0$ on $\partial\Omega_{co}$ and $\partial\Omega_{fo}$ and $P_{ex} = -P_a$ (the aspiration pressure) on $\partial\Omega_{ci}$ and $\partial\Omega_{fi}$.

The final boundary condition necessary for an specification of our aspiration model concerns the relative velocity of the solvent and the cytoskeleton at the periphery of the cell. The simplest such condition takes the form:

$$(V_i^{(n)} - V_i^{(s)})N_i = 0 \ . \tag{9d}$$

This form of the boundary condition on relative velocity is only valid if the cytoskeleton is firmly anchored to the plasma membrane and also if fluxes of solvent through the plasma membrane can be neglected.

Computational Methodology for Free Boundary Problems

Computations were carried out using a general hydrodynamics transport code based on the finite element method. This program was developed in recent years specifically for investigating the solutions of field theoretic formulations of the mechanics and chemistry of biological cells with particular emphasis on amoeboid cells. Many of its important features are not needed to analyze the simple RIF model of cell aspiration that is of immediate concern. Nevertheless, since the general method has never before been described, we will give a brief overview emphasizing philosophy and the outline of its main capabilities and limitations.

We refer to the minimum information needed to completely specify the input of a computation, as static data. Aside from the source code itself, static data is provided explicitly by editing a setup file or by running a special initialization program. It is stored in fixed arrays and is never changed by the progress of the computation itself. In the current application the values of the various physical constants (e.g., cytoplasmic viscosity, cell volume, coefficients of the constitutive law for surface tension, initial conditions, etc.) are examples of static data of a physical nature. The various pointer arrays needed for determining mesh topology and for indirect addressing of global arrays from the element level also constitute static information, but of a purely numerical sort.

Figure 4 shows a typical finite element mesh for a cell in the initial condition and also after 40% aspiration. Notice that the basic topology of the mesh is the same in both figures: only the geometric positioning of the nodes has been altered. We also note that all interior elements are bilinear isoparametric quadrilaterals; we do not use mid-side nodes or transition elements since these present difficulties for the rezoning algorithm. The quadrilaterals are either completely interior, having no direct contact with the boundary, or else they intersect the boundary along only one edge. As a result each node on the mesh boundary is a vertex of exactly two quadrilaterals. Finally, there is always a surface node precisely at the sharp corner of the nozzle. This means that the definition of the normal and tangent directions at the nozzle can be just like any other surface node and that no special procedures are needed.

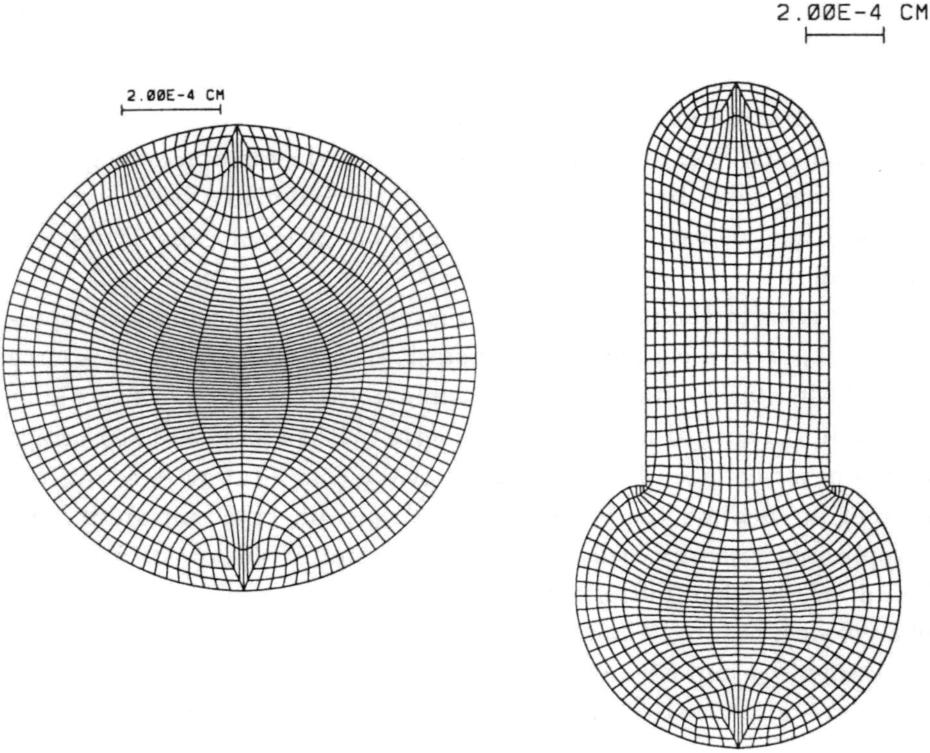

Figure 4. Adaptivity of the Computational Grid. Shown is a typical mesh of quadrilateral elements used for the computations of leukocyte aspiration. There are 592 vertex nodes, 120 surface elements, and 531 quadrilateral elements. On the left, the cell is in its initial (round) configuration, and on the right we show the state of the same mesh after the cell has been partially aspirated. This illustrates the ability of the rezoning algorithm to prevent mesh distortion despite very substantial deformations of the free boundaries. Note that the mesh adaptivity is accomplished by geometric rearrangements of the mesh nodes so as to optimize a functional measure of global distortion and resolution; the topology of the mesh is unaffected by rezoning.

Information that is contingent on the results of a computation is called dynamical information. Such information is stored in two ways: by means of surface fields and by means of interior fields. Interior fields exist in the volume Ω, and a minimum encoding of these fields is discretized by a collection of vectors living at the vertex nodes of the mesh (i.e., the so-called node vectors). The first and second positions of the node vectors are hardcoded and interpreted as either the X, Y, R, Z, or Z, R coordinates of the node, depending on the setting of a parameter. These positions of the node vectors are followed by the two components of the solvent velocity field, the pressure field, the two components of the velocity field of the cytoskeleton, and the density field of the cytoskeletion. Finally, memory is reserved for up to five auxiliary fields that can be arbitrarily programmed at the discretion of the user. Neither the cytoskeleton fields nor the auxiliary fields play a role in the current application.

The linear edges that comprise the boundary of the mesh are called surface elements (not to be confused with the kind of elements used in the boundary element method). The static data structures associated with the surface elements include pointers of the two nodes at either end of the element, a pointer of the quadrilateral of which the surface element is an exterior edge.

A minimal encoding of the dynamical fields that exist on $\partial\Omega$ (i.e., the surface fields) is stored as the elements of a vector field that lives at the midpoint of surface elements (i.e., edge vectors). The first element of the edge vector always constitutes a pointer of the boundary condition that holds on the surface element. For example, in the present application the various operative boundary conditions of the problem are assigned arbitrary numerical values during the initialization phase: $1 =$ free boundary exterior of pipet, $2 =$ slip boundary contacting wall of pipet, $3 =$ free boundary interior of pipet, etc. Memory is reserved for up to five different boundary conditions. The parameters needed to implement these boundary conditions are part of the static data. The edge vectors of the surface elements can have up to five additional components, but in the current example the membrane area density is the only additional surface field utilized.

Operations at the level of the surface elements are used in the assembly of the stiffness matrices and load vectors that describe the boundary conditions, and they are used extensively for all aspects of the chemistry and transport of surface fields. The use of surface elements is also essential for computational handling of contact dynamics (see below).

Following common practice, our computational methodology divides the overall advance of the dynamical fields with time into individual operations that are ultimately superimposed or assembled to obtain the final result (this is called operator splitting). The splitting we employ matches most of the operators with basic physical processes: diffusive mass transport, chemical reaction, transport of momentum-force balance, advective mass transport, and boundary kinematics. There is also an operation that is indispensable but that has no physical significance, namely mesh rezoning.

Diffusion Operator

The change imposed by the diffusion operator corresponds to what occurs if diffusion is the only process operating over a small time interval. For each concentration field the diffusion constant is computed at each node of the mesh. If these are all zero, the operation is terminated; otherwise the element mass matrix and the element stiffness matrix for each bilinear quadrilateral is computed using four-point Gaussian quadrature. The surface elements are used to generate penalty matrices enforcing any Direchlet conditions and or adding loads arising from the diffusive flux of mass through the boundaries. The mass matrix together with the time step is used to assemble the load vector corresponding to either a backward Euler or Crank-Nicolson time step. The final load vector together with the element stiffness matrices are then passed to a conjugate gradient solver. This operator is not used for the computation of leukocyte aspiration according to the current simple model, but it is necessary in computations like those required by the *proteus* model where diffusion of cybernetic factor is important.

Reaction Operator

The reaction operator changes the node vectors according to what occurs if all processes except chemical reaction are ignored. In order to handle arbitrary reactions with widely dispersed time scales, a second order semi-implicit scheme is utilized. The Jacobian matrix of the reaction rate at each node is computed using numerical differentiation, and the final matrix equation at each node is solved by direct LU decomposition.

Advection Operator

The advection operator takes one time step of the linear convection equation for all the dynamic fields both surface and volumetric; diffusive and reactive parts of the continuity equation are handled separately as described above. The velocity field governing advection of each field is assumed to be some linear combination of the network and cytoskeleton velocity at each node. The mesh velocity is also assigned independently at each node – the general idea being to follow the change in shape of the boundary and other "Lagrangian" features while preserving Eulerian features and minimizing the interpolation error. For the aspiration calculation, interior nodes, regular boundary nodes on the axis, and the boundary node at the nozzle are regarded as fixed or Eulerian nodes. Boundary nodes on the unaspirated surface are moved according to the normal component of fluid velocity, and boundary nodes on the aspirated surface are moved with both the normal and tangential fluid velocity (i.e., they are fully Lagrangian). Volumetric fields like the cytoskeleton density are advected using a shape preserving semi-Lagrangian interpolation scheme (Rash and Williamson, 1990). Surface fields are advected using an upwind difference scheme. In the present calculations the advection operator is used only for transport of the membrane area density. The operator is also used in order to visualize the displacement of fluid elements from their starting positions in the droplet as aspiration proceeds.

Rezone Operator

The rezone operator consists of two parts, the standard interior rezone and the boundary rezone. The goal of the interior rezoner is to compute new coordinates of interior mesh nodes so as to optimize a functional measure of resolution and distortion; the dynamical data are then interpolated from the existing mesh to the optimal mesh. The operator does not change the position of boundary nodes except in the special case of purely logical boundaries (e.g., the nodes falling on the axis of symmetry). The method we employ for optimizing mesh geometry is an adaptation of the variable diffusion method described by Thompson (1982). The original method was based on a finite difference formulation of the generator equation and was restricted to simple mesh topologies. Our adaptations remove these restrictions so as to accommodate the meshes of arbitrary topology common in finite element practice. We also use a more efficient solution routine (conjugate gradient vs. Jacobi). In our case the optimization can be constrained so that the mesh is perpendicular to certain boundary segments. The rezoning algorithm also allows for the control of mesh resolution through the specification of a weighting function. For example, a finer resolution of the elements in the immediate vicinity of the nozzle can be obtained by locally increasing the value of this function.

A collection of adjacesent surface elements all with the same value of the boundary condition pointer is called a boundary "segment." The position of the interface between two segments is physically essential and must therefore be preserved by rezoning operations. Boundary nodes that fall on the interface between two boundary segments are called "corner" nodes, and nodes that lie in the interior of segments are called "regular" boundary nodes. The first job of the boundary rezoner is to detect gradients in surface resolution within the confines of each boundary segment. Such gradients are then corrected by sliding of the regular nodes back and forth.

If a check reveals the presence of a significant resolution gradient between two disjoint boundary segments, then the boundary rezoner takes further steps. First two elements on the high resolution segment are fused together. This creates a "ghost" element (i.e., an element with zero arclength). The ghost in the element topology is then progressively shifted clockwise or anti-clockwise until it finds its "target" (i.e., the largest segment on the low resolution boundary). This target segment is divided, half of its arclength and the associated surface fields being donated to revitalize the ghost. The boundary condition pointer for the ghost element is set to the same value as that of the target segment.

Contact Operator

The "contact" operator is concerned with the kinematics of changing boundary conditions as the droplet interacts with the pipet. This operator computes the position of the midpoint of all unconstrained boundary elements and also the normal stress acting on the midpoint of all constrained boundary elements. When an unconstrained surface element comes into contact with the wall of the pipet the boundary condition pointer of this surface element is changed so as to reflect the appropriate constraint. In a similar fashion, when the normal stress on a constrained boundary element exceeds a threshold, the contact operator can remove the constraint by a simple change in the value of a boundary condition pointer. This operator is of course very important in following the interaction of a droplet and a pipet, but it is also useful for analyzing the interaction of the cytoskeleton with the inner surface of the plasma membrane and the interaction of cells with external surfaces and particles.

Momentum Operator

The momentum operator computes the network and solvent velocity fields together with the pressure field by solving the momentum equations for the network and solution phases subject to the incompressibility constraint and the boundary conditions. The procedure for doing this involves first dividing Eq. (2c) by $\theta_n\mathcal{H}$, taking the divergence and applying the incompressibility constraint. The result of these manipulations is an ecliptic equation involving the network velocity and the pressure:

$$\partial_{x_j}\left[\frac{\theta_s}{\theta_n\mathcal{H}}\partial_{x_j}P\right] = \partial_{x_j}V_j^{(n)} \quad . \tag{10a}$$

A second equation involving only $V_j^{(n)}$ and P is obtained by simply adding Eq. (2c) and (2d):

$$0 = \partial_{x_j}[\theta_n\mathcal{M}\mathbf{e}_{ij}^{(n)}] - (1/3)\partial_{x_i}[\theta_n\mathcal{M}\mathbf{e}_{kk}^{(n)}] + \partial_{x_i}[\theta_n\mathcal{C}] - \partial_{x_i}P \; . \tag{10b}$$

These two equations are discretized by the standard Galerkin finite element procedure for bilinear quadrilaterials (Hughes, 1987). The constraint boundary conditions are enforced by a penalty method. The resulting linear system is symmetric and positive definite, and the solution is obtained by means of the preconditioned conjugate gradient method. In a final step the solvent velocity is derived by back-substitution into Eq. (2c).

Some Preliminary Results

Most of the parameter values required for a realistic computation with the RIF model of leukocyte aspiration can be fixed using

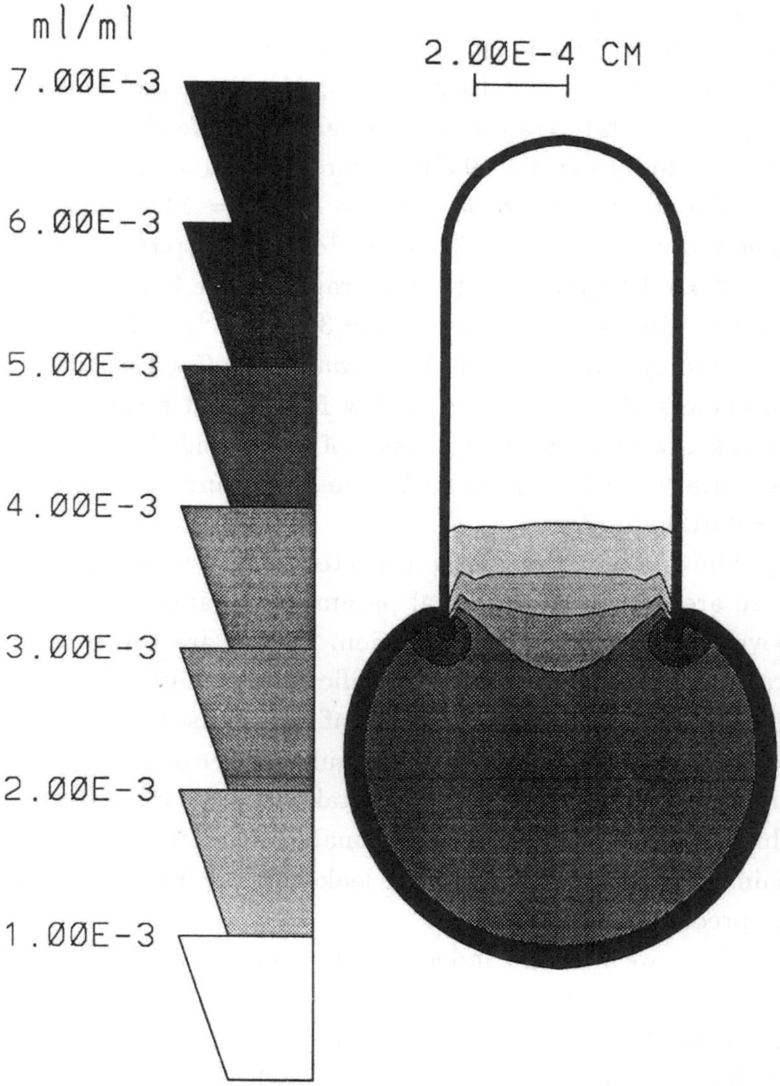

Figure 5. Phase Separation During Leukocyte Aspiration. Shading of the interior of the cell indicates network density. The thickness of the black border is proportional to the excess area density of the plasma membrane. Note that the network density in the aspirated material is very low relative to the initial condition. All parameters are as discussed in the text. Forty percent of the cell volume has been aspirated. The elapsed time to reach this stage of aspiration is 130 milliseconds.

direct measurements reported from several laboratories (Evans and Kukan, 1984; Simon and Schmidt-Schoenbein, 1988; Evans and Yeung, 1989; Yeung and Evans, 1989; Needham and Hochmuth, 1990; Dong et al., 1991; Tran-Son-Tay et al., 1991; Needham and Hochmuth, 1992; Hochmuth et al., 1993). Accordingly, in all our computations we will take the volume of the cell to be $v_\Omega = 147.5 \times 10^{-12} cm^3$; the pipet radius to be $r_p = 2.4 \times 10^{-4} cm$; the aspiration pressure to be $P_a = 2.0 \times 10^4 dynes/cm^2$; the average volume fraction of cytoskeleton in the resting leukocyte to be $\bar{\theta}_n = 3.0 \times 10^{-3}$; and the specific viscosity of the cytoplasm to be $\mathcal{M} = const. = 4.0 \times 10^5 poise$. Finally the parameters of the constitutive law for cortical tension [see Eq. (8b)] are taken as: excess area density of the rounded cell, $\rho_0 = 2.0$; resting surface tension, $\gamma_0 = 0.012 dynes/cm$; and specific area modulus, $\omega = 0.012 dynes/cm$.

Since polymerization is neglected on the time scale of aspiration, there are only two additional parameters that are needed in order to move forward with a computation. These are the specific hydraulic drag coefficient, $(\mathcal{H})$, and the specific contractility ,$(\mathcal{C})$. The former can be estimated with considerable confidence based on the theory of low Reynolds number flow through a regular lattice of filaments (Happel and Brenner, 1973). Consequently, we take $\mathcal{H} = const. = 10^{+11}$ $poise/cm^2$. This is at the upper range of reasonable estimates, but in any event the main qualitative features of the leukocyte aspiration are insensitive to the precise value of $\mathcal{H}$.

Since we have no independent means of estimating the value of $\mathcal{C}$, there is little choice but to explore the implications of a range of values for this parameter. For example, Figure 5 shows a computation in which $\mathcal{C} = -1.0 \times 10^6 dynes/cm^2$ – other parameters being fixed as discussed above. Forty percent of the cell volume has been aspirated into the pipet. The density of the shading in the interior of the cell indicates the network volume fraction (see the scale bar). The amount of wrinkling of the membrane (i.e., the value of ρ) is indicated by the thickness of the black border around the perimeter of the cell. Notice that the computation predicts a considerable smoothing of wrinkles on

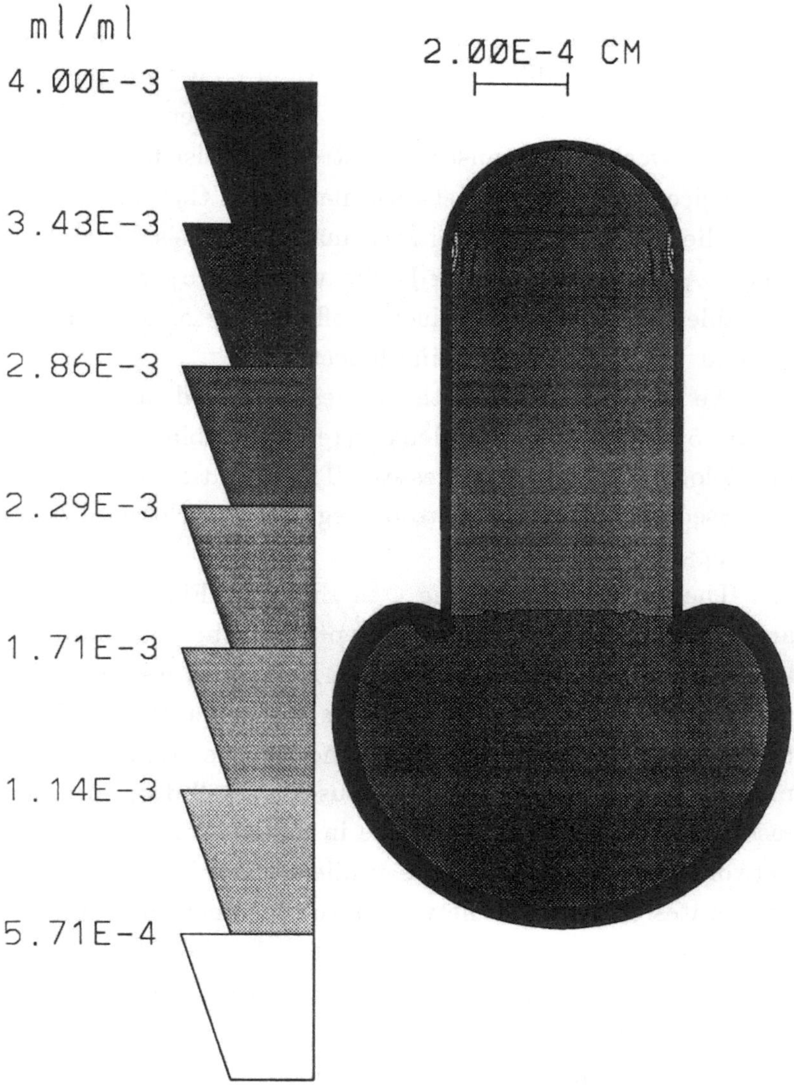

Figure 6. Effect of Increased Cytoskeletal Swelling Stress on Leukocyte Aspiration. Parameters and format are the same as in Figure 5 except that the swelling stress has been increased by an order of magnitude (i.e., the contractility parameter is more negative). Note that the network density in the aspirated material is only slightly reduced compared to the initial condition. As in Figure 5, 40% of the cell volume has been aspirated. The cell reaches this stage of aspiration in 158 milliseconds.

the aspirated portions of the cell surface. Also note the characteristic flattening of the cell against the external surface of the pipet.

The most notable prediction of Figure 5 is an extensive separation of the network and solution phases. This happens because the viscosity of the network phase causes it to strongly resist flowing inward toward the orifice of the pipet. Network material is therefore largely excluded from the pipet and the aspirated material consists mainly of a watery fluid. As aspiration proceeds the watery projection grows, becomes unstable, and eventually pinches off by way of a capillary instability (similar to an inverse dripping faucet).

We conclude that for the values of $\mathcal{C}$ used in the computation of Figure 5, our model of the leukocyte is unstable and cannot withstand physiological aspiration stresses. The situation is even worse if $\mathcal{C}$ is increased (i.e., if we take $\mathcal{C}$ to be negative but closer to zero or if we let $\mathcal{C}$ be positive).

The effect of decreasing $\mathcal{C}$ is shown in Figure 6. All parameters are the same as in Figure 5 except that the contractility is reduced by an order of magnitude; $\mathcal{C} = -1.0 \times 10^7 dynes/cm^2$. The effect of this is to increase the swelling stress of the network to the point where these stresses are comparable to the viscous stresses. When aqueous material flows into the pipet, it causes a small drop in network density inside the pipet and a small rise in density outside. This gradient of network density causes a large gradient of swelling stress, which in turn accelerates the entry of network into the pipet and retards the further entry of more aqueous material. The ultimate effect is to nullify the nascent onset of phase separation and to maintain a tight coupling between the phases as aspiration proceeds to completion.

We conclude that if the RIF model is correct in its essential elements, then the constitutive contractility of the cytoskeleton of passive leukocytes is such as to cause this network to swell and resist compaction. This raises many questions; for example, what happens when the leukocyte is activated? and what are the implications of negative contractility for the behavior of the cytoskeleton during crawling and phagocytosis? A manuscript documenting more extensive theoretical studies of leukocyte aspiration is being prepared for publication elsewhere.

Acknowledgements

This work was supported by NIAID grant RO1-AI21002 and by the United States Department of Energy. The author also wishes to thank Dr. Richard Skalak, Director of the Institute for Mechanics and Materials at the University of California, San Diego, for hospitality and support during the preparation of this manuscript.

References

Albrecht-Buehler, G. (1980). Autonomous movements of cytoplasmic fragments. *Proc. Natl. Acad. Sci. USA* 77:6639-6644.

Allen, R.D. (1961). Ameboid movement. In: *The Cell, Vol. II*, J. Brachet and A.E. Mirsky, eds. Academic Press, New York, pp. 135-216.

Alt, W. (1987). Mathematical models in actin-myosin interaction. In: *Fortschritte der Zoologie, Band 34: Nature and Function of Cytoskeletal Proteins in Motility and Transport*, Hrgs., Wholfarth-Bottermann. Gustav Fischer Verlag, Stuttgart and New York, pp. 219-230.

Bereiter-Hahn, J.M. Luck, T. Miebach, H.K. Stelzer and M. Voth. (1990). Spreading of trypinized cells: Cytoskeletal dynamics and energy requirements. *J.Cell Sci* 96:171-188.

Bowen, R.M. (1980). Incompressible porous media models by use of the theory of mixtures. *Int. J. Eng. Sci.* 18:1129-1148.

Crick, F.H.C. and A.F.W. Hughes (1950). The physical properties of cytoplasm: A study by means of the magnetic particle method. *Exp. Cell Res.* 1:37-80.

De Bary, A. (1864). *Die Mycetozen.* W. Engelmann, Leipzig.

Dellinger, O.P. (1906). Locomotion of *Amoebae* and allied forms. *J. Exp. Zool.* 3:337-358.

Dembo, M., F. Harlow, and W. Alt (1984). The biophysics of cell surface motility. In: *Cell Surface Dynamics: Concepts and Models*, C. DeLisi, A. Perelson, and F. Wiegel, eds. Marcel Dekker, New York, pp. 495-542.

Dembo, M., M. Maltrud, and F. Harlow (1986). Numerical studies of unreactive contractile networks. *Biophys. J.* 50:123-137.

Dembo, M. (1986). The mechanics of motility in dissociated cytoplasm. *Biophys. J.* 50:1165-1183.

Dembo, M. and F. Harlow (1986). Cell motion, contractile networks, and the physics of interpenetrating reactive flow. *Biophys. J.* 50:109-121.

Dembo, M. (1989). The mechanics and control of the cytoskeleton in *Amoeba proteus. Biophys. J.* 55:1053-1080.

Dong, C., R. Skalak, K-L.P. Sung, G.W. Schmidt-Schoenbein and S. Chien. (1988). Passive peformation analysis of human leukocytes. *J. Biomech. Eng.* 110:27-36.

Dong, C., R. Skalak and K-L.P. Sung. (1991). Cytoplasmic rheology of Passive Neutrophils. *Biorheology* 28:557-567.

Dong, C., and R. Skalak. (1992). Leukocyte Deformability: Finite Element modeling of large viscoelastic deformation. *J. Theor. Biol.* 158:173-193.

Drew, D.A., and L.A. Segel (1971). Averaged equations for two phase flow. *Stud. Appl. Math.* 50:205-231.

Dujardin, F. (1835). Recherches sur les organismes inferieurs. *Ann. Sci. nat. Zool.* 4:343-377.

Dujardin, F. (1838). Memoire sur l'organisation des Infusoires. *Ann. Sci. nat. Zool.* 10:230-315.

Dunn, G.A. (1980). Mechanisms of fibroblast locomotion. In: *Cell Adhesion and Motility*, A.S.G. Curtis and J.D. Pitts, eds. Cambridge Univ. Press, London, pp. 409-424.

Ehrenberg, C.G. (1830). *Organisation, Systematic und Geographisches Verhaltniss der Infusionsthierchen.* F. Dummler, Berlin.

Ehrenberg, C.G. (1832). *Zur Erkenntniss der Organisation in der Richtung des Kleinsten Raumes.* F. Dummler, Berlin.

Evans, E.A. and B. Kukan (1984). Passive material behavior of granulocytes based on large deformation and recovery after deformation tests. *Blood* 64:1028-1035.

Evans, E., and A. Yeung. (1989). Apparent viscosity and cortical tension of blood granulocytes determined by micropipet aspiration. *Biophys. J.* 50:151-160.

Evans, E. and M. Dembo (1990). *Physical Model for Phagocyte Motility: Local Growth of a Contractile Network from a Passive Body.* Nato ASI Series. Vol H42. Biomechanics of Active Movement and Deformation of Cells. N. Akkas ed. Springer Verlag Berlin, 185-214.

Ezzell, R.M., A.J. Brothers, and W.Z. Cande (1983). Phosporylation-dependent contraction of actomyosin gels from amphibian eggs. *Nature* 306:620-622.

Geiger, B. (1983). Membrane-cytoskeleton interaction. *Biochem. Biophys. Acta.* 737:305-341.

Gelfand, V.I., N.A. Glushankova, O.Y. Ivanova, L.A. Mittelman, O.Y. Pletyushkina, J.M. Vasiliev, and I.M. Gelfand (1985). Polarization of cytoplasmic fragments microsurgically detached from mouse fibroblasts. *Cell Biol. Intern. Rep.* 9:883-892.

Grebecki, A. (1984). Relative motion in *Amoeba proteus* in respect to adhesion sites. I. Behavior of monotactic forms and the mechanism of fountain phenominon. *Protoplasm* 123:116-134.

Happel, J. and H. Brenner (1973). *Low Reynolds Number Hydrodynamics.* Noordhoff Int., Leyden, p. 553.

Hartwig, J.H., R. Neiderman, and S.E. Lind (1985). Cortical actin structures and their relation to mammalian cell movements. In: *Subcellular Biochemistry, Vol. 11*, D.B. Roodyn, ed. Plenum Press, New York, pp. 1-49.

Hartwig, J.H. and H.L. Yin (1988). The organization and regulation of the macrophage actin skeleton. *Cell Motil. Cytoskel.* 10:117-125.

Heitzmann, C. (1873). Untersuchungen unber das Protoplasma. I. Bau des Protoplasmas. *Sitzber. kaiserl. Akad. Wiss., Math.-Naturwiss. Cl.* 67(III. abth.):100-115.

Hill, T.L. and M.W. Kirschner (1982). Bioenergetics and kinetics of microtubule and actin filament assembly-disassembly. *Int. Rev. Cytol.* 78:1-125.

Hiramoto, Y. (1969). Mechanical properties of the protoplasm of the sea urchin egg, I. Unfertilized egg. *Exp. Cell Res.* 56:201-208.

Hochmuth, R.M., H.P. Ting-Beall, B.B. Beaty, D. Needham and R. Tran-Son-Tay. (1993). Viscosity of passive human neutrophils undergoing small deformations. *Biophys. J.* 64:1596-1601.

Hughes, T.J.R. (1987). *The Finite Element Method.* Prentice Hall, Englewood Cliffs, New Jersey, p. 803.

Ishijima, A., T. Doi, K. Sakurada and T. Yanagida. (1991). Subpiconewton force fluctions of actomyosin *in vitro*. *Nature* 352:301-306.

Izzard, C.S. (1988). A precursor of the focal contact in cultured fibroblasts. *Cell Motil. Cytoskel.* 10:137-142.

Janmey, P.A. (1991). Mechanical properties of cytoskeletal polymers. *Curr. Opin. Cell Biol.* 2:4-11.

Jennings, H.S. (1904). Contributions to the study of the behavior of lower organisms. 6. The movements and reactions of *Amoeba*. *Carnegie Inst. Wash. Pub.* 16:129-234.

Jeon, K.W. and M.S. Jeon. (1975). Cytoplasmic filaments and cellular wound healing in *Amoeba proteus*. *J. Cell Biol.* 67:243-249.

Kamiya, N. (1986). Some motility characteristics of living cytoplasm from observations on *Physarium*. In: *Cell Motility: Mechanism and Regulation.* H. Ishikawa, S. Hatano, and H. Sato, eds. Alan R. Liss, New York, pp. 577-585.

Korn, E.D. (1982). Actin polymerization and its regulation by proteins from nonmuscle cells. *Physiol. Rev.* 62:672-737.

Kuroda, K. (1979). Movement of cytoplasm in a membrane-free system. In: *Cell Motility: Molecules and Organization* , S. Hatano, H. Ishikawa, and H. Sato, eds. Univ. Park Press, Baltimore, pp. 347-362.

Lovtrup, S. and A. Pigon (1951). Diffusion and active transport of water in the amoeba *Chaos chaos*. *L. Comptes Rendus des Travaux du Lab. Carlsberg, Ser. chim.* 28:1-36.

Lyass, L.A., A.D. Bershadsky, V.I. Gelfand, A.S. Serpinskaya, A.A. Stavrovskaya, J.M. Vasiliev, and I.M. Gelfand (1984). Multinucleation-induced improvment of the spreading of transformed cells on the substratum. *Proc. Natl. Acad. Sci. USA.* 81:3098-3102.

Mast, S.O. (1926). Structure, movement, locomotion and stimulation in *Amoeba*. *J. Morph. Physiol.* 41:347-425.

Matsudaria, P. and P. Janmey (1988). Pieces in the actin-severing protein puzzle. *Cell* 54:139-140.

McClendon, J.F. (1909). Protozoan studies. *J. Exp. Zool.* 6:265-287.
McNeil, P.L. (1991). Cell wounding and healing. *American Scientist* 79:222-235.
McNiven, M.A., M. Wang, and K.R. Porter (1984). Microtubule polarity and the direction of pigment transport reverse simultaneously in surgically severed melanopore arms. *Cell* 37:753-765.
Needham, D., and R.M. Hochmuth (1990). Rapid flow of passive neutrophils into a μm pipet and measurments of cytoplasmic viscosity. *J. Biomech. Engr.* 112:269-276.
Needham,D., and R.M. Hochmuth. (1992). A sensitive measure of surface stress in the resting neutrophil. (*Biophys. J.* 61:1664-1670.
Nowakowska, G. (1978). Twisting of suspended monotactic specimens of *Amoeba proteus. Acta Protozool.* 17:347-352.
Oosawa, F., S. Fujime, S. Ishiwatwa, and K. Mihashi (1973). Dynamic property of F-actin and thin filament. *Cold Spring Harbor. Symp. Quant. Biol.* 37:277-284.
Oster, G.F. and A.S. Perelson (1987). The physics of cell motility. *J. Cell Sci. Suppl.* 8:35-54.
Pantin C.F.A. (1923). On the physiology of amoeboid movement. I. *J. Marine Biol. Ass.* 13:24-69.
Pantin C.F.A. (1924). On the physiology of amoeboid movement. II. *Brit. J. Exp. Biol.* 1:519-538.
Peskin, C.S., G.M. Odell and G.F. Oster (1993). Cellular motions and thermal fluctuations: The Brownian ratchet. *Biophys. J.* 65:316-324.
Pollard, T.D. and S. Ito (1970). Cytoplasmic filaments of *Amoeba proteus*: I. The role of filaments in consistency changes and movement. *J. Cell Biol.* 46:267-289.
Pollard, T.D., S.C. Selden, and P. Maupin (1984). Interaction of actin with microtubules. *J. Cell Biol.* 99:33s-37s.
Pollard, T.D. and J.A. Cooper (1986). Actin and actin-binding proteins: A critical evaluation of mechanisms and functions. *Annu. Rev. Biochem.* 55:987-1035.
Porter K.R. (1984). The cytomatrix: A short history of its study. *J. Cell Biol.* 99:3s-12s.
Rash, P.J., and D.L. Williamson. (1990). On shape-preserving interpolation and semi-Lagrangian Transport. *Siam J. Sci. Stat. Comput.* 11:656-687.
Rhumbler, L. (1898). Physikalische Analyse von Lebenserscheinungen der Zelle. I. Bewegung, Nahrungsaufname, Defakation, Vacuolen-Pulsation und Gehausebau bei lobosen Rhizopoden. *Arch. Entwmech. Org.* 7: 103-198.
Sato, M., T.Z. Wong, and R.D. Allen (1983). Rheological properties of living cytoplasm: Endoplasm of *Physarium* plasmodium. *J. Cell Biol.* 97:1089-1097.

Sato, M., T.Z. Wong, D.T. Brown, and R.D. Allen (1984). Rheological properties of living cytoplasm: A Preliminary investigation of squid axoplasm *Loligo pealei*. *Cell Motility* 4:7-23.

Sato, M., W.H. Schwartz, and T.D. Pollard (1987). Dependence of the mechanical properties of actin/alpha-actinin gels on deformation rate. *Nature* 325:828-830.

Schroder, T.E. (1975). Dynamics of the contractile ring. In: *Molecules and Cell Movement*, S. Inoue and R.E. Stephens, eds. Raven Press, New York, pp. 305-334.

Schulze, F.E. (1875). Rhizopodenstudien IV. *Arch mik. Anat.* 11:329-353.

Seifriz, W. (1921). Observations on some physical properties of protoplasm by aid of microdissection. *Ann. Bot.* 35:269-296.

Simon, S.I., and G.W. Schmidt-Schoenbein. (1988). Biophysical aspects of microsphere engulfment by human neutrophils. *Biophys. J.* 53:163-173.

Stossel, T.P., C. Chaponnier, R.M. Ezzell, J.H. Hartwig, P.A. Janmey, D.J. Kwiatkowski, S.E. Lind, D. Smith, F.S. Southwick, H.L. Yin, and K.S. Zaner (1985). Nonmuscle actin binding proteins. *Annu. Rev. Cell Biol.* 1:353-402.

Stossel, T.P. (1990). How cells crawl. *American Scientist* 78:408-423.

Stossel, T.P. (1993). On the crawling of animal cells. *Science* 260:1086-1094.

Sung, K.-L.P., C.Dong, G.W. Schmidt-Schoenbein, S. Chein and R. Skalak. (1988). Leukocyte relaxation properties. *Biophys. J.* 53:331-336.

Taylor, D.L., J.S. Condeelis, P.L. Moore, and R.D. Allen (1973). The contractile basis of amoeboid movement. I. The chemical control of motility in isolated cytoplasm. *J. Cell Biol.* 59:378-394.

Taylor, D.L. and J.S. Condeelis (1979). Cytoplasmic structure and contractility in amoeboid cells. *Internat. Rev. Cytol.* 56:57-114.

Thompson, C.M. and L. Wolpert (1963). The isolation of motile cytoplasm from *Amoeba proteus*. *Exp. Cell Res.* 32:156-160.

Tilney, L.G. and S. Inoue (1982). Acrosomal reaction of *Thyone* sperm. II. The kinetics and possible mechanism of acrosomal process elongation. *J. Cell Biol.* 93:820-827.

Tran-Son-Tay, R., D. Needham, A. Yeung, and R.M. Hochmuth. 1991. Time-dependent recovery of passive neutrophils after large deformation. *Biophys. J.* 60:856-866.

Trusdell, C. and R.A. Toupin (1960). *The Classical Field Theories, Handbuch der physik III/1*, Springer-Verlag, Berlin.

Valberg, P.A. and D.F. Albertini (1985). Cytoplasmic motions, rheology, and structure probed by a novel magnetic particle method. *J. Cell Biol.* 101:130-140.

Wallich, G.C. (1863a). Further observations on an undescribed indigenous *Amoeba*, with notices on remarkable forms of *Acinophrys* and *Diffugia*. *Ann. Mag. Nat. Hist.* 11(ser. 3):365-371.
Wallich, G.C. (1863b). On the value of distinctive characters in *Amoeba*. *Ann. Mag. Nat. Hist.* 11(ser. 3:111-151.
Wang, Y-L. (1985). Exchange of actin subunits at the leading edge of living fibroblasts: Possible role of treadmilling. *J. Cell Biol.* 101:597-602.
Yagi, K. (1961). The mechanical and colloidal properties of amoeba protoplasm and their relations to the mechanism of amoeboid movement. *Comp. Biochem. Physiol.* 3:73-91.
Yeung, A. and E. Evans (1989). Cortical shell-liquid core model for passive flow of liquid-like spherical cells into micropipets. *Biophys. J.* 56:139-149.
Zheng, Q. and D.C. Chang (1991). Reorganization of cytoplasmic structures during cell fusion. *J. Cell Sci.* 100:431-442.
Zhu, C. and R. Skalak (1988). A continuum model of protrusion of pseudopod in leukocytes. *Biophys. J.* 54:1115-1138.

A MODEL OF THE LEUKOCYTE MIGRATION THROUGH SOLID TISSUE

Boguslaw A. Skierczynski, Shunichi Usami* and Richard Skalak
Institute for Biomedical Engineering
University of California, San Diego
9500 Gilman Drive
La Jolla, CA 92093-0412
USA

INTRODUCTION

Neutrophils are defensive cells that are recruited from the blood into damaged or inflamed tissues by chemotactic agents. Ligand-receptor interactions at the cell surface elicit adhesion, locomotion, granule exocytosis, and the respiratory burst [Baggiolini and Kernen, 1992]. *In vitro* analysis shows that these responses are controlled at several levels of intracellular signal transduction processes. In this introduction we first review some of the biochemical events associated with these responses.

The polymorphonuclear leukocytes (PMNs) circulate passively in the blood, and sense infection or damage through various types of "alarm" molecules that are distributed by diffusion within the local microvessels [Lloyd and Oppenheim, 1992]. These chemoattractants direct PMN migration through the vessel wall toward the injured tissue. There are several chemotactic molecules with different origins, such as complement fragment C5a, fMet-Leu-Phe, PAF (platelet-activating factor), LTB4 (leukotriene B4), IL-8 (interlukin-8), etc. [Baggiolini and Kernen, 1992, Westlin, Kiely, et al., 1992].

* Institute of Biomedical Science, Academia Sinica, Taiwan, ROC

NATO ASI Series, Vol. H 84
Biomechanics of Active Movement
and Division of Cells
Edited by N. Akkaş
© Springer-Verlag Berlin Heidelberg 1994

Chemoattractant molecules interact with specific receptors and elicit the responses that are required for the defensive functions of the PMNs.

Leukocyte adhesion is controlled by the dynamic interaction between adhesion molecules expressed by the PMNs and ligands on the endothelial cells (EC) [Kuijpers, Hakkert, et al., 1992]. The three families of adhesion molecules that participate in this process are: selectins, [L-selectin, E-selectin, P-selectin], the Immunoglobulin (Ig) superfamily; [ICAM-1, ICAM-2, VCAM-1] and the integrins, [CD11a/CD18, CD11b/CD18 , CD11c/CD18 and CD49d/CD29] [Cummings and Smith, 1992, Fuortes, Jin, et al., 1993, Hogg, 1992, Keeney, Schmalstieg, et al., 1993, Lasky, 1992, Patarroy and Makgoba, 1989, Symington, Takada, et al., 1993].

Following receptor-ligand interaction and coupling to G-protein, [Birnbaumer, 1992, Linder and Gilman, 1992, Sarndahl, Lindorth, et al., 1989, Spiegel, 1992] phosphatidylinositol-specific phosphase C is activated [Howard, Sefton, et al., 1993, Sternweis and Smrcka, 1992] and generates two second messengers, inositol 1,4,5-trisphosphate (IP_3) and diacylglycerol (DAG) [Condeelis, 1992, Cunningham, Gorlin, et al., 1992]. IP_3 induces the release of calcium from intracellular storage organelles, leading to transient rise in cytosolic free calcium [Connor, 1993]. Diacylglycerol remains associated with the membrane and participates in activation of protein kinase C and ponticulin [Chia, Shariff, et al., 1993, Downey, Chan, et al., 1992, Luna and Hitt, 1992]. External calcium concentration has been shown to affect cell pressure, shape and locomotion [Bereiter-Hahn, Strohmeier et al., 1983, Strohmeier and Bereither-Hahn, 1984,1987].

Leukocyte adhesivness is a finely regulated process, controlled by membrane receptors involved in cell-cell and cell-matrix recognition. PMN's adhesion receptors may have low- or high-affinity for their respective ligands, depending on triggering molecules delivered by the surrounding cells and the extracellular milieu. In addition to the intrinsic properties of the adhesion receptors and ligands, such as surface density and diffusitivity, both the positive and negative feedback loops affect PMNs adhesion and locomotion, through a network of intracellular signaling pathways [Schwartz, 1993]. For example, integrin not only mediates adhesion but sends signals into interior of a cell [Juliano and Haskill, 1993, Ruoslahti, 1992].

ROLES OF THE ACTIN NETWORK.

It has been understood for a long time that the cytoskeleton is a critical component in the processes of cell locomotion. A principal constituent of the cytoskeleton is actin which may occur in a globular form (G-actin) or in a polymer network (F-actin). The roles of the actin network are diverse. In addition to maintaining cell shape, actin may define cell polarity and drive cell locomotion. The actin cytoskeleton is composed of a complicated meshwork of microfilaments whose precise geometry may be regulated by more than 70 proteins [Goldschmidt-Clermont and Janmey, 1991, Hofmann, Eichinger, et al., 1992, Hug, Miller, et al., 1992, Pollard, 1986]. Regulated changes in the actin cytoskeleton occur in a large number of biological responses; however the signal transduction pathways leading to actin remodeling are still poorly understood, mainly due to the complexity of the events [Pollard and Goldman, 1993, Turley, 1992]. The actin polymerization has to be precisely controlled but the balance between monomeric and polymerized actin is influenced by a large number of the accessory actin-binding proteins [Goldschmidt-Clermont, Furman, et al., 1992, Stossel, 1989] and by ras-related GTP-binding proteins [Hall, 1992]. During activation by chemoattractants' phosphorylation of the p64 protein may play a role in the cytoskeletal reorganization [Shibata, Yamakawa, et al., 1993]. Polymerization of the G-actin can also be induced by adherence to a substratum or to other cells. Rapidly polymerizing actin fibers are assembled either into bundles of filaments in filopodia or into a crosslinked meshwork as in pseudopodia and lamellipodia.

In case of adherence-induced polymerization, the F-actin is found in small foci [Samuelsson, Luther, et al., 1993, Southwick and Young, 1990, Tawil, Wilson, et al., 1993, Woods and Couchman, 1992]. In passive state 60 to 75% of total actin in the cell is in the monomeric state at a concentration about 1000-fold above the critical concentration [Southwick, Dabiri, et al., 1989]. To prevent spontaneously polymerization the actin monomers (G-actin) is bound in complexes of G-actin and sequestering-proteins (ADF-actin depolymerizing factor, depactin, destrin, actophorin, profilin, and Tß4 -thymosin ß4). The Tß4 seems to play a crucial role in maintaining pool of unpolymerized actin in platelets and leukocytes. Tß4 does not bind to F-actin and shows no cutting, capping or nucleation activity. Tß4 binding is neither calcium sensitive nor affected by PIP_2, and phospohorylation of Tß4 has not yet been detected [Cassimeris, Safer, et al., 1992, Safer, 1992, Weber, Nachmias, et al., 1992].

In vivo, the first phase of adhesion results in the rolling of the neutrophil along the surface of the endothelial cells. PMN rolling is mediated by the P-selectin and E-selectin. A second phase involves integrins that strengthen the adhesion and permit emigration. The translocation of integrins from the intracellular storage pool onto neutrophil surface, their activation and shedding of L-selectin from the PMN's surface are part of the responses to chemoattractants [Butcher, 1990, Keeney, Schmalstieg, et al., 1993, Rot, 1992]. After adhesion, pseudopod formation contributes significantly to the processes by which cells migrate through the endothelium [Mueller and Chen, 1991].

MOTILITY MODELS OF THE PMN

During last few years, several models have been proposed to explain the motility of the PMNs, but the source of the driving force for extension of the leading lamella is unclear [Stossel, 1989, 1990, 1993]. When the PMN is locomoting through the EC layer the leading edge of the pseudopod must generate forces not only required to advance cell by itself but must also overcome the resistance of the EC layer.

One proposed mechanism of advancement of the leading edge of a pseudopod is the forward cytoplasmic flow hypothesis [Beraither-Hahn, Strohmeier et al.,1981]. A similar hypothesis is the cortical expansion model [Condeelis, 1992, Condeelis, Bresnick, et al., 1990], which invokes cycles of brief and localized osmotic swelling in transiently disrupted actin filament gel at the leading edge [Oster, 1984, 1988, Oster and Odell, 1984a, 1984b, Oster and Perelson, 1987]. Subsequent actin assembly and crosslinking, leading to reconstitution of the disrupted gel, cause solidification of the cortical gel in the region created by the osmotically driven advance.

A very different mechanism proposed, is the "Brownian Ratchet" [Peskin, Odell, et al., 1993] in which rectified Brownian motion is the key operation. An actin rod polymerizes at the membrane whose mobility is characterized by its diffusion coefficient D. The ratchet mechanism is the intercalation of monomers between the membrane and the polymer tip when sufficiently large fluctuations of the membrane position occur so as to open the gap wide enough to allow an actin monomer to polymerize onto the end of the F-actin filament. The actin proteins are assumed to be flexible, elastic structures, whose thermal fluctuations significantly affect their motions. The gap between the membrane and the tip of the polymer depends not only on the diffusion of the barrier, but Brownian motion of the gel which induces thermal "breathing" modes that contribute to the gap width. These fluctuations within the F-actin network can compress the filament sufficiently to permit polymerization even if the membrane is too large to more diffusively appreciably. In this situation the elastic compression energy generated by thermal motions is the proximal origin of the force. In this model it is not immediately clear how the energy release in the polymerization process can be utilized to generate propulsion forces.

Another proposal of a pseudopod protrusion mechanism is that membrane-associate myosin I molecules move antegrade along immobilized actin filaments and push the membrane at the leading edge [Kuijpers, Hakkert, et al., 1992]. Myosin-actin contraction may also be involved in cellular motions of amoeba. Dembo and colleagues [Dembo, 1986, 1989, Dembo and Harlow, 1986] have developed a biphasic continuum model of the ameboid motion on this basis. One phase is a filamentous network of actin and myosin, and the other is an aqueous solution; both phases are treated as continuous and interpenetrating. This model reproduces the internal flow observed in ameboid cells, but it is not immediately clear if it is suitable to explain leukocyte motility.

There are several models of PMN locomotion which involve polymerization of actin at the front face of the pseudopod. Several authors have shown that relative to the leading edge of a lamella the actin network moves away from the advancing edge of the lamella. This relative motion implies that the actin network polymerizes at the leading edge. As a cell moves forward, the actin network may be at rest relative to the substrate over which it moves. Transmembrane ligands can then connect the substrate and to the actin filaments to create an attachment site as the cell advances.

Toward the rear of the cell, disassembly of the actin network followed by disaggregation of the adhesion complex can allow the rearward portion of the cell to detach from the substrate. Contraction of membrane-bound actin filaments by myosin may be responsible for some of the internal-motion of the crawling cell's lamella [Stossel, 1993]. The models of cell motion postulating polymerization at the front at the leading edge of the advanced pseudopod allow for the active driving mechanism to be based on the chemical energy transduction to mechanical work associated with actin polymerization [Perelson and Coutsias, 1986, Zhu and Skalak, 1988].

Most of the above models allow a continuous motion of a locomoting cell. Another possibility is a cyclic motion as proposed by DiMilla [DiMilla, Barbee, et al., 1991]. In this model, a migration cycle is proposed consisting of successive stages of lamelopodia extension, cytoskeletal contraction and relaxation. In this model the cell cytoskeleton is described as a simple network of viscoelastic and contractile elements. A crucial aspect of the model is that translocation of the cell requires an asymmetry in the cell-substratum interactions; bonds must be stronger at the front end of the cell. Such asymmetry of bonds strength is generally necessary in any cell locomotion model [Lauffenburger, 1989].

A MODEL OF PMN LOCOMOTION.

In the present paper we propose a model of the cell locomotion suggested by the observations of a leukocyte actively moving through a pipette [Usami, Wung, et al., 1992b]. The observation of locomoting cells [Usami, Wung, et al., 1992a] shows that the cells develop three distinct zones indicated schematically in Fig.1. The leading zone (Zone A) is uniform and clear, devoid of granules and other organelles. The second short section (Zone B) in which granules may be present, and the rear portion of the cell (Zone C) containing the cell nucleus and organelles. There is little relative motion of the constituents within Zone C as the cell moves forward. Analyses of the records of the cell motion [Skalak, Skierczynski, et al., 1993] indicate that the length of the pseudopod is approximately constant during the locomotion.

The model suggested here is as simple as possible and still capable of describing the main mechanical events of the observed cell locomotion and in keeping with known features of biophysics and biochemistry. The main assumptions of the model are:

- Monomeric G-actin is polymerized to form F-actin filament at the front face of the pseudopod. Depolymerization of the actin network takes place at the rear part of the pseudopod region to the generate G-actin monomers. This cycle is regulated by several classes of proteins that sequester actin monomers, control nucleation of the filaments, block the barbed and pointed ends and sever the F-actin filaments.

- Actin filaments are crosslinked to form a stiff network which (Zone A) remains stationary relative to the pipette wall during locomotion due to bonding of transmembranous adhesion receptors [Otey, Pavalko, et al., 1990].

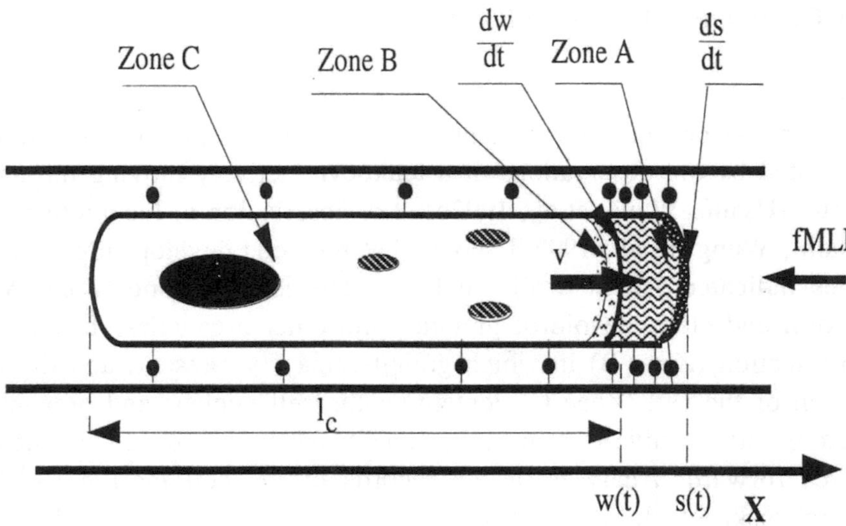

Fig.1 Sketch of the mathematical model

- The lipid bilayer of the cell membrane is passively pulled past the adhesion bonds and anchoring proteins which attach the F-actin network to the substratum [Kucik, Elson, et al., 1990].

- Forces opposing the forward motion of the cell are transmitted through the actin gel and the adhesion bonds in the pseudopod region to the pipette walls.

- Contraction forces are generated in the depolymerization region (Zone A) and pull the rear part of the cell (Zone C) forward. Depolymerization of actin gel and its contraction are assumed to occur in the same place and at the same time.

- G-actin, receptors, and other proteins involved in locomotion are passively transported through the actin gel to the front of the cell by diffusion.

-The chemotactic and the adhesion receptors are recycled. They are expressed and form bonds at the front of the cell and are internalized at the rear.

Active force-generating processes

As mentioned above we postulate that in the locomotion of the PMN cells two independent and different mechanisms of force generation are utilized: (i) contraction at the end of the pseudopod in the depolymerization are due to actin-myosin interaction and (ii) polymerization of the actin at the leading edge of the pesudopod. The chemical free energy liberated by actin polymerization may be partially converted into mechanical work.

Polymerization is known to be able to exert forces as in the sickling of red blood cells due to hemoglobin polymerization. Other examples are: the polymerization of the microtubules inside spherical liposomes [Cooper, 1991], shape changes in actin-containing vesicles [Cortese, Schwab III, et al., 1989], and polymerization of the actin utilized by bacterium *Listeria monocytogenes* to locomote through the host cell cytoplasm [Theriot, Mitchison, et al., 1992]. These examples and their experimental results support the possibility that the polymerization of actin might exert protrusive forces. Some additional information on actin polymerization and contraction is summarized in the following paragraphs.

Actin consists of a single polypeptide chain of 375 residues that binds ATP and ADP with a high affinity. ATP is hydrolyzed to ADP by both filamentous polymeric actin and by globular monometric actin. Actin consists of four domains (I-IV) with cleft between domain I-II and II-IV containing the actin bound nucleotide [Wertman and Drubin, 1992]. F-actin is a double-stranded right-handed helix with 14 actin molecules in each strand per complete turn [Aktories and Wagner, 1992]. Structural data indicate that the actin family is complex and can be divided into three classes of isoforms. Class I and II are typified by the amino-terminal domain and class III is a new class of the actin-related proteins [Herman, 1993].

Studies of the thermodynamics of actin polymerization have shown that the free energy change accompanying this process is 10-15 kcal/mol [Gordon, Yang, et al., 1976]. A recent study shows that the change of free energy during polymerization for Mg-ATP-actin is about 1 kcal/mol, while change for the polymerization of Ca-ATP-actin, Mg-ADP-actin and Ca-ADP-actin is higher, in range of 5.1-9.1 kcal/mol. The energy change in actin polymerization has been attributed to changes in the state of water molecules bound to the actin monomers [Estes, Selden, et al., 1992].

The experiments based on differential scanning calorimetry (DSC), have given new insight into the energetic transactions that accompany changes of protein conformations [Galisteo and King, 1993]. The observation of an energy release upon switching of a protein between two physiologically defined states suggests a mechanism by which protein folding or unfolding may perform mechanical work at the molecular level [Kremen, 1992, Steven, 1993].

The phosphorylation of ATP to ADP is not required for actin polymerization and can take place after the filament is polymerized. The F-actin filaments made from the ATP-G-actin are stiff rods. F-actin prepared from the ADP-G-actin produce more flexible filaments [Janmey, Hvidt, et al., 1990].

A chemo-mechanical transduction, structurally realized by the twist and stretch mode of F-actin, could correspond to transition from ATP to ADP after polymerization. The energy stored in molecular structure can by used again during depolymerization process [Schutt, Linderberg, et al., 1989]

We postulate that the free chemical energy released during polymerization of the G-actin into F-actin filament may be partially converted into mechanical work through unfolding of the G-actin at the time of incorporation into the filament (Fig.2). A similar mechanism for unfolding of a protein involving a change of the inverse temperature transition ΔT_t has been proposed as a mechanism for the free energy transduction [Urry, 1992].

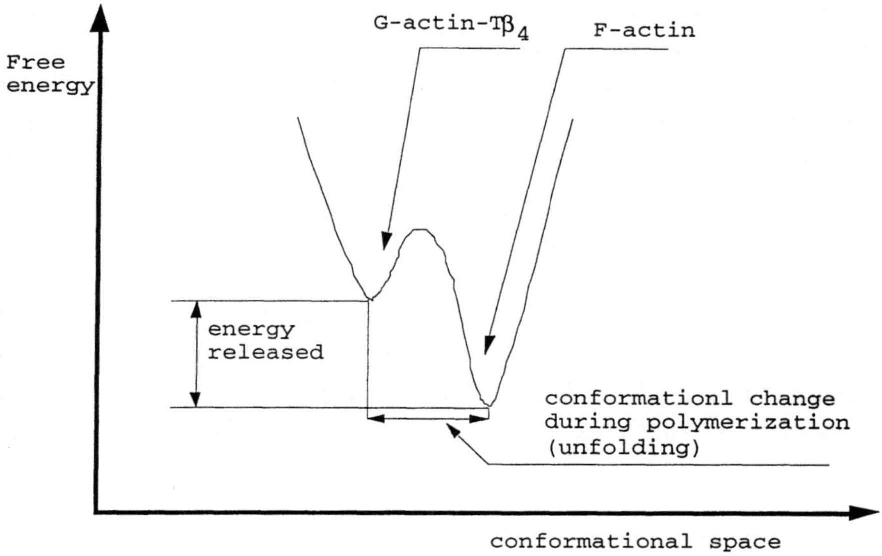

Fig. 2 Sketch of change free energy during polymerization.

For the case of actin polymerization we assume that when no external forces are acting on the filament during polymerization, the released energy is dissipated into heat and the reaction can be described by the equation [Frieden, 1985, Pollard and Mooseker, 1981].

$$\frac{dL}{dt} = l(k_+ c_a - k_-) \tag{1}$$

where $k_+\left[\frac{cm^3}{s}\right]$, $k_-\left[\frac{1}{s}\right]$ are association and dissociation rate constants respectively, $c_a\left[\frac{1}{cm^3}\right]$ is concentration of the G-actin in the region of question, $\frac{dL}{dt}\left[\frac{cm}{s}\right]$ is the rate of elongation of the filament

Rate of association between a G-actin molecule and an F-actin filament end is determined by collision theory [Pontius, 1993] as the amount of time it takes for the diffusing G-actin to encounter the F-actin tip, and the probability that a binding encounter will occur. For most macromolecules associations, the rate constants are well below the encounter limit. One reason for this is that most complementary macromolecules have specific binding sites and are not reactive over their entire surface. This introduces a steric factor. The experimental values for association and dissociation constants include this effect. When the external forces are applied during polymerization the probability of the encounter to occur will be decreased and the dependence on the association and dissociation constants on external force can be written in the form [Hill, 1981, Hill and Kirschner, 1982]:

$$k_{+} = k_{+0} \exp(-lF / 2kT) \tag{2}$$
$$k_{-} = k_{-0} \exp(lF / 2kT) \tag{3}$$

where $k_{-0} = 1.4\left[\frac{1}{s}\right], k_{+0} = 1.93 \cdot 10^{-14}\left[\frac{cm^3}{s}\right]$ are dissociation and association constants without forces present, l is the length by which F-actin filament will extend after addition one G-actin , k is the Boltzman constant, T is the absolute temperature, and F represents the opposing force.

The opposing forces in the case of locomoting cell may arise from at least two sources. First and most important, are resisting forces exerted by the tissue through which cell migrates, and second, the drag forces generated by membrane which is passively pulled past the anchoring proteins and must slide over the surrounding tissue surface.

In the micropipette experiment the locomoting cell can be subjected to a counter-pressure applied at the front of the cell which may be similar to the resisting force of surrounding tissue *in vivo*.

To evaluate membrane drag a first approximation of anchoring and transmembrane adhesion proteins can be modeled as cylinders. Then using the Saffman approach [Saffman, 1976, Saffman and Delbruck, 1975], the drag forces generated on the lipid membrane by transmembrane proteins per one F-actin filament can be described by equation:

$$F_d = A_1 \frac{ds}{dt}\left[s(t) - w(t)\right] \tag{4}$$

where

$$A_1 = \frac{8\pi\mu_s n_2}{r_c n\left(\log\frac{\mu_s}{\mu' a} - 0.0772\right)} \tag{5}$$

and μ_s is a surface viscosity of the lipid bilayer which is in range $10^{-4} - 10^{-3}\left[\frac{dyn \bullet s}{cm}\right]$ [Evans and Skalak, 1980], $n\left[\frac{1}{cm^2}\right]$ is a surface concentration of the F-actin ends at the leading edge of the pseudopod, $n_2\left[\frac{1}{cm^2}\right]$ is a surface concentration of protein generating drag, $r_c[cm]$ is a radius of the cell in the micropipette, a [cm] is a radius of cylinder (model of the protein), μ' is a viscosity of the water $10^{-2}\left[\frac{dyn \bullet s}{cm^2}\right]$, $\frac{ds}{dt}\left[\frac{cm}{s}\right]$ is velocity of the leading edge of the cell and (s(t) - w(t)) [cm] is the length of the pseudopod (Fig.1).

The opposing forces per F-actin filament arising from pressure applied to the front face of the cell is

$$F_p = \frac{p_{ext} - p_{\text{int}}}{C_{sF}} \tag{6}$$

where; $p_{ext}\left[\frac{dyn}{cm^2}\right]$ is the external pressure at the front of the cell, $p_{\text{int}}\left[\frac{dyn}{cm^2}\right]$ is the internal pressure at the front inside the pseudopod, $C_{sF}\left[\frac{1}{cm^2}\right]$ is the surface concentration of F-actin ends which are reacting with G-actin

The estimated value for the drag force F_d in locomoting PMN with velocity $0.3\left[\frac{\mu m}{s}\right]$ and length of the pseudopod 4 μm, with assumption that radius of model anchoring proteins is 3 nm and F-actin gel is isotropic with surface density of F-actin $4 \cdot 10^{10}\left[\frac{1}{cm^2}\right]$ and each end of the filament is anchored to the membrane is about $1.73 \cdot 10^{-3}[dyn]$. This value is equivalent to 9 cm H_2O applied at the front of the cell. This simple estimation showing that even in the case then no external pressure is applied to locomoting cell the polymerization of the G-actin is doing work to drag the lipid membrane forward.

The second source of active force generation is actin-myosin interaction. A large family of actin-myosin based motors are functionally defined by their ability to generate directional movement. The myosins were originally divided into two groups: conventional (having the ability to form thick filament) and unconventional. Additional recently discovered groups bring the total up to seven, and there may be more [Cheney, Riley, et al., 1993, Titus, 1993]. It is likely that the many cells contain multiple types of myosins, each of which plays a unique role in generating actin-based movements, but the exact role for each myosin is presently not known.

Several experiments have shown that the cells (Dictyostelium discoideum) without heavy chain myosin have a reduced capability of locomotion [Knecht and Loomis, 1988, Peters, Knecht, et al., 1988, Schwyter, Kron, et al., 1990, Wessels, Soll, et al., 1988]. Conventional myosin is necessary for cytokinesis [Egenhoff, Brown, et al., 1991, Fukui,

Lozanne, et al., 1990]. Immunofluorescence localization of myosin II in fixed cells, and microinjection of fluorescent analog of myosin II into living cells have shown that the myosin II is absent from initial protrusion at the leading edge of migrating fibroblast [Conrad, Nederlof, et al., 1989]. Experiments during wound healing responses of fibroblasts have shown that the myosin II is also absent from initial protrusions and is concentrated at the base of the pseudopod as well as in more distant transverse fibers. Initial protrusions may contain actin and myosin I [Wang, Butler, et al., 1993]. Conventional myosin is absolutely required for normal cytokinesis, capping, and generation of the cortical tension [Pasternak, Spudich, et al., 1989, Titus, Wessel, et al., 1993]

At a molecular level, actin-myosin interactions proceed in a step-wise manner. The step size of myosin associated with single ATP hydrolysis has been estimated to be 8±2 nm [Toyoshima, Kron, et al., 1990], and 10-15 nm by Ishijima [Ishijima, Doi, et al., 1991]. The actin/myosin II force generation is regulated by calcium [Janson, Sellers, et al., 1992]. Using a remote-sensitive, isometric force transducer, the forces generated by locomoting PMN have been measured directly by ferrite probes [Guilford, Lantz, et al., 1993]. A neutrophil can generate forces on the order of 10 nN in contractions.

The activation of the myosin is a first order reaction and can be described by the equation:

$$\frac{dc_{ma}}{dt} = k_{+m} c_m c_{Cca} - k_{-m} c_{ma} \tag{7}$$

where : $c_{ma}\left[\frac{1}{cm^3}\right]$ is concentration of activated myosin, $c_m\left[\frac{1}{cm^3}\right]$ is concentration of no active myosin, $c_{Ca}\left[\frac{1}{cm^3}\right]$ is concentration of calcium and k_{+m}, k_{-m} are the association and dissociation constants respectively.

The activation of the myosin is very fast compare to other processes taking place during locomotion. Hence this reaction can be treated as quasi-state, and the concentration of activated myosin can be written in the form:

$$c_{ma} = \frac{c_{mT}\left(c_{Ca} - c_{CaT}\right)\chi\left(c_{CaT}\right)}{\left(c_{Ca} - c_{CaT}\right) + K_{mD}} \tag{8}$$

where, $c_{mT}\left[\frac{1}{cm^3}\right]$ is the total concentration of myosin, $c_{CaT}\left[\frac{1}{cm^3}\right]$ is the threshold concentration of calcium and

$$K_{mD} = \frac{k_{-m}}{k_{+m}}\left[\frac{1}{cm^3}\right] \tag{9}$$

$$\chi\left(c_{CaT}\right) = \begin{cases} 0 & for \quad c_{Ca} < c_{CaT} \\ 1 & for \quad c_{Ca} > c_{CaT} \end{cases} \tag{10}$$

The force generated by contraction of myosin in a volume V_{cm} is

$$F_{cm} = f_m c_{ma} V_{cm} \tag{11}$$

where: f_{cm} is the force generated by one myosin filament, V_{cm} is volume of the reaction chamber.

Signal generation at the leading edge of the cell

Cell-surface receptors play a variety of roles in PMN functions. A schematic sketch of the events which take place at the cell membrane at the front of the cell in response to fMLP is shown in Fig.3. Binding of the fMLP to its receptor on the cell membrane surface induces dynamic changes inside the neutrophil. One aspects of these involves G-protein. The fMLP also stimulates the appearance and reappearance of new binding sites for the fMLP, and such cycles are necessary to sustain chemotaxis over long periods of time [Norgauer, Eberle, et al., 1991].

The first steps of the PMN activation require an assembly of ternary complex (Fig. 3). The ligand (L) binds to the receptor (R), then the complex (LR) interact with G-protein (G) and releases the endogenous GDP to form the stable ternary complex [Fay, Posner, et al., 1991]. The G-proteins mediate regulation of several effectors molecules and constitute a large family of high homologous proteins. They are heterotrimers consisting of α- β- and γ-subunits [Sternweis and Smrcka, 1992]. The α-subunit appear to be most diverse and can also account for the primary activity of the G-proteins. The α-subunits (G_1) activate the phospholipase C (PLC). The complex ligand-receptor-βγ-subunit can then dissociate to

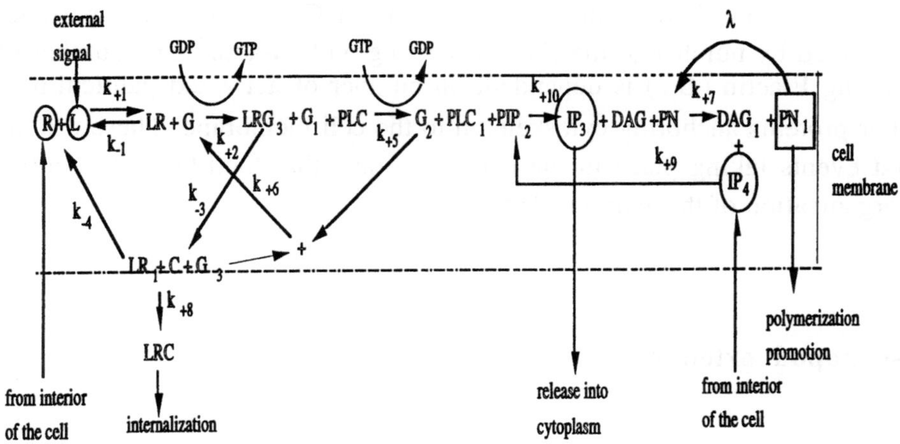

Fig 3. Simplified sketch of the generation of the secondary messengers during activation of the cell by external signals. (Adapted from Oster (1987), Aderen (1992), Liscovitch (1992), Sklar (1990))

LR complex and βγ-subunit (G_3) and the ligand-receptor complex can return to the L and R or form a slowly dissociating desensitized receptor LR_1 which interacts with cytoskeleton (C) and is internalized [Johansson, Wymann, et al., 1993, Sklar and Omann, 1990, Spiegel, 1992]. The βγ-subunit binds to the α-subunits to reestablish G-protein. It is clear that many more G protein-regulated pathways exist [Asaoka, Nakamura, et al., 1992, Hepler and Gilman, 1992].

In further steps shown in Fig.3 activation of the PLC stimulates the phosphatidylinositol 4,5 bisphosphate (PIP_2) break down generating the IP_3 and DAG. The IP_3 is released to the cytoplasm and seems to be responsible for release calcium from internal granules [Connor, 1993], converting itself into IP_4 which is transported to the leading edge reacting with the DAG forming IP_3. The DAG remains at the membrane surface and is responsible for activation of PKC and panticulin (PN) - which in turn promotes polymerization of the G-actin. The velocity of the front of the cell is controlled by the polymerization of G-actin which is partially controlled by number of the F-actin reacting with G-actin. The number of reacting F-actin ($\boldsymbol{C_{sF}}$) is dependent on number of activated panticulin or other proteins anchoring cytoskeleton to the cell membrane. These are the first events taking place in the cell membrane that facilitate and control reorganization of the actin inside the cell.

Pseudopod extension

The rate of extension of the pseudopod is the apparent cell forward velocity. the mechanisms of the pseudopod extension suggested here follows the model of Zhu and Skalak (1988). The pseudopod is assumed to be a porous medium with constant porosity ϕ. The fluid phase in Zone A and Zone B is considered to be incompressible. The G-actin is transported to the front of pseudopod as complex G-actin-Tβ4, by diffusion. This transport is discribed by the equation:

$$\frac{\partial c_G}{\partial t} = v_f \frac{\partial c_G}{\partial x} + D_G \frac{\partial^2 c_G}{\partial x^2} \qquad w(t) < x < s(t) \qquad (12)$$

where, D_G is diffusion coefficient for complex G-actin-Tβ4, c_G is the concentration of the complex, v_f is the fluid velocity inside the pseudopod and s(t) and w(t) are the coordinates of the leading edge of the cell and depolymerization front respectively (Fig.1).

It is assumed that the fluid is incompressible and that the fluid does not escape at the front of the cell. Then the fluid velocity inside the pseudopod is equal to the extension of the F-actin network; because the volumetric fraction of the solid network is small ($\phi_s \approx 0.01$) the fluid velocity is described by expression:

$$v_f = \frac{ds}{dt} = \frac{C_{sF}}{C_{sT}} l \left(k_{+0} e^{\frac{-lF_d - A_1 \frac{ds}{dt}(s(t)-w(t))}{2kT}} c_G(x=s,t) - k_{-0} e^{\frac{lF_d + A_1 \frac{ds}{dt}(s(t)-w(t))}{2kT}} \right) \qquad (13)$$

where, C_{sT} is the surface concentration of the F-actin filament ends.

The transport of the IP_3, IP_4, $T\beta_4$, R_{ad}, R_{fM} inside pseduopod are also described by convection-diffusion equations;

$$\frac{\partial c_i}{\partial t} = v_f \frac{\partial c_i}{\partial x} + D_i \frac{\partial^2 c_i}{\partial x^2} \qquad w(t) < x < s(t) \qquad (14)$$

where, $c_i, D_i \left(i = IP_3, IP_4, T\beta_4, R_{ad}, R_{fM}\right)$ are concentrations and diffusion coefficients of the IP_3, IP_4, $T\beta_4$, adhesion receptors (R_{ad}), and fMLP receptors (R_{fM}) respectively. In the model proposed here adhesion receptors and fMLP receptors are recycled. The adhesion receptors are internalized outside the pseudopd i.e. in the depolymerization zone and in the main cell body The fMLP receptors after association with the cytoskeleton at the leading edge of the pseudopod become internalized as the pseudopod advances forward and are released from the actin gel in depolymerization zone, released from fMLP molecules and passively transported to the front of the cell.

The conservation of the G-actin at the front of the cell leads to,

$$\frac{\partial c_G}{\partial x}=-\frac{C_{sF}}{\phi D_G}\left(k_{+0}e^{\frac{-lF_d-A_1\frac{ds}{dt}(s(t)-w(t))}{2kT}}c_G-k_{-0}e^{\frac{lF_d+A_1\frac{ds}{dt}(s(t)-w(t))}{2kT}}\right)\qquad x=s(t)$$

(15)

Depolymerization takes place at x=w(t) and serves as a source of the G-actin molecules, which are instantaneously sequestered by the Tβ4 molecules forming the complex G-actin-Tβ4. Thus

$$c_G+\frac{C_{sT}}{l}\chi=c_{Gb}$$

$$\chi=\begin{cases}0 & if\quad \frac{dw}{dt}=0\\ 1 & if\quad \frac{dw}{dt}\neq 0\end{cases}\qquad x=w(t)\qquad (16)$$

and

$$c_{Gb}\frac{ds}{dt}-D_G\frac{\partial c_{Gb}}{\partial x}-\frac{dw}{dt}\left(c_{Gb}-\phi c_G\right)+\frac{C_{sT}}{l}\frac{dw}{dt}=\phi\left(c_G\frac{ds}{dt}-D_G\frac{\partial c_G}{\partial x}\right)$$

$$x=w(t)\qquad (17)$$

The concentration of the G-actin-Tβ4 complex in the zone C $\left(c_{Gb}\right)$ is described by the equation:

$$\frac{\partial c_{Gb}}{\partial t}+\frac{ds}{dt}\frac{\partial c_{Gb}}{\partial x}=D_G\frac{\partial^2 c_{Gb}}{\partial x^2}\qquad w(t)-l_c<x<w(t)$$

(18)

where l_c is the length of the main cell body, with the boundary condition,

$$\frac{\partial c_{Gb}}{\partial x}=0 \qquad\qquad x=w(t)-l_c \qquad (19)$$

The initial condition for the both equations are:

$$\text{At} \quad t=0 \qquad c_G=c_{Gb}=c_0 \qquad \textit{and} \qquad w=s=0 \qquad (20)$$

where the c_0 is the initial concentration of the G-actin complexes (the concentration of the complexes in the resting cell).

The depolymerization zone

The chemistry that effects depolymerization is a complicated series of molecular events that are only partially understood. The actin can react with numerous other proteins and molecular factors which can affect polymerization and depolymerization rates, cross-link filaments or cut actin chains into small units. When the concentration of calcium is sufficiently high, cutting protein will be activated and fragment actin chains. The activation of cutting protein is a first order reaction;

$$\frac{dc_{wa}}{dt}=k_{+w}c_w c_{Caw}-k_{-w}c_{wa} \qquad (21)$$

where, $c_{wa}\left[\frac{1}{cm^3}\right]$ is the concentration of the active cutting protein, $c_w\left[\frac{1}{cm^3}\right]$ is the concentration of the nonactive cutting protein, $c_{Caw}\left[\frac{1}{cm^3}\right]$ is the concentration of the calcium above threshold value, $k_{+w}\left[\frac{cm^3}{s}\right], k_{-w}\left[\frac{1}{s}\right]$ are association and dissociation constant for this reaction.

The time scale of the cutting process is much faster that overall time of the cell locomotion, and in steady-state the concentration of active cutting protein can by expressed;

$$c_{wa} = \frac{c_{wT} c_{Caw}}{c_{Caw} + K_{Dw}} \tag{22}$$

where, c_{wT} is the total concentration of the cutting protein and

$$K_{Dw} = \frac{k_{-w}}{k_{+w}} \left[\frac{1}{cm^3} \right] \tag{23}$$

The cutting process for one filament of F-actin, is described by the probability of binding of the cutting protein to a site on the F-actin chain and is proportional to the number of structural units in the chain. Thus the cutting process can be described by [Nossal, 1988]

$$\frac{dK}{dt} = k_{+cu} c_{wa} \left(r_0 - 1 - K \right) - k_{-cu} K \tag{24}$$

where: K is the number cuts per F-actin filament, r_0 is the average number of unbounded sites on one F-actin filament and k_{+cu}, k_{-cu} are association and dissociation constants for the cutting reaction.

In quasi steady-state the number cuts per F-actin filament is expressed by:

$$K = \frac{c_{wa} \left(r_0 - 1 \right)}{K_{cu} + c_{wa}} \tag{25}$$

where $K_{cu} = \dfrac{k_{-cu}}{k_{+cu}}$ and the total number of the F-actin ends in the control volume is:

$$n_x = n \cdot K \tag{26}$$

where n is the number of the F-actin filaments before cutting occurs.

The depolymerization velocity is described by the equation:

$$\frac{dw}{dt} = l\left(n_x k_{-dep} - c_{Gf} k_{+dep} \right) \qquad x = w\ (t) \tag{27}$$

where c_{Gf} is concentration of the free G-actin.

The rise of the cytosolic calcium produces a number of important effects on the actin network. The elastic modulus of the gel is regulated by solation or gelation of the gel which depends on calcium concentration. Calcium also initiates phosphorylation of the myosin light chain, which leads to generation contraction forces.

The adhesion process

Adhesion is required for the locomotion of the cell on a substrate or through a solid tissue. Adhesion can be regulated by several different pathways; (i) modulation of the number of receptors molecules expressed on the cell surface (ii) modification or removal of the molecules by activation process or enzymes (iii) specific activation of the receptors to affect their ligand binding affinity (iv) changes in the organization of the cytoskeleton and of adhesion receptors at the cell surface (v) change in concentration of ligands.

Under normal circumstances, neutrophils flow through the vascular system and make a few contacts with the endothelial cells. But within minutes of an exposure to the an inflammatory stimulus, PMNs roll along the wall, adhere firmly and then emigrate through the endothelial cell layer [Diamond and Springer, 1993].

In the adhesion phenomenon, the principle of apparent redundancy appears to hold, as reflected by the variety of actin-binding proteins, the diverse cytoplasmic motors, and probably also the mechanisms responsible for coupling cytoskeletal proteins to receptors [Isenberg and Goldmann, 1992]. Cell adhesion usually involves ligands binding to the receptors on the cell membrane which are also anchored to the cytoskeleton . Association of the ligand and the cell cytoskeleton is a dynamic process in which individual receptors are continuously forming and breaking the association [Miettinen, Matter, et al., 1992, Ward and Hammer, 1993].

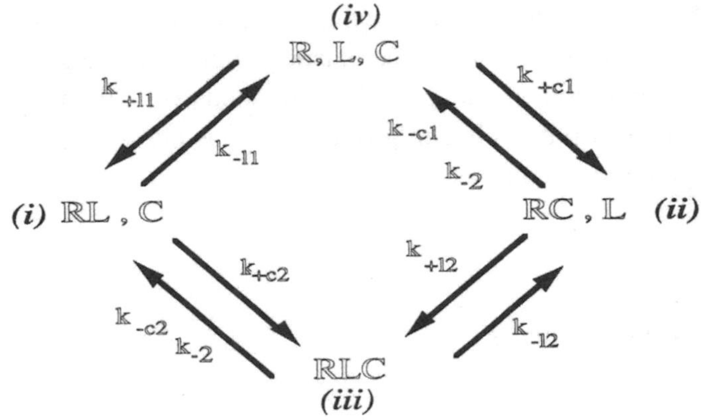

Fig.4 Schematic representation of the adhesion process.
R = adhesion receptor, L = ligand, C = cytoskeleton.
A comma indicates no connection.

The schematic representation on this process is shown in Fig.4. The adhesion receptors can be in four states: (i) receptor bound to the ligand but mobilized in respect to the cell cytoskeleton, (ii) receptor bound to the cytoskeleton but not bound to the ligand, (iii) receptor bound to ligand and cytoskeleton, generating the true adhesion bonds which capable of transmitting forces, and (iv) with no bonds, the receptor is free to diffuse in the cell membrane. The rate equations take the form:

$$\frac{dc_{RL}}{dt} = k_{+l1} c_R c_L + \left(k_{-c2} + k_{-2}\right) c_{RLC} - k_{-l1} c_{RL} - k_{+c2} c_{RL} c_C \tag{28}$$

$$\frac{dc_{RC}}{dt} = k_{+c1} c_c c_R + k_{-l2} c_{RCL} - \left(k_{-2} + k_{-c1}\right) c_{RC} - k_{+l2} c_L c_{RC} \tag{29}$$

$$\frac{dc_{RCL}}{dt} = k_{+l2} c_L c_{RC} + k_{+c2} c_{RL} c_C - \left(k_{-l_2} + k_{-c_2} + k_{-2}\right) c_{RCL} \tag{30}$$

where, $c_{RCL}\left[\frac{1}{cm^2}\right]$ is surface concentration of the complex receptor-ligand-cytoskeleton , $c_L\left[\frac{1}{cm^2}\right]$ is surface concentration of the ligand, $c_C\left[\frac{1}{cm^2}\right]$ is surface concentration of the attachment sites on the cytoskeleton, $c_{RL}\left[\frac{1}{cm^2}\right]$ is surface concentration of the complex receptor-ligand, $c_{RC}\left[\frac{1}{cm^2}\right]$ is surface concentration of the complex receptor-cytoskeleton, $k_{-2}\left[\frac{1}{s}\right]$ is the rate of depolymerization of the cytoskeleton.

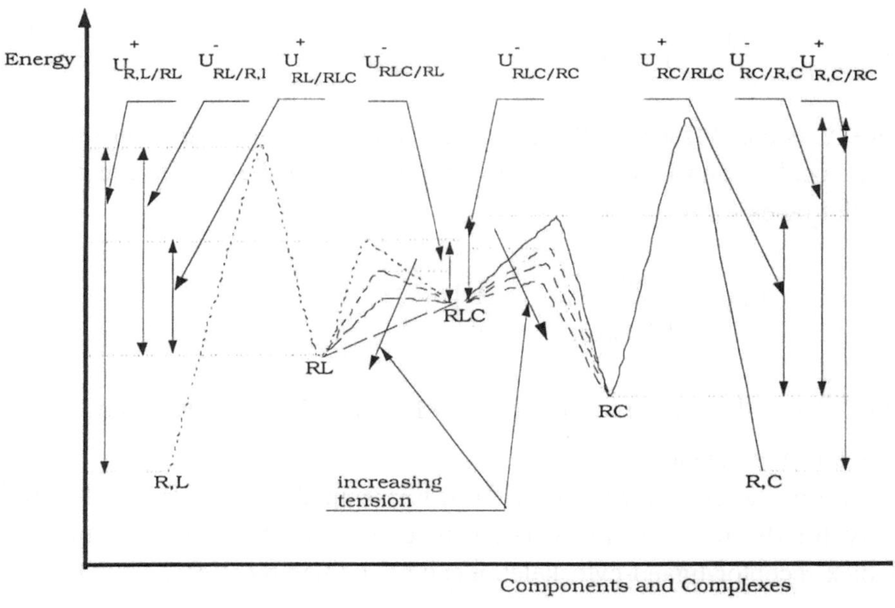

Fig.5 Sketch of energy potentials of complexes and components involved in adhesion. The effects of separation stresses (debonding tension) are shown as reducing energy barriers.

Binding of the ligand to the receptor can induce changes in receptor conformation and affinity. Similar effects can occur after binding of the receptor to the cytoskeleton. The sketch of the potential energy distribution in adhesion process is shown in Fig.5. We propose that the bariers to formation of the RL, RC, RLC and dissociation of the RC have energies which do not depend on changes of potential energy delivered to the system through the stretching by tension, as shown in Fig.5. The complex interactions that determine the detailed energetic of adhesion is not very well understood, but general qualitative forms involved are suggested in Fig.5.

The electrostatic and steric repulsion forces may be modeled by a simple phenomenological form suggested by Bell [1984] [Bell, Dembo, et al., 1984],

$$\Gamma(y)=\frac{\gamma}{y}\exp\left(-\frac{y}{\tau}\right) \tag{31}$$

where γ is the stiffness of the glycocalyx (10^{-6} dyn), τ is the thickness of the glycocalyx and Γ is the repulsion potential. For the single receptor the potential takes form:

$$\Gamma(y)=\frac{\gamma A_c}{y A_2 c_{RLC}}\exp\left(-\frac{y}{\tau}\right) \tag{32}$$

where, A_2 is the cross-section of the RLC $\left(5.15\cdot 10^{-11}\left[cm^2\right]\right)$ and A_c is the area of the contact.

From the energy point of view it is assumed that the Hemholtz free energy for the receptor-ligand (RL), receptor-cytoskeleton (RC) and the complex receptor-ligand-cytoskeleton can be defined by [Ben-Naim, 1992]

$$A_i=-kT\log q_i \qquad (i=RL,RC) \tag{33}$$

$$A_{RLC}=-kT\log q_{RLC} \tag{34}$$

where q is the partition function.

The equilibrium constant can then be defined as

$$K_{eqi} = \frac{q_i}{q_{RLC}} = \exp\left[-\frac{A_i - A_{RLC}}{kT}\right] \qquad (i = RL, RC) \tag{35}$$

Now it is assumed that stretching of the bounds delivers potential energy to the bounds increasing the rate of dissociation. The effect on external forces on the equilibrium constant have been also proposed by several authors [Bell, 1979, Cozens-Roberts, Lauffenburger, et al., 1990]

At the very beginning complex RLC behave as a freely joined random coil which have been modeled by Bell et al. (1984) as a linear spring with spring constant

$$k_s = \frac{kT}{L^2} \tag{36}$$

where L is the length of the complex.

In course of stretching, the bounds tend to behave as a α-helical protein segment. The first approximation model should included this phenomenon. As simple model we propose that the complex behaves as a nonlinear spring with potential energy changing according to the relation:

$$E_p = \kappa x^4 \tag{37}$$

where $\kappa = 2.38 \cdot 10^{12}\left[\frac{dyn}{cm^3}\right]$.

The dissociation coefficients for RLC can be expressed as

$$k_i = k_{oi}\, exp\left(\frac{\kappa x^4}{kT}\right) \qquad \left(i = -l2, -c2\right) \tag{38}$$

where k_{oi} are the dissociation constant without stretching, and x is the stretched length.

Actin gel contraction

As indicated above, it is assumed in the present model that there is an active contraction in the depolymerizing zone at the rear end of the pseudopod. Such contraction has been discribed for amoeba by Dembo (1987) on the basis of a two phase theory of the contracting gel. The theory of the mixture is based on the idea that the individual components of the mixture can be modeled as interpenetrating continua , such theories have been used extensively in soil science, [Biot, 1955, Biot, 1962, Biot, 1972, Kenyon, 1976]. These theories have been recently extended to study the compressive behavior of cartilage [Lai, Hou, et al., 1991, Mow, Ratcliffe, et al., 1992], deformation of porous material by unsteady flow [Barry and Aldis, 1991, Pascal, 1991] cardiac muscle [Yang and Taber, 1990], and brain tissue [Basser, 1992].

It is assumed that the two components (the solid network and the fluid) of the pseudopod are immiscible and each phase is intrinsically incompressible. It is also assumed there are no vesicles or bubbles. Then

$$\phi_s + \phi = 1 \tag{39}$$

where ϕ_s is the fraction of volume occupied by solid and ϕ is the fluid fraction. The continuity equations have the form

$$\frac{\partial \phi}{\partial t} = -\frac{\partial\left(\phi V_j^s\right)}{\partial x_j} - \frac{d\phi_p}{dt} \tag{40}$$

and

$$\frac{\partial}{\partial x_j}\left[\phi\left(V_j^f - V_j^s\right) + V_j^f\right] = 0 \tag{41}$$

where $V_j^s = \dfrac{du_j}{dt}$ is the velocity of the network, u_j is the displacement of the network, V_j^f is the fluid phase velocity and $\dfrac{d\phi_p}{dt}$ is rate at which fibers are subtracted from the gel.

The equations of motion for the fluid and solid phases are:

$$\frac{\partial}{\partial x_i}\left(\frac{\partial(2\mu_s E_{ij})}{\partial x_j}\right)+\phi\kappa\left(V_i^s-V_i^f\right)-\frac{\partial p}{\partial x_i}-\frac{\partial \pi}{\partial x_i}=0 \tag{42}$$

$$\frac{\partial \sigma_{ij}^e}{\partial x_j}+\phi\kappa\left(V_i^f-V_i^s\right)+\frac{\partial \sigma_{ij}^c}{\partial x_j}=0 \tag{43}$$

where μ_s is the viscosity of the fluid phase, E_{ij} is the strain rate in the fluid:

$$E_{ij}=\frac{1}{2}\left[\frac{\partial V_i^f}{\partial x_j}+\frac{\partial V_j^f}{\partial x_i}\right] \tag{44}$$

κ is the permeability of the porous medium, σ_{ij}^c is the contraction stress generating by activation of the myosin. The is the elastic stress σ_{ij}^e is:

$$\sigma_{ij}^e=\lambda e_{kk}\delta_{ij}+2\mu e_{ij} \tag{45}$$

λ,μ are Lame constants and e_{ij} is the strain in the solid:

$$e_{ij}=\frac{1}{2}\left[\frac{\partial u_i}{\partial x_j}+\frac{\partial u_j}{\partial x_i}\right] \tag{46}$$

p is the hydrodynamic pressure and π is the osmotic pressure.

To apply the above theory to the locomotion of a cell, the constitutive relations for the elastic moduli of the network, active stress and rate of fiber subtraction must be defined in terms of the variables ϕ_s and concentration of myosin, Ca^{++} and other constituents which control the contraction process. These relations are missing links at present in the sense that no quantitative relations or measurements for the PMN contraction zone are available. Studies of *in vitro* rheology of actin networks have shown that the active stress can be a very complicated function of the deformation and rate of deformation [Bremer, Milloning, et al., 1991, Evans, 1993, Howard, Chaponnier, et al., 1991], and also is strongly dependent of biochemical processes taking place inside the cell which are likely to be quite complex.

For proposes of the present model we adopt and simplify the theory of elasticity of polymer networks [Pearson and Graessley, 1978]. The Lame coefficients are assumed to have the forms:

$$\lambda = \lambda_0 e^{-(n_0 - n_f)} \tag{47}$$

$$\mu = \mu_0 e^{-(n_0 - n_f)} \tag{48}$$

where λ_0, μ_0 are elastic coefficients before activation or cutting of filaments, n_0 and n_f are the number actin filaments per unit volume before and after activation of the cutting proteins.

The active stress generated by the myosin can be described

$$\sigma_{ij}^{c} = \delta_{ij} \xi c_{ma} f_m \tag{49}$$

where $f_m[dyn]$ is force generated by one myosin filament, c_{ma} is concentration of activated myosin described by Eq.(8), ξ is the coefficient describing the fraction of the myosin heads not interacting with F-actin filament. If all myosin interact with F-actin and there is not a preferential polarization the coefficient ξ i s equal to one. The active stress and elastic coefficients are calcium dependent. Thus, there may be spatial gradients of the elastic coefficients due to spatial distribution of calcium:

$$\frac{\partial \lambda}{\partial x_i} = \frac{\partial \lambda}{\partial c_{Ca}} \frac{\partial c_{Ca}}{\partial x_i} \tag{50}$$

$$\frac{\partial \mu}{\partial x_i} = \frac{\partial \mu}{\partial c_{Ca}} \frac{\partial c_{Ca}}{\partial x_i} \tag{51}$$

and

$$\frac{\partial \sigma_{ij}^{c}}{\partial x_i} = \delta_{ij} \frac{\partial c_{ma}}{\partial c_{Ca}} \frac{\partial (\xi c_{Ca})}{\partial x_i} \tag{52}$$

The rate at which fibers of F-actin are depolymerized and subtracted from the gel can be expressed as,

$$\frac{d\phi_p}{dt} = \frac{V_{Gact}}{\phi V_c}\left(n_x k_{-dep} - c_{Gf} k_{+dep}\right) \tag{53}$$

where V_{Gact} is the volume of the one G-actin monomer, V_c is the volume of the reaction chamber and the rest of symbols are described by Eqs. (26), (27).

The equations (1-53) together with equation describing generation of the secondary messengers (Fig.3), distribution of the complexes ligand-receptor-cytoskeleton over the cell surface and appropriate boundary conditions constitute a complete model for the cell locomotion.

DISCUSSION

The extravasation and migration of the leukocytes through solid tissue involves complex series of processes including the detection of the chemoattractant, activation of the PMN, rolling, firm adhesion, pseudopod formation, gelation and contraction, penetration the endothelium cell layer. Some of these steps are not well understood as yet. In this paper we have proposed formulations of some of the processes of the PMN locomotion. This model is based on the observation of PMN locomotion in a micropippete experiment [Usami, Wung, et al., 1992b].

The general features of the proposed model are supported by various experiments reported in literature. Some of the proposed mechanisms require further experimental verification. The model proposed here is as simple as possible but retaining the main features of the process. It is proposed that in the cell locomotion of the PMN there are two different force generating mechanisms. One is the polymerization of actin at the front face of the cell: the other is contraction within the pseudopod region Fig.6. The fact that these processes can be independently controlled results in the possibility that the protrusion of the pseudopod can be different than the rate of depolymerization and hence the length of the pseudopod may vary [Skalak, Skierczynski, et al., 1993, Zigmond, 1993].

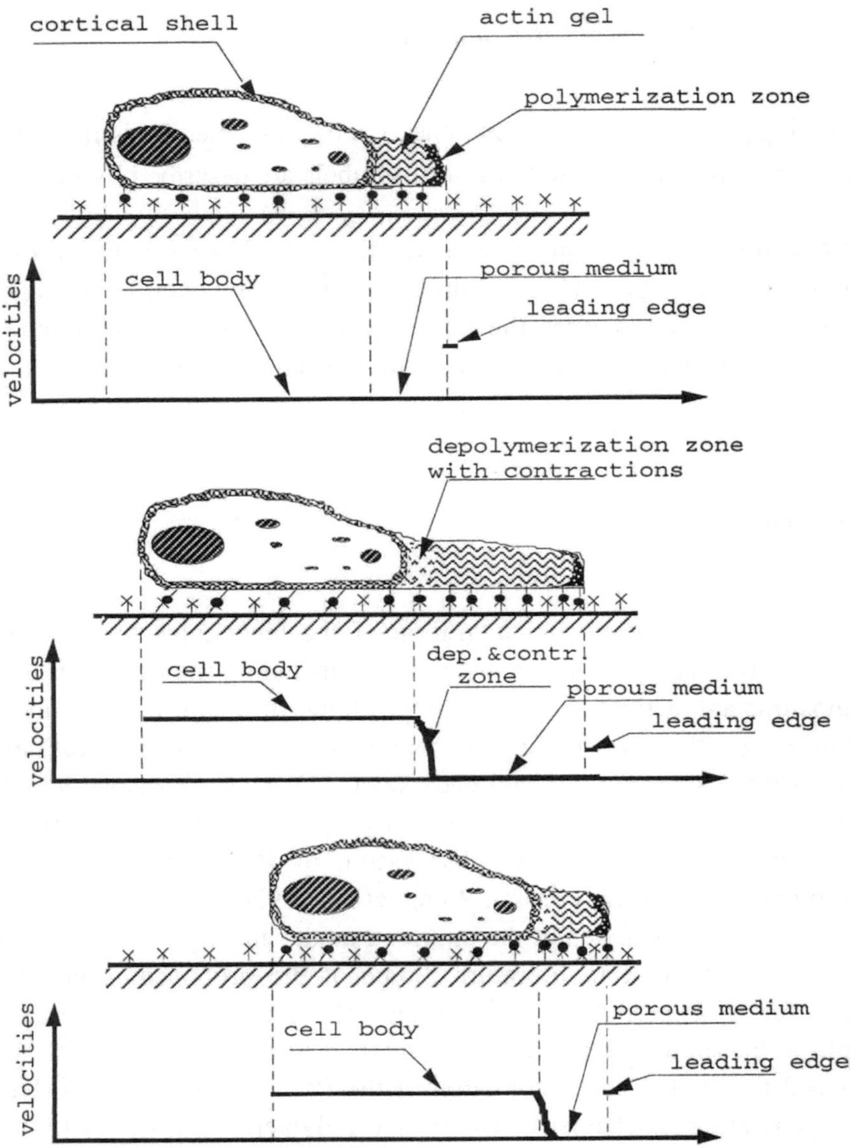

Fig. 6. Sketch of the proposed events taking place during locomotion of the cell.

The proposition that contraction is generated by myosin in the rear part of the pseudopod between cortical shell is supported by recent publications [Coates, Watts, et al., 1992, Kocks, Hellio, et al., 1993, Watts and Howard, 1992].

The proposed mechanism for converting chemical energy released during polymerization into mechanical work through the unfolding G-actin may require involvement of some specialized membrane proteins [Gabai, Kabakow, et al., 1992]. This is an alternative to the "Brownian ratchet". The polymerization force mechanism has also been proposed as the explanation of the translation of the bacterium *Listeria monocytogenes* in the cytoplasm of the host cell, protrusion of a pseudopod and the transport of protein through the cell membrane in recent experiments [Kocks, Hellio, et al., 1993].

While the present discussion is motivated by the experiment of Usami et al. (1992b), demonstrating locomotion of PMN's in a capillary, the same kinds and sequences of events are expected to pertain to locomotion of PMN's through a solid tissue or along a vascular wall. The interpretation suggested here for the case of cell locomotion along a planar substrate is shown schematically in Fig.6. In Fig. 6A, the pseudopod is extending but the cell body has not yet moved. In Fig. 6B, the cell body is moving faster then the leading (polymerizing) edge so the length of the pseudopod decreasing. In Fig. 6C, the velocity of the cell body is equal to the rate of polymerization at the leading edge and to the rate of depolymerization so the cell moves at constant velocity with no change in pseudopod length.

In Fig. 6A,B,C, the actin network (gel) of the psudopod is assumed to have zero velocity because of adhesion bonds which fix the actin cytoskeleton relative to the planar substrate. In Fig. 6B and C, the cell body is assumed to be moving. Thus there is zone in which the is a velocity gradient $\left(\frac{\partial V^s}{\partial x} < 0\right)$ which indicates contraction on purely kinematics grounds, i.e. a requirement due to the qualitative nature of the velocity distribution. It is possible that the contraction zone is wider then shown in Fig. 6B and C and may be two-dimensional rather than one -dimensional as assumed in Fig.6.

Some solutions of the governing equations (Eg.1-53) are available, but a simultaneous solution of the entire set of equation will require an extensive numerical (finite difference or finite elements) computation. Such comprehensive computations have not yet been completed.

ACKNOWLEDGMENT

This research was supported by U.S. National Institute of Health Grant HL-43026 (Dr. Shu Chien, Director)

REFERENCES

Aktories K. and Wagner A., (1992) Mechanisms of the Cytopathic Action of Actin-ADP-ribosylating Toxins., Molecular Microbiology, 6(20): 2905-2908.

Asaoka Y., Nakamura S.-i., Yoshida K. and Nishizuka Y., (1992) Protein Kinase C, Calcium and Phospholipid Degradation., Trends in Biochemical Sciences, 17(10): 414-417.

Baggiolini M. and Kernen P., (1992) Neutrophil Activation: Control of Shape Change, Exocytosis, and Respiratory Burst., News In Physiological Sciences, 7(October): 215-219.

Barry S.I. and Aldis G.K., (1991) Unsteady Flow Induced Deformation of Porous Material., Int. J. Non-linear Mechanics, 26(5): 687-699.

Basser P.J., (1992) Interstital Pressure, Volume, and Flow During Infusion into Brain Tissue., Microvascular Research, 44: 143-165.

Bell G.I., (1979) Theoretical Models for Cell-Cell Interaction in Immune Responces., in Physical Chemical Aspects of Cell Surface Events in Cellular Regulation Ed. C.Delisi, R.Blumenthal, : 371-386.

Bell G.I., Dembo M. and Bongrand P., (1984) Cell Adhesion., Biophysical Journal, 45: 1051-1064.

Ben-Naim A.,(1992), Statistical Thermodynamics for Chemists and Biochemists., Plenum Press: New York and London.

Bereiter-Hahn J., Strohmeier R., Beck K., (1983) Bestimmung des Dickenprofils von Zellen mit dem Reflexionskontrastmikroskop, Leitz-Mitt. Wiss. u, Techn. Bd VIII(5):147-150.

Bereiter-Hahn J., Strohmeier R., Beck K.,Kunzenbacher I., Beck K. and Voth M., (1981)Locomotion of *Xynopus* Epidermis Cells In Primary Culture, J.Cell Sci. 52:289-311.

Biot M.A., (1955) Theory of Elasticity and Consolidation for a Porous Anisotropic Solid., Journal of Applied Physics, 26(3): 182-185.

Biot M.A., (1962) Mechanics Deformation and Acoustic Propagation in Porous Media., Journal of Applied Physics, 33(4): 1482-1498.

Biot M.A., (1972) Theory of Finite Deformations of Porous Solids., Indiana University Mathematics Journal, 21(7): 597-620.

Birnbaumer L., (1992) Receptor-to-Effector Signaling through G Proteins: Roles for $\beta\gamma$ Dimers as well as α Subunits., Cell, 71: 1069-1072.

Bremer A., Milloning R.C., Sutterlin R., Engel A. and Pollard T.D., (1991) The Structural Basis for the Intrinsic Disorder of the Actin Filament: The "Lateral Slipping" Model., The Journal of Cell Biology, 115(3): 689-703.

Butcher E.C., (1990) Cellular and Molecular Mechanisms that Direct Leukocyte Traffic., American Journal of Pathology, 136(1): 311.

Cassimeris L., Safer D., Nachmias V.T. and Zigmond S.H., (1992) Thymosin β_4 Sequesters the Majority of G-actin in Resting Human Polymorghonuclear Leukocytes., The Journal of Cell Biology, 119: 1261-1270.

Cheney R.E., Riley M.A. and Mooseker M.S., (1993) Phylogenetic Analysis of the Myosin Superfamily, Cell Motility and the Cytoskeleton, 24: 215-223.

Chia C.P., Shariff A., Savage S.A. and Luna E.J., (1993) The Integral Membrane Protein, Ponticulin, Acts as a Monomer in Nucleating Actin Assembly., The Journal of Cell Biology, 120(4): 909-922.

Coates T.D., Watts R.G., Hartman R. and Howard T.H., (1992) Relationship of F-actin Distribution to Development of Polar Shape in Human Polymorphonuclear Neutrophils., The Journal of Cell Biology, 117: 765-774.

Condeelis J., (1992) Are All Pseudopods Created Equal?, Cell Motility and the Cytoskeleton, 22: 1-6.

Condeelis J., Bresnick A., Demma M., Dharmawardhane S., Eddy R., Hall A.L., Sauterer R. and Warren V., (1990) Mechanisms of Amoeboid Chemotaxis: An Evaluation of the Cortical Expansion Model., Developmental Genetics, 11: 33-340.

Connor J.A., (1993) Intracellural Calcium Mobilization by Inositol 1,4,5-triphosphate: Intracellular Movments and Compartmentalization., Cell Calcium, 14: 185-200.

Conrad P.A., Nederlof M.A., Herman I.M. and Taylor D.L., (1989) Correlated Distribution of Actin, Myosin, and Microtubules at the

Leading Edge of Migrating Swiss 3T3 Fibroblasts., Cell Motility and the Cytoskeleton, 14: 527-543.

Cooper J.A., (1991) The Role of Actin Polymerization in Cell Motility., Annu. Rev. Physiol., 53: 585-605.

Cortese J.D., Schwab III B., Frieden C. and Elson E.L., (1989) Actin Polymerization Induces a Shape Change in Actin-containing Vesicles., Proc. Natl. Acad. Sci. USA, 86: 5773-5777.

Cozens-Roberts C., Lauffenburger D.A. and Quinn J.A., (1990) Receptor-mediated cell attachment and detachment kinetics. I. Probabalistic model and analysis., 58: 841-856.

Cummings R.D. and Smith D.F., (1992) The Selectin Family of Carbohydrate-Binding Proteins: Structure and Importance of Carbohydrate Ligands for Cell Adhesion., BioEssays, 14(12): 849-856.

Cunningham C.C., Gorlin J.B., Kwiatkowski D., Hardwig J.H., Janmey P.A., Byers H.R. and Stossel T.P., (1992) Actin-binding Protein Requirement for Cortical Stability and Efficient Locomotion., Science, 255: 325-327.

Dembo M., (1986) The Mechanics of Motility in Dissociated Cytoplasm., Biophysical Journal, 50: 1165-1183.

Dembo M., (1989) Mechanics and Control of the Cytoskeleton in *Amoeba proteus*., Biophysical Journal, 55: 1053-1080.

Dembo M. and Harlow F., (1986) Cell Motion, Contractile Networks, and the Physics of Interpenetrating Reactive Flow., Biophysical Journal, 50: 109-121.

Diamond M.S. and Springer T.A., (1993) A Subpopulation of Mac-1 (CD11b/CD18) Molecules Mediates Neutrophil Adhesion to ICAM-1 and Fibrinogen., The Journal of the Cell Biology, 120(2): 545-556.

DiMilla P.A., Barbee K. and Lauffenburger D.A., (1991) Mathematical Model for the Effects of Adhesion and Mechanics on Cell Migration Speed., Biophysical Journal, 60: 15-37.

Downey G.P., Chan C.K., Lea P., Takai A. and Grinstein S., (1992) Phorbol Ester-Induced Actin Assembly in Neutrophils: Role of Protein Kinase C., The Journal of Cell Biology, 116: 695-706.

Dufort P.A. and Lumsden C.J., (1993) Cellular Automaton Model of the Actin Cytoskeleton., Cell Motility and the Cytoskeleton, 25: 87-104.

Egenhoff T.T., Brown S.S. and Spudich J.A., (1991) Spatial and Temporal Control of Nonmuscle Myosin Localization: Identification of a Domian That Is Necessary for Myosin Filament Disassembly In Vitro., The Journal of Cell Biology, 112(4): 677-688.

Estes J.E., Selden L.A., Kinosian H.J. and Gershman L.C., (1992) Tightly-bound Divalent Cation of Actin., Journal of Muscle Research and Cell Motility, 13: 272-284.

Evans E., (1993) New Physical Concepts for Cell Amoeboid Motion., Biophysical Journal, 64: 1-16.

Evans E.A. and Skalak R.,(1980), Mechanics and Thermodynamics of Biomembranes., CRC Press Inc.: Boca Raton, Florida

Fay S.P., Posner R.G., Swann W.N. and Sklar L.A., (1991) Real-time Analysis of the Assembly of Ligand, Receptor, and G Protein by Quantitative Fluorescene Flow Cytometry., Biochemistry, 30: 5066-5075.

Frieden C., (1985) Actin and Tubulin Polymerization: The Use of kinetic Methods to Determine Mechanisms., Ann. Rev. Biophys. Biophys. Chem., 14: 189-210.

Fukui Y., Lozanne A.D. and Spudich J.A., (1990) Structure and Function of the Cytoskeleton of a *Dictyostelium* Myosin-defective Mutant., The Journal of Cell Biology, 110: 367-378.

Fuortes M., Jin W.-W. and Nathan C., (1993) Adhesion-dependent Protein Tyrosine Phosphorylation in Neutrophils Treated with Tumor Necrosis Factor., The Journal of Cell Biology, 120(3): 777-784.

Gabai V.L., Kabakow A.E. and Mosin A.F., (1992) Association of Blebbing with Assembly of Cytoskeletal Proteins in ATP-Depleted EL-4 Ascites Tumour Cells., Tissue and Cell, 24(2): 171-177.

Galisteo M.L. and King J., (1993) Conformation Transformations in the Protein Lattice of Phage P22 Procapsids., Biophysical Journal, 65: 227-235.

Goldschmidt-Clermont P.J., Furman M.I., Wachsstock D., Safer D., Nachmias V.T. and Pollard T.D., (1992) The Control of Actin Nucleotide Exchange by Thymosin β_4 and Profilin. A Potential Regulatory Mechanisms for Actin Polymerization in Cells., Molecular Biology of the Cell, 3: 1015-1024.

Goldschmidt-Clermont P.J. and Janmey P.A., (1991) Profilin, a Weak CAP for Actin and RAS., Cell, 66: 419-421.

Gordon D.J., Yang Y.-Z. and Korn E.D., (1976) Polymerization of *Acanthamoeba* Actin: Kinetics, Thermodynamics, and Co-polymerization with Muscle Actin., J. Biol. Chem., 251: 7474-7479.

Griffiths S.K. and Nilson R.H., (1992) Freezing Flow in a Subcooled Permeable Medium., Transactions of the ASME, 114: 1036-1041.

Guilford W.H., Lantz R.C. and Gore R.W., (1993) The Forces Generated by Single Neutrophils and Macrophages Migrating *In Vivo* and *In Vitro*., University of Arizona, College of Medicine, : March 30,1993.

Hall A., (1992) Ras-Related GTPase and the Cytoskeleton., Molecular Biology of the Cell, 3: 475-479.

Hepler J.R. and Gilman A.G., (1992) G Proteins., Trends In Biochemical Sciences, 17(10): 383-387.

Herman I.M., (1993) Actin Isoforms., Current Opinion in Cell Biology, 5: 48-55.

Hill T.L., (1981) Microfilament or Microtubile Assembly or Disassembly Against a Force., Proc. Natl. Acad. Sci. USA, 78(9): 5613-5617.

Hill T.L. and Kirschner M.W., (1982) Subunit Treadmilling of Microtubules or Actin in the Presence of Cellular Barriers: Possible Conversion of Chemical Free Energy Into Mechanical Work., Proc. Natl. Acad. Sci. USA, 79: 490-494.

Hofmann A., Eichinger L., Andre E., Reiger D. and Schleicher M., (1992) Cap 100, a Novel Phosphatidylinositol 4,5-Biophosphate-regulated Protein that Caps Actin Filaments but Does Not Nucleate Actin Assembly., Cell Motility and the Cytoskeleton, 23: 133-144.

Hogg N., (1992) Roll, Roll, Roll Your Leucocyte Gently Down the Vein., Immunology Today, 13(4): 113-115.

Howard P.K., Sefton B.M. and Firtel R.A., (1993) Tyrosine Phosphorylation of Actin in *Dictyostelium* Associated with Cell-Shape Changes., Science, 259: 241-246.

Howard T., Chaponnier C., Yin H. and Stossel T., (1991) Gelsolin-Actin Interaction and Polymerization in Human Neutropils., The Journal of Cell Biology, 110: 1983-1991.

Hug C., Miller T.M., Torres M.A., Casella J.F. and Cooper J.A., (1992) Identification and Characterization of an Actin-binding Site of CapZ., The Journal of Cell Biology, 116(4): 923-931.

Isenberg G. and Goldmann W.H., (1992) Actin-membrane Coupling: A Role for Talin., Journal of Muscle Research and Cell Motility, 13: 587-589.

Ishijima A., Doi T., Sakurada K. and Yanagida T., (1991) Sub-piconewton Force Fluctuations of Actomyosin *in Vitro*., Nature, 352: 301-306.

Janmey P.A., Hvidt S., Oster G.F., Lamb J., Stossel T.P. and Hartwig J.H., (1990) Effect of ATP on Actin Filament Stiffness., Nature, 347: 95-99.

Janson L.W., Sellers J.R. and Taylor D.L., (1992) Actin-Binding Proteins Regulate the Work Performed by Myosin II Motors on Single Actin Filaments., Cell Motility and the Cytoskeleton, 22: 274-280.

Johansson B., Wymann M.P., Holmgren-Peterson K. and Magnusson K.-E., (1993) N-Formyl Peptide Receptors in Human Neutrophils Display Distinct Membrane Distribution and Lateral Mobility when Labeled with Agonist and Antagonist., The Journal of Cell Biology 121(6): 1281-1289.

Juliano R.J. and Haskill S., (1993) Signal Transduction from the Extracellular Matrix., The Journal of Cell Biology, 120(3): 577-585.

Keeney S.E., Schmalstieg F.C., Palkowetz K.H., Rudloff H.E., Le B.-M. and Goldman A.S., (1993) Activated Neutrophils and

Neutrophils Activators in Human Milk: Increased Expresion of CD11b and Decreased Expresion of L-selectin., Journal of Leukocyte Biology 54: 97-104.

Kenyon D.E., (1976) The Theory of an Incompressible Solid-Fluid Mixture., Arch .Rat. Mech. Anal., 62: 131-147.

Knecht D.A. and Loomis W.F., (1988) Developmental Consequences of the Lack of Myosin Heavy Chain in *Dictyostelium discoideum.*, Developmental Biology, 128: 178-184.

Kocks C., Hellio R., Gounn P., Ohayon H. and Cossart P., (1993) Polarized Disstribution of *Listeria Monocytogenes* Surface Protein ActA at the Site of Directional Actin Assembly., Journal of Cell Science 105: 699-710.

Kremen A., (1992) Biological Molecular Energy Machines as Measuring Devices., Journal of Theoretical Biology, 154: 405-413.

Kucik D.F., Elson E. and Sheetz M.P., (1990) Cell Migration Does Not Produce Membrane Flow., The Journal of Cell Biology, 11: 1617-1622.

Kuijpers T.W., Hakkert B.C., Hart M.H.L. and Roos D., (1992) Neutrophil Migration Across Monolayers Of Cytokine-Prestimulated Endothelial Cells: A Role For Platelet-Activating Factor and IL-8., The Journal of Cell Biology, 117(3): 565-572.

Lai W.M., Hou J.S. and Mow V.C., (1991) A Triphasic Theory for the Swelling and Deformation Behaviors of Articular Cartilage., Journal of Biomechanical Engineering, 113: 245-258.

Lasky L.A., (1992) Selectins: Interpreters of Cell- Specific Carbohydrate Information During Inflammation., Science, 258: 964-969.

Lauffenburger D., (1989) A Simple Model for the Effects of Receptor-mediated Cell-substratum Adhesion on Cell Migration., Chemical Engineering Science, 44(9): 1903-1914.

Linder M.E. and Gilman A.G., (1992) G Proteins., Scientific American, July: 56-65.

Liscovitch M., (1992) Crosstalk Among Multiple Signal-activated Phospholipases., Trends In Biochemical Sciences, 17: 393-399.

Lloyd A.R. and Oppenheim J.J., (1992) Poly's Lament, the Neglected Role of the Polymorphonuclear Neutrophil in the Afferent Limb of the Immune Response., Immunology Today, 13(5): 169-172.

Luna E.J. and Hitt A.L., (1992) Cytoskeleton-Plasma Membrane Interactions., Science, 258: 955-963.

Miettinen H.M., Matter K., Hunziker W., Rose J.K. and Mellman I., (1992) Fc Receptor Endocytosis Is Controlled by a Cytoplasmic Domian Determinant That Actively Prevents Coated Pit Localization., The Journal of Cell Biology, 116(4): 875-888.

Mow V.C., Ratcliffe A. and Poole A.R., (1992) Cartilage and Diarthordial Joints as Paradigme for Hiererchical Materials and Structures., Biomaterials, 13(2): 67-98.
Mueller S.C. and Chen W.-T., (1991) Cellular Invasion Into Matrix Beads:Localization Of β_1 Integrins And Fibronectin To The Invadopodia., Journal of Cell Science, 99: 213-225.
Norgauer J., Eberle M., Fay S.P., Lemke H.D. and Sklar L.A., (1991) Kinetics of N-formyl Peptide Receptor Up-regulation During Stimulation in Human Neutrophils., The Journal of Immunology, 146(3): 975-980.
Nossal R., (1988) On the Elasticity of Cytoskeletal Network., Biophysical Journal, 53: 349-359.
Oster G., (1988) Biophysics of the Leading Lamella., Cell Motility and the Cytoskeleton, 10: 164-171.
Oster G.F., (1984) On the Crawling of Cells., J. Embryol. exp. Morph, 83 Supplement: 329-364.
Oster G.F. and Odell G.M., (1984a) Mechanics of Cytogel I: Oscilation in Physarum., Cell Motility, 4: 469-503.
Oster G.F. and Odell G.M., (1984b) The Mechanochemistry of Cytogels., Physica, 12D: 333-350.
Oster G.F. and Perelson A.S., (1987) The Physics of Cell Motility., Journal of Cell Science, Suppl. 8: 35-54.
Otey C.A., Pavalko F.M. and Burridge K., (1990) An Interaction Between α-Actinin and the β_1 Integrin Subunits In Vitro., The Journal of the Cell Biology, 111: 721-729.
Pascal H., (1991) On Non-linear Efects in Unsteady Flows Through Porous Media., Int. J. Non-linear Mechanics, 26(2): 251-261.
Pasternak C., Spudich J.A. and Elson E.L., (1989) Capping of Surface Receptors and Concomitant Cortical Tension are Generated by Convetional Myosin., Nature, 341: 549-551.
Patarroy M. and Makgoba M.W., (1989) Leucocyte Adhesion to Cells., Scand. J. Immunol., 30: 129-164.
Pearson D.C. and Graessley W.W., (1978) The Structure of Rubber Networks with Multifunctional Joints., Macromolecules 11: 528-533.
Perelson A.S. and Coutsias E.A., (1986) A Moving Boundary Model for Acrosomal Elongation., Journal of Mathematical Biology, 23: 361-379.
Peskin C.S., Odell G.M. and Oster G.F., (1993) Cellular Motion and Thermal Fluctuations: The Brownian Ratchet., Biophysical Journal, 65: 316-324.
Peters D.J.M., Knecht D.A., Loomis W.F., Lozanne A.D., Spudich J. and Haastert P.J.M.V., (1988) Signal Transduction,Chemotaxis, and Cell Aggregation in *Dictyostelium*

discoideum Cells without Myosin Heavy Chain., Developmental Biology, 128: 158-163.

Pollard T.D., (1986) Actin and Actin-Binding Proteins. A Critical Evaluation of Mechanisms and Functions., Ann. Rev. Biochem., 55: 987-1035.

Pollard T.D. and Goldman R.D., (1993) Cytoplasm and Cell Motility., Current Opinion in Cell Biology, 5: 1-2.

Pollard T.P. and Mooseker M.S., (1981) Direct Measurements of Actin Polymerization Rate Constants by Electron Microscopy of Action Filaments Nucleated by Isolated Microvillus Cores., The Journal of Cell Biology, 88: 654-659.

Pontius B.W., (1993) Close Encounters: Why Unstructured, Polymetric Domains Can Increase Rates of Specific Macromolecular Association., TIBS, 18(May):

Redmond T. and Zigmond S.H., (1993) Distribution of F-Actin Elongation Sites in Lysed Polymorphonuclear Leukocytes Parallels the Distribution of Endogenous F-Actin., Cell Motility and Cytoskeleton, 26: 7-18.

Rot A., (1992) Endothelial Cell Binding of NAP-1/IL-8: Role in Neutrophil Emigration., Immunology Today, 13(8): 291-294.

Ruoslahti E., (1992) Control of Cell Motility and Tumor Invasion by Extracellular Matrix Interactions., British Journal of Cancer, 66: 239-242.

Safer D., (1992) The Interaction of Actin with Thymosin β_4., Journal of Muscle Research and Cell Motility, 13: 269-271.

Saffman P.G., (1976) Brownian Motion in Thin Sheets of Viscous Fluid., Journal of Fluid Mechanics, 73(4): 593-602.

Saffman P.G. and Delbruck M., (1975) Brownian Motion in Biological Membranes., Proc. Nat. Acad. Sci. USA, 72(8): 311-3113.

Samuelsson S.J., Luther P.W., Pumplin D.W. and Bloch R.J., (1993) Structures Linking Microfilaments Bundles to the Membrane st Focal Contacts., The Journal of Cell Biology, 122(2): 485-496.

Sarndahl E., Lindorth M., Bengtsson T., Fallan M., Gustavsson J., Stendahl O. and Andersson T., (1989) Association of Ligand-Receptor Complex with Actin Filaments in Human Neotrophils: A Possible Regulatory Role for the G-Protein., The Journal of Cell Biology, 109(No.6,Pt.1): 2791-2799.

Schutt C.E., Linderberg U., Myslik J. and Strauss N., (1989) Molecular Packing in Profilin: Actin Crystals and its Implications., Journal of Molecular Biolology, 209: 735-746.

Schwartz M.A., (1993) Spreading of Human Endothelium Cells on Fibronectin or Vitronectin Triggers Elevation of Intracellular Free Calcium., The Journal of Cell Biology, 120(4): 1003-1010.

Schwyter D.H., Kron S.J., Toyoshima Y.Y., Spudich J.A. and Reisler E., (1990) Subtilisin Cleavage of Actin Inhibits In Vitro Sliding Movment of Action Filaments Over Myosin., The Journal of Cell Biology, 111: 465-470.

Shibata M., Yamakawa Y., Ohoka T., Mizuno S. and Suzuki K., (1993) Characterization of a 64-kd Protein Phosphorylated During Chemotactic Activation with Il-18 and fMLP of Human Polymorphonuclear Leukocytes. II. Purification and Amino Acid Analysis of Phosphorylated 64-kd protein., Journal of Leukocyte Biology, 54: 10-15.

Skalak R., Skierczynski B.A., Wung S.-L., Chien S. and Usami S., (1993) Mechanical Models of Pseudopod Formation., Blood Cells, (in press):

Skierczynski B.A., Usami S., Chien S. and Skalak R., (1993) Active Motion of Polymorphonuclear Leukocyte in a Micropipette., : (in press).

Sklar L.A. and Omann G.M., (1990) Kinetics and Amplification in Neutrophil Activation and Adaptation., Seminars in Cell Biology, 1: 115-123.

Southwick F.S., Dabiri G.A., Paschetto M. and Zigmond S.H., (1989) Polymorphonuclear Leukocyte Adherence Induces Actin Polymerization by a Transduction Pathway wich Differs from that Used by Chemoattractants., The Journal of Cell Biology, 109: 1561-1569.

Southwick F.S. and Young C.L., (1990) The Actin Released From Profilin-Actin Complex Is Insufficient to Account for the Increase in F-Actin in Chemoattractant-stimulated Polymorphonuclear Leukocytes., The Journal of Cell Biology, 110: 1965-1973.

Spiegel A.M., (1992) G Proteins in Cellular Control., Current Opinion in Cell Biology, 4: 203-211.

Sternweis P.C. and Smrcka A.V., (1992) Regulation of Phospholipase by G Proteins., Trends in Biochemical Sciences, 17(12): 502-506.

Steven A.C., (1993) Conformational Change- An Alternative Energy Source?: Exothermic Phase Transition in Phase Capsid Maturation., Biophysical Journal, 65: 5-6.

Stossel T.P., (1989) From Signal to Pseudopod., The Journal of Biological Chemistry, 264(N0.31): 18261-18264.

Stossel T.P., (1990) How Cells Crawl., American Scientist, 78: 408-423.

Stossel T.P., (1993) On the Crawling of Animal Cells., Science, 260: 1086-1094.

Strohmeier R., Bereiter-Hahn J.,(1987) Hydrostatic Pressure in epidermal Cells is Dependent on Ca-mediated Contractions, J.Cell Sci. 88:631-640.

Strohmeier R., Bereiter-Hahn J.,(1984) Control of Cell Shape and Locomotion by External Calcium, Expermental Cell Research 154:412-420.

Symington B.E., Takada Y. and Carter W.G., (1993) Interaction of Integrins $\alpha_3\beta_1$ and $\alpha_2\beta_1$: Potential Role in Keratinocyte Intercellular Adhesion., The Journal of Cell Biology, 120(2): 523-535.

Tawil N., Wilson P. and Carbonetto S., (1993) Integrins in Point Contacts Mediate Cell Spreading: Factors That Regulate Integrin Accumulation in Point Contacts vs. Focal Contacts., The Journal of Cell Biology, 120(1): 261-271.

Theriot J.A., Mitchison T.J., Tilney L.G. and Portnoy D.A., (1992) The Rate of Actin-based Motility of Intracellular *Listeria monocytogenes* Equals the Rate of Actin Polymerization., Nature, 357: 257-260.

Titus M.A., (1993) Myosins.,Curr. Opinion in Cell Biol. 5: 77-81.

Titus M.A., Wessel D., Spudich J.A. and Soll D., (1993) The Unconventional Myosin Encoded by the *myo* A Gene Plays a Role in *Dictyostelium* Motility., Molecular Biology of the Cell, 4: 233-246.

Toyoshima Y.Y., Kron S.J. and Spudich J.A., (1990) The myosin step size: Measurement of the unit displacment per ATP hydrolized in an *in vitro* assay., Proc. Natl. Acad. Sci. USA, 87(Biophysics): 7130-7134.

Turley E.A., (1992) Molecular Mechanisms of Cell Motility., Cancer and Methastasis Revieves, 11: 1-3.

Urry D.W., (1992) Free Energy Transduction in Polypeptides and Proteins Based on Inverse Temperature Translocations., Progress in Biophysics & Molecular Biology, 57(1): 23-57.

Usami S., Wung S.-L., Skierczynski B.A., Skalak R. and Chien S. (1992a), Leukocyte Locomotion in Chemotaxis *Eight International Congerss of Biorheology,* Biorheology 29(1)

Usami S., Wung S.-L., Skierczynski B.A., Skalak R. and Chien S., (1992b) Locomotion Forces Generated by a Polymorphonuclear Leukocyte., Biophysical Journal, 63: 1663-1666.

Wachsstock D.H., Schwarz W.H. and Pollard T.D., (1993) Afinity of α-Actinin for Actin Determines the Structure and Mechanical Properties of Actin Filament Gels., Biophysical Journal, 65: 205-214.

Wang N., Butler J.P. and Inberg D.E., (1993) Mechanotransduction Across the Cell Surface and Trough the Cytoskeleton., Science, in press:

Ward M.D. and Hammer D.A., (1993) A Theoretical Analysis for the Effect of Focal Contact Formation on Cell-substrate Attachment Strength., Biophysical Journal, 64: 936-959.

Watts R.G. and Howard T.H., (1992) Evidence for a Gelsolin-Rich, Labile F-Actin Pool in Human Polymorphonuclear Leukocytes., Cell Motility and the Cytoskeleton, 21: 25-37.

Weber A., Nachmias V.T., Pennise C.R., Pring M. and Safer D., (1992) Interaction of Thymosin β_4 with Muscle and Platelet Actin: Implication for Actin Sequestration in Resting Platelets., Biochemistry, 31: 6179-6185.

Wertman K. and Drubin D.G., (1992) Actin Constitution: Guaranteeing the Right to Assemble., Nature, 258: 759-760.

Wessels D., Soll D.R., Knecht D., Loomis W.F., Lozanne A.D. and Spudich J., (1988) Cell Motility and Chemotaxis in *Dictyostelium* Amebae Lacking Myosin Heavy Chain., Developmental Biology, 128: 164-177.

Westlin W.F., Kiely J.-M. and Gimbrone_Jr. M.A., (1992) Interleukin-8 Induces Changes in Human Neutrophil Actin Conformation and Distribution: Relationship to Inhibition of Adhesion to Cytokine-activated Endothelium., Journal of Leukocyte Biology, 52: 43-51.

Woods A. and Couchman J.R., (1992) Protein Kinase C Involvment in Focal Adhesion Formation., Journal of Cell Science, 101: 277-290.

Yang M. and Taber L.A., (1990) The Possible Role of Poroelasticity in the Apparent Viscoelestic Behavior of Passive Cardiac Muscle., Journal of Biomechanics, 24(7): 587-597.

Zhu C. and Skalak R., (1988) A Continuum Model of Protrusion of Pseudopod in Leukocytes., Biophysical Journal, 54: 1115-1137.

Zigmond S.H., (1993) Recent Quantitative Studies of Actin Filament Turnover During Cell Locomotion., Cell Motility and the Cytoskeleton 25: 309-316.

ADHESION AND TRACTION FORCES IN MIGRATION: INSIGHTS FROM MATHEMATICAL MODELS AND EXPERIMENTS

Paul A. DiMilla
Department of Chemical Engineering and Center for Light Microscope Imaging and Biotechnology
Carnegie Mellon University
Pittsburgh, PA 15213-3890

Introduction

The migration of mammalian tissue and white blood cells plays a critical role in a diverse array of physiological and pathological phenomena, including the proper development and repair of organs, inflammation, angiogenesis, and metastasis (Trinkaus, 1984). The phenomenon of migration is complex, depending on coordinated interactions among many underlying biochemical and biophysical processes. Much information has been obtained at the molecular level about the identity and properties of components involved in these processes, especially receptor-mediated adhesion (Buck and Horwitz, 1987; Hynes, 1992) and cytoskeletal force generation (Stossel *et al.*, 1985; Bray and White, 1988), as well as at the cell behavioral level (Lackie, 1986; Singer and Kupfer, 1986; Devreotes and Zigmond, 1988; Heath and Holifield, 1991).

Our studies are motivated by a desire to understand the relationship between molecular-level components and mechanisms and the functional characteristics of cell adhesion and migration. In particular, the types of questions that we wish to pose include: How do changes in molecular-level adhesive properties, such as the number of adhesion receptors on the cell surface or the density of adhesion ligand on the substratum, affect the rate of migration? What is the role of cellular mechanics in active movement? Does an optimal adhesiveness exist for cell movement? We have addressed these questions through a variety of mathematical modeling (DiMilla *et al.*, 1991; DiMilla and Lauffenburger, 1991) and experimental (DiMilla *et al.*, 1992a; DiMilla *et al.*, 1993) techniques designed to both provide predictive and interpretive information.

NATO ASI Series, Vol. H 84
Biomechanics of Active Movement
and Division of Cells
Edited by N. Akkaş
© Springer-Verlag Berlin Heidelberg 1994

The Cell Biology of Adhesion-Mediated Movement on 2-D Substrata

Migration over adhesive surfaces involves the dynamic formation and breakage of attachments with the underlying substratum (Regen and Horwitz, 1992). These attachments are mediated primarily by specific reversible interactions between heterodimeric cell-surface adhesion receptors known as integrins (Albelda and Buck, 1990; Hemler, 1990; Hynes, 1992) and substratum-bound extracellular matrix (ECM) proteins (adhesion ligands), such as fibronectin and laminin (Straus *et al.*, 1989; Bauer *et al.*, 1992). Cytoskeletal contraction is achieved by the reversible polymerization of actin monomers into filaments (Taylor and Condeelis, 1979). Although many of the components involved in the transmission of force from the contractile apparatus to the underlying substratum have not been identified, integrins likely are part of a chain of transient molecular linkages (Mueller *et al.*, 1989; Heidemann *et al.*, 1990; Wang *et al.*, 1993): integrins bind both ECM proteins and cytoskeletal elements with low affinity ($K_d \approx 10^{-6}$ M for both integrin/fibronectin (Akiyama and Yamada, 1985) and integrin/talin interactions (Horwitz *et al.*, 1986)). Further, both short time-scale interactions -- such as binding between individual receptors and ligands -- and longer time-scale events (*e.g.*, clustering of adhesion receptors into focal contacts (Burridge *et al.*, 1988) and the formation of stress fibers, bundled filaments of actin, that terminate at focal contacts (Dejana *et al.*, 1987)) may be relevant to movement occurring over periods of many hours. The relative contributions of these different time-scale adhesive interactions to cell locomotion presently are poorly understood. However, cytoskeletal and cell surface organization appears to be important: in stationary cells integrins tend to aggregate into focal contacts and stress fibers are abundant. In contrast, actively migrating cells tend to exhibit an absence of focal contacts and stress fibers, with a more diffuse, mesh-like cytoskeleton (Duband *et al.*, 1988).

Intuitively, cell migration should depend on the strength of the transient cell-substratum attachments (Lackie and Wilkinson, 1984). Three regimes of motile, morphological, and adhesive behavior can be envisioned for a cell interacting with a surface (Fig. 1). Cells on surfaces in which adhesive interactions are much weaker than contractile forces can not generate net traction forces at the cell/substratum interface; no locomotion is possible and the cell spreads poorly. Conversely, cells on surfaces in which adhesive interactions are much stronger than contractile forces can not dynamically disrupt cell/substratum attachments; locomotion is not allowed but spreading does occur. Cells on surfaces in which adhesive and contractile forces balance, however, can move if the internal resistance to deformation can be overcome by intracellular contractile forces.

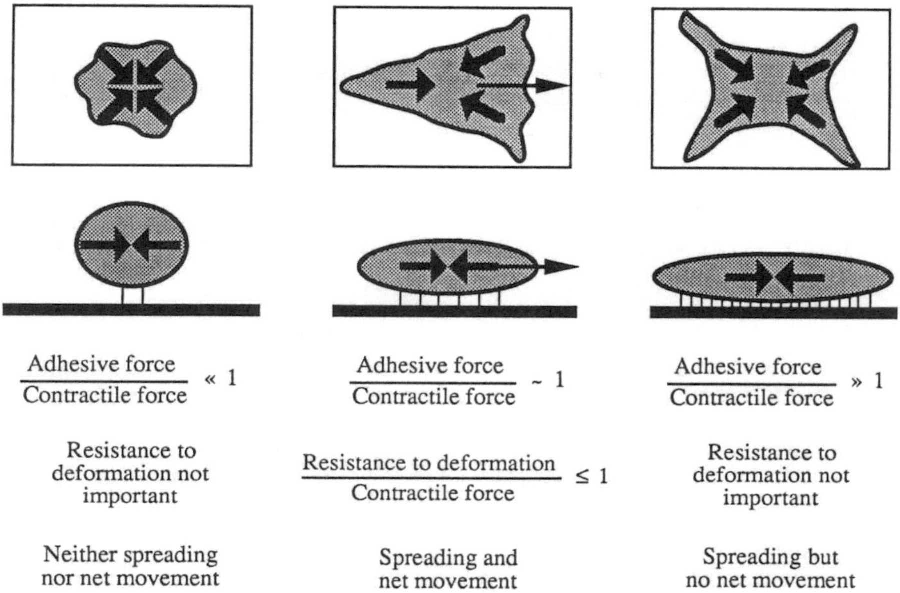

Fig. 1. Schematic of hypothetical relationships between cell-substratum adhesiveness, intracellular contractile force, cellular resistance to deformation, cell morphology, and migration.

Several recent experimental studies have demonstrated that variations in the density of substratum-bound adhesion ligands can affect motility. For example, we have studied the migration of human smooth muscle cells (HMSCs) on surfaces coated with different densities of the adhesion ligands fibronectin and collagen type IV using time-lapse videomicroscopy and image analysis (DiMilla *et al.*, 1993). The paths of individual HMSCs provide a qualitative picture of how migration varied with the composition of the substratum (Fig 2a), but we also can obtain rigorous information regarding the speed of individual cells by applying a persistent random walk model to data for the mean-squared displacement as a function of time for each individual cell path (Gail and Boone, 1970; Dunn, 1983; DiMilla *et al.*, 1992b). Fig. 2b demonstrates that HSMC speed reached a maximum at intermediate densities of each adhesion ligand; the optimal densities for maximum migration -- 10^3 molecules/μm^2 for fibronectin and 10^4 molecules/μm^2 for collagen -- differed, however, by an order of magnitude.

The addition of soluble competitors, such as adhesion ligands, analogs to these ligands, or antibodies to adhesion receptors, can affect how motility depends on the density of substratum-bound adhesion ligands. For example, Wu *et al.* (1993) recently have shown that

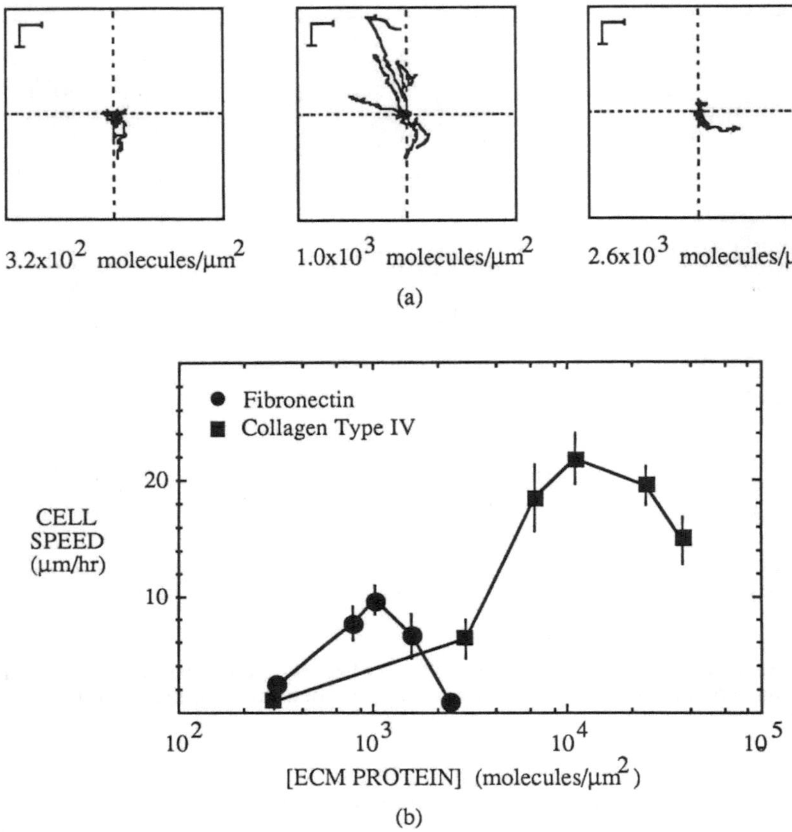

Fig. 2. Experimental data from DiMilla *et al.* (1993) for the relationship between the migration of individual human smooth muscle cells (HSMCs) and the surface density of adhesion ligand. (a) Typical cell paths over 24 hours, superimposed to a common origin in "wind-rose" displays, for cells migrating on three concentrations of fibronectin. Bars, 100 μm. (b) Biphasic relationships between cell speed, S, plotted as means and standard errors, and the surface densities of fibronectin and collagen type IV.

the random motility coefficient -- a measure of the ability of a population of cells to move on a uniform surface -- for rat microvessel endothelial cells (RMVECs) decreased with increasing density of adsorbed fibronectin but increased over the same range of density of fibronectin in the presence of soluble echistatin, a disintegrin protein which binds with high affinity ($K_d \approx 10^{-7}$ M) to integrins (Savage *et al.*, 1990) (Fig. 3).

Only a limited amount of experimental data currently exists regarding the role of a cell's mechanical properties on migration. One particular interesting example is provided by

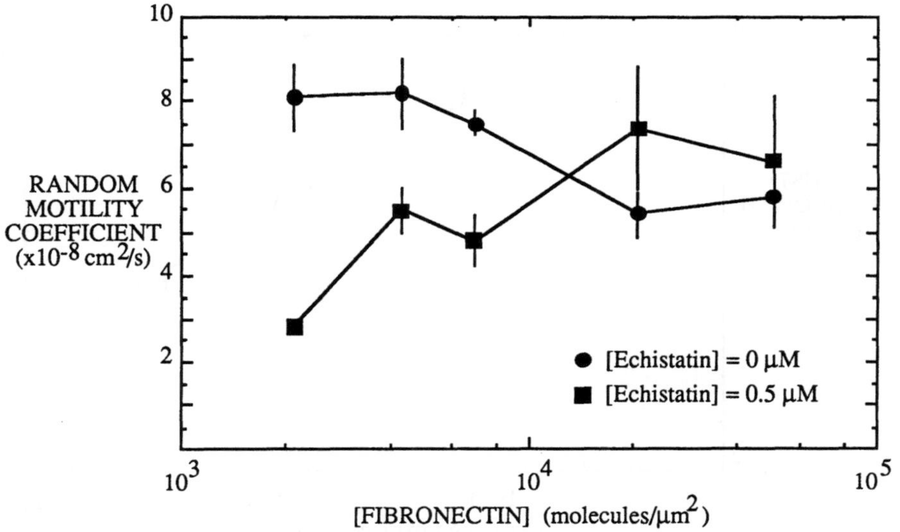

Fig. 3. Experimental data from Wu *et al.* (1993) for the random motility coefficient of rat microvessel endothelial cells (RMVECs) as a function of the surface density of fibronectin in the presence of different soluble concentrations of echistatin, an inhibitor of binding between integrin adhesion receptors and fibronectin.

Young and Herman (1985), who observed that the rate of outgrowth of wounded monolayers of endothelial cells was inversely proportional to the density of stress fibers within the migrating cells (Fig. 4). Nerem and coworkers have found that the formation of stress fibers correlated with an increase in the mechanical stiffness of endothelial cells (Sato *et al.*, 1987; Theret *et al.*, 1988), but no explanation has been suggested for why migration might be inversely proportional to cell stiffness. This observation, along with the data for the effects of variations in density of adhesion ligand on migration, directs the particular types of predictions we wish to make with our models.

Developing a Model for Migration Based on a Chronological Cycle

Any model for predicting the rate of migration must incorporate three general requirements identified by cell biologists for persistent cell movement over adhesive substrata (Trinkaus, 1984; Singer and Kupfer, 1986; Bershadsky and Vasiliev, 1988): (1) contraction of cytoskeletal elements generates intracellular forces; (2) dynamic adhesion processes transmit intracellular forces as traction forces at the cell/substratum interface; and, (3)

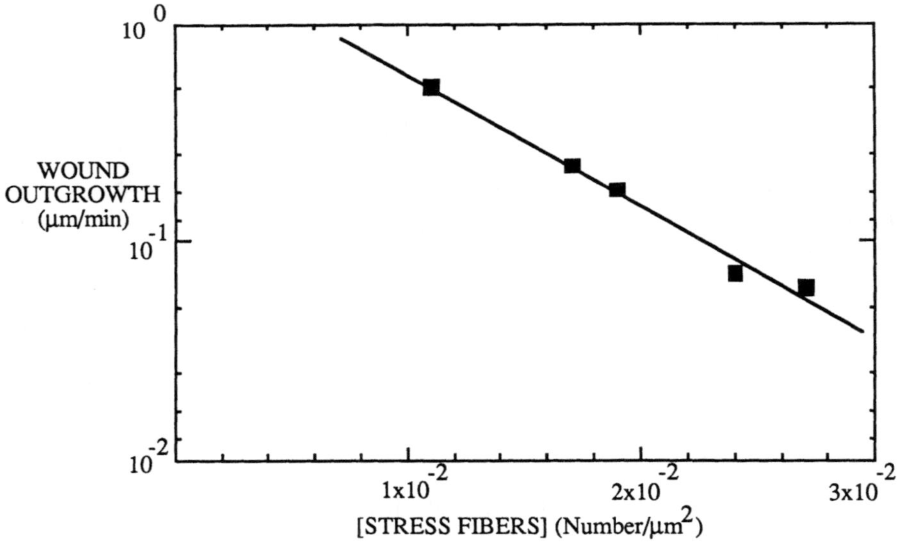

Fig. 4. Experimental data from Young and Herman (1985) for the relationship between the rate of outgrowth of a wounded monolayer of endothelial cells and the density of intracellular stress fibers.

morphological polarization promotes a spatial asymmetry in the traction force. We can conceptualize these characteristics by visualizing a moving cell as an automobile. Cytoskeletal contraction is the "motor", dynamic adhesion bonds are the "tires", and the mechanism responsible for asymmetry in traction is the "driver".

Observed with time-lapse microscopy over a time scale of minutes, the movement of an individual tissue cell appears to follow a saltatory cycle of extension at the cell's front (lamellipod) and retraction of the cell's tail (uropod) (Trinkaus, 1984, Harris, 1990; DiMilla *et al.*, 1993). During lamellipodal extension the body of the cell remains relatively stationary. These descriptions motivate modeling unidirectional cell movement as a temporal cycle of three distinct stages (Fig. 5). During the first stage of this cycle, a period t_l, the cell extends a lamellipod in the direction of motion and adhesion receptors reversibly bind with immobilized adhesion ligands on the underlying substratum until equilibrium is reached (the kinetics of adhesion receptor/ligand binding are much faster than the dynamics of lamellipodal extension). Our treatment ignores angular direction -- this characteristic is the subject of a recent model by Dickinson and Tranquillo (1993) -- and does not incorporate detailed descriptions of the biophysics of this extension (*i.e.*, Oster and Perelson (1987), Zhu and Skalak (1988), and Evans and Dembo (1990)).

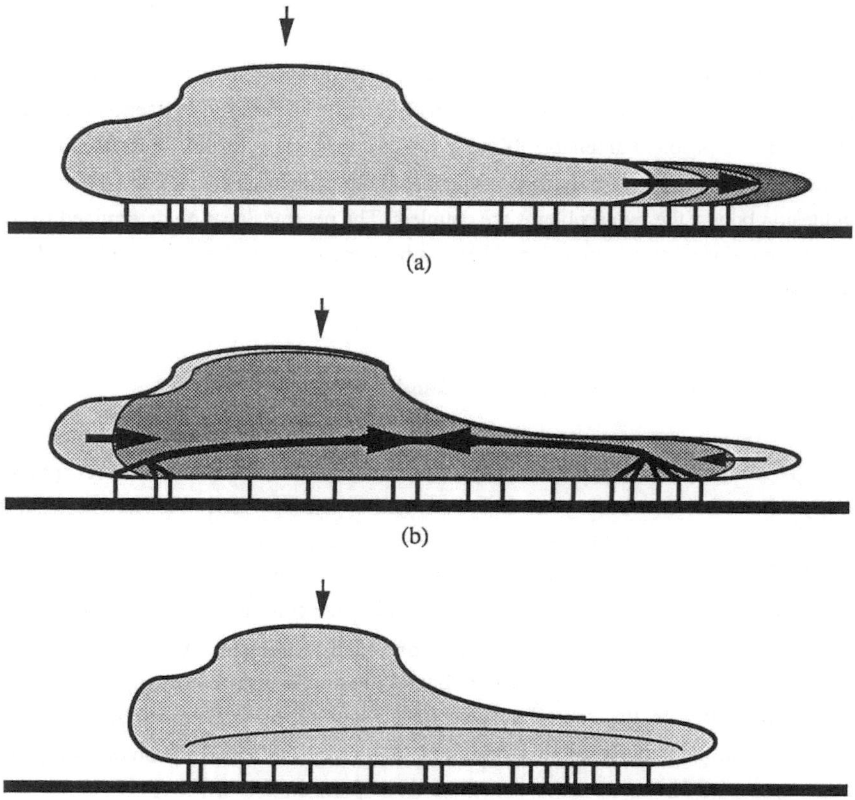

(a)

(b)

(c)

Fig. 5. Schematic of chronological cycle of events in cell migration over an adhesive substratum. (a) In first stage of cycle the cell extends a lamellipod. (b) In the second stage of the cycle the cytoskeleton contracts. (c) In the last stage of the cycle the cytoskeleton relaxes. Arrows depict the location of the cell's center of mass at each point in the cycle.

During the second stage of the movement cycle, a period of duration t_c, the cell's cytoskeleton contracts, pulling on both ends of the cell body. The cytoskeleton pulls at connected adhesion bonds at both the lamellipod and uropod, and some bonds at each end break (we assume that the attachments at either end are most important in locomotion (Chen, 1981; Marks *et al.*, 1991)). If an asymmetry exists in the cytoskeleton/substratum interaction between the lamellipod and uropod, a net traction force on the substratum is generated, and the cell's center of mass moves. Finally, during the third stage of the cycle lasting a period t_r, the cytoskeleton relaxes, and adhesion bonds reestablish unstressed equilibrium distributions.

The events of lamellipodal extension, cytoskeletal contraction, and cytoskeletal relaxation can be described mathematically by dividing the problem into two parts. The dynamics of adhesion bonds and trafficking of adhesion receptors is a classical reaction-diffusion problem. Similarly, the deformation of the cell and exertion of traction forces during contraction is a problem amenable to modeling as a viscoelastic solid. Because the traction generated at the lamellipod and uropod depend on both cellular mechanics and the distribution of adhesion bonds, the two problems are coupled. The net translocation determined from cell displacement during the contraction phase can be averaged over the entire cycle to obtain an observable speed.

Receptor/Ligand-Mediated Adhesion Bond: Molecular "Tires"

To simplify the description of the dynamic distribution of adhesion receptors and bonds on the cell surface during lamellipodal extension and cytoskeletal relaxation, we restrict ourselves to situations in which ligand is immobilized on the substratum at a uniform density, n_s, and soluble competitors are present at a uniform concentration, n_c. By further requiring both of these species to be present in excess, n_s and n_c are constant. We model the dorsal and ventral surfaces of the cell as two flat sheets of length L and width W connected along all edges. Free receptors, $n_r{}^d$, and receptors bound to soluble competitors, $n_{bc}{}^d$, exist on the dorsal surface, but on the ventral surface free receptors, $n_r{}^v$, receptors bound to ligand immobilized on the underlying substratum, n_b, and receptors bound to soluble competitors, $n_{bc}{}^v$, are present. The total number of adhesion receptors on the surface is R_T:

$$\int_0^L \int_{-\frac{W}{2}}^{\frac{W}{2}} \left(n_r{}^d + n_{bc}{}^d + n_r{}^v + n_{bc}{}^v + n_b\right) dy\, dx = R_T \qquad [1]$$

Fig. 6 depicts the range of events governing receptor dynamics and trafficking. Adhesion bonds formed reversibly by association between free receptors on the ventral surface and immobilized ligand bind at a rate proportional to k_f and dissociate at a rate proportional to k_{ro}; free receptors on both surfaces and soluble competitors similarly associate with rate constant k_i and dissociate with rate constant k_j. Free receptors and receptor/competitor complexes diffuse at a rate D_r and are allowed to move freely between dorsal and ventral surfaces. Free receptors also are internalized at a rate k_e; reinsertion into the cell surface balances internalization. Finally, adhesion bonds drift backwards relative to the cell's center of mass at the cell's overall velocity of v. To model the effects of focal contacts

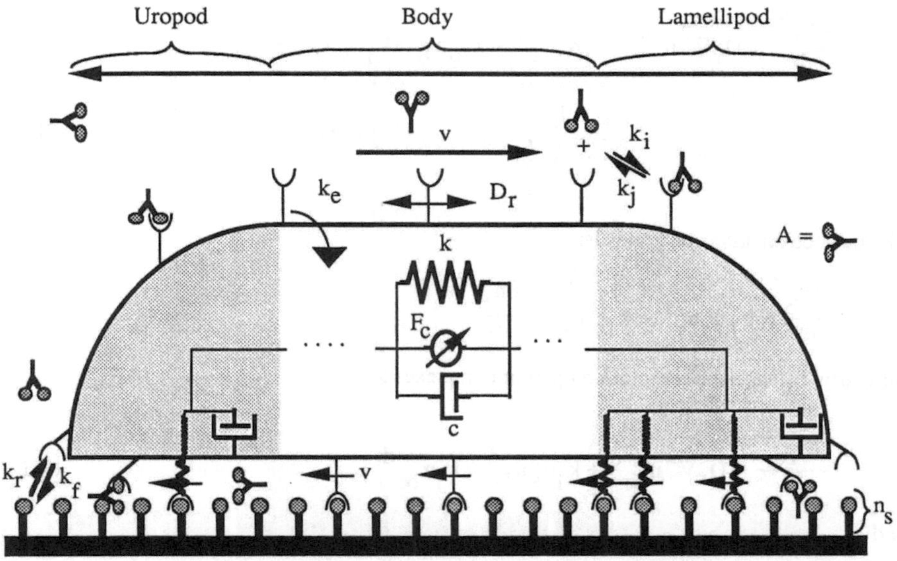

Fig. 6. Schematic of dynamics and trafficking events for adhesion receptors and the viscoelastic-solid model for cellular mechanics during migration over adhesive substrata. Free receptors reversibly bind immobilized adhesion ligand and soluble competitors, diffuse, and are internalized; bound receptors drift backwards with respect to the cell's forward motion. The cell body is divided into four compartments, which each contain Hookean springs and viscous dashpots in parallel with contractile elements; the lamellipodal and uropodal compartments are modeled with sets of springs and a dashpot in parallel.

we also can allow adhesion bonds to dynamically organize into clusters (DiMilla and Lauffenburger, 1991).

Mass balances for free receptor on the dorsal and ventral surfaces, respectively, are written as:

$$\frac{\partial n_r^d}{\partial t} = D_r \nabla^2 n_r^d - k_i n_c n_r^d + k_j n_{bc}^d - k_e n_r^d + \frac{k_e}{2LW} \int_0^L \int_{-\frac{W}{2}}^{\frac{W}{2}} \left(n_r^d + n_r^v \right) dy dx \quad [2a]$$

and

$$\frac{\partial n_r{}^v}{\partial t} = D_r \nabla^2 n_r{}^v - k_f n_s n_r{}^v + k_{ro} n_b - k_i n_c n_r{}^v + k_j n_{bc}{}^v$$

$$- k_e n_r{}^v + \frac{k_e}{2LW} \int_0^L \int_{-\frac{W}{2}}^{\frac{W}{2}} \left(n_r{}^d + n_r{}^v \right) dydx. \qquad [2b]$$

The balance on adhesion bonds is:

$$\frac{\partial n_b}{\partial t} = k_f n_s n_r{}^v - k_{ro} n_b + v \frac{\partial n_b}{\partial x}. \qquad [3]$$

Similarly, balances on receptor/competitor complexes are:

$$\frac{\partial n_{bc}{}^d}{\partial t} = D_r \nabla^2 n_{bc}{}^d + k_i n_c n_r{}^d - k_j n_{bc}{}^d \qquad [4a]$$

and

$$\frac{\partial n_{bc}{}^v}{\partial t} = D_r \nabla^2 n_{bc}{}^v + k_i n_c n_r{}^v - k_j n_{bc}{}^v. \qquad [4b]$$

Relevant boundary conditions include continuity for free receptor and receptor/competitor complex density and continuous total receptor fluxes along the "seams" between dorsal and ventral surfaces, along with conservation of total receptors (Eqn. [1]).

Cytoskeletal Contraction: Cellular "Motors"

We base our description of cellular mechanics on a viscoelastic-solid model because, in contrast to the dynamics for adhesion receptors, many of the molecular components and mechanisms responsible for cytoskeletal mechanics currently are ill-defined -- although this problem is being studied intensely (*e.g.*, Kolega *et al.*, 1991). Cells are complex rheological materials which exhibit both elastic-like and fluid-like properties (Dong *et al.*, 1988; Hochmuth *et al.*, 1993). We model elasticity with linear Hookean springs, for which stress is proportional to strain, and fluidity with linear viscous dashpots, for which stress is proportional to the rate of strain (Fung, 1977).

We divide the contracting cell into six compartments of equal length, limit deformation to the axis of motion, and assume that the mechanical properties of each compartment are spatially and temporally homogeneous (relaxing this assumption for the contractile force and rheological parameters may produce interesting dynamics and merits future attention). Four middle compartments, representing the cell body, are identical and individually represented by a spring, k, dashpot, c, and contractile element, F_c, in parallel (Fig. 6). The anterior and posterior compartments, representing the lamellipod and uropod, also are modeled with springs and dashpots (Fig. 7). In these compartments, however, the dashpot, c, is in parallel with both a spring describing the intrinsic stiffness of each pseudopod in the absence of adhesion bonds, k', and R_{bl} and R_{bu} springs (for the lamellipod and uropod, respectively) in parallel, with Hookean constant k_c, representing the mechanics of the connections between the cytoskeleton and substratum (the R_{bl} and R_{bu} adhesion bonds in the pseudopodia each are associated with individual connections). If we treat the adhesion bonds themselves as springs with Hookean constants k_{spring} (Dembo *et al.*, 1988) and assume that each adhesion bond is stiffer than its cytoskeletal connection, the bond displacements u_l and u_u are much smaller than (u_5-u_l) and (u_1-u_u), respectively. These arrangements of viscoelastic elements are limiting cases of the standard viscoelastic-solid model used by Schmid-Schonbein *et al.* (1981); here we remove an additional spring in series with the dashpot in each compartment, which is reasonable because the dynamics of contraction and deformation during movement are slow for tissue cells.

The boundaries at either end of the cell are fixed, but the motion of the free nodes between each compartment, u_i (i = 1-5), is described by a series of force balances on the nodes:

$$c\frac{du_1}{dt} + k'u_1 + k_c R_{bu} u_1 = c\frac{d(u_2 - u_1)}{dt} + k(u_2 - u_1) + F_c, \qquad [5a]$$

$$\begin{aligned} c\frac{d(u_i - u_{i-1})}{dt} + k(u_i - u_{i-1}) = \\ c\frac{d(u_{i+1} - u_i)}{dt} + k(u_{i+1} - u_i) \quad (i = 2, 3, 4), \end{aligned} \qquad [5b]$$

and

$$\begin{aligned} c\frac{d(u_5 - u_4)}{dt} + k(u_5 - u_4) + F_c = \\ c\frac{d(-u_5)}{dt} + k'(-u_5) + k_c R_{bl}(-u_5). \end{aligned} \qquad [5c]$$

We define a frame of reference for each node and constrain them to remain within the cell:

$$u_i(t = 0) = 0 \qquad (i = 1, ..., 5) \qquad [6a]$$

and

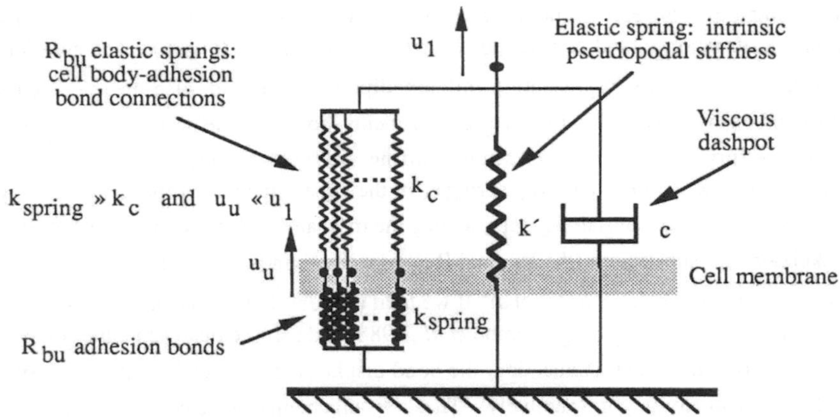

Fig. 7. Schematic of arrangement of viscoelastic elements in the uropodal compartment. The lamellipod is described by an identical arrangement.

$$-\frac{iL}{6} \leq u_i(t) \leq \frac{(6-i)L}{6} \qquad (i = 1, ..., 5). \qquad [6b]$$

The cell's overall velocity is simply the average velocity of the nodes over a full cycle of extension, contraction, and relaxation:

$$v = \frac{\sum_{i=1}^{5} u_i(t = t_c)}{5(t_1 + t_c + t_r)} \qquad [7]$$

We assume that adhesion bonds within the lamellipodal and uropodal compartments are distributed uniformly, are at pseudo-steady state (Lauffenburger, 1989), and are stressed homogeneously during contraction. We also assume that the level of stress can affect the kinetics of binding. We write mass balances on R_{bl} and R_{bu} as a function of the number of free receptors in each compartment, R_{rl} and R_{ru}, respectively, as:

$$\frac{dR_{bl}}{dt} = k_{fu} n_s R_{rl} - k_{rl} R_{bl} = 0 \qquad [8a]$$

and

$$\frac{dR_{bu}}{dt} = k_{fu} n_s R_{ru} - k_{ru} R_{bu} = 0 \qquad [8b]$$

Similar balances also govern the numbers of receptor/competitor complexes in each pseudopod, R_{bcl} (lamellipod) and R_{bcu} (uropod), except that the kinetics relevant for these complexes are independent of stress:

$$\frac{dR_{bcl}}{dt} = k_f n_c R_{rl} - k_{ro} R_{bcl} = 0 \qquad [9a]$$

and

$$\frac{dR_{bcu}}{dt} = k_f n_s R_{ru} - k_{ro} R_{bcu} = 0 \qquad [9b]$$

We also assume that the equilibrium receptor/ligand dissociation constants for the lamellipod and uropod, K_{dl} and K_{du}, respectively, increase exponentially with the energy applied by contractile forces, scaled by the number of bonds in the compartment and thermal energy (Bell, 1978):

$$K_{dl} = \frac{k_{ro} \exp\left(\frac{(-k_c u_5)(-u_l)}{k_b T}\right)}{k_f} = \frac{k_{ro} \exp\left(\frac{k_c^2 u_5^2}{k_{sping} k_b T}\right)}{k_f} \qquad [10a]$$

and

$$K_{du} = \frac{k_{ro} \exp\left(\frac{(k_c u_1) u_u}{k_b T}\right)}{k_f} = \frac{k_{ro} \exp\left(\frac{k_c^2 u_1^2}{k_{sping} k_b T}\right)}{k_f}, \qquad [10b]$$

where $k_{spring} u_l = k_c u_5$ and $k_{spring} u_u = k_c u_1$. This treatment is based on an analysis by Zhurkov (1965) for the effects of stress on the kinetics of material fracture. The total numbers of bonds in each compartment, then, are:

$$R_{bl}\{u_5\} = \frac{\int_0^{\frac{L}{6}} \int_{\frac{-W}{2}}^{\frac{W}{2}} \left(n_r^{\,v} + n_{bc}^{\,v} + n_b\right) dy\,dx}{1 + \left(\frac{k_{ro}}{k_f n_s}\right)\left(1 + \frac{k_i n_c}{k_j}\right) \exp\left(\frac{k_c^2 u_5^2}{k_{spring} k_b T}\right)} \qquad [11a]$$

and

$$R_{bu}\{u_1\} = \frac{\int_{\frac{5L}{6}}^{L} \int_{\frac{-W}{2}}^{\frac{W}{2}} \left(n_r^{\ v} + n_{bc}^{\ v} + n_b\right) dy\,dx}{1 + \left(\frac{k_{ro}}{k_f n_s}\right)\left(1 + \frac{k_i n_c}{k_j}\right) \exp\left(\frac{k_c^{\ 2} u_1^{\ 2}}{k_{spring}\, k_b T}\right)}. \qquad [11b]$$

R_{bl} and R_{bu} decrease and the deformation of the lamellipodal and uropodal compartments, respectively, increase with the application of strain .

A Spatial Asymmetry in Traction: The Cellular "Driver"

In addition to the reaction-diffusion model for the dynamics and trafficking of adhesion receptors and the viscoelastic-solid model for the mechanics of a moving cell, we must specify a mathematical description for a mechanism which can generate a spatial asymmetry in traction across the cytoskeleton/substratum interface between the lamellipod and the uropod. The primary roles of the lamellipod and uropod are supported by the data of Marks *et al.* (1991), who have demonstrated that cell speed can be reduced by either limiting uropodal detachment or lamellipodal extension. A general molecular-level mechanism responsible for asymmetry in all tissue cells has not been identified, but at least four plausible types of mechanisms have been proposed (Fig. 8), and we previously have considered two in particular (DiMilla *et al.*, 1991). It also is possible that multiple mechanisms may be active simultaneously in a motile cell.

(1) *A spatial asymmetry in the affinity of adhesion bonds between lamellipod and uropod may be generated by a number of biochemical events (Dustin and Springer, 1989).* At the molecular level, affinity can be decreased by phosphorylation (Hirst *et al.*, 1986; Tapley *et al.*, 1989) or proteolysis (Grinnell, 1986) of adhesion receptors; conversely, the exchange of Ca^{2+} with Mn^{2+} increases the affinity of adhesion bonds (Gailit and Ruoslahti, 1988). Schmidt *et al.* (1993) recently have demonstrated that there is an spatial asymmetry in the strength of interactions between β_1 integrins and the cytoskeleton in motile but not immotile cells. Using laser optical trapping and nanometer-level motion analysis these investigators measured the binding of dorsal-surface integrins to the cytoskeleton at the lamellipod and uropod for murine 3T3 fibroblasts transfected with chicken β_1 integrin cDNAs containing different mutations in the cytoplasmic domain. They found that the ratio of integrin-cytoskeletal attachments at the lamellipod to the uropod was approximately 2 for migrating cells but near unity for stationary cells; they also observed that the membrane of motile cells was more deformable than the membrane of corresponding immotile cells.

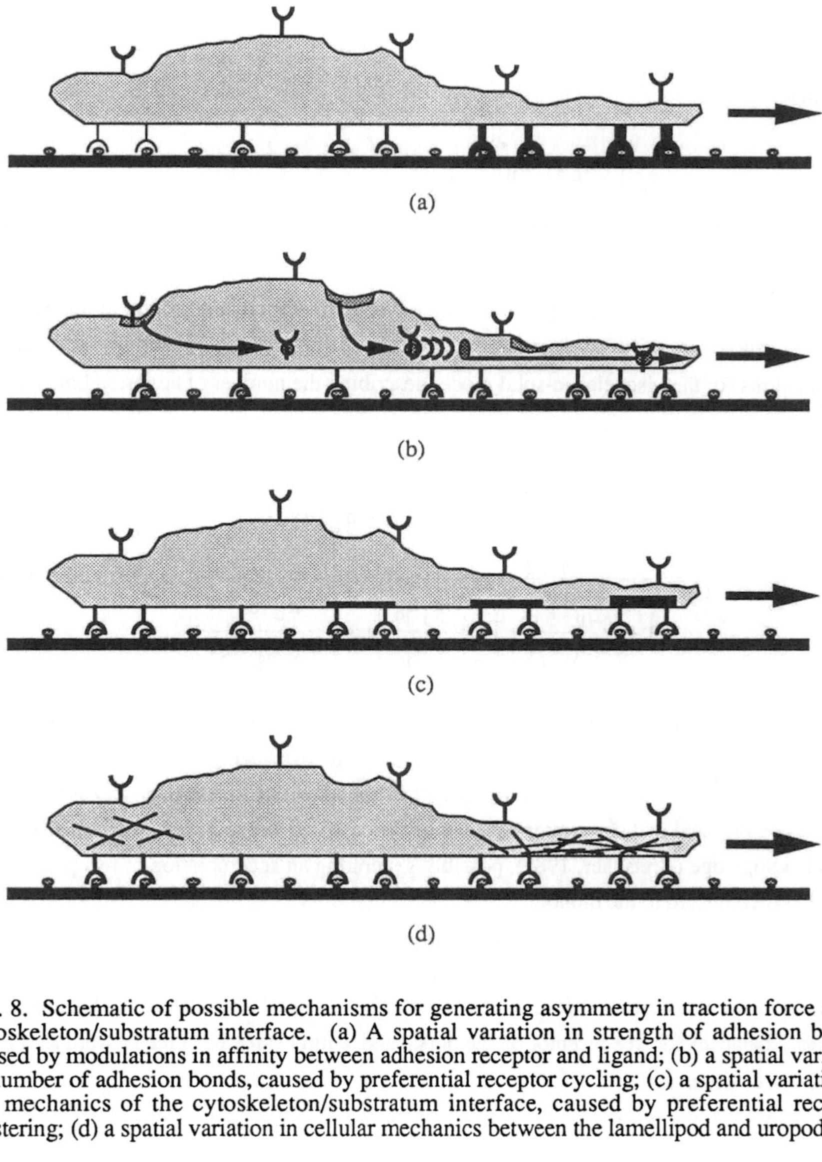

Fig. 8. Schematic of possible mechanisms for generating asymmetry in traction force at the cytoskeleton/substratum interface. (a) A spatial variation in strength of adhesion bonds, caused by modulations in affinity between adhesion receptor and ligand; (b) a spatial variation in number of adhesion bonds, caused by preferential receptor cycling; (c) a spatial variation in the mechanics of the cytoskeleton/substratum interface, caused by preferential receptor clustering; (d) a spatial variation in cellular mechanics between the lamellipod and uropod.

To model a spatial variation in the strength of adhesion bonds we define the dimensionless parameter ψ as the ratio of the equilibrium dissociation constant between adhesion receptor and ligand in the absence of applied stress for the lamellipod to the corresponding constant for the uropod:

$$\psi = \frac{K_{dl}\{F=0\}}{K_{du}\{F=0\}}, \qquad [12]$$

such that:

$$K_{du} = \frac{k_{ro}\exp\left(\dfrac{(k_c u_1)u_u}{k_b T}\right)}{\psi k_f} = \frac{k_{ro}\exp\left(\dfrac{k_c^2 u_1^2}{k_{sping} k_b T}\right)}{\psi k_f}. \qquad [13]$$

ψ is a measure of the asymmetry in traction between lamellipod and uropod in this scheme. We will assume that this mechanism only functions during contraction, and introduce ψ into the equations for the viscoelastic-solid model describing the number of adhesion bonds in the uropod:

$$R_{bu}\{u_1\} = \frac{\displaystyle\int_{\frac{5L}{6}}^{L}\int_{\frac{-W}{2}}^{\frac{W}{2}}\left(n_r^{\,v} + n_c^{\,v} + n_b\right)dy\,dx}{1 + \psi^{-1}\left(\dfrac{k_{ro}}{k_f n_s}\right)\left(1 + \dfrac{k_i n_c}{k_j}\right)\exp\left(\dfrac{k_c^2 u_1^2}{k_{spring} k_b T}\right)}. \qquad [14]$$

(2) *A spatial asymmetry in the number of adhesion receptors between lamellipod and uropod may be generated by preferential endocytic trafficking of adhesion receptors (Bretscher, 1984; Bretscher, 1992).* Recent experiments with Chinese hamster ovary (CHO) cells have suggested that some integrins may be endocytosed and recycled preferentially to the cell's leading edge (Bretscher, 1989), possibly generating an accumulation of receptors at the lamellipod compared to the uropod.

To model a spatial variation in the number of adhesion receptors we assume that adhesion receptors are inserted homogeneously -- at a rate which balances internalization -- on both the dorsal and ventral surface over a fraction λ of the cell length from the leading edge. The mass balances for free receptor on the dorsal and ventral surfaces in the forward λL of the cell now are written as:

$$\frac{\partial n_r^d}{\partial t} = D_r \nabla^2 n_r^d - k_i n_c n_r^d + k_j n_{bc}^d - k_e n_r^d + \frac{k_e}{2\lambda LW} \int_0^L \int_{-\frac{W}{2}}^{\frac{W}{2}} \left(n_r^d + n_r^v\right) dy dx \qquad [15a]$$

and

$$\frac{\partial n_r^v}{\partial t} = D_r \nabla^2 n_r^v - k_f n_s n_r^v + k_{ro} n_b - k_i n_c n_r^v + k_j n_{bc}^v - k_e n_r^v + \frac{k_e}{2\lambda LW} \int_0^L \int_{-\frac{W}{2}}^{\frac{W}{2}} \left(n_r^d + n_r^v\right) dy dx. \qquad [15b]$$

The equations for the rear $(1-\lambda)L$ of the cell become:

$$\frac{\partial n_r^d}{\partial t} = D_r \nabla^2 n_r^d - k_i n_c n_r^d + k_j n_{bc}^d - k_e n_r^d \qquad [16a]$$

and

$$\frac{\partial n_r^v}{\partial t} = D_r \nabla^2 n_r^v - k_f n_s n_r^v + k_{ro} n_b - k_i n_c n_r^v + k_j n_{bc}^v - k_e n_r^v. \qquad [16b]$$

Because the value of the rate of endocytosis controls the rate of recycling and, thus, the distribution of receptors between lamellipod and uropod, k_e is the functional asymmetry parameter for this mechanism (although a true parameterization of asymmetry for this mechanism is more complicated and requires numerical computations).

(3) *A spatial asymmetry in the organization of adhesion bonds between lamellipod and uropod may be generated if adhesion bonds preferentially form clusters, such as focal contacts (Burridge et al., 1988), at the lamellipod.* Clustered bonds may withstand stronger transmitted forces than isolated bonds (Ward and Hammer, 1993), possibly because they cooperatively form stiffer structures than a collection of single bonds. This type of scheme is the subject of ongoing work in our group.

(4) *A spatial asymmetry in the mechanical properties of the cytoskeleton between lamellipod and uropod may be generated if there are no differences in the affinity, number, or organization of adhesion bonds between lamellipod and uropod but the mechanical properties of these two regions of the cell are different.* For example, Barbee *et al.* (1993) have found that the magnitude and orientation of strains within bovine vascular smooth muscle cells

cultured on mechanically-deformed substrata varied spatially and temporally, although they did not identify distinct lamellipodia and uropodia. Measurements and modeling of spatial and temporal differences in the rheological properties of migrating tissue cells deserves attention in future studies.

Analysis and Solution: "Assembling" the Model Cell from its Parts

The fundamentals of scaling and solving model equations have been presented previously (DiMilla *et al.*, 1991). Several dimensionless parameters deserve particular attention. Our efforts focus on predicting the dependence of key dimensionless parameters on the dimensionless cell speed, $\nu = \frac{v}{k_{ro}L}$. Key dimensionless parameters describing molecular adhesive interactions include dimensionless adhesion ligand density, $\kappa = \frac{k_f n_s}{k_{ro}}$ (the product of ligand density and stress-free affinity), dimensionless asymmetry in affinity of adhesion bonds, ψ (Eqn. [12]), dimensionless rate of endocytosis, $\theta = \frac{k_e}{k_{ro}}$, and dimensionless concentration of soluble competitors, $\kappa_c = \frac{k_i n_c}{k_j}$. Definition of dimensionless parameters for cellular mechanics requires the assumptions that Hookean spring constants are proportional to a Young's modulus, E, and dashpot coefficients to a viscosity, μ. Based on the intrinsic rheological properties E and μ, key dimensionless parameters for mechanics include dimensionless contractile force, $f_c = \frac{F_c}{\sqrt{k_{spring}\, k_b T}}$, dimensionless cytoskeletal stiffness, $\alpha = \frac{E}{\sqrt{k_{spring}\, k_b T}}$, and dimensionless cytoskeletal viscosity, $\omega = \frac{\mu}{\sqrt{k_{spring}\, k_b T}}$. We define $\tau_c = k_{ro} t_c$ as the dimensionless contraction time

The particular set of equations to solve depends on which mechanism for asymmetry in traction is chosen. For the two examples we have outlined here, Eqns. [1]-[9], [10a], [11a], and [12]-[14] are valid when asymmetry is generated by a spatial variation in the strength of adhesion bonds; similarly, Eqns. [1]. [2]-[11], [15], and [16] describe a cell in which asymmetry is generated by a spatial variation in number of adhesion receptors. Values and ranges for key dimensionless parameters appropriate for tissue cells, such as fibroblasts and endothelial cells, have been presented in DiMilla *et al.* (1991) based on reported literature values and experimental observations.

Effects of Molecular Adhesive Properties

Our most important predictions focus on the dependence of dimensionless cell speed, ν, on dimensionless adhesion ligand density, κ (Fig. 9). For non-trivial values of the asymmetry parameter ψ (*i.e.*, $\psi < 1$) or moderate levels of receptor recycling ($10^{-5} \leq \theta \leq 10^{-3}$), we predict a biphasic relationship between speed and ligand density (Fig. 9a). At low κ cytoskeletal contraction dissociates the few receptors that have bound at the lamellipod and uropod and no net movement occurs. Increasing κ allows the formation of more bonds at both ends of the cell, but the lamellipod has proportionally more bonds to withstand contractile forces than the uropod, and the cell experiences a net forward translocation. With further increases in κ enough bonds form at the uropod to withstand the contractile force without dissociating, and net translocation of the cell decreases. Increasing the value of ψ leads to the generation of greater adhesion-bond asymmetry, and cell movement occurs over a larger range of values of κ. Dimensionless speed is independent of dimensionless adhesion ligand density, however, for large dimensionless rates of endocytosis ($\theta > 10^{-3}$): ν does not decrease significantly as κ is increased because even if all the receptors in the uropod form adhesion bonds there are not enough to resist contraction. The critical value of θ at which monophasic behavior is predicted depends on the total number of receptors, R_T. Finally, maximum speed -- which occurs when cytoskeletal contraction disrupts all uropodal adhesion bonds while leaving some lamellipodal adhesion bonds -- does not depend significantly on ψ or θ for relatively large degrees of asymmetry ($\psi \geq 0.1$ or $\theta \geq 10^{-4}$).

Because dimensionless adhesion ligand density is the product of the density of adhesion ligand and the affinity between adhesion receptor and ligand, we can apply our model predictions to variations in either ligand density or affinity. In Fig. 9 we plot -- in dimensionless form -- the experimental data for cell speed for HSMCs on fibronectin and collagen as depicted in Fig. 2b. The correspondence between experimental data and model predictions clearly is apparent, and our data suggest that a relatively low level of asymmetry ($\psi = 0.5$-1 or $\theta = 10^{-5}$-10^{-4}) is necessary for the movement of these cells on these surfaces, consistent with the low levels of asymmetry observed directly by Schmidt *et al.* (1993). (The values of the asymmetry parameter measured by Schmidt *et al.*, the ratio of cytoskeletally-attached integrins at the lamellipod to uropod, is the reciprocal of our ψ.) We can not distinguish between the two alternative mechanisms for asymmetry with this data. Others also have observed biphasic relationships between rate of migration and the density of adhesion ligands (*e.g.*, Goodman *et al.*, 1989).

Duband *et al.* (1991) have examined the effect of variations in both the concentration of adhesion ligands and the affinity of adhesion ligands for receptors using anti-integrin antibodies with different affinities as adhesion ligands They found that increasing the

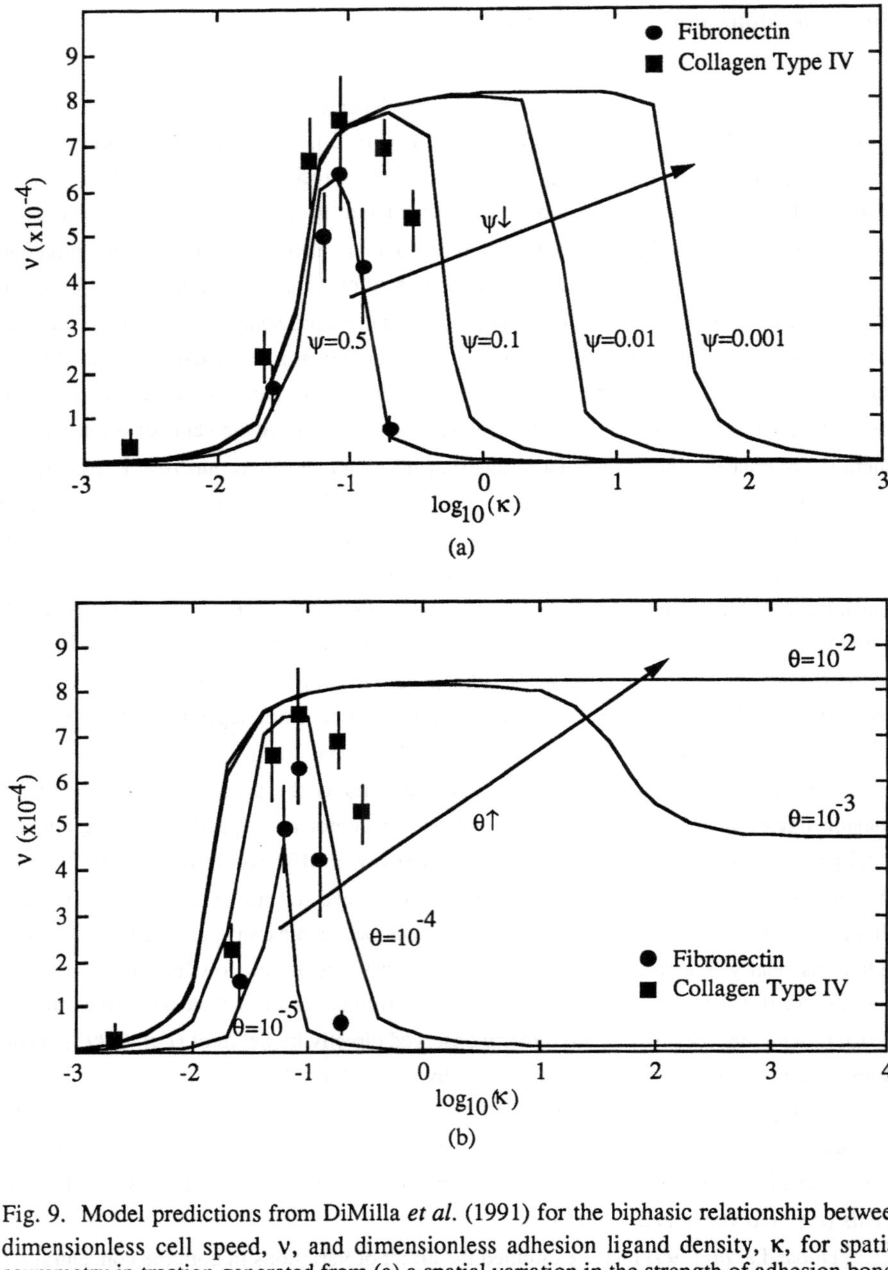

Fig. 9. Model predictions from DiMilla *et al.* (1991) for the biphasic relationship between dimensionless cell speed, ν, and dimensionless adhesion ligand density, κ, for spatial asymmetry in traction generated from (a) a spatial variation in the strength of adhesion bonds and (b) a spatial variation in number of adhesion bonds through preferential receptor cycling. Dedimensionalized experimental data for HSMCs from DiMilla *et al.* (1993) (see Fig. 2b) has been superimposed.

concentration of low-affinity antibodies increased migration, but increasing the concentration of high-affinity antibodies over the same range of concentrations decreased migration. These observations are consistent with the predictions in Fig. 9: because the key parameter is the product of density and affinity, the data for the low-affinity ligands represent the increasing portion of the curve for speed versus dimensionless adhesion ligand density, and the data for the high-affinity ligands represent the decreasing portion of the same curve.

Fig. 10 summarizes predictions for the effects of soluble competitors on cell speed. Increasing the dimensionless concentration of competitors -- experimentally accomplished by either varying competitor concentration or receptor/competitor affinity -- shifts the optimal range of dimensionless adhesion ligand densities for movement to larger values (Fig 10a). We can rationalize this effect by noting that in Eqns. [11] and [14] the effective dimensionless adhesion ligand density, κ_{eff}, is the actual dimensionless adhesion ligand density, κ, reduced by the dimensionless competitor concentration, κ_c:

$$\kappa_{eff} = \frac{\kappa}{1 + \kappa_c}. \qquad [17]$$

Fig. 10b demonstrates that with this new parameterization the predictions for different values of κ_c collapse to a single curve. This prediction is applied successfully in Fig. 11 to the data of Wu *et al.* (1993) presented in Fig 3. (Note that their data is for the random motility coefficient, μ, which is directly related to cell speed, S, and directional persistence time, P, as $\mu = \frac{S^2P}{2}$ (Othmer *et al.*, 1988; Rivero *et al.*, 1989). We recently have shown that directional persistence time can vary with the density of adhesion ligand (DiMilla *et al.*, 1993), a caveat which must be acknowledged when applying our model predictions to the data for Wu *et al.*) The observation that migration can be inhibited by a soluble competitor is not unique (*e.g.*, Marks *et al.*, 1991), but previous studies have not noted that migration can be promoted by adding soluble competitors.

We previously have shown that the dependence of cell speed on receptor number depends on the specific type of mechanism responsible for generating asymmetry in traction (DiMilla *et al.*, 1991). For the differential bond affinity mechanism with $\psi = 10^{-2}$, the dependence of ν on κ is biphasic regardless of R_T, and the range of κ permitting substantial migration is independent of R_T; for the polarized endocytic trafficking mechanism with $\theta = 10^{-3}$, the dependence of ν on κ is biphasic for small R_T but monophasic for large R_T, and the range of κ permitting substantial migration increases with increasing R_T. However, because these calculations were based on levels of asymmetry an order of magnitude greater than what we have observed experimentally, we have reexamined the relationship between speed and receptor number at lower levels of asymmetry consistent with our data for HSMCs (Fig. 12). The plots in Fig. 12 demonstrate that similar behavior is expected for movement

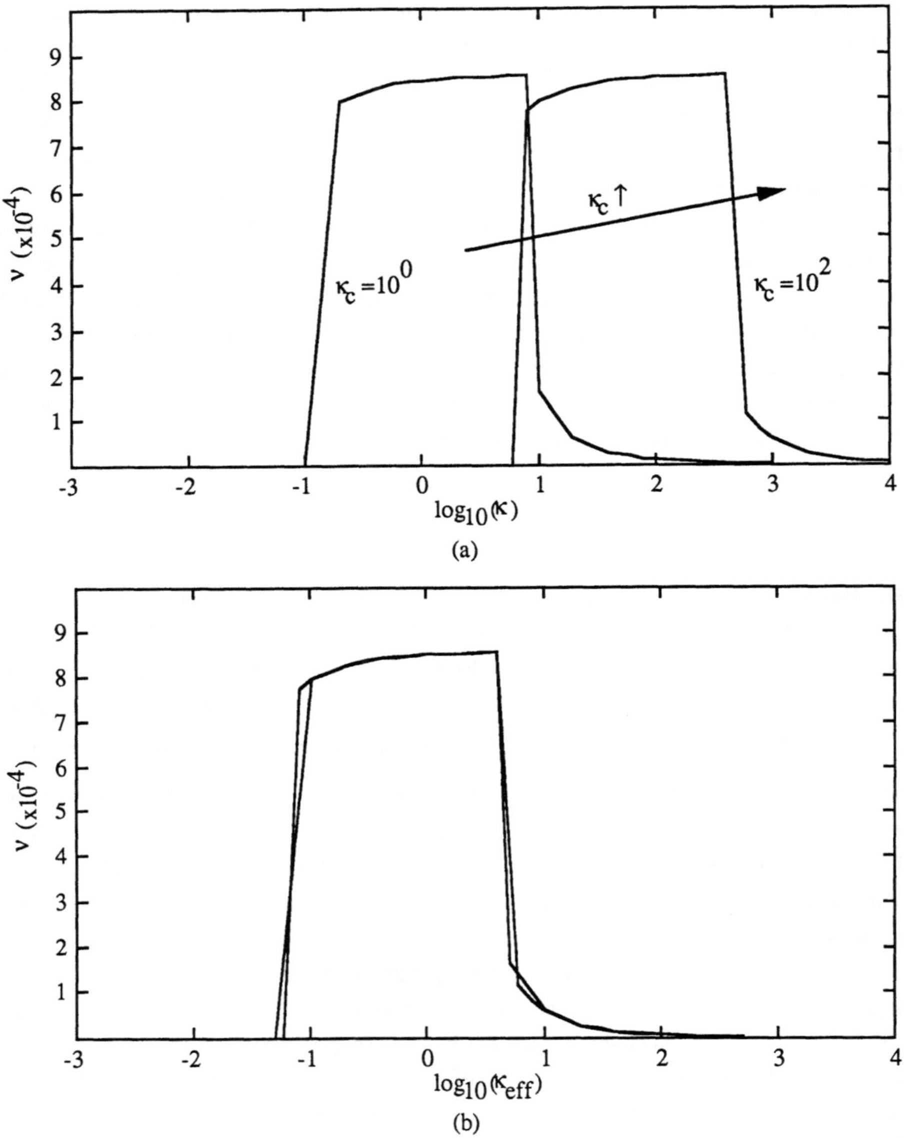

Fig. 10. Model predictions for the relationship between dimensionless cell speed, ν, dimensionless adhesion ligand density, κ, and dimensionless concentration of soluble competitors, κ_c. (a) Plot of ν as a function of κ for two values of κ_c. (b) Plot of ν as a function of effective dimensionless adhesion ligand density, $\kappa_{eff} = \frac{\kappa}{1 + \kappa_c}$.

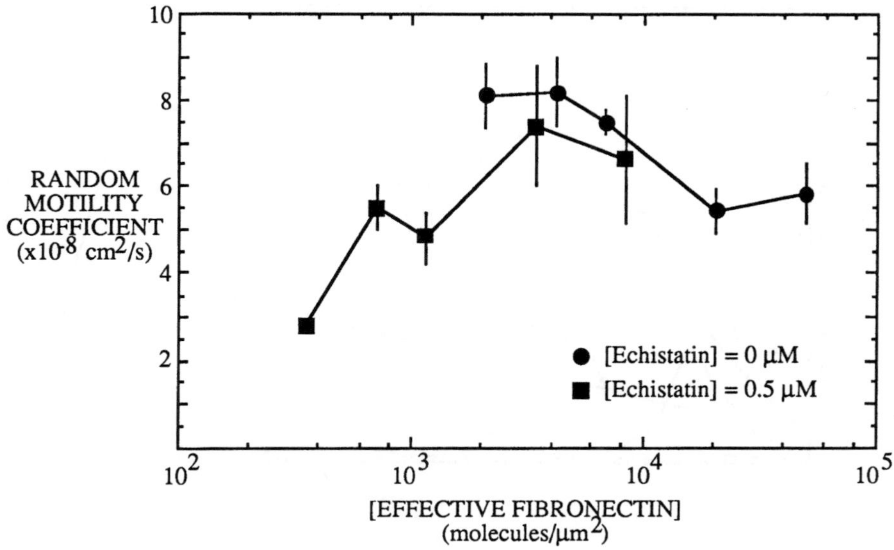

Fig. 11. Experimental data from Wu *et al.* (1993) plotted as random motility coefficient as a function of effective adhesion ligand density, $\text{[effective fibronectin]} = \frac{\text{[fibronectin]}}{1 + \text{[echistatin]}/K_c}$, assuming $K_c = 10^{-7}$ M (Savage *et al.*, 1990).

with asymmetry parameters $\psi = 0.1$ or $\theta = 10^{-4}$: a biphasic relationship between ν and R_T is predicted, and the optimum value of R_T for maximum movement decreases linearly with κ for $\kappa < 10^2$. Maximum cell speed also is smaller for $\kappa = 10^2$ than for smaller values of κ for both mechanisms because at the optimal value of $R_T \approx 5 \times 10^3$ for maximum speed not enough receptors are present to sufficiently withstand contractile forces at the pseudopodia.

We have examined experimentally the effect of variations in receptor number on cell speed using a set of CHO cell clones which had been transfected with increasing levels of the integrin $\alpha_5\beta_1$, a receptor for fibronectin (Giancotti and Ruoslahti, 1990). Migration, including cell speed, decreased with increasing levels of expression of the fibronectin receptor (Fig. 13). Although it is difficult to compare these observations with our model predictions, our overall approach should be useful in future studies in which substratum properties, including the density and type of adhesion ligand, are varied.

Ongoing efforts in our group are focusing on the role of adhesion receptor and bond organization in the regulation of cell migration. We predict that the dynamics of receptor clustering can affect motility significantly: increasing rates of clustering decrease the optimal dimensionless adhesion ligand density for movement, and decreasing the rate of dissociation

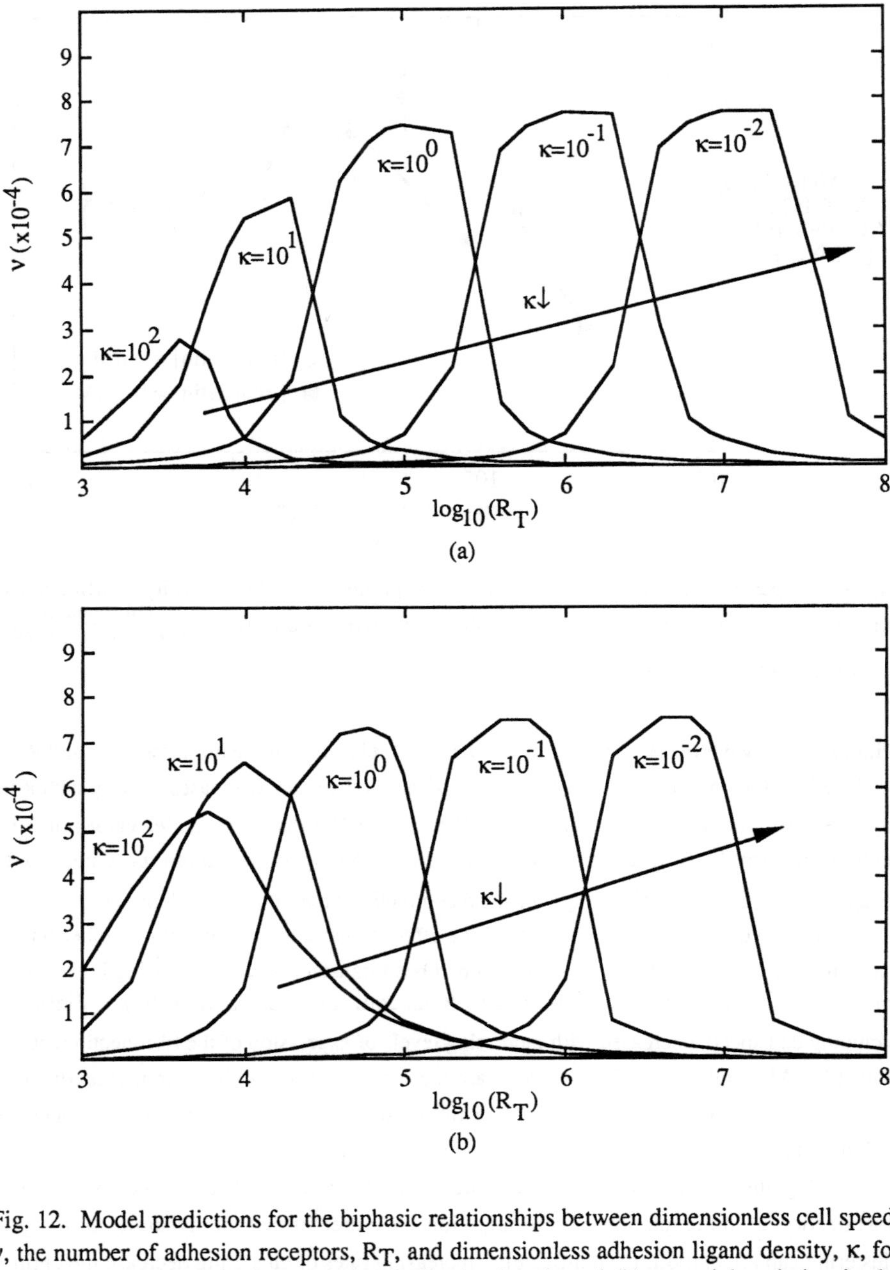

Fig. 12. Model predictions for the biphasic relationships between dimensionless cell speed, ν, the number of adhesion receptors, R_T, and dimensionless adhesion ligand density, κ, for (a) a spatial variation in the strength of adhesion bonds and (b) a spatial variation in the number of adhesion bonds through preferential receptor cycling.

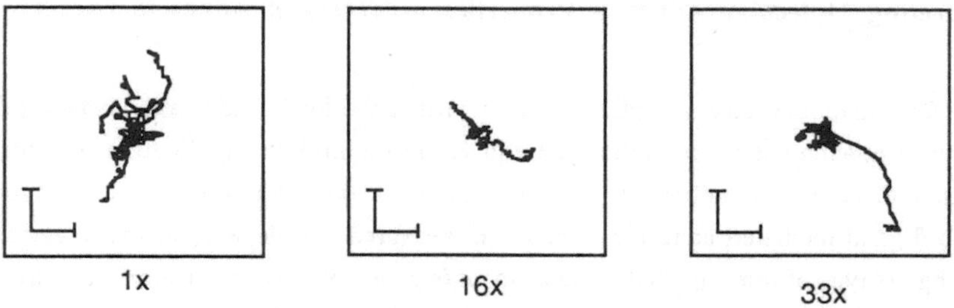

(a)

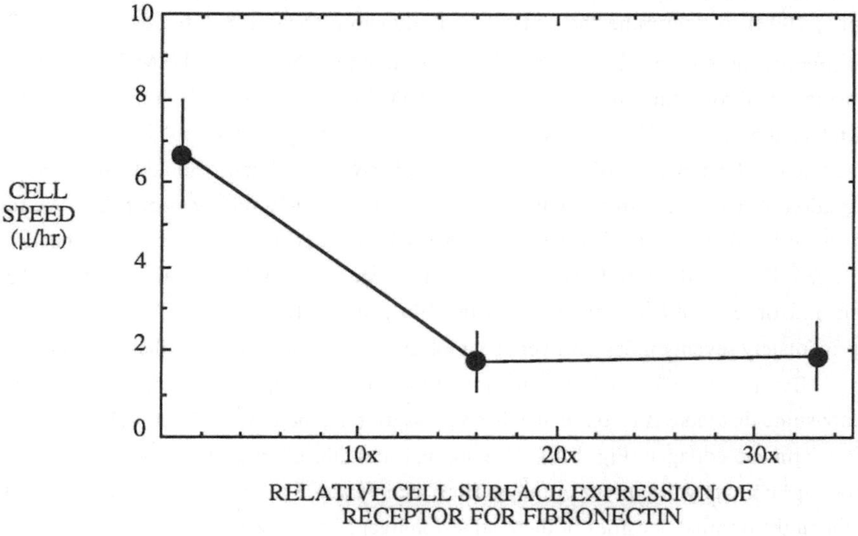

(b)

Fig 13. Experimental data for the migration of individual Chinese hamster ovary cells which have been transfected to express different levels of a receptor for fibronectin. Clonal populations were developed by Giancotti and Ruoslahti (1990). (a) Typical cell paths over 24 hours, superimposed to a common origin in "wind-rose" displays, for cells expressing different levels of the receptor for fibronectin. Bars, 100 μm. (b) Relationship between cell speed, S (plotted as means and standard deviations), and the level of expression of the receptor for fibronectin.

of clusters can increase the range of dimensionless adhesion ligand densities permitting movement (DiMilla and Lauffenburger, 1991).

Interpreting Molecular Adhesive Properties in Terms of Adhesive Forces

Determining whether an optimal dimensionless adhesion ligand density and an optimal strength of attachment for migration are equivalent requires the application of rigorous, quantitative assays for adhesion. Further, even for well-defined assays the type of receptor/ligand-mediated adhesive interaction measured can depend on many variables, including the type of force applied in detachment (e.g., hydrodynamic shear versus tensile by micromanipulation), the temporal characteristics of the applied force, the presence of nonspecific interactions (Bongrand and Bell, 1984), and the experimental conditions chosen, such as attachment time (Stone, 1993). As one approach to this problem we have applied a radial-flow detachment assay (RFDA) (Cozens-Roberts *et al.*, 1990; DiMilla *et al.*, 1992a; Stone, 1993) to determine the critical hydrodynamic shear stress for detachment, S_c, a quantitative measure of the strength of cell/substratum adhesion, for HSMCs on a range of densities of fibronectin and collagen type IV (DiMilla *et al.*, 1993). By limiting the time for initial attachment to 30 minutes we measured the strength of initial receptor-mediated attachments; the dynamics of short time-scale adhesive interactions should be relevant to the migration of tissue cells moving at rates on the order of 10-20 microns per hour. We found that the critical shear stress increased linearly with the density of adsorbed fibronectin and collagen by $3.6\pm0.2\text{x}10^{-9}$ dynes/adsorbed fibronectin molecule and $5.0\pm1.4\text{x}10^{-11}$ dynes/adsorbed collagen molecule (Fig. 14a), suggesting that HSMC attachment is approximately seventy times stronger on a molecular basis for fibronectin than for collagen.

By plotting cell speed as a function of the critical shear stress for detachment, we can demonstrate that there is an optimal adhesiveness for maximum migration for HSMCs on both fibronectin and collagen (Fig. 14b). The optimal strengths of attachment for maximum speed were approximately 3.8 μdynes/μm^2 on fibronectin and 1.5 μdynes/μm^2 on collagen. Thus, although the optimal densities of these adhesion ligands for migration differed by an order of magnitude, the optimal strengths of initial attachment for migration were similar. Further, the same minimum strength of initial attachment of about 1 μdyne/μm^2 was required for movement on either protein.

Mathematical models are instructive for elucidating why the rate of migration is correlated better with the strength of cell/substratum attachments than simply dimensionless adhesion ligand density. Cozens-Roberts *et al.* (1990) have predicted that the S_c is linearly proportional to the number of adhesion receptors and density of adhesion ligands:

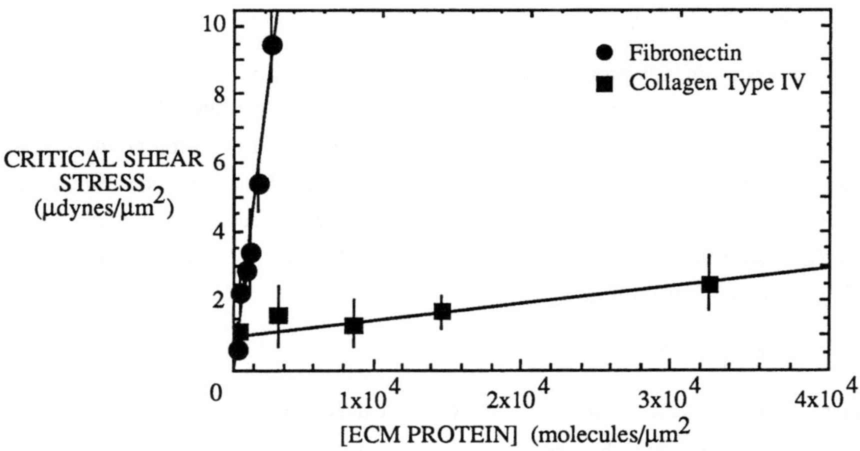

(a)

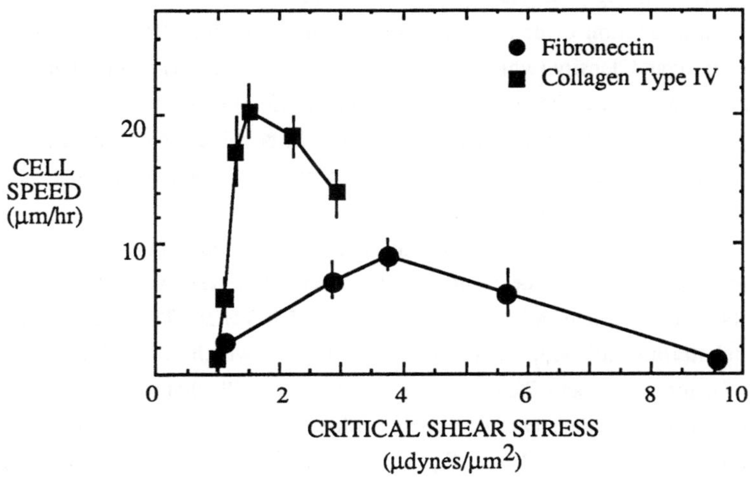

(b)

Fig. 14. Interpreting experimental data for the biphasic relationship between density of adhesion ligand and cell speed for HSMCs on fibronectin and collagen from Fig. 2b in terms of adhesion forces (DiMilla *et al.*, 1993). (a) Experimental data for the critical shear stress for cell detachment, S_C, as linear functions of the surface densities of fibronectin and collagen type IV, as measured in a radial-flow detachment assay. (b) Plot of cell speed, S, as a function of strength of initial cell-substratum attachment, S_C.

$$S_C = \beta R_T n_S + \lambda, \qquad [18]$$

where β is both linearly proportional to receptor/ligand affinity and a function of geometric parameters, such as contact area and cell diameter, and λ is a measure of nonspecific forces. In addition to applying the fracture model for the kinetics of adhesion bonds, this model makes a number of critical assumptions regarding the effect of applied fluid shear on adherent cells; Eqn. [18] has explained satisfactorily, however, several sets of experimental results to date (Cozens-Robert *et al.*, 1990; DiMilla *et al.*, 1992a; Stone, 1993). We can rewrite Eqn. [18] in dimensionless form as:

$$\alpha_C = R_T \kappa + \Lambda, \qquad [19]$$

where α_C is the dimensionless critical force for detachment and Λ is the dimensionless nonspecific force. Thus, the steeper slope observed in Fig. 14a for fibronectin compared to collagen may be due to a greater number of adhesion receptors for fibronectin than for collagen or a stronger receptor/ligand affinity for fibronectin than for collagen. Thus, plotting speed as a function of detachment stress incorporates the effects of both dimensionless adhesion ligand density (which encompasses dimensional density and affinity) and receptor number in a single variable. Ward and Hammer (1993) recently have provided a model for receptor/ligand-mediated adhesion incorporating the possible strengthening of adhesion by the formation of focal contacts which should be useful in conjunction with models for the effects of receptor clustering on migration.

Fig 15 demonstrates how we also can apply the concept of an optimal relationship between migration and adhesiveness to interpret the observations by Wu *et al.* (1993) for the effects of soluble echistatin on the migration of RMVECs. These investigators measured cell/substratum adhesion as the fraction of cells which resisted detachment using a centrifugation assay and found that the percentage of cells that remained attached increased monotonically with density of fibronectin; adhesion was reduced, however, by the addition of 0.5 μM echistatin (Fig 15a). Fig 15b demonstrates that plotting the random motility coefficient as a function of the fraction of cells attached for each experimental condition superimposes the data roughly onto a single curve, as expected from our modeling and the analysis from Fig. 11. The scatter in the data is reasonable considering experimental constraints limited measurements of migration to the random motility coefficients of populations rather than the speeds of individual cells and adhesion to the centrifugation assay at a constant *g* force rather than a range of forces.

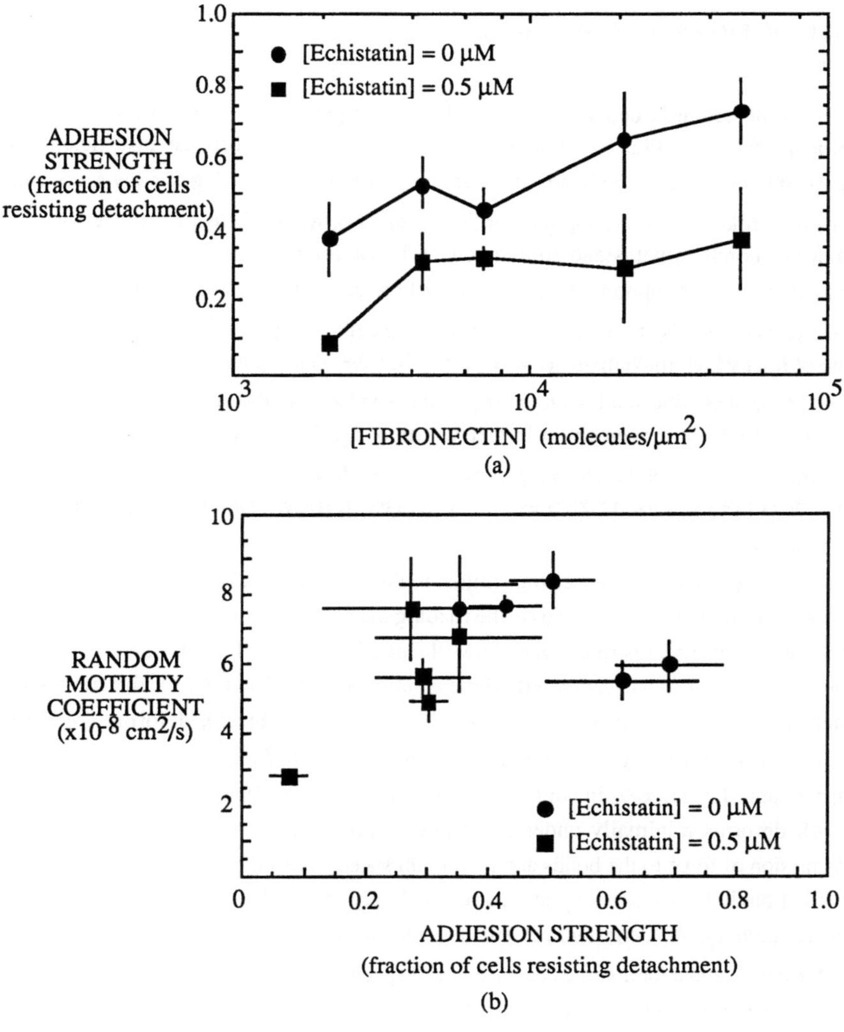

Fig. 15. Interpreting experimental data for the biphasic relationship between effective density of fibronectin and random motility coefficient for RMVECs from Fig. 3 in terms of adhesion forces (Wu *et al.*, 1993). (a) Experimental data for %attached cells as a function of the surface density of fibronectin and the concentration of soluble echistatin, as measured using a centrifugation assay. (b) Plot of random motility coefficient as a function of %attached cells.

Effects of Intracellular Mechanics

Dimensionless cell speed also exhibits a biphasic dependence on dimensionless contractile force, f_c (Fig. 16). For low f_c no movement occurs because insufficient force is generated to break adhesion bonds at either end of the cell. Increasing f_c allows movement as bonds are broken at the uropod but not the lamellipod, where there are stronger or a greater number of bonds to withstand contraction. ν is maximum at the value of f_c for which the bonds at the uropod support a trivial fraction of the transmitted force. Further increases in f_c break adhesion bonds at both pseudopodia and the cell merely contracts inwards. The critical value of f_c at which movement ceases occurs when the lamellipodal bonds break and depends on dimensionless adhesion ligand density: increasing κ produces more bonds at both ends of the cell and allows movement at higher values of f_c. For large κ and moderate α and ω, however, the cell can not effectively transmit enough force to break all the bonds in the front, and ν does not decrease to zero as f_c is increased further. Maximum speed generally is independent of κ.

Cell speed and the forces transmitted to the bonds and substratum depend not only on internal contractile forces but also on the rheological properties of the cell and the contraction time the forces are exerted over (DiMilla *et al.*, 1991). Fig. 17 demonstrates how dimensionless cell speed varies with dimensionless cell stiffness, α, for different values of dimensionless cell viscosity, ω. For small α the cell is flaccid, deforms greatly under contraction, little force is transmitted to the bonds at either pseudopod, and movement is insignificant. For large α, in contrast, the cell becomes excessively rigid and ν decreases as the cell deforms minimally under contraction. Intermediate values of α allow efficient transmission of force to the bonds and maximum speed. The effect of small α is equivalent to the effect predicted for small f_c, and correspondingly effects for large α are the equivalent of those for large f_c. At a constant cell stiffness increasing viscosity results in greater resistance to deformation and a monotonic decrease in cell speed.

The duration of the contraction phase of the movement cycle also can have dramatic effects on maximum speed: dimensionless contraction time, τ_c, has little influence on cell speed for small f_c but has a great biphasic effect as f_c is increased (DiMilla *et al.*, 1991). At short contraction times little movement occurs because the contractile force is not applied long enough to break a sufficient number of uropodal bonds. Likewise, at long contraction times the contractile force breaks all the bonds at both ends of the cell, and no net translocation occurs.

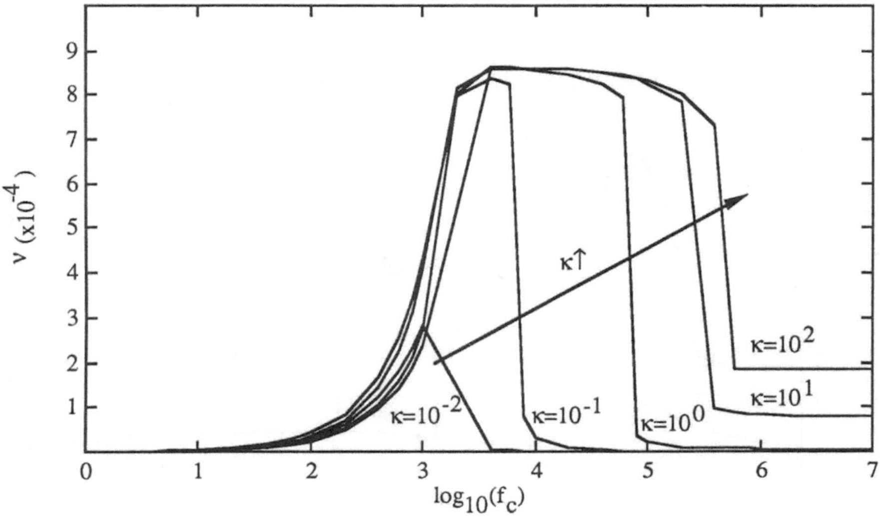

Fig. 16. Model predictions from DiMilla *et al.* (1991) for the biphasic relationship between dimensionless cell speed, v, and dimensionless intracellular contractile force, f_c, for surfaces with different dimensionless adhesion ligand densities, κ.

The Linkage Between Cellular Mechanics, Traction Forces, and Migration

Although our model predictions demonstrate that cytoskeletal contractile force *per se* is important in regulating the rate of migration over adhesive surfaces, the key parameter of interest should be one characterizing the actual forces exerted by the cell on the supporting substratum, namely the traction force. We define the dimensionless total traction force, f_T, as the sum of the dimensionless total forces exerted at the lamellipodal and uropodal compartments of our model cell at the end of the contraction period, t_c (the choice of t_c is arbitrary; see DiMilla *et al.* (1991) for a complete description of this definition). The transmitted total traction force depends implicitly on the intracellular contractile force and explicitly on the molecular adhesive and rheological properties of the cell which govern the dissipation of the contractile force. Harris *et al.* (1980) have demonstrated how it is possible to indirectly measure these forces using elastomeric silicone rubber substrata.

We can examine the relationship between cell speed and total traction force by comparing calculations for the effects of dimensionless contractile force and dimensionless cell

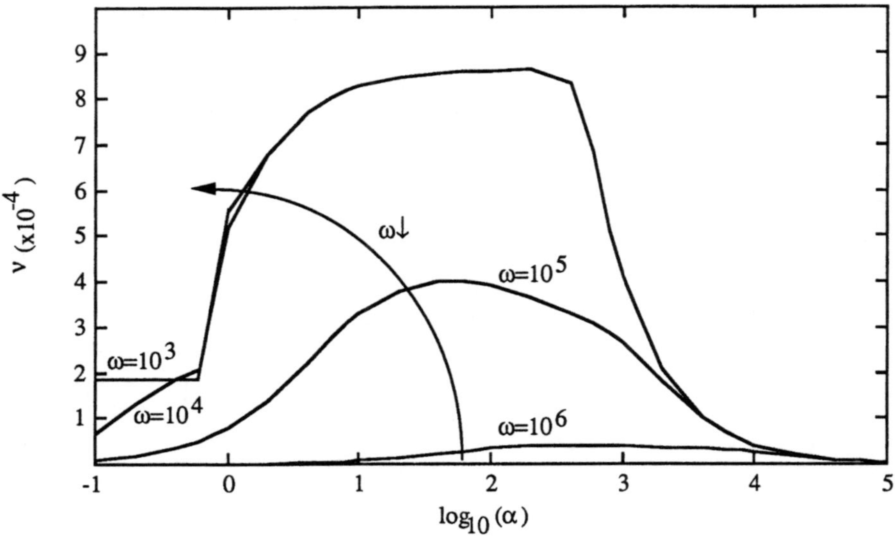

Fig. 17. Model predictions from DiMilla *et al.* (1991) for the biphasic relationship between dimensionless cell speed, ν, and cell rheology. α is dimensionless cell stiffness and ω is dimensionless cell viscosity.

stiffness on both dimensionless speed and dimensionless total traction force. f_T increases with f_C for small f_C and is relatively independent of α, up to a maximum total traction force that is exerted at complete cell deformation. Resistance to deformation, however, can limit the total traction force transmitted to the substratum.

Fig. 18 demonstrates that ν also exhibits a biphasic dependence on f_T, for reasons analogous to those for Fig. 16. In this plot dimensionless cell viscosity is defined such that changes in stiffness are matched by changes in viscosity. Because a stiffer cell dissipates and transmits greater force without fully contracting, increases in α shift the optimal f_T for maximum movement to larger f_T. Our predictions offer one possible explanation for why the rate of migration observed by Young and Herman (Fig. 4) decreased with increasing densities of stress fibers: at a constant level of contractile force (and net traction force) cell speed decreases with increasing stiffness. It is difficult to compare our predictions directly with this and other experimental data, however, because these studies did not distinguish between several possible variations in cellular mechanics (*i.e.*, changes in contractile force versus changes in stiffness). Further studies in which the effects of molecular alterations in the composition and structure of the cytoskeleton on cellular mechanics and cell behaviorial properties are measured in tandem are necessary to rigorously test our predictions.

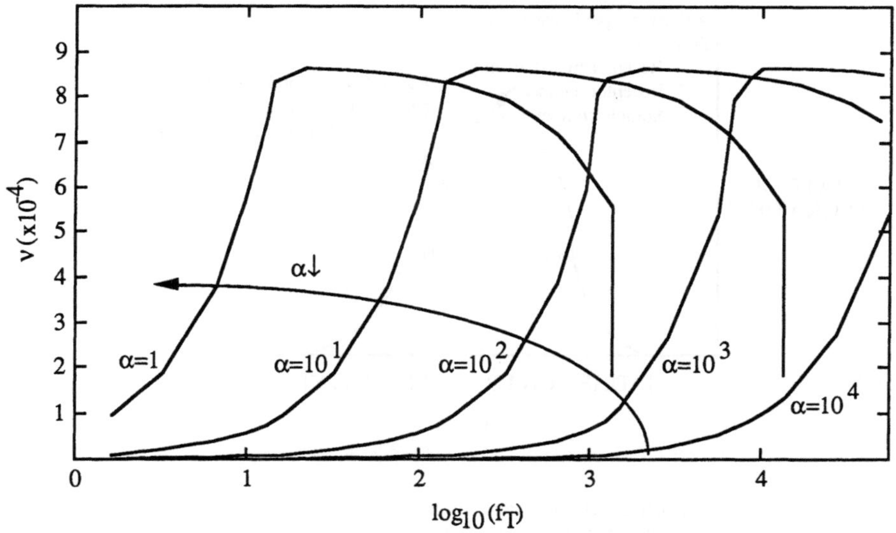

Fig. 18. Model predictions for the biphasic relationship between dimensionless cell speed, ν, dimensionless total traction force, f_T, and cell rheology. α is dimensionless cell stiffness; in this plot dimensionless cell viscosity, ϖ, is proportional to α.

Conclusions and Perspectives for Future Research

Our experimental and modeling studies have demonstrated that, although cell movement over adhesive surfaces is indeed a complex process involving dynamic interactions between cytoskeletal, cell surface, and extracellular components, adhesive interactions play a primary role in regulating the speed of migration. Our conclusions are summarized in Fig. 19:

(1) *An optimal density of adhesion ligand exists for maximum cell speed* (compare Fig. 19a with Figs. 2b and 9). The specific density of adhesion ligand for maximum speed depends on the affinity between adhesion receptor and ligand, receptor number, and the concentration and affinity of soluble competitors for receptors. Also, the range of densities of adhesion ligand allowing movement depends on the magnitude of the asymmetry in the cell-substratum traction between front and rear of the cell; levels of asymmetry in which the lamellipod is approximately twice as adhesive for the substratum as the uropod are consistent with both cellular- and molecular-level observations. Maximum cell speed is primarily a function of cellular mechanics, although the type of adhesion ligand also appears to have an

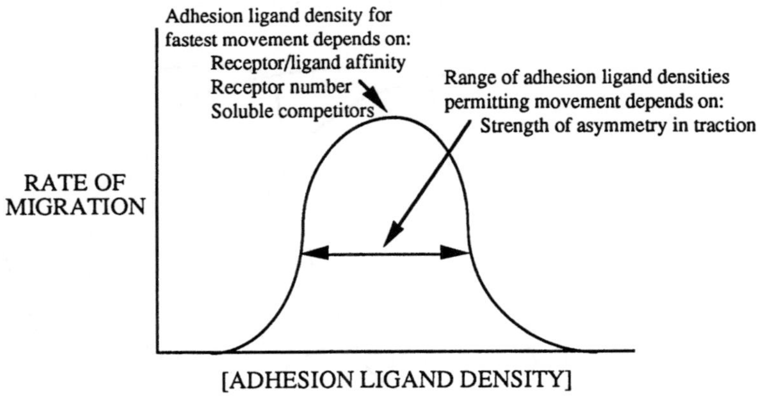

(a)

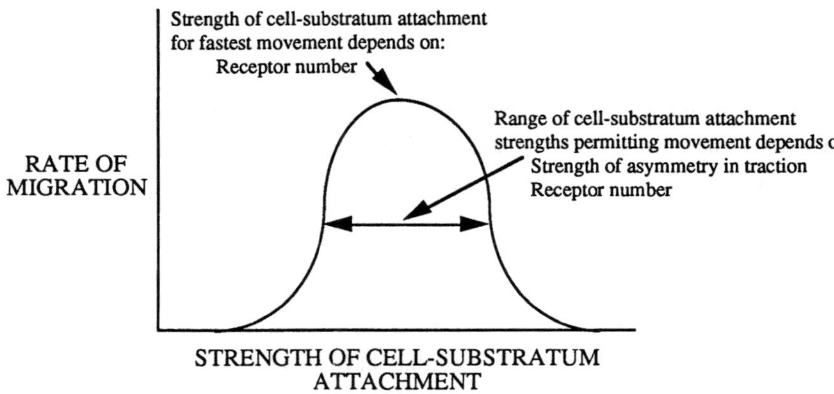

(b)

Fig. 19. Schematic of the relationships between cell speed, surface density of adhesion ligand, and strength of cell-substratum attachment. (a) An optimal surface density of adhesive ligand exists for maximum cell speed (compare with Figs. 2b and 9). (b) An optimal strength of cell-substratum attachment exists for maximum cell speed (compare with Figs. 14b and 15b). These relationships hold only when cellular resistance to deformation -- controlled by cell rheology -- does not limit the effective traction forces generated by intracellular contractile forces (Fig. 18).

effect on this property, possibly due to differences in receptor/cytoskeletal interactions or receptor signaling.

(2) *An optimal strength of transient cell-substratum attachments exists for maximum cell speed* (compare Fig. 19b with Figs. 15b and 16b). Measured as a critical stress or force

for detachment, the strength of adhesion incorporates the effects of adhesion ligand density, adhesion receptor/ligand affinity, and adhesion receptor number into a single functional parameter. The strength of adhesion for maximum speed depends on receptor number but not on the receptor/ligand affinity or the properties (*i.e.*, concentration or affinity) of soluble competitors. Both the asymmetry in traction and receptor number, however, affect the range of adhesiveness allowing significant movement, although a minimum strength of adhesion of approximately 1 $\mu dyne/\mu m^2$ appears necessary for movement. These results suggest that variations in the strength of adhesive bonds are more critical to movement than the density of the adhesion ligand or affinity for adhesion receptors *per se*.

(3) *An optimal balance of cytoskeletal contractile and dissipative forces exists for maximum cell speed* (Figs. 16-18). Biphasic relationships exists between cell speed and intracellular contractile force, cell stiffness, and total traction force. Further, the strength of the adhesive interaction -- in this work evaluated as the dimensionless adhesion ligand density -- can affect the optimal intracellular contractile force (and correspondingly the optimal total traction force) for movement, but movement can only occur under conditions in which the cell can efficiently deform and exert traction forces on the supporting substratum.

A number of problems deserve further attention in future modeling and experimental studies and are outlined below.

(1) *Models for the effects of molecular adhesiveness on the speed of white blood cells.* Because the equations for the dynamics and trafficking of adhesion receptors and cellular mechanics are coupled but solved separately, the current form of our model can not predict the speed of faster-moving white blood cells, such as neutrophils and macrophages. Solving the full transient equations for each system simultaneously is straightforward but computationally expensive.

(2) *Dynamics of the molecular linkages between cytoskeleton and substratum as transient attachments during migration.* Identifying the mechanisms of detachment and, in particular, the point in the mechanical linkage that is disrupted upon contraction, remains a critical problem, especially for situations in which focal contacts are important, ligand is poorly immobilized, or the substratum is deformable. Regen and Horwitz (1992) have shown that migration involves both dissociation of adhesion-receptor/ligand linkages and distraction of adhesion receptors from the cell membrane. Two types of approaches are critical considerations for future experiments: the application of sensitive *in situ* assays -- such as surface plasmon resonance or total internal reflection microscopy -- which can identify changes in substratum composition, including removal of ligand and deposition of receptor-laden membrane fragments, and the development of methods which can quantitatively and unambiguously distinguish transient receptor/ligand-mediated events from adhesion governed by organized receptor-cytoskeletal interactions (these processes currently can be addressed only by qualitative examination of cell morphology and the structure of molecular linkages).

(3) *Quantitative measurements of traction forces as a function of molecular adhesive and mechanical properties of cells and supporting substrata.* Opas (1987) has suggested that migration occurs when traction forces are greater than adhesive forces. To evaluate this hypothesis requires direct comparison of adhesion and traction forces in the same system. Measurements of adhesion are relatively well-advanced today, and techniques based on culturing cells on elastomeric silicone rubber substrata (Harris *et al.*, 1980) and measuring the strain histories of microbeads -- either on the substratum, the cell surface, or injected within the cell -- using image processing (Simon and Schmid-Schönbein, 1990; Barbee *et al.*, 1993) present a powerful set of technologies for investigating traction forces and the mechanics of motile cells in general. Coupling these approaches, which examine cell behavioral properties, with molecular biology that can alter the molecular adhesive properties of cells (*e.g.*, Giancotti and Ruoslahti, 1990; Shih *et al.*, 1993) is a promising direction for future research.

Acknowledgments

Some of the modeling and experiments described here were initiated as part of my doctoral thesis with Doug Lauffenburger and John Quinn at the University of Pennsylvania, and I am grateful to them for their support and encouragement. I also appreciate many helpful discussions with Julie Stone, Ken Barbee, Steve Albelda, and Clayton Buck, and I thank Doug Lauffenburger, Julie Stone, and Ken Barbee for sharing submitted manuscripts with me. This work has been supported by grants from the Whitaker Foundation, the NIH, and the NSF Center for Light Microscope Imaging and Biotechnology at Carnegie Mellon University.

References

Akiyama, SK, and KM Yamada (1985) The interaction of plasma fibronectin with fibroblastic cells in suspension. *J. Biol. Chem.* **260**: 4492-4500.

Albelda, SM, and CA Buck (1990) Integrins and other cell adhesion molecules. *FASEB J.* **4**: 2868-2880.

Barbee, KA, EJ Macarak, and LE Thibault (1993) Strain measurements in cultured vascular smooth muscle cells subjected to mechanical deformation. *Ann. Biomed Eng.* (in press).

Bauer, JS, CL Schreiner, FG Giancotti, E Ruoslahti, and RL Juliano (1992) Motility of fibronectin receptor-deficient cells on fibronectin and vitronectin: collaborative interactions among integrins. *J. Cell. Biol.* **116**: 477-487.

Bell, GI (1978) Models for the specific adhesion of cells to cells. *Science* **200**: 618-627.

Bershadsky, AD, and JM Vasiliev (1988) Cytoskeleton. Plenum Press, New York.

Bongrand, P, and GI Bell. (1984) Cell-cell adhesion: parameters and possible mechanisms. In: Cell Surface Dynamics: Concepts and Models. AS Perelson, C DeLisi, and F Wiegel, (eds) Marcel Dekker, Inc., New York. 459-493.

Bray, D, and JG White (1988) Cortical flow in animal cells. *Science* **239**: 883-888.
Bretscher, M (1984) Endocytosis: relation to capping and cell locomotion. *Science* **224**: 681-686.
Bretscher, MS (1989) Endocytosis and recycling of the fibronectin receptor in CHO cells. *EMBO J.* **8**: 1341-1348.
Bretscher, MS (1992) Circulating integrins: a_5b_1, a_6b_4 and Mac-1, but not a_3b_1, a_4b_1 or LFA-1. *EMBO J.* **11**: 405-410.
Buck, CA, and AF Horwitz (1987) Cell surface receptors for extracellular matrix molecules. *Ann. Rev. Cell Biol. 1987* **3**: 179-205.
Burridge, K, K Faith, T Kelly, G Nuckolls, and C Turner (1988) Focal adhesions: transmembrane junctions between the extracellular matrix and the cytoskeleton. *Ann. Rev. Cell Biol. 1988* **4**: 487-525.
Chen, WT (1981) Surface changes during retraction-induced spreading of fibroblasts. *J. Cell. Sci.* **49**: 1-13.
Cozens-Roberts, C, DA Lauffenburger, and JA Quinn (1990) Receptor-mediated adhesion phenomena. Model studies with the radial-flow detachment assay. *Biophys. J.* **58**: 107-125.
Dejana, E, S Colella, LR Languino, G Balconi, GC Corbascio, and PC Marchiso (1987) Fibrinogen induces adhesion, spreading, and microfilament organization of human endothelial cells. *J. Cell Biol.* **104**: 1403-1411.
Dembo, M, DC Torney, K Saxman, and D Hammer (1988) The reaction-limited kinetics of membrane-to-surface adhesion and detachment. *Proc. Roy Soc. Lond. B* **234**: 55-83.
Devreotes, PN, and SH Zigmond (1988) Chemotaxis in eukaryotic cells: a focus on leukocytes and *Dictyostelium. Ann. Rev. Cell Biol. 1988* **4**: 649-686.
Dickinson, RB, and RT Tranquillo (1993) A stochastic model for cell random motility and haptotaxis based on adhesion receptor fluctuations. *J. Math. Biol.* (in press).
DiMilla, PA, K Barbee, and DA Lauffenburger (1991) Mathematical model for the effects of adhesion and mechanics on cell migration speed. *Biophys. J.* **60**: 15-37.
DiMilla, PA, and DA Lauffenburger (1991) Models for integrin clustering dynamics, anti-integrin antibody binding, and cell migration speed. *J. Cell Biol.* **115**: 114a.
DiMilla, PA, JA Stone, SM Albelda, DA Lauffenburger, and JA Quinn (1992a) Measurement of cell adhesion and migration on protein-coated surfaces. In: Tissue-Inducing Biomaterials. LG Cima and ES Ron, (eds) Mater. Res. Soc. Proc. Vol. 252, Pittsburgh, PA. 205-212.
DiMilla, PA, SM Albelda, DA Lauffenburger, and JA Quinn (1992b) Measurement of individual cell migration parameters for human tissue cells. *AIChE J.* **37**: 1092-1104.
DiMilla, PA, JA Stone, JA Quinn, SM Albelda, and DA Lauffenburger (1993) Maximal migration of human smooth muscle cells on type IV collagen and fibronectin occurs at an intermediate initial attachment strength. *J. Cell Biol.* **122**: 729-737.
Dong, C, R Skalak, K-LP Sung, GW Schmid-Schonbein, and S Chien (1988) Passive deformation analysis of human leukocytes. *J. Biomech. Eng.* **110**: 27-36.
Duband, J-L, GH Nuckolls, A Ishihara, T Hasegawa, KM Yamada, JP Thiery, and K Jacobson (1988) Fibronectin receptor exhibits high lateral mobility in embryonic locomoting cells but is immobile in focal contacts and fibrillar streaks in stationary cells. *J. Cell Biol.* **107**: 1385-1396.
Duband, J-L, S Dufour, SS Yamada, KM Yamada, and JP Thiery (1991) Neural crest cell locomotion induced by antibodies to β_1 integrins. A tool for studying the roles of substratum molecular avidity and density in migration. *J. Cell Sci.* **98**: 517-532.
Dunn, GA (1983) Characterizing a kinesis response: time averaged measures of cell speed and directional persistence. *Agents and Actions [Suppl.]* **12**: 14-33.
Dustin, ML, and TA Springer (1989) T-cell receptor cross-linking transiently stimulates adhesiveness through LFA-1. *Nature (London)* **341**: 619-624.

Evans, E, and M Dembo (1990) Physical model for phagocytic motility: local growth of a contractile network from a passive body. In: N Akkas, (ed) Biomechanics of Active Movement and Deformation of Cells. Springer-Verlag, Berlin. 185-214.

Fung, YC (1977) A First Course in Continuum Mechanics. Prentice-Hall, Inc., Englewood Cliffs, NJ.

Gail, MH, and CW Boone (1970) The locomotion of mouse fibroblasts in tissue culture. *Biophys. J.* **10**: 980-992.

Gailit, J, and E Ruoslahti (1988) Regulation of the fibronectin receptor affinity by divalent cations. *J. Biol. Chem* **263**: 12927-31292.

Giancotti, P, and E Ruoslahti (1990) Elevated levels of the α5β1 fibronectin receptor suppress the transformed phenotype of chinese hamster ovary cells. *Cell* **60**: 849-859.

Goodman, SL, G Risse, and K von der Mark (1989) The E8 subfragment of laminin promotes locomotion of myoblasts over extracellular matrix. *J. Cell Biol.* **109**: 799-809.

Grinnell, F (1986) Focal adhesion sites and the removal of substratum-bound fibronectin. *J. Cell Biol.* **103**: 2697-2706.

Harris, AK (1990) Protrusive activity of the cell surface and the movements of tissue cells. In: N Akkas, (ed) Biomechanics of Active Movement and Deformation of Cells. Springer-Verlag, Berlin. 249-294.

Harris, AK, P Wild, and D Stopak (1980) Silicone rubber substrata: a new wrinkle in the study of cell locomotion. *Science* **208**: 177-179.

Heath, JP, and BF Holifield (1991) Cell locomotion: new research tests old ideas on membrane and cytoskeletal flow. *Cell Motil. Cytoskel.* **18**: 245-257.

Heidemann, SR, P Lamoureux, and RE Buxbaum (1990) Growth cone behavior and production of traction forces. *J. Cell Biol.* **111**: 1949-1957.

Hemler, ME (1990) VLA proteins in the integrin family: structures, functions, and their roles on leukocytes. *Ann. Rev. Immunol. 1990* **8**: 365-400.

Hirst, R, A Horwitz, C Buck, and L Rohrschneider (1986) Phosphorylation of the fibronectin receptor complex in cells transformed by oncogenes that encode tyrosine kinases. *Proc. Natl. Acad. Sci. USA* **83**: 6470-6474.

Hochmuth, RM, HP Ting-Beall, BB Beaty, D Needham, and R Tran-Son-Tay (1993) Viscosity of passive human neutrophils undergoing small deformations. *Biophys. J.* **64**: 1596-1601.

Horwitz, AF, K Duggan, C Buck, MC Beckerle, and K Burridge (1986) Interaction of plasma membrane fibronectin receptor with talin. A transmembrane linkage. *Nature (London)* **320**: 531-532.

Hynes, RO (1992) Integrins: versatility, modulation, and signaling in cell adhesion. *Cell* **69**: 11-25.

Kolega, J, L Janson, and D Taylor (1991) The role of solation-contraction coupling in regulating stress fiber dynamics in nonmuscle cells. *J. Cell Biol.* **114**: 993-1003.

Lackie, JM (1986) Cell Movement and Cell Behavior. Allen and Unwin, London.

Lackie, JM, and PC Wilkinson (1984) Adhesion and locomotion of neutrophil leukocytes on 2-D substrata and 3-D matrices, In: White Blood Cell Mechanics: Basic Science and Clinical Aspects. HJ Meiselman, MA Lichtman, and PL LaCelle, (eds) Alan R. Liss, New York. 237-254.

Lauffenburger, D (1989) A simple model for the effects of receptor-mediated cell-substratum adhesion on cell migration. *Chem. Eng. Sci.* **44**: 1903-1914.

Lotz, MM, CA Burdsal, HP Erickson, and DR McClay (1989) Cell adhesion to fibronectin and tenascin: quantitative measurements of initial binding and subsequent strengthening response. *J. Cell Biol.* **109**: 1795-1805.

Marks, PW, B Hendley, and FR Maxfield (1991) Attachment to fibronectin or vitronectin makes human neutrophil migration sensitive to alterations in cytosolic free calcium concentration. *J. Cell Biol.* **112**: 149-158.

Mueller, SC, T Kelly, M Dai, H Dai, and W-T Chen (1989) Dynamic cytoskeleton-integrin associations induced by cell binding to immobilized fibronectin. *J. Cell Biol.* **109**: 3455-3464.

Opas, M (1987) The transmission of forces between cells and their environment. In: Cytomechanics. The Mechanical Basis of Cell Form and Structure. J Bereiter-Hahn, OR Anderson, and W-E Reif, (eds.) Springer-Verlag, Berlin. 273-285.

Oster, GF, and AS Perelson (1987) The physics of cell motility. *J. Cell Sci.* **8**: 35-54.

Othmer, HG, SR Dunbar, and W Alt (1988) Models of dispersion in biological systems. *J. Math. Biol.* **26**: 263-298.

Regen, CM, and AF Horwitz (1992) Dynamics of b_1 integrin-mediated adhesive contacts in motile fibroblasts. *J. Cell Biol.* **119**: 1347-1359.

Rivero, MA, RT Tranquillo, HM Buettner, and DA Lauffenburger (1989) Transport models for chemotactic cell populations based on individual cell behavior. *Chem. Eng. Sci.* **44**: 2881-2897.

Sato, M, MJ Levesque, and RM Nerem (1987) Micropipette aspiration of cultured bovine aortic endothelial cells exposed to shear stress. *Atherosclerosis* **7**: 276-285.

Savage, B, UM Marzee, BH Chao, LA Harker, JM Maraganore, and ZM Ruggeri (1990) Binding of the snake venom-derived proteins applaggin and echistatin to the arginine-glycine-aspartic acid recognition site(s) on platelet glycoprotein IIb.IIa complex inhibits receptor function. *J. Biol. Chem.* **265**: 11766-11772.

Schmid-Schonbein, GW, K-LP Sung, H Tozeren, R Skalak, and S Chien (1981) Passive mechanical properties of human leukocytes. *Biophys. J.* **36**: 243-256.

Schmidt, CE, AF Horwitz, DA Lauffenburger, and MP Sheetz (1993) Integrin-cytoskeletal interactions in migrating fibroblasts are dynamic, asymmetric, and regulated. *J. Cell Biol.* (accepted).

Shih, D-T, JM Edelman, and CA Buck (1993) Structure/function analysis of the integrin β_1 subunit by epitope mapping. *J. Cell Biol.* (in press).

Simon, SI, and GW Schmid-Schönbein (1990) Cytoplasmic strains and strain rates in motile polymorphonuclear leukocytes. *Biophys. J.* **58**: 319-332.

Singer, S, and A Kupfer (1986) The directed migration of eukaryotic cells. *Ann. Rev. Cell Biol. 1986* **2**: 337-365.

Stone, JA (1993) Quantitative analysis of receptor-mediated adhesion of mammalian cells. Ph. D. Thesis, University of Pennsylvania.

Stossel, T, C Chapponnier, R Ezzell, J Hartwig, and P Janney (1985) Nonmuscle actin-binding proteins. *Ann. Rev. Cell Biol. 1985* **1**: 353-402.

Straus, AH, WG Carter, EA Wayner, and S-I Hakomori (1989) Mechanism of fibronectin-mediated cell migration: dependence or independence of cell migration susceptibility on RGDS-directed receptor (integrin). *Exp. Cell Res.* **183**: 126-139.

Tapley, P, A Horwitz, C Buck, K Duggan, and . Rohrschneider (1989) Integrins isolated from Rous sarcoma virus-transformed chicken embryo fibroblasts. *Oncogene* **4**: 325-333.

Taylor, DL, and JS Condeelis (1979) Cytoplasmic structure and contractility in amoeboid cells. *Int. Rev. Cytol.* **56**: 57-143.

Theret, DP, MJ Levesque, M Sato, RM Nerem, and LT Wheeler (1988) The application of a homogeneous half-space model in the analysis of endothelial cell micropipette measurements. *J. Biomech. Eng.* **110**: 190-199.

Trinkaus, JP (1984) Cells Into Organs. The Forces that Shape the Embryo. Prentice-Hall, Inc., Englewood Cliffs, NJ.

Wang, N, JP Butler, and DE Ingber (1993) Mechanotransduction across the cell surface and through the cytoskeleton. *Science* **260**: 1124-1127.

Ward, MD, and DA Hammer (1993) A theoretical analysis for the effect of focal contact formation on cell-substrate attachment strength. *Biophys. J.* **64**: 936-959.

Wu, P, JB Hoying, SK Williams, BA Kozikowski, and DA Lauffenburger (1993) Soluble integrin-binding peptide can either inhibit or enhance endothelial cell migration, in manner predictable from cell adhesion effort. In preparation.

Young, WC, and IM Herman (1985) Extracellular matrix modulation of endothelial cell shape and motility following injury *in vitro*. *J. Cell Sci.* **73**: 19-32.
Zhu, C, and R Skalak (1988) A continuum model of protrusion of pseudopod in leukocytes. *Biophys. J.* **54**: 1115-1137.
Zhurkov, SN (1965) Kinetic concept of strength of solids. *Int. J. Frac. Mech.* **1**: 311-323.

The Active Response of an Endothelial Cell to the Onset of Flow

Gabriel Helmlinger, Olivier Thoumine, Theodore F. Wiesner, and Robert M. Nerem
Biomechanics Laboratory
Georgia Institute of Technology
Atlanta, GA 30332-0405 U.S.A.

Introduction

Over the past 25 years, there has been considerable interest in the role of hemodynamics in vascular biology and in those pathobiological events leading to arterial disease. Early on much of the effort was focused on *in vivo* studies and on experimental investigations using excised tissue; however, more recently there has been a significant emphasis on the use of cell culture studies as will shortly be discussed (Nerem and Girard, 1990). In both the *in vivo* and *in vitro* studies much of the focus has been on the vascular endothelial cell (EC). *In vivo* the vascular endothelium serves as the interface between the flowing blood and the underlying vessel wall. As such it might be expected to serve as a mediator of blood-associated effects on the biology of the vascular wall. This would include hemodynamic-related effects.

As blood flows through an artery, it imparts a force on the vessel wall. This force can be decomposed into (i) a normal component, pressure, which acts directly on the endothelium, but also distends the basement membrane on which the endothelial monolayer 'rides,' and (ii) a tangential component, shear stress, which is the frictional force per unit area imposed by the viscous nature of blood. The shear stress acting on the vessel wall is orders of magnitude smaller than the pressure, yet there is evidence that shear stress is an important regulator of endothelial function.

NATO ASI Series, Vol. H 84
Biomechanics of Active Movement
and Division of Cells
Edited by N. Akkaş
© Springer-Verlag Berlin Heidelberg 1994

Although there is some evidence for this from *in vivo* experiments, it is difficult to study quantitatively the influence of blood flow on the endothelium *in vivo*. Not only is it not possible to accurately define the details of the *in vivo* hemodynamic environment, but the biologic end points are ones which require changes over a period of days, weeks, even months, and during that time the blood flow and the associated hemodynamic environment will be varying significantly. Thus, for more than a decade, a number of laboratories have been involved in the use of cell culture to study the effect of flow and the associated shear stress on vascular EC (Dewey et al., 1981; Eskin et al., 1984; Levesque and Nerem, 1985). Cell culture, even with the inclusion of flow, does not represent a physiologic model of the vascular endothelium, but it is a biologic model and one where the flow environment is known and can be controlled.

The classic cell culture experiment on the effect of flow involves placing an EC monolayer in a flow apparatus, either a parallel-plate flow chamber or a cone-plate apparatus (Tran-Son-Tay, 1993), and then "turning" on the flow. What EC see is a sudden onset of flow, and with this a variety of changes in cell structure and function are observed. Although as will be discussed such an "onset of flow" experiment is not physiologic, it has proved itself to be a fascinating model, one in which the endothelial cells respond to flow and adapt to their new environment. This process of adaptation appears to be an active one, and in the next few sections the nature of this response for cultured large vessel EC will be summarized. This will be followed by a discussion of what we know about signaling mechanisms involved in this response and our initial attempts to mathematically model these signaling processes. Finally, possible future directions for work in this area will be discussed.

Cell Shape and the Cytoskeletal Structure

As noted earlier, both the parallel-plate flow chamber and cone-plate viscometer have been employed for cultured EC flow experiments. In the study of large vessel EC, most of the available data is either for bovine aortic EC (BAEC) or human umbilical vein EC (HUVEC), although recently some studies have been carried out using human adult EC. The medium used in such flow experiments generally is Dulbecco's modified Eagle's medium (DMEM) with approximately ten percent calf serum (CS) or fetal calf serum (FCS) and supplemented with antibiotics. The substrate to which the cells are grown and anchored is plastic or glass and may be covered by an extracellular matrix protein layer, e.g. a single protein such as fibronectin or a commercially available mixture. Most studies have been carried out for steady state, laminar flow conditions.

With the onset of flow there are a variety of changes in structure and function which are observed. The most visible and quite dramatic is the elongation in cell shape and the alignment with the direction of flow which takes place. In regard to the former, the monolayer changes from one with a cobblestone appearance, where each endothelial cell has a rather polygonal shape, to one where the EC are elongated into almost an ellipsoid shape. The time course of events for BAEC on a relatively adhesive substrate (Levesque and Nerem, 1985) is such that at early times there is no measurable change in cell shape; however, by two hours the process of elongation has been initiated. This continues with the final shape being attained at a time in the range of 24-48 hours. Concommitant with this is an orientation of the cell with the direction of flow. There is a suggestion in the data that the initiation of the process of elongation slightly precedes that of orientation, but these results are inconclusive.

The shape and angle of orientation of an EC monolayer can be quantified using image analysis techniques where first each cell is analyzed and then population averages are determined. The shape of a cell is characterized using the shape index, S.I., which is defined as $4\pi A/P^2$ where A and P are the cell area and perimeter respectively (Nerem et al., 1981). Whereas an EC monolayer in a static culture, no flow environment is characterized by a shape index of approximately S.I. = 0.8, with elongation due to exposure to flow for 24 hours, the value decreases and is found to be in the range of 0.4 to 0.6, depending on the level of shear stress. This is similar to what is observed *in vivo* (Flaherty et al., 1972; Silkworth and Stehbens, 1975), but still represents values slightly higher than that observed *in vivo* in high flow regimes (Nerem et al., 1981; Levesque et al., 1986), i.e. EC *in vivo* may be slightly more elongated than that found in cell culture flow experiments.

Associated with the above noted changes in cell shape and orientation are changes in the cytoskeletal components (White, 1982; Eskin et al., 1985; Wechezak et al., 1985; Sato et al., 1987b). It is the F-actin microfilament network which has been studied in most detail (Wechezak et al., 1985; Sato et al., 1987b). For BAEC in static culture, the distribution of F-actin is characterized by dense peripheral bands (Wong and Gotlieb, 1986); however, with the sudden onset of flow, a dramatic reorganization occurs. In this, the dense peripheral bands first disappear, with this being initiated even before any change in cell shape can be detected (Sato et al., 1987b). This disappearance of the dense peripheral bands is followed by an increased number of centrally located stress fibers and a widening of intercellular spaces with few connecting microfilaments. These stress fibers are mostly aligned with the direction of flow; however, as the cells reach a more complete state of elongation and alignment, intercellular spaces are no longer visible and a higher concentration of microfilaments at the cell's periphery is observed, i.e. there is a reappearance of the dense peripheral bands. This alteration in F-actin localization occurs over a 24 hour period.

The course of these events is influenced by the flow rate, including the level of shear stress, and also the exact cell type and the surface to which the cells are attached. In regard to the level of shear stress as a factor, it should not be surprising that for lower shear stress levels the sequence of events is quite similar, but with an extended time course, commensurate with the increased time required for elongation and alignment.

It should be noted that this dramatic reorganization in F-actin is directly linked to the elongation in cell shape that takes place. If one compares the elongation observed over 48 hours for BAEC exposed to flow with that for BAEC also exposed to flow, but in this latter case treated with cytochalasin B which disrupts actin assembly, one can see a significant difference (Levesque et al., 1989). Cells that have had actin assembly disrupted do not elongate in any major way. The cytochalasin-B-treated BAEC, even after 48 hours of exposure to flow, are still rather round in shape, with their shape index only changing from a value of approximately S.I. = 0.8 in static culture to a value S.I. = 0.7 with flow. It thus appears that the ability to assemble F-actin is critical to the cell elongation process.

Recently our laboratory has investigated the influence of various types of pulsatile flow on BAEC shape and orientation and also on F-actin localization (Helmlinger et al., 1991). Three types of pulsatile flow were examined: (i) a nonreversing pulsatile flow, with a shear stress of 40 ± 20 dyn/cm^2; (ii) reversing pulsatile flows, including both a high-amplitude, reversing shear stress case of 20 ± 40 dyn/cm^2 and a low-amplitude, reversing shear stress case of 10 ± 15 dyn/cm^2, and (iii) two cases of purely oscillatory flow, 0 ± 20 and 0 ± 40 dyn/cm^2. Two steady flow controls were included for these pulsatile flow studies, one with a wall shear stress corresponding to the peak pulsatile flow value and one corresponding to the mean value. For nonreversing pulsatile flow it was observed that, although BAEC shape initially changed less rapidly compared to cells in a steady flow, after 28-32 hours these cells were more elongated. This confirmed earlier findings from our laboratory (Levesque and Nerem, 1989; Levesque et al., 1989). For reversing pulsatile flow, BAEC shape also was found to change less rapidly, but in this case the cells were less elongated at all times compared to steady controls. Finally, for purely oscillatory flow, BAEC remained rounded, polygonal in shape, being virtually identical to that of static culture.

These results suggest that the preferred shape for BAEC when exposed to a high mean shear stress is an elongated one which might involve some polarity. This is in contrast to the case of a purely oscillatory flow where the preferred shape is a polygonal one, but perhaps without any polarity. Obviously, this whole question of cell polarity needs to be directly investigated. However, whatever the case, it is clear that BAEC in culture can not only discriminate between different types of pulsatile flow but also transduce that recognition into alterations in cell morphology.

In these studies we also observed the distribution of F-actin (Helmlinger et al., 1991). For types I and II pulsatile flows, specifically the nonreversing and reversing cases where cell elongation occurred, the localization of F-actin was very similar to that in a laminar, steady flow. Thus, even in the reversing-flow case, well-formed stress fibers aligned with the direction of flow were seen to develop. In contrast, in the oscillatory flow cases where the cell shape remained polygonal, the F-actin localization was concentrated in dense peripheral bands and thus appeared just as in cells in static culture. These pulsatile flow studies thus reinforce our view that there is a direct relationship between cell shape and cytoskeletal localization, and these observations are consistent with *in vivo* results (Levesque et al., 1986; Kim et al., 1989).

The reorganization in cytoskeletal structure that occurs in cultured

BAEC in response to flow and the associated shear stress also is reflected in the cell's mechanical properties. In fact, it may be that the cell's ability to adapt its mechanical properties to the flow environment in which it resides is a key aspect of its response to the mechanical environment imposed by flow. Experiments have been conducted in our laboratory to study cell mechanical properties, where in these studies cell deformation was measured using the micropipette technique as applied to detached BAEC (Sato et al., 1987a, 1987b; Theret et al., 1988; Sato et al., 1990). A cell that has never been exposed to flow becomes spherical on detachment. F-Actin staining has been used to visualize the submembranous, cortical cytoskeletal layer which we believe represents the primary load-bearing component in a micropipette deformation, i.e. a stress-strain measurement. By using a membrane-like model for this cortical layer, one can determine an effective shear modulus that represents the strength of this cortical cytoskeletal layer (Sato et al., 1987a). The importance of cytoskeletal elements in the determination of the effective cortical layer shear modulus is demonstrated by using cytochalasin B to disrupt actin assembly. For BAEC so treated, the effective cortical layer shear modulus if found to be an order of magnitude less than that for control cells. Colchicine which affects microtubule assembly, also causes a reduction in the effective shear modulus, although not as great as that due to cytochalasin B.

In contrast to cells detached from a static culture condition, EC which have been exposed to a high shear stress for a long period of time and subsequently mechanically detached, retain their elongated shape (Sato et al., 1987b). With increasing flow, as characterized by the level of wall shear stress, and/or increasing exposure time, there is an increase in cell stiffness. To analyze micropipette data obtained from BAEC exposed to shear stress a model has been used that treats these elongated cells as a homogeneous material with an effective Young's modulus, E (Theret et al., 1988). Applying the results of this analysis to the micropipette data, characteristic values for E are of the order of 10,000 dyn/cm^2 for BAEC exposed to 30 dyn/cm^2 for 24 hours. This is as compared to values of E of approximately 1000 dyn/cm^2 for cells grown in static culture or on microcarrier beads under control, no-flow conditions. In this later case, the measurements were performed with the cell both attached and detached to its microcarrier bead, with comparable results being obtained. Thus, our observations of F-actin and our measurements of EC mechanical properties all support the notion that EC, in response to shear stress, adapt their cytoskeletal structure to the flow environment in which they reside.

In more recent studies, it has been shown that EC not only have an elastic mechanical nature, but also a viscoelastic behavior (Sato et al., 1990). Furthermore, it is the F-actin that has been demonstrated to dominate these viscoelastic mechanical properties. The microtubules also are important, but to a lesser extent, both in terms of the elastic and viscoelastic properties of EC.

In discussing the mechanical properties of EC, it is important to introduce the concept of tensegrity, i.e. tensional integrity, first introduced by R. Buckminster Fuller (1975) and applied to geodesic domes. Ingber and his coworkers (1985, 1989, 1993) have applied this concept to the structure of a cell. The actin microfilaments and the microtubules appear to be the major load bearing components within a cell with, as an example, the actin forming a continuous network under tension while the microtubules are in compression and, at least in part, support the actin structure. The intermediate filaments, the other major cytoskeletal system, may serve as a cellular integrator. This structure is connected to the extracellular matrix (ECM) through transmembrane receptors, and it is the ECM and these receptors which will be discussed next.

Extracellular Matrix

As part of the adaptation of cultured EC to their flow environment, mechanisms by which the cells remain anchored to the underlying surface are important. The mechanism of cell attachment is a complicated one involving the integration of the cell's cytoskeletal structure with its extracellular matrix (ECM) through focal contacts. These focal contacts are specialized regions of the plasma membrane where transmembrane cell adhesion receptors interact with ECM proteins and the actin microfilament network. They also represent sites where cellular morphological changes may be modulated by alterations in focal contact-associated proteins or by modified contacts with the ECM (for a review, see Burridge et al., 1988).

Over the 24 to 48 hour period during which EC elongation and orientation take place, there is a marked change in the composition of the extracellular matrix proteins. Immunofluorescent staining of fibronectin (FN), laminin (LN), and collagen type IV (CIV), reveal both a granular and a fibrillar pattern, the former, i.e. the granular pattern, is specifically found in the cellular layer, in a perinuclear location, and the latter, i.e. the fibrillar pattern, is localized underneath the EC layer. In contrast, vitronectin (VN) exhibits a bright labeling at the basal surface of BAEC, but no cellular distribution even in the perinuclear area nor a fibrillar pattern in the ECM. For BAEC in static culture, FN is organized into a dense matrix of randomly oriented fibrils. With the onset of flow, these fibrils progressively align over a 12-48 hour period with the direction of flow and/or group into thicker fibers. Extracellular LN and CIV in contrast in static culture form a rather diffuse network of fibrils. Exposure for 48 hours to shear stress results in a progressive grouping of LN and CIV into thick fibers, which, in contrast to FN, are randomly oriented.

In order to examine the influence of shear stress on the protein composition of the ECM, the material remaining on the plastic substratum after removal of the BAEC was analyzed by two-dimensional gel electrophoresis. Consistent increases in the levels of four acidic proteins (Mr/isoelectric pH = 52/4.9, 70/4.7, 70/5.5, and 110/4.4) were observed after 3-6 hours of exposure to flow. The relative levels of FN, LN, and VN, in the ECM fraction of sheared versus static BAEC, were determined by one-dimensional immunoblotting and scanning laser densitometry. The level of FN in BAEC appeared to decrease after 12 hours of exposure to flow, but then increased after 24 to 48 hours. The level of LN exhibited a steady increase with flow up to 48 hours, while the level of VN was not altered by shear stress. These results indicate that specific qualitative and quantitative changes occur in the ECM of BAEC exposed to shear stress. These changes, possibly linked to a reorganization of cytoskeletal elements, may support the cell elongation and orientation observed in response to flow.

To examine the role of cytoskeleton-associated focal contact proteins in the modulation of shear-stress-related EC morphological changes, a double-label immuno-fluorescence study was carried out in order to determine the localization of vinculin and actin microfilaments in BAEC in response to the sudden onset of flow (Girard and Nerem, 1993). Vinculin is known to act as a linker protein, anchoring actin microfilaments to ECM receptor proteins in the plasma membrane. Following exposure to a shear stress of 30 dyn/cm^2 for 24 hours, vinculin was found to be concentrated at the 'upstream' portions of EC, specifically at the ends of stress fibers where vinculin appeared to be anchoring the microfilament bundles upstream. This relocalization of vinculin occurs concomitant with cell elongation and alignment, and is in marked contrast to BAEC in static culture where vinculin is found to be concentrated around the periphery of most cells corresponding to the location of the dense peripheral band of actin microfilaments.

Our laboratory also has examined the role of transmembrane ECM receptors which link actin microfilaments and associated vinculin to specific ECM proteins. In BAEC grown in static culture, $\alpha_5\beta_1$ FNR was found to be present as relatively long streaks of immunofluorescence, indicating a clustering of receptors located around the periphery of each cell. After 24 hours of exposure to flow with a shear stress of 30 dyn/cm^2, the $\alpha_5\beta_1$ FNR appears to be rather evenly distributed over the entire length of each elongated cell. This was distinctly different from the upstream localization of vinculin. With the exception of the upstream portion of the cells, many of the streaks of $\alpha_5\beta_1$ FNR were not associated with vinculin, suggesting a role primarily in cell adhesion.

The $\alpha_V\beta_3$ VNR also was found to be localized around the periphery of BAECs in static culture, although not in a streaked appearance like FNR. In EC subjected to 24 hours of flow, the localization of $\alpha_V\beta_3$ VNR was strikingly different from that of $\alpha_5\beta_1$ FNR as the VNR appeared to concentrate upstream in a manner rather similar to that of vinculin. These results suggest that the VNR acts as a transmembrane adhesion protein, specifically at the upstream portion of the cells.

Clearly the ECM, including the associated focal contact proteins, are a dynamic part of the EC response to the onset of flow. In addition to EC reorganizing their F-actin microfilament structure, the process of adaptation to flow also appears to involve selected changes in ECM proteins and in the organization of the ECM itself. As important as this may be for the maintenance of cell attachment, it equally well may be critical to the signaling processes involved (Wang et al., 1993).

Signal Recognition and Transduction

A critical, still unanswered question is how do EC recognize the flow environment in which they reside, and having done so, how do they transduce this signal into the observed changes in structure and function? In regard to the latter, although not focused on in this discussion, there are alterations in the synthesis and secretion of a wide variety of substances (Frangos et al., 1985; Grabowski et al., 1985; Diamond et al., 1989; Taylor et al., 1991), including the ECM-associated proteins previously noted. Furthermore, the effects of flow extend to the level of gene expression (Diamond et al., 1990; Mitsumata et al., 1991; Resnick et al., 1991), with in some cases messenger RNA being unregulated, in others downregulated, and others yet there being no observed effect at all. There also are changes in endocytic processes due to the influence of flow (Davies et al., 1983), including receptor-mediated LDL endocytosis (Sprague et al., 1987). Most importantly, there are effects on cell proliferation, with the rate of EC proliferation being decreased due to flow (Levesque et al., 1990), this associated with an inhibition of entry into S-phase (Mitsumata et al., 1991). Thus there is an effect of flow on cell cycle.

Returning to the issue of signal recognition and transduction, there have been a number of studies of the mechanisms involved. These initially have focused on the second messengers associated with the transduction of a mechanical signal. Such studies indicate that a mechanical stimuli such as flow does stimulate the phosphoinositide system (Nollert et al., 1990; Prasad et al., 1993), cause an elevation in intracellular calcium (Ando et al., 1988; Dull and Davies, 1991; Mo et al., 1991; Shen et al., 1992; Geiger et al., 1992), and produces a translocation of protein kinase C from cytosol to membrane (Nerem and Girard, 1990; Girard and Nerem, 1993). Of interest is that these second messengers, known to be activated by chemical agonists, are also stimulated by flow.

In our own laboratory the focus has been on intracellular calcium $[Ca^{2+}]_i$, as a second messenger resulting from the flow signal. An important part of these studies recently has been their extension to pulsatile flow. In this we have conducted a systematic investigation of the $[Ca^{2+}]_i$ response to the three different types of pulsatile flow previously described, using steady flow as a control. The $[Ca^{2+}]_i$ response was measured for BAEC monolayers subjected to steady and pulsatile flows with DMEM + 10% CS as the perfusing medium. Steady flow (30 dyn/cm^2) elicited the highest amplitude response, as compared to the three types of pulsatile flows. A 1 dyn/cm^2 steady flow induced the same $[Ca^{2+}]_i$ response as a 30 dyn/cm^2 steady flow (same rise time, amplitude response, and decay time), except that the time for the occurrence of a response was more delayed. Non-reversing (20 ± 15 dyn/cm^2) and reversing (20 ± 30 dyn/cm^2) flows also induced a $[Ca^{2+}]_i$ elevation in BAEC monolayers. The initial $[Ca^{2+}]_i$ peak was followed by a slow return to basal level or a sustained plateau above the basal value. A purely oscillatory flow (0 ± 20 dyn/cm^2) did not elevate $[Ca^{2+}]_i$ in monolayers. Single cell analyses revealed interesting features, where for all flow conditions, single-cell $[Ca^{2+}]_i$ oscillations were observed, with a fast rise time to peak, and an equally fast return time to the previous basal level. Because these $[Ca^{2+}]_i$ oscillations were not synchronized among the cells within a monolayer, they were not revealed in the monolayer traces. A striking observation was that a purely oscillatory flow (0 ± 20 dyn/cm^2) induced numerous single-cell $[Ca^{2+}]_i$ oscillations, although the monolayer response remained similar to that of static, no-flow conditions. The apparent periodicity of the $[Ca^{2+}]_i$ oscillations (between 50 and 200 seconds) was far off the frequency (1 Hz) of the pulsatile flows used. Not all single cells within the field of view were induced to oscillate, and the three types of pulsatile flow induced more cells to oscillate (≈85% of all cells analyzed) than steady flow did (≈57%).

A particularly critical question is whether the elevation in $[Ca^{2+}]_i$ due to flow is purely an agonist effect, where the flow effect is due to the agonist transport being controlled by convection-diffusion coupling (Dull and Davies, 1991; Nollert and McIntire, 1992), or whether there is a more direct mechanical effect (Watson, 1991), e.g. through mechanically-activated ion channels (Lansman et al., 1987; Olesen et al., 1987), an effective "strain gauge" which senses deformation of the cell's cytoskeletal structure, and/or a "shear" receptor in the cell's membrane, perhaps coupled to a G protein (Brown, 1991; Simon et al., 1991), but where the receptor is shear sensitive.

Our results indicate that although the amplitude of the $[Ca^{2+}]_i$ response is much lower with DMEM without CS as a medium, this in comparison to DMEM with CS, there still is a definite elevation in response to flow. Furthermore, not only is there a reduced amplitude, but there also is a decrease in the frequency of occurrence of single-cell asynchronous $[Ca^{2+}]_i$ oscillations with DMEM without CS as compared to with CS. This suggests a more direct mechanical effect in the recognition of flow by EC and is consistent with results of Shen et al. (1992). It is, however, at variance with the views of Mo et al. (1991) and Dull and Davies (1991).

In concluding this discussion, our knowledge of the recognition event in the response of EC to flow is still very unclear. It is quite likely that recognition does not take place through a single pathway, but rather through multiple, parallel events, some of which represent a more direct, mechanical effect. It is even possible that recognition could take place at the abluminal surface, e.g. through the focal contact adhesion points. However, once the cell recognizes its flow environment, then various second messengers, including $[Ca^{2+}]_i$, communicate this intracellularly, and the resulting changes in structure and function occur.

A Mathematical Model of an Endothelial Cell's Response

As a complement to our experimental studies of the influence of flow on vascular EC, a mathematical model of the response of EC to flow is being developed. Our initial effort has involved the description of the agonist pathway of signal recognition and transduction into calcium as a second messenger. Thrombin was chosen as a model agonist. Thrombin is a serine protease known to cause calcium transients in EC. The thrombin receptor is activated via a two-step kinetic process which has been described by Vu et al. (1991). Experimental studies of EC responses to flow were conducted in a square capillary tube flow chamber in which a confluent HUVEC monolayer was exposed to a steady laminar flow of agonist-containing medium.

To model these experiments, the equations describing the conservation of mass and momentum are coupled to a state-space model of receptor activities and intracellular concentrations. The mass transfer of agonist is obtained by solving a species continuity balance numerically, subject to the appropriate boundary conditions. The velocity profile is obtained from the analytical solution to the Navier-Stokes equations for this geometry. The concentration of ligand at the cell surface, c_s, obtained from the above, is used to drive a model of receptor binding and degradation. The general model of receptor binding and degradation given by Lauffenburger et al. (1985) is used.

A bound receptor balance is given by :

$$\dot{R}_b = k_{on}\,(R_T - R_b)c_s - k_{off}R_b - k_dR_b$$

In this expression, the first two terms represent the net rate of binding of thrombin to its receptor, while the third term represents the rate of proteolysis. The receptor occupancy in turn drives the state-space model of the calcium oscillator described by Meyer and Stryer (1988). This model is represented conceptually by:

$$y_i = f(y_i, R_b)$$

Here, y_i is a vector of concentrations of intracellular species. These species include $[Ca^{2+}]_i$ and IP_3. The foregoing equations may be integrated numerically to obtain the temporal response of the agonist concentration and the state variables R_b and y_i. Thus, the response of $[Ca^{2+}]_i$ to changes in c_s, including its modification by flow, may be calculated.

Using parameter values that are physiologically realistic, computer calculations have been carried out, and the predictions of the model have been compared to experimental monolayer responses. The model prediction compares quite favorably to experiment in a qualitative sense. Both the experimental time course and the model prediction feature a delay time, a peak value, and a plateau phase. This is considered reasonable agreement in view of the inaccuracies in determining cellular rate and equilibrium constants.

Through a parametric analysis conducted using the model, the following was found. As the shear stress is increased from a low of 0.1 dynes/cm^2 to 30 dynes/cm^2, one sees an increase in the rate of response. Also apparent is a slight increase in the peak value at higher flow rates. These effects are due to the enhanced mass transfer of agonist to the cell's surface. Noteworthy here is the fact that there is little difference between the responses at 5 dynes/cm^2 and 30 dynes/cm^2. This indicates that the agonist-mediated calcium mobilization process is controlled by mass transfer limitations at low shear, but by the intrinsic kinetics of receptor activation and release from intracellular stores at higher shear stresses, i.e. above 5 dynes/cm^2.

Comparing the model prediction to experiment indicates that the agonist pathway contributes significantly to signal recognition and transduction in EC. This pathway is modulated by flow at low shear stresses. However, it is unlikely that the agonist pathway is the sole modality by which the cell senses flow. Future work includes investigation of the contributions of other factors such as ion channels and exogenous calcium. It is hoped that this mathematical approach, when taken in concert with experiments, can provide added insight into the actual mechanism(s) of signal recognition and transduction involved in the response of EC to flow.

Future Directions

Although considerable progress has been made in the last decade, there is still much to be done. Experimentally, there are many functional aspects which are still unstudied. Key questions that remain involve the signaling mechanisms associated with the response of the vascular endothelium to the mechanical environment imposed by flow. This is particularly true of the events involved in recognizing a mechanical stimulus, but it is also true of intracellular events, including nuclear pathways. Equally important is the identification of the regulatory mechanisms involved in specific structural and functional changes.

Even though cell culture studies of the effect of flow represent a well-defined biological model, they do not simulate the *in vivo* 'world.' The addition of flow into cell culture studies does improve the simulation, but there is still much that could be done to make cell culture systems a better model for the study of vascular biology and events associated with the initiation of disease.

As important as the addition of flow to cell culture studies has been, it still must be recognized that the flow environment used is not physiologic. Not only is the study of pulsatile flow, and not steady flow, important, but in addition EC *in vivo* are not exposed to a sudden onset of flow. To be sure flow in the human body is time-varying over periods of minutes, hours, and days. However, in general, the flow is not suddenly accelerated from a no-flow to a flow condition, but rather changed from one flow environment to another. Such studies can be done in the cell culture laboratory and have been initiated in our group.

It also should be recognized that there is some variation in shear stress over the cell surface due to the intrusion of the cell into the flow. Satcher et al. (1992) have carried out calculations using an idealized cell shape. Their results suggest that the peak shear stress acting on the cell is on the order of 25 percent higher than the nominal value calculated based on flow rate and the geometry of the flow chamber. This variation in shear stress is non-trivial.

Another aspect is the fact that *in vivo* EC reside not on a plastic surface, but in a biological milieu where vascular smooth muscle cells (SMC) are their neighbors. Thus, cell culture studies designed to study the influence of flow on EC would benefit from the use of EC/SMC co-culture models.

Finally, our effort to mathematically model EC response to flow is one which has progressed more rapidly than first anticipated; however, here also there is still much to be done. This includes modeling additional mechanisms of signal recognition and coupling the second messengers to changes in structure and function. In this the dynamics of actin will be all important, and one can even envision the day when, as part of predicting the EC responses, one will use finite element mechanics to calculate the stresses within the cell's cytoskeleton, a structure which might be of a tensegrity nature as proposed by Ingber, but one undergoing dynamic change as the cell actively adapts to its flow environment.

Acknowledgment

This work was supported by National Institutes of Health Grants HL-26890 and HL-41175 and NSF Grants ECS-8815656 and BCS-9111761. The authors thank R.W. Alexander, B.C. Berk, P.R. Girard, C. J. Schwartz, E.A. Sprague, T.M. Wick, and C. Zhu for the many discussions which have led to the ideas reflected in this paper and for their participation as collaborators in the study of this subject.

References

Ando J, Komatsuda T, Kamiya A (1988) Cytoplasmic calcium responses to fluid shear stress in cultured vascular endothelial cells. In Vitro Cell Dev. Biol. 24:871-877.

Brown, AM (1991) A cellular logic for G protein coupled ion channel pathways. FASEB J. 5:2175-2179.

Burridge K, Fath K, Kelly I, Nuckolls G, Turner C. (1988) Focal adhesions: transmembrane junctions between the extracellular matrix and the cytoskeleton. Annu. Rev. Cell. Biol. 4:487-525.

Davies PF, Dewey CF Jr., Bussolari SR, Gordon EJ, Gimbrone MA Jr. (1983) Influence of hemodynamic forces on vascular endothelial function. In vitro studies of shear stress and pinocytosis in bovine aortic endothelial cells. J. Clin. Invest. 73:1121-1129.

Dewey CF, Bussolari SR, Gimbrone MA, Jr, Davies PF (1981) The dynamic response of vascular endothelial cells to fluid shear stress. ASME J. Biomech. Engr. 103:177-181.

Diamond SL, Eskin SG, McIntire LV (1989) Fluid flow stimulates tissue plasminogen activator secretion by cultured human endothelial cells. Science 243:1483-1485.

Diamond SL, Sharefkin JB, Dieffenbach C, Frazier-Scott K, McIntire LV, Eskin SG (1990) Tissue plasminogen activator messenger RNA levels increase in cultured human endothelial cells exposed to laminar shear stress. J. Cell. Physiol. 143:364-371.

Dull RO, Davies PF (1991) Flow modulation of agonist (ATP)-response (Ca++) coupling in vascular endothelial cells. Am. J. Physiol. 261:H149-154.

Eskin SG, Ives CL, Frangos JA, McIntire LV (1985) Cultured endothelium: the response to flow. Am. Soc. Artif. Organs 8:109-112.

Eskin SG, Ives CL, McIntire LV, Navarro LT (1984) Response of cultured endothelial cells to steady flow. Microvasc. Res. 28:87-94.

Flaherty JR, Pierce JR, Ferrans VJ, Patel DJ, Tucker WK, Fry DL (1972) Endothelial nuclear patterns in the canine arterial tree with particular reference to hemodynamic events. Circ. Res. 30:23.

Frangos JA, McIntire LV, Eskin SG, Ives CL (1985) Flow effects on prostacyclin production by cultured human endothelial cells. Science 227:1477-1479.

Fuller JB (1975) Synergetics. Macmillan, New York.

Geiger RV, Berk BC, Alexander RW, Nerem RM (1992) Flow-induced calcium transients on single endothelial cells: spatial and temporal analysis. Am. J. Physiol.: Cell Physiol. 262:C1411-C1417.

Girard PR, Nerem RM (1990) Role of protein kinase C in the transduction of shear stress to alterations in endothelial cell morphology. J. Cell Biochem. 14E:210 (Abstract).

Girard PR, Nerem RM (1993) Endothelial cell signaling and cytoskeletal changes in response to shear stress. Frontiers Med. Biol. Engng, 5:31-36.

Grabowski EF, Jaffe EA, Weksler BB (1985) Prostacyclin production by cultured human endothelial cells exposed to step increases in shear stress. J. Lab. Clin. Med 105:36-43.

Helmlinger G, Geiger RV, Schreck S, Nerem RM (1991) Effects of pulsatile flow on cultured vascular endothelial cell morphology. ASME J. Biomech. Engr. 113:123-131.

Ingber DE, Folkman J (1989) Tension and compression as basic determinants of cell form and function: Utilization of a cellular tensegrity mechanism. In Cell Shape: Determinants Regulation and Regulatory Role, 1-32, Stein WD, Bronner F (eds.), Academic Press, Orlando, FL.

Ingber DE, Jamieson JD (1985) Cells as tensegrity structures: Architectural regulation of histodifferentiation by physical forces transduced over basement membrane. In Gene Expression during Normal and Malignant Differentiation, 13-32, Anderson LC, Gahmberg CG, Ekblom P (eds.), Academic Pressi, Orlando, FL.

Ingber DE, Karp S, Plopper G, Hansen L, Mooney D (1993) Mechanochemical transduction across extracellular matrix and through the cytoskeleton. In Physical Forces and the Mammalian Cell, 61-79, Francos JA (ed.), Academic Press, San Diego, CA.

Kim DW, Gotlieb AI, Langilie BL (1989) In vivo modulation of endothelial F-actin microfilaments by experimental alterations in shear stress. Arteriosclerosis 9:439-445.

Lansman JB, Hallam TJ, Rink TJ (1987) Single stretch-activated ion channels in vascular endothelial cells as mechanotransducers? Nature 325:811-813.

Lauffenburger DA, Lindermann JJ, Hammer DA (1985) Mathematical modeling of receptor-mediated mammalian cell phenomena. Biotechnology and Bioengineering Symposium 15:675-688.

Levesque MJ, Liepsch D, Moravec S, Nerem RM (1986) Correlation of endothelial cell shape and wall shear stress in a stenosed dog aorta. Arteriosclerosis 6:220-229.

Levesque MJ, Nerem RM (1985) The elongation and orientation of cultured endothelial cells in response to shear stress. ASME J. Biomech. Engr. 176:341-347.

Levesque MJ, Nerem RM (1989) The study of rheological effects on vascular endothelial cells in culture. Biorheology 26:345-357.

Levesque MJ, Sprague EA, Nerem RM (1990) Vascular endothelial cell proliferation in culture and the influence of flow. Biomaterials 11:702-707.

Levesque MJ, Sprague EA, Schwartz CJ, Nerem RM (1989) The influence of shear stress on cultured vascular endothelial cells: The stress response of an anchorage-dependent mammalian cell. Biotech. Prog. 5:1-8.

Meyer T, Stryer L (1988) Molecular model for receptor-stimulated calcium spiking. Proceedings of the National Academy of Science U.S.A. (Biochemistry), 85:5051-5055.

Mitsumata M, Nerem RM, Alexander RW, Berk BC (1991) Shear stress inhibits endothelial cell proliferation by growth arrest in the G0/G1 phase of the cell cycle. FASEB J. 5(4):A527 (Abstract).

Mitsumata M, Nerem RM, Alexander RW, Berk BC (1991) Inverse relationship in mRNA expression between c-sis and GAPDH in endothelial cells subjected to shear stress. Abstract book of workshop on Mechanical Stress Effects on Vascular Cells, Atlanta, GA, April 20-21.

Mo M, Eskin SG, Schilling WP (1991) Flow-induced changes in Ca2+ signaling of vascular endothelial cells: effect of shear stress and ATP. Am. J. Physiol. 260:H1698-H1707.

Nerem RM, Girard PR (1990) Hemodynamic Influences on vascular endothelial biology. Toxic. Path. 18:572-582.

Nerem RM, Levesque MJ, Cornhill JF (1981) Vascular endothelial morphology as an indicator of blood flow. ASME J Biomech Engr 103:172-176.

Nollert MU, Eskin SG, McIntire LV (1990) Shear stress increases inositol trisphosphate levels in human endothelial cells. Biochem . Biophys. Res. Commun. 170:281-287.

Nollert MV, McIntire LV (1992) Convective mass transfer effects on the intracellular calcium response of endothelial cells. ASME J. Biomech. Engr. 114:321-326.

Olesen SP, Clapham DE, Davies PF (1987) Hemodynamic shear stress activates a K+ current in vascular endothelial cells. Nature 331:168-170.

Prasad ARS, Logan SA, Nerem RM, Schwartz CJ, Sprague EA (1993) Flow-related responses of intracellular inosital phosphate levels in cultured aortic endothelial cells. Circulation Res. 72:827-836.

Resnick N, Dewey CF, Atkinson B, Collins T, Gimbrone MA, Jr. (1991) Shear stress induces PDGF-B promoter/CAT fusion gene expression in bovine aortic endothelial cells. FASEB J. 5(5):A1247 (Abstract).

Satcher RL Jr., Bussolari SR, Gimbrone MA Jr., Dewey CF Jr. (1992) The distribution of fluid forces on model arterial endothelium using computational fluid dynamics. ASME J. Biomech. Engr. 114:309-316.

Sato M, Levesque MJ, Nerem RM (1987a) Application of the micropipette technique to the measurement of the mechanical properties of cultured bovine aortic endothelial cells. ASME J. Biomech. Eng. 109:27-34.

Sato M, Levesque MJ, Nerem RM (1987b) Micropipette aspiration of cultured bovine aortic endothelial cells exposed to shear stress. Arteriosclerosis 7:276-286.

Sato M., Theret DP, Wheeler LT, Ohshima N, Nerem RM (1990) Application of the micropipette technique to the measurement of cultured porcine aortic endothelial cell viscoelastic properties. ASME J. Biomech. Eng. 112:263-268.

Shen J, Luscinkas FW, Connolly A, Dewey CF Jr., Gimbrone MA Jr. (1992) Fluid shear stress modulates cytosolic free calcium in vascular endothelial cells. Am. J. Physiol. 262 (Cell Physiol. 31): C384-390.

Silkworth JB, Stehbens WE (1975) The shape of endothelial cells in en face preparations of rabbit blood vessels. Angiology 26:474-487.

Simon MI, Strathmann MP, Gautam N (1991) Diversity of G proteins in signal transduction. Science 252:802-808.

Sprague EA, Steinbach BL, Nerem RM, Schwartz CJ (1987) Influence of a laminar steady-state fluid-imposed wall shear stress on the binding, internalization, and degradation of low density lipoproteins by cultured arterial endothelium. Circulation 76:648-656.

Taylor WR, Harrison DG, Nerem RM, Peterson TE, Alexander RW (1991) Characterization of the release of endothelium-derived nitrogen oxides by shear stress. FASEB J. 56(6):A1727 (Abstract).

Theret DP, Levesque MJ, Sato M, Nerem RM, Wheeler LT. (1988) The application of a homogeneous half-space model in the analysis of endothelial cell micropipette measurements. ASME J. Biomech. Eng. 110:190-199.

Tran-Son-Tay, R (1993) Techniques for studying the effects of physical forces on mammalian cells and measuring cell mechancial properties. In Physical Forces and the Mammalian Cell, 1-59, Frangos JA (ed.), Academic Press, San Diego, CA.

Vu, TH, Hung AT, Wheaton VI, Coughlin, SR (1991) Molecular cloning of a functional thrombin receptor reveals a novel proteolytic mechanism of receptor activation. Cell 64:1057-1068.

Wang N, Butler JP, Ingber DE (1993) Mechanotransduction across the cell surface and through the cytoskeleton. Science 260:1124-1127.

Watson PA (1991) Function follows form: generation of intracellular signals by cell deformation. FASEB J. 5:2013-2019.

Wechezak AR, Viggers RF, Sauvage LR (1985) Fibronectin and F-actin redistribution in cultured endothelial cells exposed to shear stress. Lab. Invest. 53:639-647.

White GE, Fujiwara K, Shefton EJ, Dewey CF Jr., Gimbrone MA Jr. (1982) Fluid shear stress influences cell shape and cytoskeletal organization in cultured vascular endothelium. Fed. Proc. 41:321 (Abstract).

Wong MKK, Gotlieb AI. (1986) Endothelial cell monolayer integrity. 1. Characterization of dense peripheral band of microfilaments. Arteriosclerosis 6:212-219.

Mechanical Aspects of Ciliary Propulsion

Michael E.J. Holwill
Department of Physics
King's College London
Strand
London WC2R 2LS
England

Introduction

Cilia and eukaryotic flagella are motile cellular organelles, usually between 5 μm and 200 μm long and 0.2 μm in diameter, used to produce relative motion between a cell and its liquid environment. Cilia tend to occur in large numbers (hundreds) on the surfaces of cells, such as *Tetrahymena* or epithelial cells in the human lung. In contrast, flagellated cells bear small numbers of the organelles, usually one (*Crithidia* and many spermatozoa) or two (*Chlamydomonas*), but sometimes four (*Trichomonas*), eight (*Hexamita*) or larger numbers (the Hypermastigida). Cilia and eukaryotic flagella have the same basic structure and motile behaviour, so that to avoid repetition, and possible confusion with the prokaryotic flagellum, which has a different structure and mode of action from its eukaryotic counterpart, in this Chapter the eukaryotic organelles will be referred to collectively as cilia.

Since the cilium is a self-contained propulsive unit, it is a convenient experimental system for the investigation of the molecular basis of motility. Being found in the reproductive, digestive and respiratory tracts of humans (and other mammals), ciliary malfunction can lead to medical disorders, so that a complete understanding of their mechanism will have considerable medical, as well as biological, significance.

In this chapter, the mechanical aspects of ciliary movement will be discussed, with only limited reference to the chemical composition and interactions of the organelles. Following a description of the wide range of bend patterns seen on cilia, the nature of the forces elicited from the liquid environment will be examined. A brief description of ciliary structure then follows, with subsequent sections on mechanisms of movement and the types of modelling used to investigate ciliary motion.

NATO ASI Series, Vol. H 84
Biomechanics of Active Movement
and Division of Cells
Edited by N. Akkaş
© Springer-Verlag Berlin Heidelberg 1994

Patterns of Movement

It is important to obtain an accurate characterisation of the shape of a cilium during movement since it reflects the macromolecular behaviour within the organelle. Observations indicate that cilia propagate two- or three-dimensional bends along their length. In the case of two-dimensional bending, the curvature and length of successive bends (with opposite curvature) can vary both in single cell types and from one species to another, to give a range of motile patterns. If successive bends have equal curvature over the same length of organelle, a symmetric wave is propagated, whereas if the curvature and/or length differ, the pattern of movement will be asymmetric. Truly symmetric two-dimensional wave propagation is rare; a number of species of flagellated cells sustains propagating waves, but the waves are asymmetric and often have a small three-dimensional component. Sea urchin sperm swimming close to a the surface of a microscope coverslip propagate virtually two-dimensional waves, but since they swim in circular paths (e.g. Gibbons and Gibbons, 1972), the waveform must be asymmetric, as is demonstrated by critical measurement (e.g. Goldstein, 1977). When swimming through regions remote from a surface, these sperm rotate about their axis of progression because the wave has a small three-dimensional component. The larger bend of an asymmetric wave is referred to as the 'principal bend', while the smaller one is called the 'reverse bend' (Gibbons and Gibbons, 1972). Some flagella, such as that borne by *Euglena*, propagate predominantly three-dimensional bends. These are not helical waves, but tend to be flat loops propagated so that the plane of a loop is roughly normal to the propagation direction. In the majority of cell types with flagella that propagate waves, the waves are initiated at the flagellar base and propagate towards the tip. Exceptions are the trypanosomes and related organisms, which propagate waves from the flagellar tip towards the base (e.g. Holwill, 1966; Sugrue et al, 1988). Examples of different motility patterns are shown on the left of Fig. 3.

Smooth flagella, as might intuitively be expected, and can be demonstrated both theoretically and experimentally, elicit a propulsive thrust from the fluid environment in the direction opposite to that of wave propagation (e.g. Holwill, 1974). Flagella of some species of organism themselves bear appendages. In some, such as *Chlamydomonas* and *Euglena*, these take the form of flexible hairs that are wrapped around the main flagellar shaft. In others, e.g. *Ochromonas*, the hairs are rigid and, to explain the observed motile behaviour, must remain roughly perpendicular to the flagellar shaft during movement, and some must lie in the plane of the undulation. In *Ochromonas*, the direction of flagellar waves is from base to tip, the flagellum is held in front of the cell, and the organism swims flagellum-first (Fig. 1). To explain this unexpected behaviour qualitatively, consider the behaviour of the mastigonemes (as the hairs are called) at

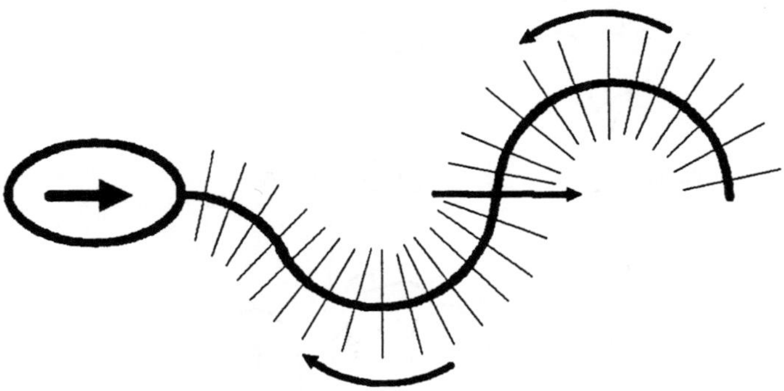

Figure 1. The action of mastigonemes in producing a propulsive force in the direction of wave propagation, indicated by the long arrow on the wave axis. The motion of the mastigonemes at a wave crest is indicated by the curved arrows, while the arrow in the cell body indicates its direction of motion.

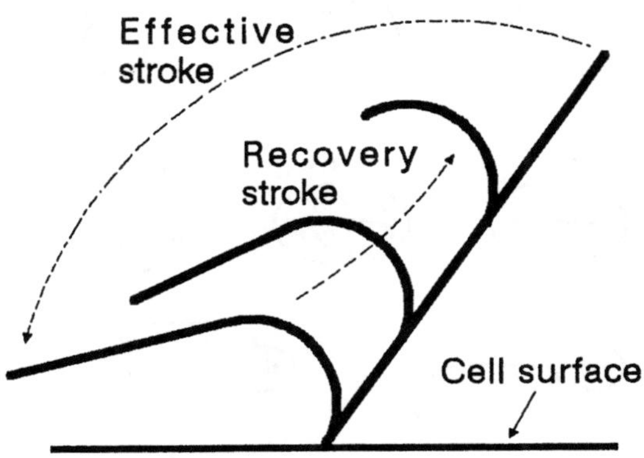

Figure 2. Showing the effective and recovery strokes of a ciliary beat. The beat is usually three-dimensional.

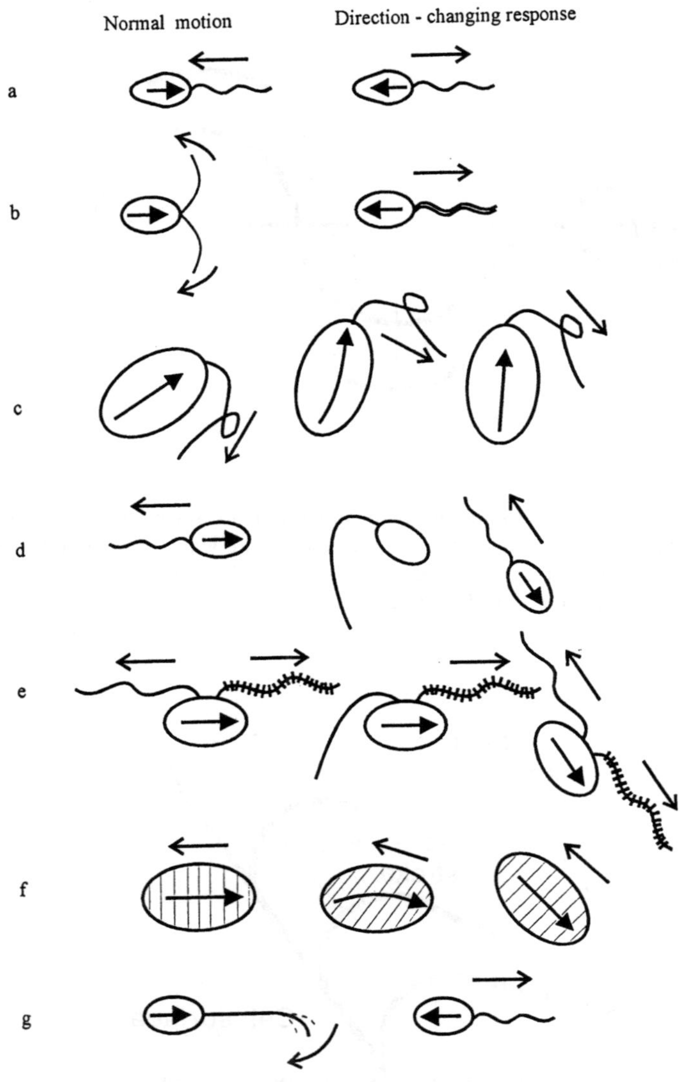

Figure 3. Showing a range of ciliary movement patterns, together with variations used to change the direction of motion of the cell. Arrows outside the body indicate ciliary motion/wave direction, while arrows in the body give its direction of motion. a. *Crithidia*, showing wave reversal; b. *Chlamydomonas*, showing a 'breast-stroke' action changing to wave propagation; c. *Euglena*, showing reorientation of the flagellum sustaining a three-dimensional bend; d. Spore of *Allomyces*, showing arrest of flagellum during cell reorientation; e. Spore of *Blastocladiella*, showing arrest of posterior flagellum while the hispid anterior flagellum continues to beat; f. *Paramecium*, showing the reorientation of the metachronal waves, the crests of which are indicated by the lines. g. *Peranema*, showing localised tip movement changing to wave propagation along the whole organelle.

the crest of a wave. As the wave passes, an individual mastigoneme at the crest moves in the manner of an oar, thereby producing a propulsive thrust in the direction opposite to that of the flagellar shaft. Provided the force generated by the mastigonemes exceeds that produced by the shaft, the net thrust will be such as to move the organism in the direction of the propagating wave (Jahn et al, 1964; Holwill and Peters, 1974).

Fields of cilia are characterised by the presence of metachronal waves, produced by the co-ordinated action of the organelles. Hydrodynamic interactions between adjacent cilia cause synchronised beating along lines on the cell surface parallel to the metachronal wave crests, while the beats are out of phase in other directions. The beat pattern of an individual cilium can be divided conveniently into two parts, an effective stroke during which it bends in the basal region with most of the organelle straight, and a recovery stroke, when the bend propagates to the tip to bring it to the beginning of the next effective stroke (Fig. 2). In practice, the two parts are not usually clearly separated, but follow one another smoothly to give a continuous cycle of activity. While the effective stroke is usually two-dimensional, and occurs in a plane roughly normal to the cell surface, the recovery stroke is three-dimensional, with the cilium remaining close to the cell surface to minimise energy expenditure. In the majority of organisms the recovery stroke occupies a greater fraction of the cycle than the effective stroke; for example, the complete cycle of a *Sabellaria* cilium occupies 60 ms, with the effective stroke taking about 40% of this period (Sleigh & Holwill, 1969). An exception is the *Mytilus* gill compound cilium, which executes the recovery stroke in 75 ms, while the effective stroke takes about twice as long (Sleigh & Holwill, 1969); these particular cilia are probably used for some purpose other than to create significant water currents.

The line of synchronised beating, and therefore the direction of the metachronal wave, depends on the length, beat form and spacing of individual cilia in the field. The direction of the effective stroke is used as a reference for the metachronal wave (Knight-Jones, 1954) . If the two are in the same direction, the metacrony is symplectic, if the directions are opposite, it is antiplectic; if the metachronal wave propagates to the right of the effective stroke it is termed dexioplectic, if to the left, it is laeoplectic.

Many cilia change their beat form in response to a change in the external environment, indicating cellular control over ciliary movement. For many organisms this may be seen in the avoidance and chemotactic responses, when the beat pattern changes to propel the cell either away from or towards a particular environment. The transient change may take different forms, such as a reversal of the wave propagation direction, as in *Crithidia*, alteration of the direction of the effective stroke, as in *Paramecium*, and complete cessation of ciliary movement with a reorientation of the cell body, as in gametes of *Blastocladiella*. These, and other examples are shown in Fig. 3.

While there is a wide range of different beat patterns, those patterns are maintained by organelles with essentially the same internal structures (see later) and it is therefore natural to enquire whether there are any features common to the various motile patterns observed. One characteristic that has been the subject of much study is the shape of a bend. An examination of a ciliary image indicates that the waves are not sinusoidal, and Brokaw (1965) showed that for sperm of several marine organisms the wave shape could be matched by a series of circular arcs separated by straight regions (an arc-line bend). An alternative shape, similar to the arc-line, is the meander, which is the curve of minimum potential energy adopted by an elastic beam when the freely-hinged ends are forced towards each other. An objective method of distinguishing between these two bend shapes was developed by Silvester and Holwill (1972), who represented the curves by a Fourier series. To avoid problems of duplicate co-ordinates on re-entrant cilia, the curve was specified in (ϕ, s) format, where ϕ is the tangent angle at position s measured along the curve. A Fourier series representation of a cilium of length S is then

$$\phi(s) = \sum_n b_n \sin\left[(2n-1)\frac{\pi s}{2S}\right], \tag{1}$$

where the coefficients b_n characterise the harmonic components of the curve and are given by

$$b_n = \frac{2}{S}\int_0^S \phi(s) \sin\left[(2n-1)\frac{\pi s}{2S}\right] ds. \tag{2}$$

A given curve has a unique set of coefficients b_n, so that cilium shapes can be described unambiguously. Application of this method to the organism *Crithidia oncopelti* showed that the waves are more closely matched by the arc-line shape than by the meander (Johnston et al, 1979). This result, and the similar conclusion reached for other organisms, suggests that the internal forces that deform the cilium may change abruptly, indicating a rapid conformational change at the macromolecular level, though the shape change in the transition region will be smoothed by the elasticity of the organelle.

Nature of the Propulsive Thrust

The motion of a fluid is described by the Navier-Stokes equation, which is an expression of Newton's second law for the system. In the equation, there are the three force terms which relate respectively to the inertia, the viscosity and the pressure associated with the fluid motion. The relative importance of the inertial and viscous forces can be assessed from a determination of the Reynolds number, Re, which is given by the relationship

$$\text{Re} = \frac{\text{Inertial force}}{\text{Viscous force}} = \frac{\rho U L}{\eta} \tag{3}$$

where ρ and η are the density and coefficient of dynamic viscosity of the fluid, while U and L are characteristic lengths associated with the motion. Where the disturbance is due to a body moving through the fluid, U is the velocity of the body relative to the fluid, while L is a particular dimension of the body. For cilia the dimension of relevance is the diameter, 0.2 μm, the velocity is about 50 μm s^{-1} and, for an aqueous solution η/ρ is about 1×10^{-6}, so that the Reynolds number is about 10^{-5}. For ciliary motion, therefore, the viscous effect dominates the inertial one, which can be neglected by comparison. Propulsion by viscous interaction with a liquid environment is not a mechanism with which we, as humans, have direct experience; our own locomotion when swimming is caracterised by a high Reynolds number, in the region of 10^5, so that inertial forces are dominant, and viscous ones can be neglected by comparison. One consequence of this is that we do not have an intuitive grasp of the nature of propulsion by ciliated cells. A simple example will illustrate this point; our experience indicates that when motive power is removed from a moving high Reynolds number system, e.g. a rowing boat, the momentum of the system will cause it to continue to move for a considerable time. At low Reynolds numbers, 10^{-1} or less, a system will come to rest quickly, in a small fraction of a second, following removal of the motive power, because of the dominant viscous force. This may be seen when observing ciliated cells swimming on a microscope slide; if the cilia stop beating, the forward motion of the cell is arrested, apparently instantaneously.

When the inertial term is neglected, the governing equation is simplified, and several authors have made theoretical studies of ciliary propulsion. A successful solution to the equation allows the force generated by the cilium to be calculated, together with an estimate of the power dissipation, both of which are relevant to an investigation of the molecular basis of ciliary movement.

A useful approach to the problem, in that it is relatively easy to apply, is the resistive-force theory developed originally by Gray and Hancock (1955), which relies on the concept of force coefficients. Two coefficients are defined with reference to a straight cylinder having a diameter equal to that of the cilium, one, C_s, for motion along the axis of the cylinder, the other, C_n, for motion perpendicular to it. A coefficient is defined as the force per unit length of the cylinder moving with unit velocity in the appropriate direction. Since a cilium does not remain straight during movement, except perhaps during the effective stroke described earlier, it is necessary to consider the force generated by short elements, ds, of the organelle, and to sum them over the whole length. The forces dF_s and dF_n produced parallel and normal to the axis of an element of the cilium are

$$dF_s = C_s V_s ds \quad \text{and} \quad dF_n = C_n V_n ds, \tag{4}$$

where V_s and V_n are the velocities of the element in the appropriate directions. The net force acting in a particular direction, making an angle θ_i with the axis of the ith element, is then

$$dF_{Ri} = dF_{Ni} \cos\theta_i + dF_{Ti} \sin\theta_i \tag{5}$$

Provided the shape of the cilium is known, θ_i can be determined, either analytically or empirically, for each point on the cilium and, by evaluating $\int_0^S dF_{Ri}$, where S is the length of the cilium, the instantaneous force in the required direction developed by the whole cilium calculated. For a free-swimming organism, propelled by a single undulating cilium, the average force generated by the organelle over a ciliary cycle can be determined and equated to the resistance offered by the cell body. Since this resistance is proportional to the cell speed u – for a sphere of radius r it is $6\pi r\eta u$ – the resulting force equation can be used to obtain u in terms of the ciliary wave velocity V_w. For example, if the undulation is assumed to be sinusoidal, of amplitude a and wavelength λ, the following relationship is obtained:

$$\frac{u}{V_w} = \frac{a^2k^2(1-(C_s/C_n))}{(2C_b r/n\lambda C_n)(1+0.5a^2k^2)^{\frac{1}{2}} + (2C_s/C_n) + 2a^2k^2}, \tag{6}$$

where $k = 2\pi/\lambda$ while C_b is the proportionality constant for the body resistance; for a sphere $C_b = 6\pi r\eta$ (Holwill and Sleigh, 1967).

To calculate the velocity ratio it is necessary to know the ratio of the force coefficients. In their original study of a smooth cilium, Gray and Hancock (1955) gave somewhat empirical values for the coefficients which did not take account of the hydrodynamic influence between neighbouring parts of the cilium or between the cilium and the cell body, and assumed $C_n/C_s = 2$, a condition only applicable to a filament of zero thickness. The reasonable agreement between experimental results and the predictions of equation (6) obtained by them and by later workers (e.g., Holwill, 1966) gave some confidence for the use of the values.

However, it was recognised that slender body theory would provide a more accurate representation of ciliary motion than the resistive force theory and several analyses using this approach have been performed (e.g. Shen et al, 1975; Lighthill, 1976). Slender body theory considers the effects of a distribution of singularities, known as stokeslets and doublets, along the centre line of the cilium. A stokeslet describes the velocity distribution which results from the application of a force $\mathbf{F}$ at a point in the fluid. The velocity $\mathbf{U}$ at a distance r from the point is given by

$$\mathbf{U} = \frac{\mathbf{F}}{6\pi\eta r} + \frac{r^2}{4}\text{grad}\left[\text{div}\left(\frac{\mathbf{F}}{6\pi\eta r}\right)\right]. \tag{7}$$

Since the force $\mathbf{F}$ will depend on position, s, along the cilium, it is convenient to represent it as a force $\mathbf{f}(s)$ per unit length of the organelle exerted on the fluid. The no-slip condition that should obtain at the cilium surface cannot be achieved by the stokeslets alone, but can be almost realised if there is an accompaying distribution of doublets of strength $\left(b^2\mathbf{f}_n(s)\right)/(4\eta)$, where b is the cilium radius and $\mathbf{f}_n(s)$ is obtained by resolving $\mathbf{f}(s)$ onto a plane normal to the centre line.

While the slender body theory is more rigorous than resistive force theory, its application to the movement of a cilium demands much more computational effort. Comparison of the two theories has indicated that improved expressions for the Gray and Hancock (1955) coefficients are possible. Lighthill (1976) showed that better values for the coefficients are

$$C_n = \frac{4\pi\eta}{\ln(2q/b) + 0.5} \quad \text{and} \quad C_s = \frac{2\pi\eta}{\ln(2q/b)}, \tag{8}$$

where q takes a value that depends on the deformation of the organelle. For undulations of wavelength λ_s measured along the cilium, $q = 0.09\lambda_s$, while for a straight cilium of length L, $q = 0.22L$. These expressions for the coefficients may be compared with those of Gray and Hancock for a waving filament:

$$C_s = \frac{2\pi\eta}{\ln(2\lambda/b) - 0.5} \quad \text{and} \quad C_n = 2C_s. \tag{9}$$

In an appraisal of the two approaches, Johnson and Brokaw (1979) showed that the resistive force theory based on the coefficients given by equation (8) agrees well with the slender body theory for wave motion on isolated organelles, or for undulating individual cilia propelling small bodies, but fails to do so if large bodies are involved. 'Small' in this context will depend on the shape as well as the dimensions of the body, and is such that the presence of the body does not significantly influence the fluid flow around the cilium; Johnson and Brokaw consider the sea urchin sperm head, shaped roughly like a cone with a height of 4 μm and a basal diameter of 1.5 μm to be sufficiently small as not to disturb significantly the flow about the cilium. The reasonable agreement between predicted and experimental velocities obtained using the coefficients in both equations (8) and (9) arises because the expression for the velocity depends on the ratio of coefficients (equation 4), which is similar in the two cases. However the power dissipation, P, depends on the absolute values of the coefficients, being given by an expression of the form

$$P = \int (C_n V_n^2 + C_s V_s^2)\, ds. \tag{10}$$

Calculations using the improved coefficients indicate that earlier studies based on the Gray and Hancock expressions underestimated the power dissipation by some 40%

(Johnson and Brokaw, 1979; Holwill and Satir, 1987). The efficiency of conversion of chemical to mechanical energy in cilia, estimated earlier (Brokaw and Gibbons, 1975) at around 30%, rises to between 40 and 45% with the revised expressions, and is similar to that of muscle (between 45% and 70%, Kushmeric and Davies, 1969; Pybus and Tregear, 1972). If some allowance is made for the energy needed to overcome elastic resistances in the cilium, the efficiency will increase further, perhaps to 80%.

Equation (6) was derived in such a way that if the right hand side takes a positive value, the propulsive velocity u and the wave velocity V_w are in opposite directions. This is the case for a smooth flagellum, since C_s/C_n is less than one and the denominator is always positive. As noted earlier, this result is in accord with experimental observation. For the flagellum of *Ochromonas*, which bears stiff mastigonemes, a crude evaluation of the force coefficients indicates that C_s/C_n is about 1.9 (Holwill and Peters, 1974), so that u and V_w are predicted to be in the same direction, as is observed. A detailed comparison of the predictions with experimental observations at a number of different viscosities shows that the two are not completely coincident, a discrepancy that might be resolved by a more rigorous hydrodynamic treatment.

The application of slender body theory was extended by Blake (1972) to include fields of cilia beating metachronously. By considering the interactions between cilia, Blake was able to construct quantitative flow fields relating to both symplectic and antiplectic metachrony. His predictions include a region of backflow between the surface of the organism and the ciliary tip. While the velocities predicted are similar to those observed in ciliated cells, there is no sufficiently detailed experimental study that would allow the predictions to be tested. Calculations of the power dissipation suggest that, for given beat frequency and ciliary length, antiplectic metachrony is a more efficient means of propulsion than symplectic beating. Blake (1971) used an alternative model of an undulating sheet surrounding the cell body for symplectic metachrony and obtained a reasonable prediction for the swimming speed of *Opalina*.

Gueron and Liron (1992) found that the equations derived by Lighthill (1976), while being an appropriate representation of the system, posed problems when attempts were made to obtain a numerical solution. These authors proposed an alternative approach, which leads to a drag-velocity equation similar to that of Gray and Hancock (1955), but with different values for the force coefficients and with correcting terms. Their arguments allow for the presence of the cell surface and other cilia. It is a more comprehensive model than described previously, in that it includes the forces generated by the internal mechanism that bends the cilium. The active bending moment will be balanced by the viscous resistance discussed earlier, and by the elasticity of the cilium. By modelling the internal active forces appropriately, and using a newly-developed representation of the time evolution of torsion and curvature on an inextensible filament, Gueron and Liron (1993) were able to generate realistic three-dimensional ciliary bend patterns.

Structure

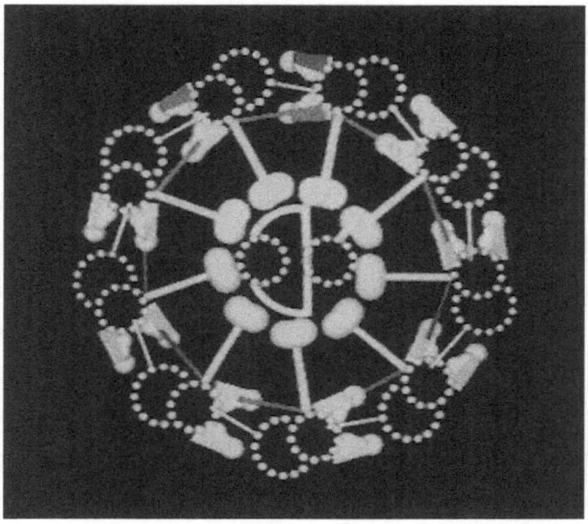

Figure 4. Diagrammatic representation of the cross section of a typical ciliary axoneme, viewed distally from the base. (Generated using the computer modelling precedure described by Sugrue et al, 1991

Cilia from all sources have the same basic architecture, consisting of nine doublet microtubules arranged cylindrically about a pair of single microtubules, referred to as the axoneme, within a membrane that is continuous with the cell membrane (Fig. 4; See Avolio et al, 1986, for references). A doublet has a complete microtubule (A), and an incomplete microtubule (B), with 13 and 11 tubulin protofilaments respectively. Attached to microtubule A of a doublet (doublet N, say), and projecting towards the neighbouring doublet (N+1), are the dynein arms; these occur on the inner and outer aspects of a doublet, and the outer arms are distributed along the doublet with a spacing of 24 nm. The inner arms may have a different periodicity in some organisms, and further structural studies are needed to investigate this. Radial spokes extend inwards from each doublet and terminate in a spoke head which abuts the central complex. The spokes usually occur in groups of three, with spokes 1 and 2 separated by 32 nm, 2 and 3 by 24 nm and spoke 3 and spoke 1 of the next group by 40 nm, giving a group repeat of 96 nm, equal to 4 dynein arm periods. The larger spoke spacing within a group occurs towards the proximal end of the cilium, thereby providing a useful directional marker on electron micrographs of a portion of the cilium containing neither base nor tip. Two sets of circumferential links connect the doublets (Witman et al, 1972, 1978; Warner, 1983; Bozkurt and Woolley, 1993). In one set the links repeat in register with the dynein arms, and may be part of the arms, while in the other set the repeat distance corresponds to that of the spoke groups.

Mechanisms of Movement

To sustain bends of consistent, and sometimes increasing amplitude, it is necessary that energy be available along the length of the cilium. This was recognised by Gray (1955) and demonstrated analytically by Machin (1958) for an elastic filament undulating in a viscous liquid. There is now a considerable body of evidence to support a sliding microtubule mechanism as the basis for ciliary movement. In a classic structural study of the tips of cilia at various stages of their cycle within a fixed metachronal wave, Satir (1965) showed that the microtubules terminated in a manner consistent with microtubule sliding. While other explanations for this observation are possible, Summers and Gibbons (1971) showed by optical microscopy that a demembranated axoneme became thinner and longer following reactivation with ATP, as would be predicted if the microtubules slid relative to each other; by monitoring this type of experiment using electron microscopy, Sale and Satir (1977) were able to confirm this interpretation. Recently, Brokaw (1991) used small gold particles as markers on the outer surfaces of the doublets of demembranated spermatozoa from marine organisms, and was able to obtain direct visualisation and measurement of sliding displacements. In *Tetrahymena* cilia and sperm of the domestic cock, relative active sliding appears to be unidirectional, with the arms on one doublet (N) pushing the neighbouring doublet (N+1) tipward (Sale and Satir, 1977; Woolley and Brammall, 1987). This unidirectional sliding characteristic is supported by studies of microtubules propelled by isolated dynein (e.g. Vale et al, 1989).

The nature and extent of microtubule sliding in the axoneme can be determined from its shape. The net amount of microtubule sliding between two points on a bent cilium is given by ϕd, where ϕ is the angle between the tangents drawn at those points, and d is the microtubule separation projected onto the plane of the bend. Appropriate adjustments in the calculation of ϕd are required for a three-dimensional bend or a twisted axoneme (Holwill et al, 1979). Since d varies according to the position of the microtubules in the axoneme, the amount of relative sliding depends on the particular pairs of microtubules examined. This geometrical relationship has been used to study the detailed nature of the sliding (e.g. Satir, 1968; Goldstein, 1976; Eshel and Brokaw, 1987; Chevrier and Dacheux, 1992). The maximum doublet displacement in a half-cycle of ciliary beating is between 0.1 and 0.4 μm (Satir, 1982), leading to an average microtubule sliding velocity of the order 10 μm s^{-1} for typical beat frequencies. This is similar to the velocity between extracted flagellar microtubules (Takahashi et al, 1982) and for the translocation of brain microtubules by outer arm dynein (Paschal et al, 1987). The study by Takahashi et al (1982) showed that sliding was independent of loading, in that the velocity remained constant as the microtubule overlap region decreased to almost zero.

The motive force for sliding is provided by the dynein arms, which undergo a cyclic interaction with the neighbouring microtubule involving attachment and detachment. The force generated by a single dynein arm in a lightly-loaded system of sliding microtubules is about 1 pN (Kamimura and Takahashi, 1981). The energy to drive the required conformational change in the dynein arms is provided by the breakdown of ATP in association with a divalent cation, usually magnesium. A sliding microtubule system in isolation will not produce the bending patterns seen in cilia; a mechanism must exist to convert the shear forces into bending moments. The mechanism will provide a resistance to free sliding between microtubules, thereby setting up a bending moment which will tend to deform the system. The generation and propagation of bends require that sliding be restricted in a controlled manner. Little or no sliding seems to occur in the basal region of the cilium, where it enters the basal body, so that active sliding in one direction along the length of the organelle will result in the formation of a single bend. Goldstein (1976) showed that in sea urchin sperm flagella, bends were formed at the base in pairs of opposite curvature, in such a way that sliding was restricted to the region of bend development. Once a pair of bends is fully formed, it can be propagated by the introduction of local sliding at the leading and trailing edges.

Studies of wave propagation in sea urchin sperm by Brokaw (1991) have shown that in addition to the localised sliding required to generate and propagate bends (referred to as metachronous sliding) sliding also occurs along the length of the organelle (synchronous sliding). This poses some mechanical problems for the system, since some microtubule sliding must be accommodated in regions of the axoneme where active sliding forces are being developed; such sliding occurs through bent regions of the axoneme and has little effect on the bend shape.

Although the geometry of a bend allows the amount of relative microtubule sliding to be determined, it is not possible to identify the regions of active sliding unequivocally from observations of bend patterns. Active sliding could occur in bending regions only, so that the region of activity would govern the length of a bend, in straight regions only or along the length of the organelle. The stiffness of a particular section of the cilium will determine the extent to which it will bend under the influence of the imposed shearing forces. As noted by Holwill (1980) for a cilium with a single bend, active sliding, and therefore shear force generation, could occur throughout the length of the organelle, with sliding localised by a reduction in stiffness at the appropriate site. Stiffness could be altered by variations in the axonemal linkages, such as the spokes, dynein arms or circumferential links. During ciliary movement, the spokes change their orientation in a well-defined way (Warner and Satir, 1974) and appear to play a significant role in bend formation and propagation. The circumferential links, which have limited elasticity (Warner, 1983) maintain the cylindrical form of the axoneme, even when it has been split so that each separate part has an incomplete set of microtubules (Holwill and Satir, 1993), and may maintain an optimum separation of the microtubules for the dynein arm interaction (Bozkurt and Woolley, 1993). By appropriate treatment it is possible to induce a non-physiological state of the axoneme where a majority of the dynein arms are attached to the appropriate neighbouring doublet (e.g. Gibbons and Gibbons, 1974). The axoneme becomes relatively stiff in this condition, the rigor state, and suggests that the arms could have a significant influence on the stiffness of the axoneme. In the rigor condition, the cilium is resistant to bending, but less so to twisting, which involves insignificant amounts of microtubule sliding. Though rigor is not a physiological condition, the ease with which twisting occurs suggests that it could be present during normal ciliary movement, as a result of the internal forces generated. The absence of twisting would indicate that the torsional forces set up by the dynein arm interaction are balanced by the provision of suitable resistances within the axoneme.

To produce a beat cycle, the microtubules in a cilium must slide synchronously. Active arms on microtubules in one half of the axoneme will produce a bend in one direction, and cause passive sliding between the microtubules in the other half; if the active and passive sets are exchanged, the bend direction is reversed. This implies, as noted by Satir and Sleigh (1990), that the dynein arm can exist in two states, cycling and noncycling. The dividing plane lies in the plane of beat and passes through the centre of the cilium. Satir (1982, 1985) has proposed that the bending cycle is controlled by a switch-point mechanism, in which arms on doublets in opposite halves of the axoneme are activated and deactivated sequentially with the necessary time delays to generate the observed bend patterns. On this basis, there should be at least two switches and one prediction of such a mechanism is that the failure of a switch to activate a set of microtubules should result in the arrest of the cilium at a preferred position during its stroke. Two arrest positions for mussel gill have been found, (Wais Staider and Satir, 1979) thus confirming this prediction. It would reasonably be expected that the switches should be sensitive to the ionic environment, and in the case of the mussel gill cilia calcium and vanadate determine the two arrest positions, which are close to the beginning of the effective and recovery strokes. According to the switch-point hypothesis, the effective strokes would be generated by active arms on one set of doublets, while activity of the complementary set would cause the recovery stroke to occur.

The split patterns observed in the axonemes of sea urchin sperm (Sale, 1986), of rat sperm (Lindemann et al 1992) and of mussel gill cilia (Satir and Matsuoka 1989) can be explained in terms of polarised sliding in the axoneme and are consistent with the switch point hypothesis. Holwill and Satir (1993) considered the bending moments developed by the interacting doublets and were able to predict the split patterns observed. An important conclusion of their study is that splitting occurs between the active and inactive sets of doublets, with the specific result that a split will occur between doublets N and N+1 if the arms on doublet N are inactive and the arms on either N+1 or N−1 are active.

According to the switch-point hypothesis, the form of a bending pattern is determined by the temporal sequence of switching arm activity on and off; symmetric waves would have equal times for the on and off states for both sets of microtubules, while asymmetric bending would result from unequal on and off periods. An alternative proposal, the biased-baseline mechanism (Brokaw, 1979; Eshel and Brokaw, 1987) suggests that the generation of bends is always a symmetric process but that the natural axis of the cilium maybe curved, thereby leading to the propagation of asymmetric waves; if the ciliary axis is straight, symmetric bends will be propagated. Evidence cited in support of this mechanism is that when the mean shear angle is subtracted from asymmetric bend patterns on sea urchin and *Ciona* sperm, the patterns become symmetric. Further, in reactivated demembranated sea urchin sperm, the addition of calcium ions caused the axoneme to curve (Okuno and Brokaw, 1981); the waves on reactivated sea urchin sperm become more asymmetric as the calcium ion concentration is raised. However, there are several situations that are inconsistent with the biased-baseline mechanism (Eshel and Brokaw, 1987), and some of the supporting evidence is not incompatible with the switch-point hypothesis described earlier.

Modelling the axoneme

In order to understand the mechanisms underlying ciliary movement, direct observations of the organelles, both in terms of their motile behaviour and biochemical properties, have been complemented by modelling studies. An early model was that of Machin (1958), described earlier, in which when the system was represented by an elastic beam deformed by active components distributed along its length; the active bending moments developed were balanced by elastic moments resulting from the deformation of the beam and viscous moments arising from the movement of the beam through its liquid environment. The resulting differential equation was solved to yield bending patterns that were dependent on the values of the parameters, such as the stiffness of the beam and the form of the active movement, supplied to the model.

While Machin's (1958) representation of the cilium was somewhat crude, it served to confirm the important detail that active deforming forces were required along the length of the organelle. With improved structural, mechanical and chemical information about cilia, and the development of computer techniques, the models have become increasingly sophisticated, though the basic principles utilised by Machin of setting up force and moment balance equations remain key elements of the modelling procedure.

Brokaw (1972) initiated a series (see Brokaw 1976 for references) of computer modelling studies based on the sliding microtubule model. He was able to show that a two-filament model incorporating feedback, based on local curvature, to the active elements was able to sustain waves of a form similar to those observed on real cilia. A two-filament model is able to generate only two-dimensional bending patterns, and is therefore limited in its general application. Hines and Blum (1983, 1984) overcame this limitation by representing the nine outer doublets and the central complex in their model, and demonstrated that both two-and three-dimensional bending shapes, similar to that seen on living organelles, can be generated. Their study was concerned with the generation of torsion and twist within the axoneme, and as such considered only equilibrium shapes, with the external viscosity assumed to be zero. In their comprehensive hydromechanical treatment, Gueron and Liron (1992, 1993) compute the general shape of a cilium moving through a liquid environment, with the active system in the cilium overcoming the resistances discussed earlier, and obtain bend patterns similar to those observed on a living organelle.

All the modelling approaches have provided useful information about ciliary motion, but have represented the structural interactions within a cilium by general parameters and not so far incorporated the detailed structural information available from electron microscopy. As the initial stage in the development of a fully functional model of the axoneme, a detailed structural computer representation has been developed (Avolio et al 1986, Sugrue et al 1991; Fig. 4). This computer model can be viewed from any direction. so that it presents a consistent image of the axoneme. It has the advantage over solid modelling that it can be changed easily in the light of new data. To create a functional model, the structures need to be assigned particular attributes; while some information is available, such as the force generated by individual dynein arms, there is a considerable amount of detail still to be acquired. The acquisition of this detail requires further theoretical and experimental study; for example, a detailed theoretical investigation of the microtubule sliding patterns has been presented (Holwill and Satir,

1990) and has indicated the types of experiment to be designed in order to gain the necessary mechanical information. A further study to investigate observed splitting of certain axonemes has established a vector representation of the development of bending moment by sliding microtubules (Holwill and Satir, 1993); this representation will be of value when bend development and propagation is modelled. While a simulation of dynein arm activity has already been presented, and confirms the value of this type of modelling in visualising macromolecular behaviour, the aim is to establish the form of the cycle and the character of the dynein tubulin interaction and incorporate them into the model, so that the effects of the shearing forces can be visualised.

The goal of a fully functional computer model of the axoneme will be achieved in stages, and will involve the successful representation of the behaviour of a range of preparations derived from, and simpler than, the intact organelle. One of the simplest preparations is that in which isolated microtubules in the presence of ATP are propelled across a surface coated with dynein molecules (Hamasaki et al, 1991). The forces developed by the dynein molecules to propel the microtubules are balanced by the viscous resistances to motion, and this equilibrium will be used to obtain an understanding of the observed relationship between length and velocity. At an increased level of complexity are demembranated cilia treated with proteolytic enzymes to allow disintegration of the system by microbule sliding without the formation of bends; more detailed information will be required than is currently available about the nature and character of the interacting structures in this type of preparation, so that the analysis of sliding by Holwill and Satir (1990) can be utilised. The next modelling stage will include resistances to sliding, such that bending moments are developed, as in the split axonemes prepared by Satir and Matsuoka (1989). The active bending moments will be opposed by the elasticity of the system, with contributions from the microtubules, links and spokes, as well as by the viscous resistance experienced by the moving groups of microtubules. The bending moments generated to cause the splits are those which, in more complete axonemes, will lead to bend formation and propagation. Discussion earlier in this Chapter of the hydrodynamic interactions of cilia indicate that these are reasonably well understood, and can be modelled successfully. Structural information is also readily available, but little detail is known about the elastic properties of the structures and their contribution to the passive elasticity of the system, and a critical study of this area is needed. Computer modelling is not a substitute for experimental investigation, but provides a convenient and informative visual framework for the presentation of data; it is also predictive, and a valuable element in the development of a complete understanding of ciliary movement.

References

Avolio J, Glazzard AN, Holwill MEJ, Satir P (1986) Structures attached to doublet microtubules of cilia:computer modelling of thin-section and negative- stained stereo images. Proc Nat Acad Sci USA 83:4804-4808

Blake JR (1971) A spherical envelope approach to ciliary propulsion. J Fluid Mech 46:199-208

Blake JR (1972) A model for micro-structure in ciliated organisms. J Fluid Mech 55:1-23

Bozkurt HH, Woolley DM (1993) Morphology of nexin links in relation to interdoublet sliding in the sperm flagellum. Cell Motil Cytoskel 24:109-118

Brokaw CJ (1965) Non-sinusoidal bending waves of sperm flagella. J Exp Biol 43:155-169

Brokaw CJ (1972) Computer simulation of flagellar movement. I. Demonstration of stable bend propagation and bend initiation by the sliding filament model. Biophys J 12:564-586

Brokaw CJ (1976) Computer simulation of flagellar movement. IV. Properties of an oscillatory two-state cross bridge model. Biophys J 16:1029-1041

Brokaw CJ (1979) Calcium-induced asymmetrical beating of Triton-demembranated sea urchin sperm flagella. J Cell Biol 82:401-411

Brokaw CJ, Gibbons IR (1975) Mechanisms of movement in cilia and flagella. In:Wu TY-T, Brokaw CJ, Brennan C (eds) Swimming and Flying in Nature, Vol 1. Plenum, New York, p89

Chevrier C, Dacheux J-L (1992) Evolution of the flagellar waveform of ram spermatozoa in relation to the degree of epididimal maturation. Cell Motil Cytoskel 23:8-18

Eshel D, Brokaw CJ (1987) New evidence for a "Biased Baseline" mechanism for calcium-regulated asymmetry of flagellar bending. Cell Motil Cytoskel 7:160-168

Gibbons BH, Gibbons IR (1972) Flagellar movement and adenosine triphosphatase activity in sea urchin sperm extracted with Triton X-100. J Cell Biol 54:75-97

Gibbons BH, Gibbons IR (1974) Properties of flagellar "rigor waves" formed by abrupt removal of adenosine tryphosphate from actively swimming sea urchin sperm. J Cell Biol 63:970-985

Goldstein SF (1976) Form of developing bendsends in reactivated sperm flagella. J Exp Biol 64:173-184

Goldstein SF (1977) Asymmetric waveforms in Echinoderm sperm flagella. J Exp Biol 71:157-170

Gray J (1955) The movement of sea urchin spermatozoa. J Exp Biol 32:775-801

Gray J, Hancock GJ (1955) The propulsion of sea urchin spermatozoa. J Exp Biol 32:802-814

Gueron S, Liron N (1993) Simulations of threee-dimensional ciliary beats and cilia interactions. Biophys J 65:499-507

Gueron S, Liron N (1992) Ciliary motion modelling, and dynamic multicilia interactions. Biophys J 63:1045-1058

Hamasaki T, Barkalow K, Richmond J, Satir P (1991) A cAMP-stimulated phosphorylation of an axonemal polypeptide that copurifies with the 22S dynein arm regulates microtubule translocation velocity and swimming speed in *Paramecium*. Proc Nat Acad Sci USA 88:7918-7922

Hines M, Blum JJ (1983) Three-dimensional mechanics of eukaryotic flagella. Biophys J 41:67-69

Hines M, Blum JJ (1984) On the contribution of moment-bearing links to bending and twisting in a three-dimensional sliding filament model. Biophys J 41:67-79

Holwill MEJ (1965) The motion of *Strigomonas oncopelti*. J Exp Biol 42:125-137

Holwill MEJ (1974) Hydrodynamic aspects of ciliary and flagellar movement. In: Sleigh MA (ed) Cilia and Flagella, Academic Press, London, p143

Holwill MEJ (1980) Movement of cilia. Symp Soc Gen Microbiol 30:273-300

Holwill MEJ, Cohen HJ, Satir P (1979) A sliding microtubule model incorporating axonemal twist and compatible with three-dimensional ciliary bending. J Exp Biol 78:265-280

Holwill MEJ, Peters PD (1973) Dynamics of the hispid flagellum of *Ochromonas danica*. J Cell Biol 62:322-328

Holwill MEJ, Satir P (1987) Generation of propulsive forces by cilia and flagella. In: Bereiter-Hahn J, Anderson OR, Reif W-E (eds) Cytomechanics. Springer-Verlag, Berlin, p120

Holwill MEJ, Satir P (1990) A physical model of microtubule sliding in ciliary axonemes. Biophys J 58:905-917

Holwill MEJ, Satir P (1993) A physical model of axonemal splitting. Cell Motil Cytoskel In Press

Holwill MEJ, Sleigh, MA (1967) Propulsion by hispid flagella. J Exp Biol 47:267-276

Jahn TL, Landman MD, Fonseca JR (1964) The mechanism of locomotion of flagellates.II. Function of the mastigonemes of *Ochromonas*. J Protozool 11:291-

Johnston DN, Silvester NR, Holwill, MEJ (1979) An analysis of the shape and propagation of waves on the flagellum of *Crithidia oncopelti*. J Exp Biol 80:299-315

Johnson RE, Brokaw CJ (1979) Flagellar hydrodynamics: A comparison between resistive-force theory and slender-body theory. Biophys J 25:113-127

Kamimura S, Takahashi K (1981) Direct measurement of the force of microtubule sliding in flagella. Nature 293:566-5681

Knight-Jones EW (1954) Relationships between metachronism and the direction of ciliary beat in Metazoa. Quart J Microsc Sci 95:503-521

Kurimoto E, Kamiya R (1991) Microtubule sliding in flagellar axonemes of *Chlamydomonas* mutants missing inner or outer arm dynein: Velocity measurements on new types of mutants by an improved method. Cell Motil Cytoskel 19:275-281

Kushmeric MJ, Davies RE (1969) The chemical energetics of muscular contraction.II. Proc Roy Soc Lond B Biol Sci 174:315-353

Lighthill J (1976) Flagellar hydrodynamics. Soc Ind Appl Math Rev 18:161-229

Lindemann CB, Orlando A, Kanous KS (1992) The flagellar beat of rat sperm is organised by the interaction of two functionally distinct populations of dynein bridges with a stable central axonemal partition. J Cell Sci 102:249-260

Machin KE (1958) Wave propagation along flagella. J Exp Biol 25:796-806

Okuno M, Brokaw CJ (1981) Calcium-induced change in form of demembranated sea urchin sperm flagella immobilised by vanadate. Cell Motil 1:349-362

Paschal BM, King SM, Moss AG, Collins CA, Vallee RB, Witman GB (1987) Isolated flagella outer arm dynein translocates brain microtubules *in vitro*. Nature 330:672-674

Pybus J, Tregear R (1972) Estimates of force and time of actomyosin interaction in an active muscleand the number interacting at one time. Cold Spring Harbor Symp Quant Biol 37:655-660

Sale WS (1986) The axonemal axis and Ca^{2+}-induced asymmetry of active microtubule sliding in sea urchin sperm tails. J Cell Biol 102:2042-2052

Sale WS, Satir P (1977) Direction of active sliding of microtubules in *Tetrahymena* cilia. Proc Nat Acad Csi USA 74:2045-2049

Satir P (1965) Studies on cilia. II. Examination of the distal region of the ciliary shaft and the role of the filaments in motility. J Cell Biol 26:805-834

Satir P (1968) Studies on cilia. III. Further studies on the cilium tip and a "sliding filament" model of ciliary motility. J Cell Biol 39:77-94

Satir P, (1982) Mechanisms and controls of microtubule sliding in cilia. Symp Soc Exp Biol 35:172-201

Satir P, (1985) Switching mechanisms in the control of ciliary motility. Mod Cell Biol 4:1-46

Satir P, Matsuoka T (1989) Splitting the ciliary axoneme: Implications for a switch-point model of dynein arm activity in ciliary motion. Cell Motil Cytoskel 14:345-358

Satir P, Sleigh MA (1990) The physiology of cilia and mucociliary interactions. Ann Rev Physiol 52:137-155

Shen JS, Tam PY, Shack WJ, Lardner TJ (1975) Large amplitude motion of self-propelling slender filaments at low Reynolds numbers. J Biomech 8:229-241

Silvester NR, Holwill MEJ (1972) An analysis of hypothetical flagellar waveforms. J Theor Biol 35:505-523

Sleigh MA, Holwill MEJ (1969) Energetics of ciliary movement in *Sabellaria* and *Mytilus*. J Exp Biol 50:733-743

Sugrue P, Avolio J, Satir P, Holwill MEJ (1991) Computer modelling of Tetrahymena axonemes at macromolecular resolution: Interpretation of electron micrographs. J Cell Sci 98:5-16

Summers KE, Gibbons IR (1971) Adenosine triphosphate-induced sliding of microtubules in trypsin treated flagella of sea urchin sperm. Proc Nat Acad Sci USA 68:3092-9096

Takahashi K, Shingyoji C, Kamimura S (1982) Microtubule sliding in reactivated flagella Symp Soc Exp Biol 35:159-177

Vale RD, Soll DR, Gibbons IR (1989) One-dimensional diffusion of microtubules bound to flagellar dynein. Cell 59:915-925

Wais-Steider J, Satir P (1979) Effect of vanadate on gill cilia: Switching mechanism in ciliary beat. J Supramolec Struct 11:339-347

Warner FD (1983) Organisation of interdoublet links in *Tetrahymena* cilia. Cell Motil 3:321-332

Warner FD, Satir P (1974) The structural basis of ciliary bend formation. Radial spoke positional changes accompanying microtubule sliding. J Cell Biol 63:35-63

Witman GB, Carlson K, Berliner J, Rosenbaum J (1972) *Chlamydomonas* flagella. I. Isolation and electrophoretic analysis of microtubules, matrix, membranes and mastigonemes. J Cell Biol 54:507-539

Witman GB, Plummer J, Sander R (1978) *Chlamydomonas* flagellar mutants lacking radial spokes and central tubules. J Cell Biol 76:729-747

Woolley DM and Brammall A (1987) Direction of sliding and relative sliding velocities within trypsinized sperm axonemes of *Gallus domesticus*. J Cell Sci 88:361-371

Local Bending Fluctuations of the Cell Membrane

Rafi Korenstein, Shmuel Tuvia, Leonid Mittelman and Shlomo Levin
Department of Physiology and Pharmacology
Sackler Faculty of Medicine
Tel-Aviv University
Ramat Aviv, 69978 Tel-Aviv
Israel

Submicron mechanical fluctuations of the cell membrane are a newly recognized dynamical activity of the living cell (Krol et al., 1990; Levin and Korenstein,1991; Mittelman et al., 1991; Tuvia et al., 1992a). These fluctuations consist of 300-20nm reversible displacements of the cell membrane in the frequency range of 0.3-30Hz, correspondingly. Submicron cell membrane fluctuations (CMF) were observed in different types of cells including red blood cells (Krol et al., 1990, Levin and Korenstein, 1991) monocytes, lymphocytes, 3T6 fibroblasts, cardiomyocytes (Krol et al., 1990) and murine lymphoma cells (Mittelman et al., 1991).

The observation of membrane fluctuations was rendered possible following the development of an optical method based on *Point Dark Field Microscopy* (Krol et al.,1990). Using cells, attached to a coverglass, we illuminate a very small area (1-0.25 μm^2) at the cell edge and record cell membrane displacements by monitoring the time-dependent changes of light reflection and scattering. The relative fluctuation of the light intensity (given in terms of $\delta I/I$, where δI is the time-dependent changes of light reflection and scattering and I is the time-independent intensity of scattered and reflected light) depends on the changes of the membrane area moving in and out of the focused light spot near the cell edge. A linear dependence between the relative change in the scattered and reflected light from the cell surface and the amplitude of the cell edge displacement was achieved by moving a coverglass, with attached glutaraldehyde fixed cells, by a calibrated vibrator (Krol et al., 1990). Linearity of $\delta I/I$ with displacement was observed over distances as long as 300nm (a relative change in light scattering of 1% corresponds to a cell membrane displacement of ≈ 17nm) (Krol et al., 1990).

NATO ASI Series, Vol. H 84
Biomechanics of Active Movement
and Division of Cells
Edited by N. Akkaş
© Springer-Verlag Berlin Heidelberg 1994

The local character of cell membrane fluctuations

CMF in the red blood cell (RBC) possess the highest amplitude of displacement (≈300nm), among the different studied cells, which makes it an ideal model system to study this phenomenon. CMF involve local (≤0.25μ^2) bending deformation of the cell membrane. The power spectrum of CMF of RBCs fits to a function of the type: $y=(a/f^b)+c$. Thus, the power spectra is composed of the 1/f component and the frequency independent Johnson noise. The local character of CMF emerges from the observed dependence of the fluctuation amplitude on the area size from which the fluctuation is monitored. Thus, we measured a maximal fluctuation amplitude of 8.4±2.5% when illuminating an area of 0.25μm^2 at the cell edge of B lymphocytes whereas, a lower fluctuation amplitude of 4.6±2.1% was obtained when we illuminated an area of 2μm^2. Similarly, a maximal amplitude of 16.9±2.8% was measured when illuminating an area of 0.25μm^2 at the cell edge of rat erythrocytes whereas, a lower fluctuation amplitude of 8.0±1.4% was obtained when we illuminated an area of 2μm^2. A comparable attenuation of CMF was obtain upon identical increase of the illumination area on the surface of human RBCs. This amplitude dependence of CMF on the area of illumination may be attributed to the growing interference of light scattering changes due to a larger number of independent sites of incoherent displacements operating when registering larger areas. The increased number of incoherent local fluctuations, when expanding the registered area, leads to the mutual loss in the amount of the changes in the scattered light, yielding an apparent decrease of the registered amplitude. Thus, we may conclude that the oscillators underlying cell membrane displacements are confined to microdomains (≤0.2μm^2), excluding the possibility that these fluctuations are due to the movements of large cell regions, such as pseudopods or result from the displacement of the whole cell. Moreover, CMF are characterized by a frequency range of 0.3-30Hz, which is one to two orders of magnitude faster than the cell edge displacement during fibroblast locomotion (Felder and Elson, 1990).

Cell membrane fluctuations are driven by metabolic energy

The primary mode of RBC deformation in narrow blood capillaries is a folding or bending about the longitudinal axis of the capillary (Bagge et al., 1980; Gaehtgens et al., 1980). The preference for folding in capillaries indicates that it requires less energy than that needed for shearing to occur, because the bending modulus of the RBC membrane, measured at the center of the cell, is very small (10^{-13}-10^{-12} dyne.cm; Brochard and Lennon, 1975; Evans, 1983). On the side surface of the RBC, where the curvature is large (1-2μ^{-1}), the bending modulus is expected to be much higher. In view of the large amplitude of CMF and their occurrence in regions of high positive curvature of the cell membrane, it may be supposed that membrane fluctuations are driven not only by thermal energy. In an attempt to verify this possibility we studied the effect of MgATP on CMF in three types of RBC ghosts (Levin and Korenstein, 1991). We could demonstrate MgATP-driven restoration of fluctuations in the RBC ghosts. Measurements of fluctuations in membrane skeleton shells prepared from RBC by Triton X-100 treatment (Shen et al., 1986; Vertessy and Steck, 1989), in the presence of 1mM MgATP, led to increase in fluctuation amplitude, before the occurrence of an irreversible contraction of the skeletal network. Also, a short heat treatment of the intact RBC, under conditions known to induce dissociation and denaturation of spectrin, led to a three-fold decrease in amplitude of fluctuations (Levin and Korenstein, 1991). These results imply the involvement of the membrane skeleton in the fluctuation process. These fluctuations occur in the semi-expanded state of the membrane skeleton, where expansion and contraction forces are balanced (Shen et al., 1986; Vertessy and Steck, 1989; Bennett, 1989,1990). These findings may suggest that cell membrane fluctuations are associated with a MgATP dependent mechano-chemical dynamic assembly of the skeletal proteins (Levin and Korenstein, 1991).

Cell membrane fluctuations may be related to the flickering phenomenon. Flickering is revealed as a shimmering at the rim of the center of the RBC and was attributed to thickness changes in the cell. Quantitative measurements of flickering was performed by Brochard and Lennon (1975). This study was corroborated by later studies (Fricke and Sackmann, 1984; Fricke et al., 1986; Zeman et al., 1990) where phase contrast and reflection interference microscopic methodologies were applied. Data of flickering was obtained mainly by an indirect method, measuring oscillation of transmitted light that passes through a small area at the RBC center. The flickering was examined as function of several parameters: osmotic swelling, temperature, ethanol, α-agonists and some other drugs (Fricke and Sackmann, 1984; Fricke et al., 1986; Zeman et al., 1990). The bending elasticity coefficient (K_c) of the membrane was obtained from flickering spectra of these studies yielding a value of 0.3-0.5.10^{-12} dyn.cm. These values are 5-7 times smaller than the value of K_c=1.8-2.0.10^{-12} dyn.cm obtained by the micropipette method (Evans, 1983). The authors conclude that interpretation based on pure physical factors (Brownian motion) is sufficient to explain the flickering phenomenon. The flickering phenomenon occurs in cell membrane regions of small or negative curvature, i.e. under conditions of excess of membrane surface (high surface/volume ratio), while fluctuations occur in a region of high positive curvature of the cell membrane. Thus, it is expected that more energy is needed for fluctuations to occur than that needed for flickering. The interrelationship between the fluctuation and the flickering phenomena is still not established.

Oxygenation-deoxygenation cycle of erythrocytes modulates cell membrane fluctuations

In an attempt to examine the physiological significance of mechanical fluctuations in erythrocytes, we have investigated the effect of oxygenation-deoxygenation cycle on cell membrane displacements (Tuvia et al., 1992a). Oxygenated RBC reveals a maximal relative change of light scattering ($\delta I/I$) of (mean ± S.D.; n - number of cells) 17.1±2.9% (n=92). Deoxygenation results in a decline of these fluctuations down to 9.6±1.9% (n=131). These changes correspond to the attenuation of membrane displacements from 290±49nm in the oxygenated state to a level of 160±32nm in the deoxygenated state). When deoxygenated cells, revealing an amplitude of displacement of 9.0%±2.6 (n=65), were reoxygenated their amplitude of fluctuation was restored to the level of 15.8%±3.7 (n=37). Thus, a cyclic change of oxygenation-deoxygenation is accompanied by reversible amplitude changes of cell membrane fluctuations. The power spectrum of the deoxygenated RBC differs from the oxygenated one, indicating a shift of the energy distribution of mechanical fluctuations towards lower frequencies, around 2Hz.

In intact RBC the effect of oxygenation-deoxygenation on cell membrane fluctuations may be attributed either to a variation in hemoglobin binding to the membrane-skeletal proteins and/or to a change in the intracellular concentration of metabolites during the oxygenation-deoxygenation cycle. The oxygenation state of hemoglobin is a major effector of the intracellular levels of 2,3 diphosphoglycerate (DPG) and Mg^{+2} (Bennett, 1989, 1990), which may be responsible for the modulation of membrane fluctuations. To evaluate this possibility, we have examined the effects of 2,3 DPG and Mg^{+2} on the displacement amplitudes in RBC ghosts. We have chosen the characteristic concentrations of 2,3 DPG and Mg^{+2} for the oxygenated and the deoxygenated states of the RBC (Rapoport et al. 1976; Petersen et al., 1990; Gupta et al., 1978; Flatman, 1980, 1988).

Non-resealable saponin ghosts, perfused with a solution of KCl, reveal a basal fluctuation of 6.5%±1.9 (n=40), 2.3 fold lower than the observed displacement level in the corresponding intact RBC. However, the fluctuation amplitudes were reconstituted to a level of 10.9%±3.5 (n=82), ($P<0.01$), by perfusing the ghosts with a solution, whose composition of metabolites mimics the oxygenated state of RBC. When the same ghosts were perfused again with a composite solution of metabolites which mimics the deoxygenated state of RBC (Rapoport et al. 1976; Petersen et al., 1990; Gupta et al., 1978; Flatman, 1980, 1988), the fluctuation amplitude decreased to a level of 6.8%±1.7 (n=54), ($P<0.01$). Furthermore, when the same ghosts were subjected to an additional set of perfusions with the "oxygenated" and "deoxygenated" composite solutions, the fluctuation level at first increased to 8.9%±2.1 (n=35; $P<0.01$) and then decreased to a level of 6.1%±1.5 (n=27) ($P<0.01$), correspondingly.

These results show that the observed changes in membrane fluctuations may be accounted for by change in concentration of 2,3 DPG and Mg^{+2} during the oxygenation-deoxygenation cycle of RBC. Moreover, they demonstrate that cyclic changes in metabolite concentrations leads to reversible changes in the amplitude of displacement of the cell membrane. The MgATP level does not change much between the oxygenated and the deoxygenated states (Petersen et al., 1990). In an attempt to evaluate whether a single metabolite is dominantly influencing the fluctuation amplitude we have examined the effects of DPG and Mg^{+2} separately in RBC ghosts. DPG alone (6mM in KCl solution) had no effect on the amplitude of fluctuations as compared to the amplitudes in the KCl solution. Mg^{+2} alone had a bimodal effect on displacement amplitudes: upon increasing Mg^{+2} concentration from 0.1mM to 0.4mM, the amplitude of fluctuation increased steeply to a maximal level of 11.0±2.7 (n=40), but higher concentrations of Mg^{+2} resulted in a decrease of the displacement amplitudes to the basal level of 6.1%±1.4 (n=37). The highest level of fluctuation was observed at 0.4 mM Mg^{+2}. Hence, the variation of intracellular Mg^{+2} during the oxygenation-deoxygenation cycle is suggested to be a major factor affecting cell membrane displacements.

Mg^{+2} is known to be required for the association of tropomyosin with actin (Fowler and Bennett, 1984; Bennett, 1989). Therefore, under deoxygenating conditions where free Mg^{+2} is high (0.5-0.75mM) (Flatman, 1980, 1988), tropomyosin is expected to be associated with actin, while in the fully oxygenated state tropomyosin would not be able to bind actin (Bennett, 1989). It may be expected that binding of tropomyosin to actin will lead to the increase of the mechanical stability of the membrane skeleton with a consequent decrease of cell-membrane fluctuations. Indeed, a decrease in fluctuation amplitudes is observed in saponin ghosts at Mg+2 concentrations higher or equal to 0.6mM Mg^{+2}. This range of Mg^{+2} concentration coincides with the minimal concentration of Mg^{+2} at which tropomyosin was found to bind F-actin from muscle (Bennett, 1989). The restoration of fluctuation in the oxygenated state may reflect the abolishment of the Mg^{+2} dependent association of tropomyosin with actin. Thus, the binding of Mg^{+2} seems to modulate the dynamic mechanical properties of the cell membrane. Thus, the oxygenation-deoxygenation process provides a potential mechanism for the dynamic control of membrane-skeleton proteins interactions.

The physiological significance of cell membrane fluctuations - Correlation with cell filterability

The occurrence of metabolically driven mechanical membrane fluctuations is expected to diminish the bending stiffness of the RBC and enables it to fold more efficiently upon entrance into blood capillaries. This suggestion is supported by the experimental verification of the direct correlation between the amplitude of cell membrane fluctuations and cell filterability (Tuvia et al., 1992b; Zamir et al., 1992).

We examined cell membrane fluctuations and filterability of human red blood cells over a wide range of medium osmolalities (181-675 mosmole/kg H_2O)(Tuvia et al., 1992b). The amplitude of cell membrane displacements of 13.2% (≈225nm) does not change in osmolality range of 307 - 470 mosmole/kg H_2O. However, when medium osmolality varies beyond this range, a strong decrease in amplitude of displacement down to 8.8% was observed (≈150nm at 220-181 mosmole/kg H_2O and 140-170nm at 675-550 mosmole/kg H_2O). This decline of displacement amplitude was reversible when resuspending the RBC in an isosmotic medium. The observed dependence of the displacement amplitude on medium osmolality may be attributed to the stretching and compression of the membrane skeleton in hypoosmotic and hyperosmotic conditions, correspondingly. Since the membrane skeleton is semiexpanded under isosmotic conditions (Vertessy and Steck, 1989), both stretching and compression of the skeleton network may lead to the attenuation of displacement amplitude.

The modulation of RBC filterability by osmotic forces was carried out at a constant pressure by recording the initial flow rates through 5μm pores (Tuvia et al.,

1992b). Filterability was evaluated as the ratio of the initial rate filtration of a RBCs suspension to the initial rate filtration of the solution devoid of RBCs. In the range of 300-470 mosmole/kg H_2O (isosmotic and hyperosmotic conditions), a maximal filterability was maintained (filterability ≈ 0.4). When medium osmolality changed beyond this range limits, a strong decline in filterability was observed (i.e. filterability ≈ 0.36-0.11 at 220-181 mosmole/kg H_2O and filterability of ≈ 0.35-0.17 at 550-675 mosmole/kg H_2O). The similar dependence of RBC filterability and of local cell membrane displacements on medium osmolality indicates the existence of a positive correlation between local cell membrane displacements and cell filterability.

The possible regulation of RBC deformability under physiological conditions is supported by our recent study of the modulation of membrane fluctuations and cell filterability by the atrial natriuretic peptide (Zamir et al., 1992). The atrial natriuretic peptide (ANP) is a polypeptide hormone which is synthesized, stored and secreted predominantly by atrial myocytes in response to increase in atrial pressure (Brenner et al., 1990). It was shown to be secreted by cardiac atria under hypoxia (Baertchi et al., 1988). The ANP was found to increase the amplitude of membrane fluctuations in a doze dependent manner (Zamir et al., 1992). The specificity of ANP in eliciting its effects on cell membrane fluctuations emerges from the failure of a number of other peptide hormones (vasoactive intestinal peptide, endothelin I, angiotensin II and ß-endorphin) to induce similar effects, even at concentrations of 10^{-7}M (Zamir et al., 1992).

In view of our previous findings that the amplitude of membrane fluctuations of RBCs positively correlates with cell filterability (Tuvia et al., 1992b), we also studied the effect of ANP on RBCs filterability (Zamir et al., 1992). Preincubation with ANP ($4x10^{-8}$M) increased membrane filterability by 36%. Typically, ANP exerts its action on cells via specific cell membrane receptors (ANP-R_1 and ANP-R_2). Since the effects of ANP on RBC deformability were mimicked by membrane permeable analogues of cyclic GMP, it may be concluded that the ANP exerts its effect on cell membrane fluctuations and cell filterability through ANP-R_1 receptors which act through particulate guanylate cyclase - cyclic GMP mechanisms. Based on its ability to affect the bending stiffness of the RBC it may be suggested that the secretion of ANP under conditions of hypoxia is a physiological defense response, where the ANP induced reduction of the RBC bending stiffness increases the ability of erythrocytes to enter narrow blood capillaries and thereby improve oxygen delivery to the tissue.

Based on the correlation between cell membrane fluctuations and cell filterability by osmotic or hormonal control we would like to suggests that the occurrence of metabolically driven mechanical membrane fluctuations on the side surface of the RBC diminishes the bending stiffness of the RBC and enables it to fold more efficiently upon entrance into blood capillaries. Thus, higher displacements amplitudes of the RBC are expected to lead to a higher filterability of the RBC with a concomitant higher ability of an erythrocyte to penetrate into blood capillaries with a consequent more efficient oxygen delivery to the tissue.

Concluding remarks

It seems that the dynamic mechanical properties of the red blood cell (RBC), reflected by membrane fluctuations, are determined to a large extent by metabolic driven mechanical fluctuations of the membrane skeleton network. These properties, are modulated via intracellular changes of MgATP, Mg^{+2} and cyclic GMP Moreover, the occurrence of metabolically driven mechanical membrane fluctuations diminishes the bending stiffness of the RBC and enables it to fold more efficiently upon entrance into blood capillaries.

Acknowledgements: This study was supported by the Basic Research Foundation, The Israel Academy of Sciences and Humanities and the United States-Israel Binational Science Foundation.

References

Bagge, U., Braneerman, P.I., Karlson, R. and Skalak, R. (1980) Three dimensional observations of red blood cell deformation in capillaries. Blood Cells 6:231-237.
Baertschi, A.J., Adams, J.M. and Sullivan, M. (1988) Acute hypoxia stimulate atrialnatriuretic factor secretion in vivo. Am. J. Physiol. 255:H295-H300.
Bennett, V. (1989) The spectrin-actin junction of erythrocyte membrane skeletons. Biochim. Biophys. Acta 988:107-121.
Bennett, V. (1990) Spectrin-based membrane skeleton: a multipotential adaptor between plasma membrane and cytoplasm. Physiol. Rev. 70:1029-1065.
Brenner, B.M., Ballerman, B.J., Gunning, M.E. and Ziedel, M.L. (1990) Diverse biological actions of atrial natriuretic peptide. Physiol. Rev. 70:665-699.
Brochard, F., and Lennon, J.F. (1975) Frequency spectrum of the flicker phenomenon in erythrocytes. J. Phys. (Paris) 36:1035-1047.
Evans, E.A., (1983) Bending elastic modulus of red blood cell membrane derived from buckling instability in micropipet aspiration tests. Biophys. J. 43:27-30.
Felder, S., and Elson, E.L. (1990) Mechanics of fibroblast locomotion: quantitative analysis of forces and motions of the leading lamellae of fibroblast. J. Cell Biol. 111:2513-2526.

Flatman, P.W., (1980) The effect of buffer composition and deoxygenation on the concentration of ionized Magnesium inside human red blood cells. J. Physiol. 300:19-30.

Flatman, P.W., (1988) The control of red cell magnesium. Magnesium. Res 1:5-11.

Fowler, V.M. and Bennett, V. (1984) Erythrocyte membrane tropomyosin: purification and properties. J. Biol. Chem. 259:5978-5989.

Fricke, K., and Sackmann E. (1984) Variation of frequency spectrum of the erythrocyte flickering caused by aging, osmolarity, temperature and pathological changes. Biochim. Biophys. Acta 803:145-152.

Fricke, K., K. Wirthensohn, R. Laxhuber and Sackmann, E. (1986) Flicker spectroscopy of erythrocytes. A sensitive method to study subtle changes of membrane bending stiffness. Eur. Biophys. J. 14:67-81.

Gaehtgens, P., Duhssen, C. and Albrecht, K.H., (1980) Motion, deformation and interaction of blood cells and plasma during flow through narrow capillary tubes. Blood cells 6:799-812.

Gupta, R.K., Benovic, J.L. and Rose, Z.B. (1978) The determination of the free Magnesium level in human red blood cell by ^{31}NMR. J. Biol. Chem. 253:6172-6176.

Krol, A.Ju., Grinfeldt, M.G. and Levin, S.V. and Smilgavichus, A.D. (1990) Local mechanical oscillations of the cell surface within range 0.2-30 Hz. Eur. Biophys. J. 19:93-99.

Levin, S.V., and Korenstein, R. (1991) Membrane fluctuations in erythrocytes are linked to MgATP dependent dynamic assembly of the membrane skeleton. Biophys. J. 60:733-737.

Mittelman, L., Levin, S.V., and Korenstein, R. (1991) Fast cell membrane displacements in lymphocytes:modulation by dihydro-cytochalasin and colchicine. FEBS Letters 293:207-210.

Petersen, A., Kristensen, S.R., Jacobsen, K.P. and Horder, M. (1990) ^{31}P-NMR measurements of ATP, ADP, 2,3-diphosphoglycerate and Mg^{2+} in human erythrocytes. Biochim. Biophys. Acta. 1035:165-174.

Rapoport, I., Berger, H., Rapoport, S.M., Elsner, R. and Gerbes, G. (1976) Response of the glycolysis of human erythrocytes to the transition from oxygenated to the deoxygenated state at constant intracellular pH. Biochim. Biophys. Acta. 428:193-204.

Shen, B.V., Josephs, R., and Steck T.L. (1986) Ultrastructure of the intact skeleton of the human erythrocyte membrane. J. Cell Biol. 102:997-1006.

Tuvia, S., Levin, S., and Korenstein, R. (1992a) Oxygenation - Deoxygenation cycle of erythrocytes modulates the submicron cell membrane fluctuations. Biphys. J. 63:599-602.

Tuvia, S., Levin, S., and Korenstein, R. (1992b) Correlation between local cell membrane displacements and filterability of human red blood cells. FEBS Lett. 304:32-36.

Vertessy, B.G., and Steck, T.L. (1989) Elasticity of the human red cell membrane skeleton. Effect of temperature and denaturants. Biophys. J. 55:255-262.

Zamir, N., Tuvia, S., Riven-Kreitman, R., Levin, S., and Korenstein, R. (1992) Atrial natriuretic peptide: Direct effects on human red blood cell dynamics. Biochim. Biophys. Res. Commun. 188:1003-1009.

Zeman, K., Engelhart and Sackman, E. (1990) Bending undulations and elasticity of erythrocyte membrane: effect of cell shape and membrane organization. Eur. Biophys. J. 18:203-219.

Simulation of Morphogenetic Shape Changes Using the Finite Element Method

G. Wayne Brodland[1] and David A. Clausi[2]
University of Waterloo
Waterloo, Ontario
N2L 3G1
Canada

Morphogenesis, the process by which organs and other embryonic structures are formed, involves a variety of physical phenomena. Complex chains of biochemical events regulate the construction and behaviour of organelles and cytoskeletal components, electric fields are produced, patterns are formed, and forces act to bring about crucial shape changes (Alberts *et al.*, 1989).

Transformation of a zygote, a ball of cells, into a functional embryo involves a myriad of successive shape changes to produce the necessary complex geometry, and a simultaneous progression of cell differentiations to produce cells of distinct functional types. In many cases, the morphogenetic shape changes involve in-plane movements, invaginations or evaginations that occur in sheets of epithelial cells on inside or outside surfaces of the embryo. Often, and perhaps always, these shape changes are driven by forces generated within the epithelial cell sheet. Possible sources of these forces include a variety of cytoskeletal components such as microfilaments, microtubules and intermediate filaments (Fig. 1A); intercellular adhesions; and cortical tractoring (Alberts *et al.*, 1989; Jacobson *et al.*, 1986). How these possible force generators act to bring about specific embryonic shape changes is the topic of much debate.

Partial answers have been provided by ultrastructural studies that determine which structural components are available to drive specific shape changes, and by teratogen studies in which the structure or action of specific structural components are disrupted and the resulting patterns of shape change observed and compared with normal patterns (Burnside, 1973; Brun and Garson, 1983).

[1]Departments of Civil Engineering and Biology
[2]Department of Systems Design Engineering

NATO ASI Series, Vol. H 84
Biomechanics of Active Movement
and Division of Cells
Edited by N. Akkaş
© Springer-Verlag Berlin Heidelberg 1994

As powerful digital computers became accessible, they proved to be a useful tool for studying the mechanics of development (Jacobson and Gordon, 1976; Odell *et al.*, 1981). Recent studies that use the finite element method (FEM) have made it possible to determine the shape changes that specific combinations of cytoskeletal components and other force generators would produce in sheets of cells (Selker *et al.*, 1992; Clausi and Brodland, 1993; Brodland and Clausi, 1993). Such computer simulations provide an additional means to test hypotheses about the forces that drive morphogenetic shape changes.

The Finite Element Method

The finite element method is a powerful, general-purpose methodology for modelling physical systems. As a result of its extensive application during the last several decades to physical problems, especially those that arise in physics and engineering, it has become highly developed (Zienkiewicz and Taylor, 1989). The potential of the method for analysis and solution of biology-related problems, however, remains largely unutilized (Brodland, 1993).

The basic approach of the method is to break the object of interest into finite-sized, "elements" (Fig. 1B) in such a way that a tractable mathematical description of the mechanical behaviour of each element can be written. For example, to model cytoskeletal mechanics, a bundle of microfilaments might constitute a single element. The force f_{MF} produced by it could be described in terms of its original length L_0, its current length L, its current rate of lengthening $\dot{L}$, and C, the concentration of surrounding Ca^{++}; i.e.,

$$f_{MF} = f(L_0, L, \dot{L}, C)\,. \tag{1}$$

Additional equations that embody conditions of dynamic equilibrium and inter-element compatibility are then constructed. The result is a large system of simultaneous equations that can, when organized according to the "stiffness" approach, be written as

$$\mathbf{f} = \mathbf{K}\mathbf{x}\,, \tag{2}$$

where $\mathbf{x}$ is the vector of displacements that result from the vector of applied forces $\mathbf{f}$ acting on a mechanical system with stiffness (i.e., mechanical properties) $\mathbf{K}$.

Morphogenetic shape changes are typically geometrically and materially nonlinear. Thus Equation (2) must be applied iteratively over small time steps (Brodland and Clausi, 1993).

The FEM provides an ideal tool for the analysis and simulation of aspects of morphogenesis because complex geometries and mechanical properties can be accommodated and because arbitrarily large changes in shape can be modelled reliably. When the finite element method is used as the basis for computer simulations of morphogenesis, real or hypothesized cytoskeletal morphologies and mechanical properties can be assigned, and the behaviour of a "virtual" embryo that has user-specified properties investigated. In this way, the FEM makes it possible to test the mechanics of hypotheses about the forces that drive specific morphological shape changes and, thus, provides an important complement to ultrastructure and teratogen studies.

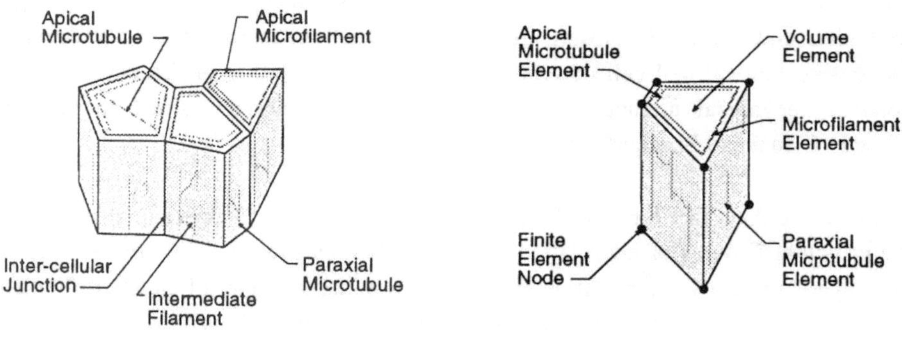

Fig. 1. A. Schematic of part of an epithelial cell sheet. **B**. Finite element model of part of the sheet shown in Fig. 1A.

Mechanics of Neurulation

Neurulation is a critical embryonic process that requires a sheet of embryonic tissue called the neural plate to undergo in-plane reshaping and subsequently roll up to form a sealed tube. Because complex three-dimensional shape changes are involved in neurulation, intuitive inferences about the mechanics of the process can be in error. Computer simulations that predict the shape changes that specific sets of driving forces would produce are, therefore, of interest.

Figure 2 shows a simulation of neural tube formation in axolotl (*Ambystoma mexicanum*). The cell cytoplasm and intermediate filament network are modelled using 8-noded, hexahedral blocks that have viscosity μ. Microfilaments act at the apical surface of each cell and exert a contractile force as described in the figure caption. During normal neurulation this strip of cells would become wider as the neural plate elongates under the action of the notochord, and its medial edge would pull the adjacent non-neural epithelium medially. Details of how these features are incorporated into the simulation through the boundary conditions, and other details relevant to the simulations reported here, are given in Clausi and Brodland (1993) and Brodland and Clausi (1993).

The simulation shows that the forces produced by contraction of apical circumferential microfilament bundles (CMB's) coupled with cephalo-caudal (axial) elongation of the embryo are sufficient to cause normal neural tube closure. Many of the salient features of neurulation, including formation of distinct neural ridges, narrowing and thickening of the neural plate, skewing of cells just medial to the ridges, "hinge" formation, and neural tube closure are evident. No cell- or region-specific "programs" or non-mechanical cell-cell communications are employed. A discussion of the detailed mechanical interactions that give rise to this sequence of shape changes are given in Clausi and Brodland (1993).

Is it possible that other biologically-reasonable sets of driving forces can also produce the principal features of neurulation? The simulations shown in Figures 3, 4 and 5 investigate, respectively, the influence of changes in the character of the microfilament driving force, the effect of microtubules acting simultaneously with microfilaments, and the possible role of medially-directed forces at the edge of the neural plate. Compared to the simulation shown in Fig. 2, none of the latter simulations correspond as well to typical normal patterns of neurulation. This suggests that the patterns of shape change produced are sensitive to the nature of the driving forces and that the driving forces acting in the first simulation are likely very close to those that act in the live embryo. That simulations of in-plane motions and dorsal surface flattening driven by the same system of microfilament forces (Clausi and Brodland, 1993) also match live embryo development supports this conclusion.

The formulation used here also provides a means to investigate other mechanical phenomena such as wave propagation, three-dimensional tissue invagination and mechanically-based pattern formation that occur in epithelial cell sheets (Brodland and Clausi, 1993). A variety of experiments in which segments of the neural plate have

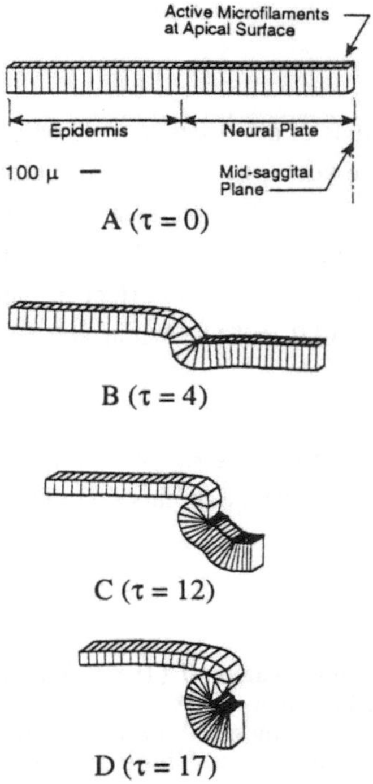

Fig. 2. Shape changes produced by microfilaments that exert a force $f_{MF} = \alpha\ L_0/L$. Part A shows the starting configuration for all of the simulations reported here. After Clausi and Brodland (1993).

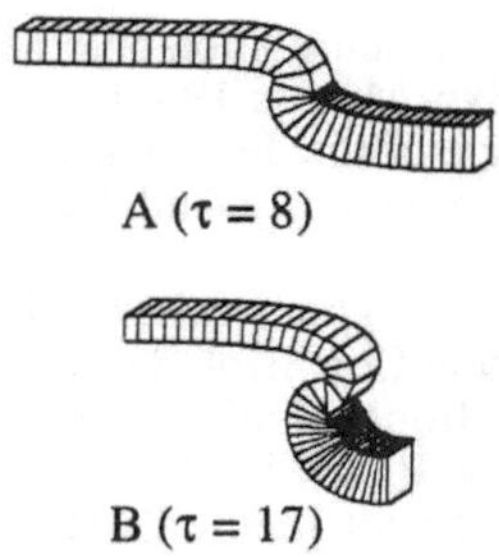

Fig. 3. Shape changes produced by microfilaments that exert a force $f_{MF} = \alpha$.

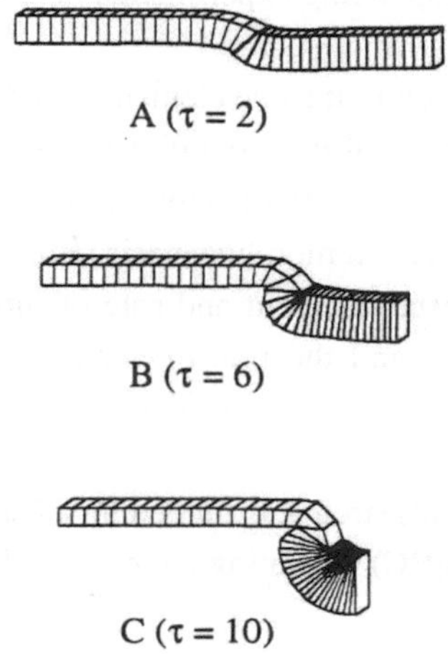

Fig. 4. Shape changes produced by microfilaments that exert a force $f_{MF} = \alpha\ L_0/L$ and paraxial microtubules.

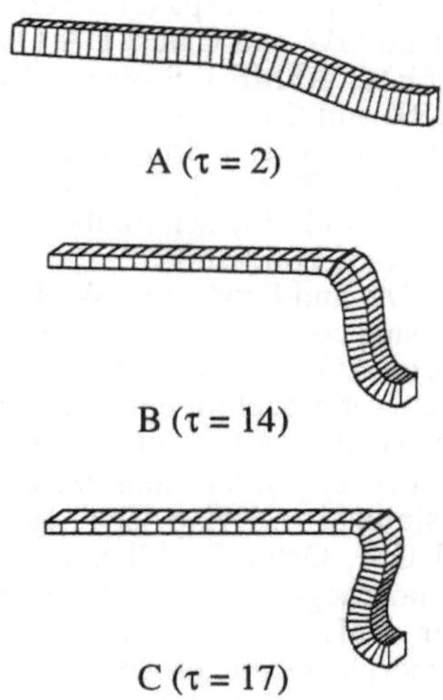

Fig. 5. Shape changes produced by medial forces located at the lateral edge of the neural plate.

been excised have also been simulated (unpublished). The simulations agree well with corresponding experimental results.

Most current finite element-based computer simulations of development deal only with mechanical considerations. The potential exists, however, to use the finite element method to simultaneously or independently investigate other physical phenomena involved in morphogenesis (Brodland, 1993). These might include cell-cell signalling, the establishment and role of morphogen fields, the formation and effect of electric fields, and the role of various substances such as ATP and Ca^{++} that are directly associated with force generation.

This research was funded by a Natural Sciences and Engineering Research Council (NSERC) Operating Grant to G.W.B and an NSERC Graduate Scholarship to D.A.C.

References

Alberts B, Bray D, Lewis J, Raff M, Roberts K and Watson J (1989) Molecular biology of the cell. 2nd ed. Garland Publishing Inc., New York
Brodland GW (1993) Finite element methods for developmental biology. Gordon R and Goodwin B (eds) The cytoskeleton in developmental biology. Int. Rev. Cytology. Accepted for publication
Brodland GW and Clausi DA (1993) Embryonic tissue morphogenesis modelled by FEM. ASME J. Biomech. Eng.: Accepted for publication
Brun RB and Garson JA (1983) Neurulation in the Mexican salamander (Ambystoma mexicanum): a drug study and cell shape analysis of the epidermis and the neural plate J. Embryol. Exp. Morph. 74:265-295
Burnside B (1973) Microtubules and microfilaments in amphibian neurulation. Am. Zool. 13:989-1006
Clausi DA and Brodland GW (1993) Mechanical evaluation of theories of neurulation using computer simulations. Development 118:1013-1023
Jacobson AG, Oster GF, Odell GM and Cheng LY (1986) Neurulation and the cortical tractor model for epithelial folding. J. Embryol. Exp. Morph: 19-49
Jacobson AG and Gordon R (1976) Changes in the shape of the developing vertebrate nervous system analyzed experimentally, mathematically and by computer simulation. J. Exp. Zool. 197:191-246
Odell GM, Oster G, Alberch P and Burnside B (1981) The mechanical basis of morphogenesis, I. Epithelial folding and invagination Dev. Biol. 85:446-462
Selker JML, Steucek GL and Green PB (1992) Biophysical mechanisms for morphogenetic progressions at the shoot apex. Dev. Biol. 153:29-43
Zienkiewicz OC and Taylor RL (1989) The finite element method McGraw-Hill, Maidenhead

LOCOMOTING BLEBBING CELLS: A NEW MODEL TO TEST WHETHER FORMATION OF PROTRUSIONS IS PRIMARILY DUE TO HYDROSTATIC PRESSURE OR TO ACTIN ELONGATION

H.U. Keller
Institute of Pathology, University of Bern, Murtenstrasse 31, 3010 Bern / Switzerland

SUMMARY

Locomoting blebbing cells appear to be a useful tool to analyse the mechanisms generating protrusions (Keller and Niggli, 1993). The following findings suggest that, in this model, hydrostatic pressure may be the primary force pushing the membrane forward:

1. In blebbing cells the membrane is pushed forward faster (about 2.4 μm /sec) than known rates of actin elongation.
2. Immunofluorescent studies indicate that blebs show either diffuse staining for F-actin and α-actinin or only a linear staining along the cell membrane. These two patterns may represent different stages of bleb formation.
3. Formation of blebs or ruffles as well as the development of a polarized shape in neutrophils is suppressed in hypertonic media (0.5 M sorbitol).

MAJOR THEORIES EXPLAINING PROTRUSION FORMATION

The view that actin elongation is the direct force pushing the membrane forward to form pseudopods has gained much support. It is compatible with the finding that the motor of neutrophils is localized at the front (Keller and Bessis, 1975) where F-actin accumulates in polarized cells (Fechheimer and Zigmond, 1983; Howard and Oresajo, 1985) and where actin elongation due to polymerization (Wang, 1985; Sanger et al., 1992) or by filament sliding (Umemoto and Sellers, 1990; Sheetz et al., 1992) occurs. While earlier data suggested a close correlation between actin movement and surface particle transport, which may be important for locomotion, recent studies suggest that there are disparities (Theriot and Mitchison, 1991). A close association between membrane protrusion and actin polymerization is not sufficient evidence that actin polymerization is the direct driving force. At least in electric fields, formation of protrusions can take place under the conditions where actin polymerization is inhibited, indicating that local actin polymerization can be a secondary event (Popov et al., 1991). A major alternative explanation for formation of protrusions is that the actin cytoskeleton produces internal hydrostatic pressure either locally or generally, and thereby elicits pseudopod formation at points where the membrane is weak or where the interaction of the membrane with the cytoskeleton is weak. There is evidence that this mechanism of protrusion formation may work in various cell types (Oster, 1984; Oster and Perelson, 1987; Bereiter-Hahn and Strohmeier, 1987; Strohmeier and Bereiter-Hahn, 1987; DiPasquale, 1975; Harris, 1990). A third possibility (lipid flow model) is that pseudopods are formed by reinsertion of recycled membrane (Bretscher, 1989). Other studies suggest that lipid flow is forward rather than rearward (Lee et al., 1990) and that the model of rearward flow of plasma membrane is incorrect (Heath and Holifield, 1991).

HYDROSTATIC PRESSURE RATHER THAN ACTIN POLYMERIZATION MAY BE THE DIRECT FORCE PRODUCING PROTRUSIONS IN BLEBBING NEUTROPHILS

The following findings are difficult to reconcile with the concept that actin elongation is the primary force driving the membrane forward in this particular model (Keller and Niggli, 1993): 1. We have measured forward expansion of surface membrane in areas where blebs are formed and observed that the rate of expansion (about 2.4 μm/sec) is much faster than known rates of actin elongation (Wang, 1985; Umemoto and Sellers, 1990; Theriot and Mitchison, 1991; Sheetz et al., 1992). 2. Whereas F-actin is consistently present throughout ruffles of neutrophils, blebs formed in the same cell type are different.

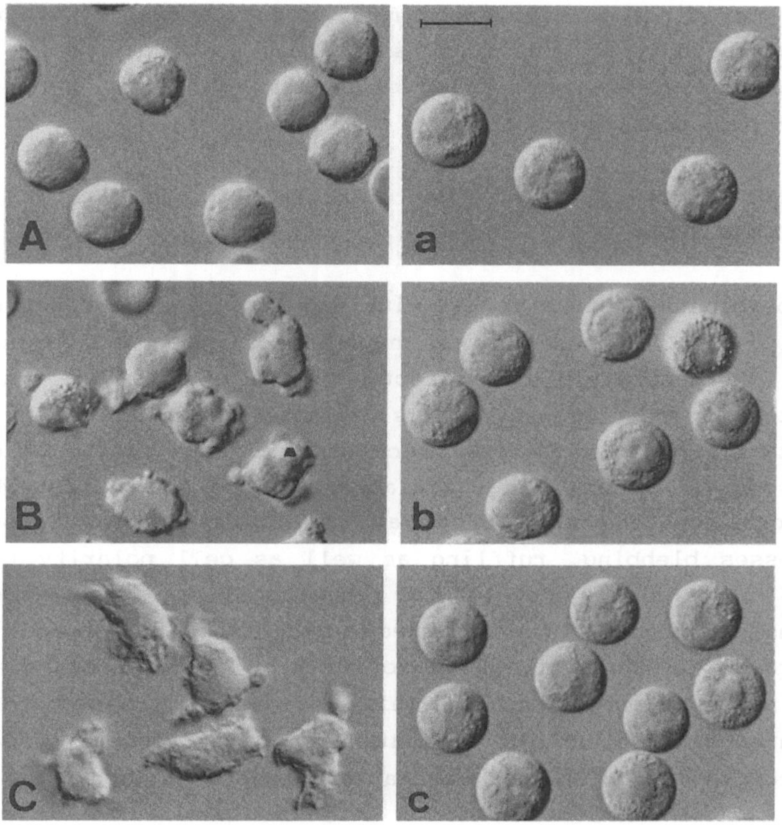

<u>Figure 1: Effect of 0.5M sorbitol on neutrophil shape</u>
A/a plain medium; B/b: 10^{-5}M colchicine; C/c: 10^{-9}M fNLPNTL in isotonic media (left; A, B, C) or hypertonic media containing 0.5 M sorbitol (right; a, b, c). Initially unstimulated cells were incubated in Gey's solution without sorbitol (310 mOsm) or with 0.5 M sorbitol (780 mOsm) at 37°C for 30 minutes and fixed in glutaraldehyde. Scale bar: 10 μm.

Sometimes they appear to contain no F-actin or α-actinin at all, or there is a linear accumulation along the membrane or a fairly diffuse but not very intense staining of the entire bleb. These findings fit the description of bleb formation of stationary cells given by Price (1967) and Harris (1990) who suggested that the expansion of the membrane is the primary event and the reorganisation and polymerization of cytoskeletal components in the area of the bleb a secondary event. Our findings have to be further substantiated with more sophisticated techniques such as confocal laser microscopy and by electron microscopy, studies on membrane flow and actin turnover in areas where the cell is blebbing.

SUPPRESSION OF RUFFLING AND BLEBBING IN NEUTROPHILS BY HYPERTONIC MEDIA

In fibroblasts protrusive activity and locomotion but not cell polarity are suppressed in hypertonic media (DiPasquale, 1975). We determined the effect of hypertonic media (0.5 M sorbitol) on resting neutrophils, neutrophils polarized by the chemotactic peptide fNLPNTL which show ruffles at the leading front and on neutrophils stimulated by 10^{-5}M colchicine which are polarized and show blebs at the leading edge. Sorbitol suppresses blebbing, ruffling as well as cell polarity under these particular experimental conditions (Fig.1). Dose-response studies (unpublished experiments) showed no difference in the suppressive activity of hypertonic media on ruffling vs blebbing cells. This indicates that the forces generated are similar but the data are not sufficient evidence that ruffles and blebs are generated by the same mechanisms.

CONCLUSIONS

Numerous studies suggest that actin elongation is the primary force driving the membrane forward in locomoting cells (for review see Heath and Holifield, 1991). Experiments suggesting that hydrostatic pressure rather than actin elongation protrudes the membrane forward in locomoting cell are relatively rare (Strohmeier and Bereiter-Hahn, 1987; Keller and Niggli, 1993). Our model of blebbing locomoting cells may prove to be suitable to test existing theories in more detail.

ACKNOWLEDGEMENTS

The work was supported by the Swiss National Science Foundation. I thank M. Kilchenmann for technical assistance.

REFERENCES

Bereiter-Hahn J, Strohmeier R (1987) Biophysical aspects of motive force generation in tissue culture cells and protozoa. Fortschritte der Zoologie 34:1-15
Bretscher MS (1989) Endocytosis and recycling of the fibronectin receptor in CHO cells. Embo J 8:1341-1348
DiPaquale A (1975) Locomotion of epithelial cells. Exptl Cell Res 95:425-439
Fechheimer M, Zigmond SH (1983) Changes in cytoskeletal proteins of polymorphonuclear leukocytes induced by chemotactic peptides. Cell Motil 3:349-361
Harris AK (1990) Protrusive activity of the cell surface and the movements of tissue cells. In: Biomechanics of active movement and deformation of cells. Edited by N. Akkas Springer-Verlag Berlin Heidelberg p. 249-294
Heath JP, Holifield BF (1991) Cell locomotion: new research tests old ideas on membrane and cytoskeletal flow. Cell Motil Cytoskel 18:245-257
Howard TH, Oresajo CO (1985) The kinetics of chemotactic peptide-induced change in F-actin content, F-actin distribution, and the shape of neutrophils. J Cell Biol 101:1078-1085
Keller HU, Bessis M (1975) Migration and chemotaxis of anucleate cytoplasmic leukocyte fragments. Nature 258:723-724
Keller HU, Niggli V (1993) Colchicine-induced stimulation of PMN motility related to cytoskeletal changes in actin, α-actinin, and myosin. Cell Motil Cytoskel 25:10-18
Lee J, Gustafsson M, Magnusson KE, Jacobson K (1990) The direction of membrane lipid flow in locomoting polymorphonuclear leukocytes. Science 247:1229-1233
Oster GF (1984) On the crawling of cells. J Embryol Exp Morph Suppl 83:329-364
Oster GF, Perelson AS (1987) The physics of cell motility. J Cell Sci Suppl 8:35-54
Popov SV, Svitkina TM, Margolis LB, Tsong TY (1991) Mechanism of cell protrusion formation in electrical field: the role of actin. Biochim Biophys Acta 1066:151-158
Price ZH (1967) The micromorphology of zeiotic blebs in cultured human epithelial (HEp) cells. Exptl Cell Res 48:82-92
Sanger JM, Sanger JW, Southwick FS (1992) Host cell actin assembly is necessary and likely to provide the propulsive force for intracellular movement of Listeria monocytogenes. Infect Immunity 60:3609-3616
Sheetz MP, Wayne DB, Pearlman AL (1992) Extension of filopodia by motor-dependent actin assembly. Cell Motil Cytoskel 22: 160-169
Strohmeier R, Bereiter-Hahn J (1987) Hydrostatic pressure in epidermal cells is dependent on Ca-mediated contractions. J Cell Science 88: 631-640
Theriot JA, Mitchison TJ (1991) Actin microfilament dynamics in locomoting cells. Nature 352:126-131
Umemoto S, Sellers JR (1990) Characterization of in vitro motility assays using smooth muscle and cytoplasmic myosins. J Biol Chem 265:14864-14869
Wang YL (1985) Exchange of actin subunits at the leading edge of living fibroblasts: possible role of treadmilling. J Cell Biol 101:597-602

Dynamic Morphology of Leukocytes: Statistical Analysis and a Stochastic Model for Receptor-Mediated Cell Motion and Orientation

Robert T. Tranquillo
Dept. of Chem. Engin. & Material Science
University of Minnesota
Minneapolis, MN 55455
U.S.A.

Oana Brosteanu and Wolfgang Alt
Div. of Theoretical Biology
University of Bonn
Kirschallee 1, D-53115 Bonn
Germany

Summary

Motility of leukocytes, like that of many other similar blood and tissue cells, can be stimulated by external chemicals which are either diffusing (chemotactic polypeptides, for example) or fixed to a substratum (adhesion proteins like fibronectin, for example). Depending on their adhesive strength, cells respond by more or less spreading, irregular protrusion of lamellipods around the cell periphery and eventually changing polarity and translocation of the cell body. Moreover, in a spatial gradient of a chemoattractant (CA) the direction of these motile activities is biased; however, cells tend to meander rather than migrate directly up-gradient, and occasionally even transiently migrate down-gradient. Thus, the inherent randomness of locomotion exhibited in a homogenous environment seems to be manifested during chemotaxis and, therefore, it should play a central role in modeling and explaining the underlying mechanisms of cell locomotion and orientation.

To this aim, we first describe a statistical approach for quantifying "dynamic shape changes" using spatial-temporal auto-correlation functions. Then we briefly describe the simplest prototype of a "stochastically perturbed morphogenetic dynamical system". It considers the (deterministic) cytomechanics of an idealized cortical actin network around the cell periphery under the influence of local signals induced by the (stochastic) distribution of bound chemotactic receptors in the cell membrane surrounding the cortex. As in earlier models, an account is made for statistical fluctuations in receptor binding, entirely determined by binding rate constants and the local CA concentration field, but generalized here to include statistical fluctuations in the spatial distribution of receptors, entirely determined by membrane diffusion coefficients. Then, as one particular case study, we assume the local density of bound receptors to determine the local F-actin assembly rate in cortical cytoplasma. Using this simple model we characterize via bifurcation analysis, stochastic simulations, and correlation functions the morphogenetic dynamics of a cortical F-actin layer and the resulting dynamic cell morphology. Animation of the simulations reveal various modes of changes in cell polarity as well as orientation in a CA gradient, whereby the degree of directional turning depends on the strength of lateral forces exerted between contracting actin network and adhesion receptors as well as the gradient steepness.

1. Statistical auto-correlation analysis of 2-dimensional shape changes

Based on two-dimensional cell contour data obtained by digitizing the outlines of migrating "amoeboid" cells (neutrophils on glass, for example), we describe the characteristics of cell

NATO ASI Series, Vol. H 84
Biomechanics of Active Movement
and Division of Cells
Edited by N. Akkaş
© Springer-Verlag Berlin Heidelberg 1994

deformation and translocation by quantifying peripheral extensions and retractions. These are thought to be induced by lamellipodial protrusions, their possible adherence to the substratum and subsequent force transduction onto the cell body. Since from phase contrast pictures we cannot estimate the mass distribution of cytoplasma, a simple hypothetical model is implemented computing position and shape of the cell body from the changing outline data. Generalizing previous methods (Soll 1988, Dunn & Brown 1987, 1990) weighted geometric moments are calculated to determine a "basic ellipse" describing mean position, orientation and elongation of a cell. Normal deviations from this elliptic shape are then plotted, classified as extensions or retractions, and eventually characterized as "significant lamellipods", see plot of "level lines" in Figure 1(a). Moreover, temporal auto-correlations of quantities as "cell area", mean protrusion activity or maximal extension are investigated, as are correlations between the directions of lamellipodial activity and cell translocation. Finally, the computation of spatial-temporal auto-correlations of peripheral extensions, see Figure 1(b), provide averaged measures of number, width and persistence of significant lamellipods as well as their frequency in space and time, for more details see (Brosteanu, 1993).

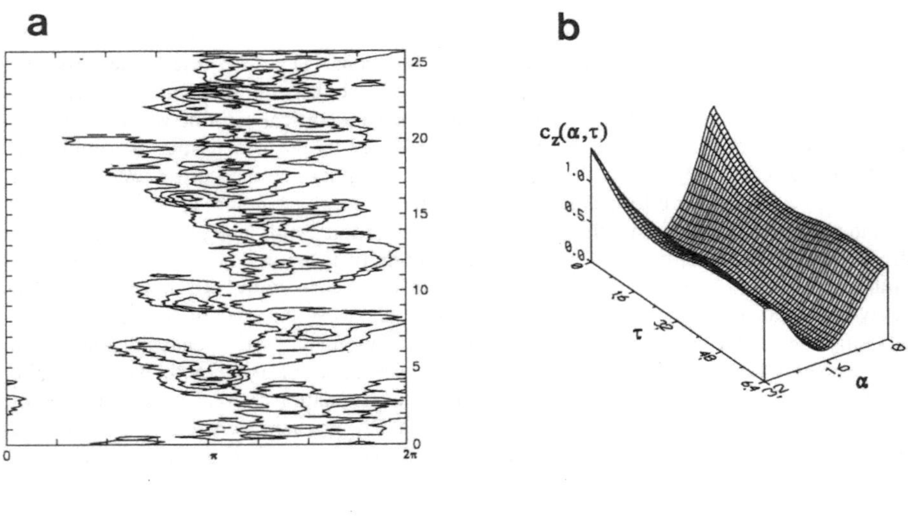

Figure 1. Evaluation of empirically recorded lamellipodial activity of (PMN) leukocytes. (a) "Level lines", i.e. lines of equal value, of peripheral extensions plotted over angular coordinate from 0 to 2π, time (min) running from bottom to top; (b) spatial-temporal auto-correlation function of data in (a), plotted vs. correlation angle α and correlation time τ. For $\alpha=0, \pi$ the mean decay time in τ measures persistence of a lamellipod; relative maximum at $\tau\approx2.4$min for $\alpha\approx2\pi/3$ corresponds to temporal-angular characteristics of the undulating waves seen in (a), namely period and mean distance between alternating lamellipods.

2. Simple cytomechanical model for the dynamic morphology of an F-actin cortex

We propose a model to simulate in a most simple manner the dynamic morphology of leukocytes that are (more or less) adherent to a two-dimensional substratum. We idealize the outline of the deformable cell cortex, $R_0 = R_0(t,\phi)$ denoting the radial extension in angular direction ϕ, to be entirely determined by the local concentration of cortical F-actin $a = a(t,\phi)$. Neglecting any changes in the radial thickness, this cortical actomyosin layer itself is modeled as a highly viscous and contractile fluid-like network, cf. (Dembo 1989, Alt 1990). Since more dense actomyosin network at a cortex position would induce stronger inward forces onto the cell body, we simply describe its extension by a monotone decreasing relationship $R_0 = R_0(a)$, using here $R_0 = \exp(-a/a_0)$.

The resistance to tangential cortical flow, defined by the *mean cortex velocity* $v(t,\phi)$, will be a combination of possible friction forces with the cytoskeleton and substratum. Both can be subsumed under the term on the left hand side of the following *tangential force balance* equation:

$$(1) \qquad \Phi \cdot a \cdot v \;=\; \partial_\phi(\, \sigma(a) + T_R \,)$$

Here the right hand side represents counteracting forces that are due to changes in cortical stress: contraction and swelling of the actomyosin network, $\sigma(a)$ as defined by Alt & Kaiser (this volume), and a stress induced by local curvature of the F-actin layer. Assuming the apparent surface tension of the actin cortex to be $\tau(a) = \tau_0\,[a-a_{min}]^+$, for instance, then the local stress can be approximated by $T_R = -\,\partial/\partial\phi\,(\,\tau(a)\,\partial R_0/\partial\phi\,)$, meaning that an outward (or inward) curved cortex leads to a compressive (or dilatational) stress. By adding the following *mass balance* equation for F-actin, cf. Alt & Kaiser (this volume) or Alt (1990),

$$(2) \qquad \partial_t\, a + \partial_\phi\,(a v) \;=\; \eta \cdot (a^* - a)$$

and inserting equation (1) therein, we obtain the following "morphogenetic" fourth order parabolic differential equation for the dynamics of cortical F-actin distribution around the idealized cortex:

$$(3) \qquad \partial_t\, a + \frac{1}{\Phi}\,\partial_\phi^2\,[\,\sigma(a) - \partial_\phi(\tau(a)\cdot \partial_\phi(R_0(a)))\,] \;=\; \eta \cdot (a^* - a)$$

Depending on the parameter values, this model produces typical morphogenetic behavior, i.e. appearance of steady state solutions describing one or more stable cortical contraction centers. In particular, one stable contraction center would simulate a polarizing cell with relatively high F-actin concentration at one side (where later the "uropod" may form) and relatively low concentration in the other cortical regions (where lamellipods appear), this being accompanied by a retrograde flow of cortical F-actin towards the "uropod". This is consistent with experimental observations of leukocytes (Coates et al. 1992), keratinocytes (Bereiter-Hahn, this volume) or even amoebae (Grębecki, this volume).

Since this model does not account for the dynamics of protrusion and retraction at the cell periphery (for dynamical models cf. Skalak or Alt & Kaiser, this volume), here we use a rather crude description of the "mean lamellipodial extension" to depend directly on local F-actin concentration via $L = L(a)$. Several functional dependencies can be considered. We take here $L(a) = a \exp(-(a/a_L))$ in order to reflect the observation that polar cells show most active protrusions not at the center of the leading lamella, but more so lateral to it, whereas regions of condensed F-actin network towards the uropod are mostly free of lamellipods. Thus, for each time instance, two polar curves describe the shape, $R_0 = R_0(a(t,\phi))$ of the cortex, and $R = R_0 + L(a(t,\phi))$ of the mean cell periphery, see Figure 3(a) at times $t \geq 0.2$, for example.

Finally, translocation or reorientation of the whole cell body is achieved through the fact that cortical actin network is connected via accessory proteins to both the cytoskeleton and adhesion receptors bound to the substratum. Thus, local contraction in the cortex induces an outward force, in direction of the outer normal vector $e(\phi)$. Integration around the periphery then gives the *net translocation force* onto the cell body

$$(4) \qquad F(t) = \oint \frac{\psi}{2}\, a(t,\phi)\, L(a(t,\phi))\, A(a(t,\phi))\, e(\phi)\, d\phi$$

where $A = A(a)$ denotes an adhesion strength, possibly depending on actin concentration, using here $A = \exp(-a/a_A)$. Similarly, tangential retrograde flow of the cortical actomyosin network induces a *net torque* onto the cell body

$$(5) \qquad \Omega(t) = -\oint a(t,\phi)\, v(t,\phi)\, A(a(t,\phi))\, d\phi$$

Balance with frictional drag forces on the adherent cell body then yields translocation and rotation speed of the model cell.

3. Stochastic stimulation of F-actin assembly by chemotactic receptors

In this cytomechanical model, both cell shape and cell movement are entirely determined by the dynamics of the cortical F-actin layer, the fourth order parabolic equation (3) only yielding steady cell shape patterns. Our main thesis is that spatial-temporal fluctuations in the local concentration of bound chemotactic receptors in the cell plasma membrane, stimulating (for example) assembly of the cortical actomyosin network, are able to induce varying dynamic morphology and orientation of our model cell resembling the observed behavior of living cells.

Thus, we have formulated a system of stochastic differential equations modeling binding kinetics and (a discretized version of) diffusion of membrane receptors that bind CA (chemoattractant). Then the concentration of bound receptors, $b^*(t,\phi_i)$, over the discretized cortical compartments constitutes a spatio-temporally correlated stochastic process, see simulated realizations in Figure 2(b). Proposing the F-actin equilibrium $a^*(t,\phi_i)$ to be locally dependent on $b^*(t,\phi_i)$, the "morphogenetic" differential equation (3) is now driven by this fluctuating stimulation process, see Figure 2(a). The main results of extensive stochastic simulations can be summarized as follows, cf. Tranquillo & Alt (1993). Depending on the choice of the mean F-actin assembly/disassembly equilibrium $<a^*>$ we obtain:

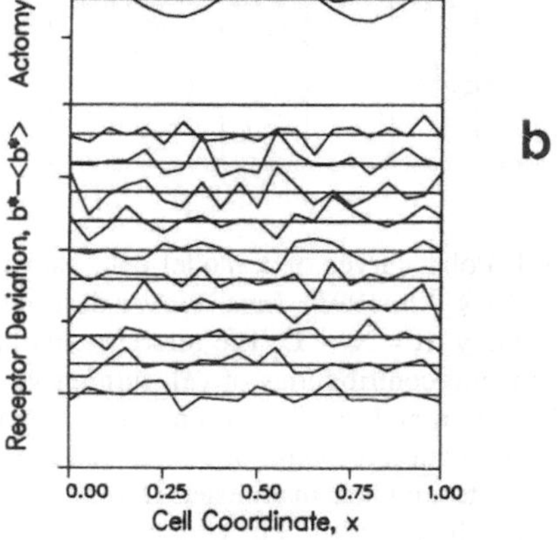

Figure 2. Stochastic simulation of (a) actomyosin, and (b) bound receptor distribution around an idealized cell cortex in a uniform CA concentration without cell rotation, according to eq.(3), with $a^* - <a^*>$ proportional to $b^* - <b^*>$. Plots are drawn vs. angular cell coordinate, $0 < x=\phi/2\pi < 1$, for a series of 10 consecutive time instances, time being counted from top to bottom in both diagrams. Note that in (a) the two maxima of actomyosin distribution slowly vary in both amplitude and position.

-- in an isotropic environment (uniform CA concentration) --

1) nonpolar cells show stochastic small amplitude shape changes, with angular-temporal auto-correlations similar to observed ones, e.g. those in Figure 1(b), cf. (Alt 1993);

2) polarizing cells show stochastically driven lamellipodial activity, analogous to Figure 2(a), with associated stochastic turning as in observed random walks, see Figure 3(a);

-- and in a constant CA gradient --

3) polar cells with relatively high drag resistance (due to strong cell-substratum adhesion) show slow orientation down-gradient, resembling observations of keratinocytes slowly turning away from a pipette releasing phalloidin (Bereiter-Hahn, this volume); in this case that means, locally enhancing actin polymerisation leads to a *phobic response*;

4) polar cells with low drag resistence (weak adhesion) show fast orientation response up-gradient with overshooting, Figure 3(b), leading to zig-zag meandering paths that are commonly found in *chemotaxis* of leukocytes.

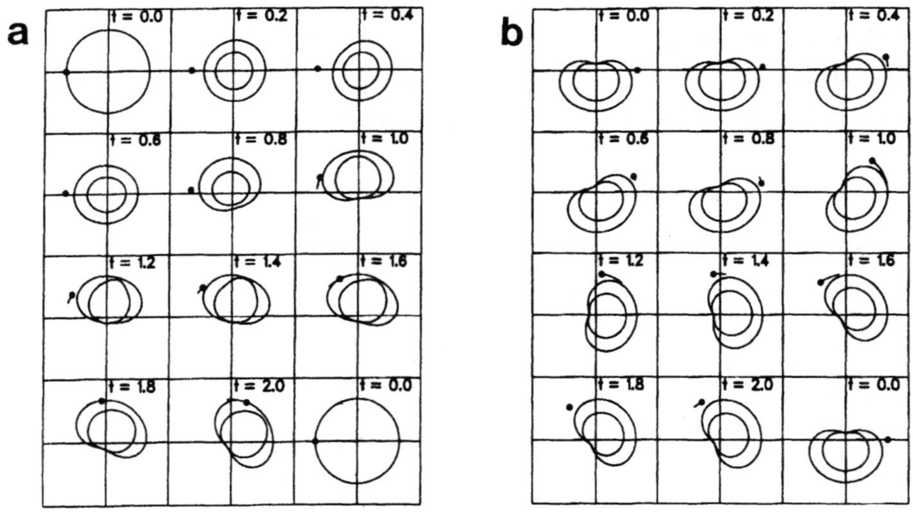

Figure 3. Polar curves of a model cell, showing typical results from computer animations (Tranquillo & Alt, 1993). *Inner curves* describe the cortex radius R_0, *outer curves* the mean cell periphery $R = R_0 + L$. The reference marker shows the position of a reference compartment allowing contributions of cell turning vs. cell repolarization to be discriminated.
(a) Polarization and random turning of an initially rounded cell in an isotropic environment; (b) Fast orientation response of a polar cell in a CA gradient (in positive x-direction); random elements in turning are manifested (t=0.8), mean response yields turning towards the side of enhanced actin assembly, but with overshooting (maximum at t=1.8).

4. Future work

Within the framework of this "most simple morphogenetic model" and its generalizations reproducing the temporal lamellipodial dynamics (Alt & Kaiser, this volume), we are going to investigate other and more detailed "pathways" of signal transduction from receptor stimulation to the cytomechanical response of the cortical actomyosin fluid. One question involves phenomena of local adaptation (e.g. via G-proteins or second messangers). Another idea is that signals do not only influence F-actin assembly, but also contractivity via myosin phosphorylation, disassembly or (dis-)connection to the plasma membrane via actin associated proteins, and receptor-mediated adhesion.

Furthermore, a detailed cytomechanical model of cell body deformation is required that includes flow properties of the interior bulk cytoplasma (Simon & Schmid-Schönbein 1990), anisotropic organization of the cortical F-actin system (Civelucoğlu, this volume), and the dynamics of receptor-mediated substrate adhesion (DiMilla, this volume).

REFERENCES

Alt W (1990) Mathematical models and analysing methods for the lamellipodial activity of leukocytes. In: (N.Akkaş ed.) Biomechanics of Active Movement and Deformation of Cells. Nato ASI Ser. H42. Springer Berlin et al. pp 403-422

Alt W (1993) Cell Motion and Orientation. In: (S.A.Levin ed.) Frontiers of Mathematical Biology. Lect. Notes in Biomath. Vol. 100. Springer Berlin et al. (to appear)

Brosteanu O (1993) Methoden zur Analyse der Lamellipodienaktivität von Leukozyten. Dissertation, Univ. Bonn

Coates TD, Watts RG, Hartmann R, Howard TH (1992) Relationship of F-actin distribution to development of polar shape in human polymorphonuclear neutrophils. J.Cell Biol. 117 : 765-774

Dembo M (1989) Field theories of cytoplasma. Comments on Theor. Biol. 1 : 159-177

Dunn GA, Brown AF (1987) A unified approach to analysing cell motility. J. Cell Sci.Suppl 8 : 81-102

Dunn GA, Brown AF (1990) Quantifying cellular shape using moment invariants. In: (W.Alt, G.Hoffmann eds) Biological Motion. Lect. Notes in Biomath. Vol. 89. Springer Berlin et al. pp 10-34

Simon SI, Schmid-Schönbein GW (1990) Cytoplasmic strains and strain rates in motile polymorphonuclear leukocytes. Biophys. J. 58 : 319-332

Soll D (1988) DMS, a computer-assisted system for quantitating motility, the dynamics of cytoplasmic flow, and pseudopod formation: its application to *dictyostelium* chemotaxis. Cell Motil. Cytoskeleton 10 : 91-106

Tranquillo RT, Alt W (1993) Simulation of chemotactic receptor-mediated leukocyte motility. Model Analysis (to appear) and Video Animation. Minnesota Supercomp. Institute

Observation, Modeling and Simulation of Keratinocyte Movement

Wolfgang Alt
Div. of Theoretical Biology
University of Bonn
Kirschallee 1, D-53115 Bonn
Germany

Hans Wilhelm Kaiser
Dept. of Dermatology
University of Bonn
Sigmund-Freud-Str. 25, D-53127 Bonn
Germany

Summary

Migration of human epithelial keratinocytes (HEK) on glass in culture medium has been observed by video-microscopy. Depending on the concentrations of epidermal growth factor (EGF) and Ca-ions, cells show a variety of characteristic shape deformations. Single cells exhibit (often pulsative) protrusions and retractions of lamellipods, mostly accompagnied by a steady retrograde flow of visible "ruffles", and dynamic changes in concentration and orientation of the cortical cytoskeleton surrounding the "cell body". Both processes induce continuing shape changes of cells, which are either freely migrating or moving as pairs or in smaller groups. Keratinocytes have obvious functions during wound healing, namely to cover free space, to establish effective cell-cell contacts and to form connected cell tissues under high tension. Responsible for these effects is the *dynamic (re-)distribution of a contractile "machinery"* consisting of actin filaments, crosslinking proteins and adhesion receptors, cf. Kaiser et al. (1989, 1993).

In this contribution we restrict mathematical modeling and simulation to the case of a single freely moving cell in order to clarify and quantify the basic biomechanical and -chemical processes underlying the lamellipodial dynamics and cell shape changes. We investigate a "circular" model describing the dynamics of radial extension of the cell periphery, cortical distribution of actin filaments and their peripheral translocation. Numerical simulations provide characteristic *"morphogenetic behavior" of this model system*, showing spatio-temporal patterns which typically resemble those observed in video-micrographs.

1. Derivation of simplified balance equations for the dynamics of an F-actin cortex

In order to describe the basic cytomechanics of keratinocyte movement let us consider the most simple and most regular case of a single cell adhering to a flat substratum, showing a circular cell body of constant radius R_0 and a lamellipodial fringe around the whole cell periphery, see examples presented by Bereiter-Hahn (this volume). Introducing polar coordinates, the radial extension $R(t,\phi) = R_0 + L(t,\phi)$ of the cell periphery in angular direction ϕ is determined by the *radial length* $L(t,\phi)$ *of the lamella region.* Assuming, for simplicity, that along this length the *cortical actin filament system* has a constant *density* $a(t,\phi)$, then we obtain the following simple

NATO ASI Series, Vol. H 84
Biomechanics of Active Movement
and Division of Cells
Edited by N. Akkaş
© Springer-Verlag Berlin Heidelberg 1994

mass balance equation for cortical F-actin:

$$(1) \qquad \partial_t(L\cdot a) + \partial_\phi(L\cdot a\cdot v) = \eta\cdot L\cdot(a^*-a)$$

where $v(t,\varphi)$ denotes the *mean angular transport velocity of cortical F-actin,* a^* *its chemical equilibrium density* and η *its (dis-)assembly rate.* Typically, assuming monomeric G-actin and other actin binding proteins to be present in excess, the time constant is in the range of 10 - 30 sec. With respect to this time scale, protrusion and translocation of the actin cortex can be viewed as a (two phase) creeping flow, with highly viscous F-actin network and less viscous cytosol pouring through it, the dynamics of which is then determined by a pseudo-steady-state force balance, cf. (Dembo, this volume). Supposing a relatively small lamella region, $L << R_0$, we are allowed to split the force balances, first, into a tangential force balance equation for F-actin network:

$$(2) \qquad \Phi\cdot a\cdot v = \partial_\phi(\ \mu\cdot a\cdot\partial_\phi v + \sigma(a) + T(a,L)\ )$$

balancing friction forces at the "cell body" (coefficient Φ), shear stress within the lamella (μ), contractile and swelling stresses $\sigma(a) = \psi/2\ a^2\ \exp(4a/a_{sat}) + s\ \ln(1-a/a_{max})$ and, finally, a compression/dilatation stress due to negative/positive lamella "curvature"

$$(2') \qquad T(a,L) = -\ \partial_\phi(\tau\cdot a\ \partial_\phi L)$$

Here the *cortical surface tension coefficient* (τ) arises as a phenomenological quantity relating the "lumped" tension in the F-actin/plasma-membrane cortex to the F-actin density. Secondly we obtain a radial force balance equation for cytosol flow:

$$(3) \qquad \Phi_1\cdot a\cdot\ \partial_t L = \beta(a) - \gamma(a)\cdot L - \delta\cdot T(a,L)$$

balancing pressures onto the cytosol at the tip of the lamella, namely

a) a pressure drop due to *drag* (coefficient Φ_1) of cytosol flow (speed $\partial_t L$) through the F-actin network (concentration a)

b) a *protrusion pressure* from the "cell body", $\beta(a) = \beta$, or $\beta(a) = \beta\ a\ \exp(-a/a_L)$

c) a *retraction force* $\gamma(a) = \gamma\ a$, e.g. due to retrograde flow of F-actin and thereby induced radial drag force onto the extending plasma membrane

d) a *curvature force*, due to cortical surface tension, same term T(a,L) as eq. (2'), but with additional coefficient (δ)

2. Dynamic protrusion-retraction properties of "model cells" and of keratinocytes

In case of no tangential flow or inhomogeneities, equations (1) and (3) provide a coupled system of ordinary differential equations describing the purely radial dynamics of lamellipodial protrusion (length L) and cortex assembly (total F-actin $A = L\, a$), for constant β :

$$(4) \qquad d_t A = \eta \, (L \cdot a^* - A)$$

$$(5) \qquad d_t L = (\frac{\beta}{A} - \gamma) \cdot L$$

This simple dynamical system always has a stable equilibrium ($a = a^*$, $L = L^* = \beta/\gamma a^*$), but perturbations lead to damped oscillations because, if protrusion and retraction (with rate $\gamma > \eta/4$) is fast compared to F-actin assembly, the lamellipod first extends at low F-actin concentration, before assembly of F-actin keeps up and yields a sufficiently strong retraction force to withdraw the lamellipod. In the contracted lamellipod the condensed F-actin network is disassembled and the cycle of extension and retraction might be repeated. In a model presented at the previous meeting (Alt 1990) similar oscillatory dynamics have been described and simulated in more detail, also including the possibility that sudden protrusion is induced by disruption of the F-actin cortex from the plasma membrane. In the mean time, this phenomenon has been experimentally observed in pseudopodial tips of amoebae, cf. Grębecki (1990) as well as the contribution by A. Grębecki in this volume.

In flat lamellipods as those of human keratinocytes, however, we do not observe such drastic disruption events, but a rather continuous extensions of a broad lamellipodial front, which from time to time starts retracting, over a large peripheral region, until the lamellipodial fringe at this site of the cell periphery is condensed into a narrow cortical border. During this process, or sometimes afterwards, new lamellipodial protrusion arise at other sites, where the F-actin cortex is less dense (cf. Bereiter-Hahn, this volume). In order to desribe these phenomena, the mathematical model presented here concentrates on possible effects that the spatial (re-)distribution of F-actin around the cell periphery has on the dynamics of lamellipodial protrusion and retraction. In the following pictures we summarize some of the characteristic spatial-temporal patterns produced by simulations of the dynamical system of partial differential equations (1)-(3).

2.1 F-actin dynamics and lamellipodial protrusion: Evolution of a steady state.

The following standard parameters $\beta = 1$, $a^*=a_L=1$, $\gamma = 0.36$ $\tau = 2$, $\delta =0.05$, $\mu = 1$, $\psi = 10$ give a "dispersion diagram" for the "initial growth rate" of modes, yielding morphogenetic instability for wave numbers m = 1 to 5.

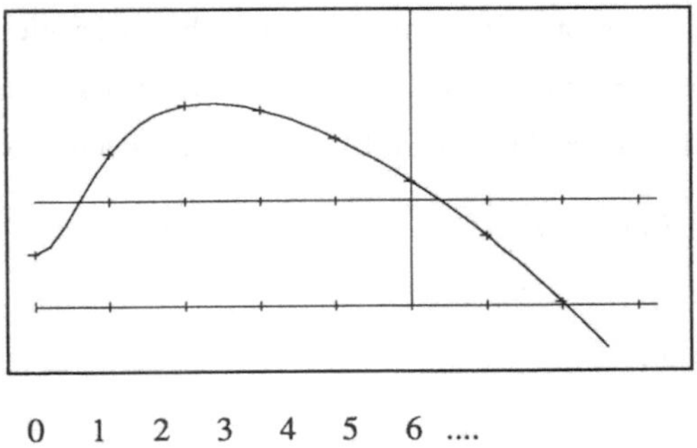

Inner circular plots show outlines of cell body and cell periphery (successive pictures drawn every minute). Plots over middle circle (---) show cortical F-actin distribution rapidly concentrating at position NE (North East) and inducing retraction of the lamella there. Plots over the outer circle (– –) depict velocity of peripheral F-actin transport, here from the remaining lamellipod at SW over both lateral regions towards NE.

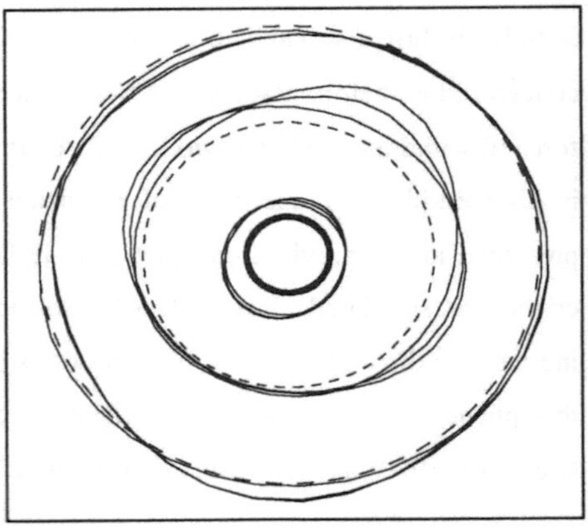

Within the following 2 minutes the F-actin maximum at NE broadens and leads to a complete lamella retraction there, while due to depletion of actin (left and right) two lateral lamella grow out. Simultaneously F-actin concentrates at SW and induces a retraction of the formerly leading lamella there. In this stable bipolar pattern cortical actin, assembled in the two lamella regions, is constantly pulled into both retraction areas (NE and SW).

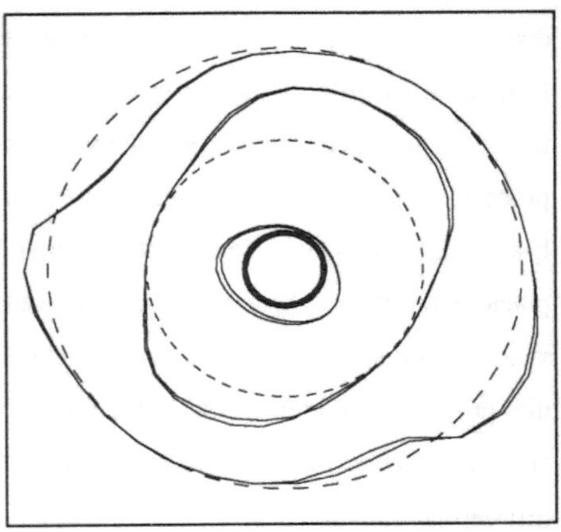

2.2 Reorientation of a polarized cell after protrusion of a competing lamellipod

For even larger protrusion rate, β=3, but lower cortical surface tension, τ=1, (otherwise unchanged parameters) the bipolar pattern changes due to one lamellipodial protrusion forming at the left hand side. After rescaling the plot units we obtain this steady pattern of a polarized cell. However, the leading lamellipod slowly decreases in length and size (while the F-actin concentration increases laterally), until ....

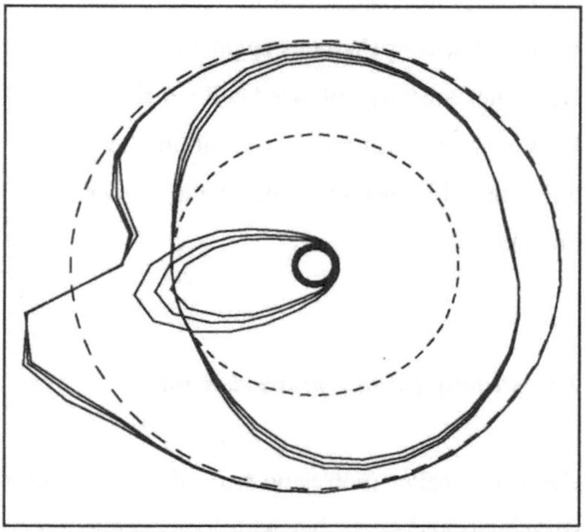

... suddenly a new lamella is formed on the opposite (right hand) side. Reason: There (at W) cortical actin network had been weakened (due to the two strong lateral contraction centers). Then this continues as a self-reinforcing process: lamellipodial protrusion yields dilution of the F-actin concentration, in particular to the left of the growing lamella (where the cortical surface experiences a stronger "negative" curvature), thus .....

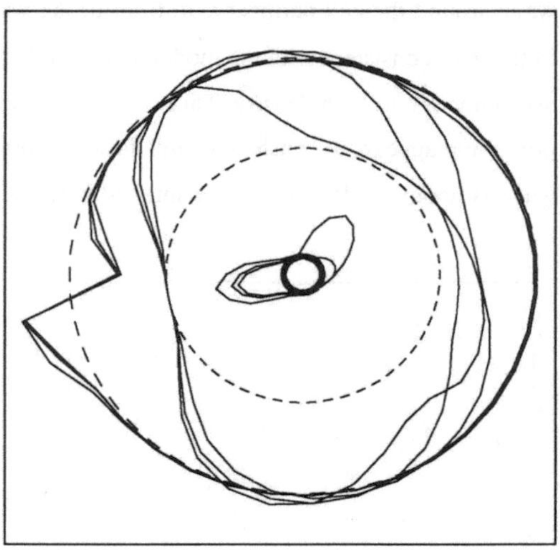

.... inducing a wave of lamellipodial protrusion towards direction N. While enlarging it touches the old lamella and merges into a unique protrusion, which first grows as a broad front, later sharpens again and, finally, creates a newly polarized cell similar as before, about 12 minutes ago, but now in a direction (NW) perpendicular to the former one (W).

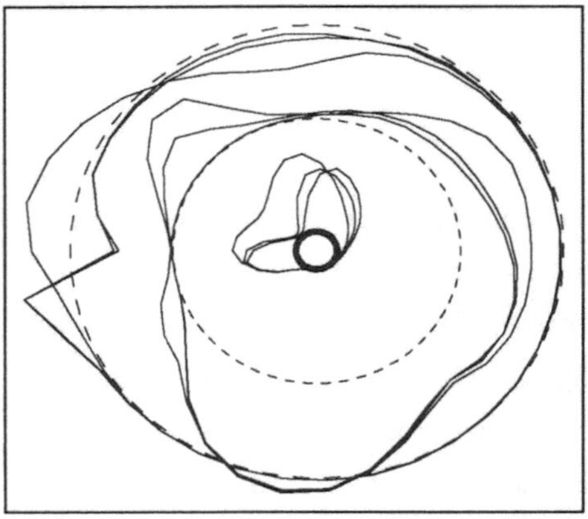

2.3 Steadily circling protrusion waves

For a bit smaller protrusion rate, $\beta = 2$, but doubled cortical surface tension, $\tau = 4$, (other parameters the same, but with lower retraction coefficient $\gamma = 0.27$) we obtain, after some intermediate perturbations, a situation in which one lamellipod forms a travelling protrusion wave around the cell periphery: In front of the wave the F-actin concentration is lower than in the rear, consistent with the modeling principle that protrusion occurs at low cortex density and retraction at high density. Thus, the F-actin concentration wave behind the lamellipodial protrusion appears to "push" the lamella wave forward (clockwise). The speed of the circling wave is about 60 deg/min (Time interval between plots is 1.5 min).

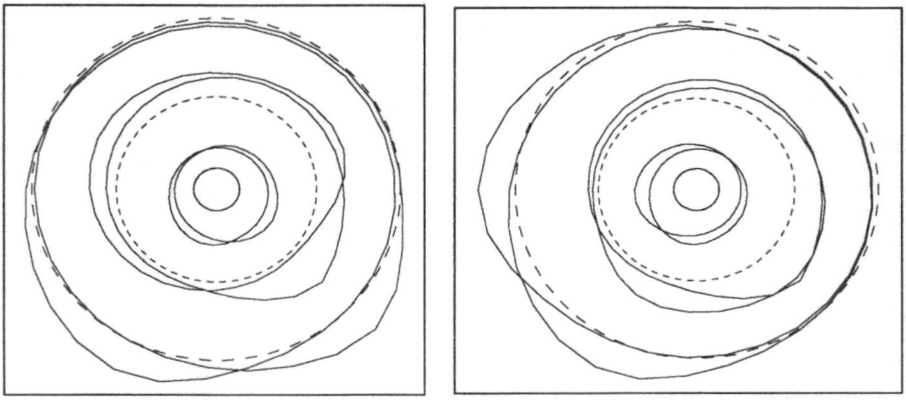

3. Discussion

In principle, all these temporal-spatial patterns of shape changes have been observed by video microscopy of single human kernatinocytes in standard medium [0.1μM Ca^{2+} in MCDB 153, supplements as discribed by Kaiser & O'Keefe (1989)]. The mathematical model, however, only assumes that the fluid-like actomyosin-cortex forms local regions of contraction and thereby, at intermediate sites of lower F-actin concentration, induces pressure driven lamellipodial protrusions (*autonomous cortex weakening hypothesis*); that in an extending lamellipod assembly of F-actin is promoted just by creating new space (into which G-actin can pour); and that the newly assembled actin meshwork induces a retraction force onto the tip of the lamellipod, dependent on its geometry and visco-elastic properties. The model does not require any dynamics in the distribution of G-actin, regulatory proteins or diffusible second messengers as Ca-ions. Although their spatial/temporal alterations would certainly change the protrusion patterns, according to our model hypotheses and simulations we can explain the observed *shape changes and lamella movements* as a purely *autonomous process of dynamic redistribution of the contractile "F-actin machinery"*.

Successive improvement and generalization of the current model should be envisaged in order to desribe some detailed phenomena that are not yet reproduced, by implementation of stochastic perturbations, cf. Tranquillo & Alt (this volume); by additional modeling of pressure balance, (even anisotropic) arrangement of actin filaments and resulting shape deformations at the border between lamella and cell body, and by a detailed analysis of pressure drop, cytosol flow and height changes in extended lamellipods.

REFERENCES:

Alt W (1990) Mathematical models and analysing methods for the lamellipodial activity of leukocytes. In: (N. Akkas ed) Biomechanics of Active Movement and Deformation of cells. Nato ASI Ser. H42. pp 403 - 422

Grębecki A (1990) Dynamics of the contractile system in the pseudopodial tips of normally locomoting amoebae, demonstrated in vivo by video-enhancement. Protoplasma 154 : 98-111

Kaiser HW, O'Keefe E (1989) Adducin: Ca-dependent association with sites of cell-cell contact. J. Cell Biol. 109 : 557-565

Kaiser HW, Ness W, Jungblut I, Kreysel HW (1993) Adherence junctions: Demonstration in human epidermis. J. Invest. Dermatol. 100 : 180-185

Strain Transduction in Non-Excitable Cells.

David B. Jones & Gunnar Leivseth
Laboratory for Cell Biology, Experimental Orthopaedics
University of Münster, Domagkstr. 3 48129 Münster, Germany,

Introduction

Nearly all types of cells, prokaryotic and eukaryotic, can respond to the mechanical environment. This environment may be due to environmental stresses (e.g. wind, or activity, perhaps also gravity) or vibrations (e.g. sound), the difference being merely the amplitudes and the frequencies involved. The response of so called non-excitable cells will be discussed here, since the biophysics of various sensory cells have been well described in the literature. Several transduction mechanisms have been proposed, listed in table 1. Of these hypotheses the first, strain related potentials, will not be discussed here, due to lack of space and for lack of solid support.

Table 1. **Proposed Mechanisms of Mechanical Transduction**

- Strain Related Potentials (SRP)
- Activation of ion channels
  membrane tension
  cytoskeleton connected
- Cytoskeleton coupled PLC
- Membrane tension effects
- Mechanical lever (primary cilium)

It is thought that the loads on the cells cause distortions of the cells or some parts of the cell and this distortion might be transduced into a electrophysiological and/or biochemical response. In some protists, organelles have been adapted into vibration sensors and can elicit evasion or aggression responses. In many multicellular organisms groups of cells can be adapted as vibration sensing organs. For instance in coelenterates special cells act as sensors to release a trigger mechanism controlling the nematocysts for

NATO ASI Series, Vol. H 84
Biomechanics of Active Movement
and Division of Cells
Edited by N. Akkaş
© Springer-Verlag Berlin Heidelberg 1994

prey capture (Golz & Thurm, 1991). Plants can adapt to environmental wind forces and also to gravity (Gehring et al. 1990). Specialised organs exist in animals to transduce small mechanical loads, such as vibrations, and link this transducer to the nervous system. Strains induced by stresses in the swim bladder of fish are detected which can aid in prey capture or predator evasion (Canfield & Eaton, 1990). The sensory hairs of vestibular ampullae (similar to the cochlea) are an example of an organ where small vibrations are amplified by exploiting the production of shear between two surfaces (Nagel et al. 1991). Interaction with the mechanical environment leading to adaptation of tissues is also necessary in non-specialised cells to allow the organism to maintain the correct functional balance of its tissues. This is especially so in skeletal and heart muscle, blood vessels, bone, skin and lung epithelium. Having too much tissue in a high functional state is wasteful. Too little tissue which has low performance means that the organism cannot meet the demands of the environment. A sensing system with feedback hence operates in these tissues to maintain the optimal functioning tissue which is balanced between breakdown and build up. In kidney tubules, the primary cilium is thought to play a role in flow sensing, and might also be the case for endothelial cells, however the primary cilium does not appear to be a flow sensor in osteoblasts or osteocytes since they are extremely rare (SD Doty, personal communication), and hardly protrude from the cell.

Many theories have been advanced as to the biochemical mechanisms by which cells can sense mechanical loading (as opposed to physical mechanisms above). In 1987 a mechanism by which a membrane bound phospholipase C was activated in osteoblasts by defined uniaxial strains was first described (Jones & Scholübbers, 1987), which now appears to be a general mechanism of transduction of mechanical forces into biological response. Using the biochemical responses it is possible to isolate in time, which response is occuring first, and thus gain insight into the physical mechanism. This approach has led us to propose a cytoskeleton-linked PLC transduction mechanism (Jones et al. 1991).

Mechanical loading of bone results in deformation. These deformations vary depending on the type of bone and the direction and amount of load placed upon it. Some people suggest that the piezoelectric and streaming potentials induced in bone during deformation result in physiological responses (Bassett & Becker, 1962, Davidovitch et al. 1984, Gjeslvik, 1973, Lee et al. 1981, Liboff et al., 1987) and some people suggest that hydrostatic forces can elicit cellular responses (Bagi & Burger 1989, Hall, Urban & Gehl, 1991, Burger (1993),in this volume), although at low pressures it is difficult to conceive of a physical mechanism to account for a biological response.

Analysis of the Strain Sensor

What is the nature of the strain sensor in the osteoblast? Is there more than one sensor? What biochemical pathways are induced by strain? These questions are at present unanswered. Many authors now agree that one of the quickest responses to physiological levels of strain, albeit of different amplitudes for different types of cells, is an increase in intracellular free calcium as mentioned above. Jones et al. (1991) and Jones and Bingmann (1992) described the strain related activation of phospholipase C (PLC) and rapid increase in intracellular free calcium (IFC) within 180 milliseconds of applied 3,000μstr in osteoblasts and similar responses have been noted in heart muscel cells (von Harsdorf et al. 1989), in endothelial cells (Ando et al. 1988) and in lung epithelial cells (Wirtz and Dobbs, 1990). Although the exact time response was resloved to within 12 seconds in these last reports. Sadoshima and Izumo (1993) apart from measuring PLC activation also noticed a stretch activation of phospholipase D, (PLD) but 30 minutes after applying strain. The stimulation of other second messengers, such as tyrosine kinases, phospholipase D and $p21^{Ras}$ mitogen activated protein were parameters also not resolved in time. The chain of events therefore remains to be established to deduce the existence of either 1 or more simultaneous pathways. If simultaneous pathways exist, then the question is whether they are linked to the same transducing mechanism. Stretch induced promotion of *c-fos* message reported by Komuro et al. (1990) is most likely linked to PLC-PKC activation (Sadoshima and Izumo, 1993). Hence it is not yet clear whether the transduction mechanism involves one or more mechanisms and one or more pathways. Since Guharay and Sachs (1984) and Lansman, Hallam and Rink (1987) described stretch activated ion channels, many groups have also also investigated this phenomenon and described not only potassium chammels, but also calcium channels. Duncan & Misler (1989) and Davidson, Tatakis & Aurerbach (1990) have investigated mechanosensitive ion channels in osteoblasts. Stretch activated ion channels have also been suggested to play a role in muscle response to stretch (Kirber, Walsh & Singer, 1988). As Morris & Horn (1991) did not find evidence to support the existence of such channels *in vivo* and taking into account the quick release of IFC, the evidence for a direct effect of strain on channel activation perhaps needs reviewing. An indirect effect on potassium channels and calcium channels through IFC release is supported by our data. However other data suggests that in endothelial cells, the ion channels are activated by fluid flow before a change in PLC activity. This point needs further investigation. The possible cascade mechanisms are showm in table 2.

Table 2. **Signal Cascade in Mechanotransduction**

- 1) Transducer
  - one
  - multiple
- 2) 2nd Messenger
  - one
  - multiple (this is most likely)
- Immediate effects (within 1 second).
  - PLC activation
  - Gap junction communiction
  - Ion channel activation
  - Release of signalling molecules
- Intermediate effects within 1-2 minutes
  - Arachidonic acid metabolism
- Late effects. (2 minutes-6 hours)
  - Gene activation
- Secondary effects
  - Effects due to release of translated genes

After the cells have sensed the strain by the 'primary' mechanism(s) a cascade of responses are set in motion.

Conclusions

Strains appropriate to the physiological strain environment of the cells can be reproduced in culture. Different machines that apply homogeneous strains of different frequencies and amplitudes may be constructed to investigate different aspects of the strain response. Analysis of the sequence of events set in motion after strain can help to deduce the nature of the strain sensing mechanism and the biological consequences. The first event, occuring at least 160 milliseconds after the application of strain appears to be the activation of PLC, which might set in motion a sequence of biological mechanisms which co-ordinates many of the physiological responses to the mechanical environment. Other mechanisms, such as the release of immediate signalling molecules, the induction of transcription of growth factor mRNA and their translation into protein is a long term response. Gap junctional communication might co-ordinate the response over the tissie as a whole. The strain sensor appears to be located in the cytoskeleton. No evidence for a significant role of streaming potentials has been discovered. Strains higher than those normally experienced in the body might evoke a specific change in phenotype and induce a type of pathological response.

Acknowledgements

This work was supported by the International BIOSIS society and by the Bertelsmann foundation. We are grateful to Prof. P. Brinkmann. Inst. f. Biomechanik, University of Münster, for help in understanding physical concepts and to Herrn W. Rück in the Biomechanik workshop for translating many of the ideas for equipment into reality.

References

Ando J, Komatsuda T, Kamiya A. (1988). Cytoplasmic calcium response to fluid shear stress in cultured vascular endothelial cells. *In Vitro* **24**,871-877

Bagi, C., & Burger, E.H. (1989). Mechanical stimulation by intermittent compression stimulates sulfate incorporation and matrix mineralization in fetal mouse long-bone rudiments under serum-free conditions. *Calcif Tissue Int* **45**,342-347

Bassett, C.A.L. & Becker, R.O. (1962). Generation of Electric Potentials by Bone in Response to Mechanical stress. *Science* **137**,1063-1064

Binderman, I. Zor, U. Kaye, A.M. Shimshoni, Z. Harell, A. & Somjen D. (1988). The transduction of mechanical force into biochemical events in bone cells may involve activation of phospholipase A2. *Calc. Tiss. Int.* **42**, 261-266

Canfield, J. & Eaton, C. (1990). Swimbladder acoustic pressure transduction initiates Mauthner-mediated escape Nature, 347:760-762, 1990

Davidovitch, Y. Shanfeld, J. L. Montgomery, P. C. Lally, E. Laster, L. Furst, L. & Korostoff E. (1984). Biochemical Mediators of the Effects of Mechanical Forces and Electric Currents on Mineralised Tissues *Calif. Tissue Int.* **36**,86-97

Davidson, R.M. Tatakis, D.W. & Auerbach, A.L. (1990). Multiple forms of mechanosensitive ion channels in osteoblast-like cells. *Eur. J. Physiol. (Pflügers Archiv)* **416**, 646-651

Duncan, R. & Misler, S. (1989). Voltage-activated and stretch-activated Ba++ conducting channels in an osteoblast-like cell line (UMR 106) *FEBS Lett.* **251**,17-21

Gehring, C.A., Williams, D.A. Cody, S.H. & Parish R.W. (1990). Phototropism and geotropism in maize coleoptiles are spatially correlated with increases in cytosolic free calcium *Nature* **345**, 528-530

Gjelsvik A. (1973). Bone Remodeling and Piezoelectricity-I. *J.Biomechanics.* **6**,69-77

Golz, R. & Thurm, U. (1991). Cytoskeleton-membrane interactions in the cnidocil complex of hydrozoan nematocytes *Cell Tiss. Res.* **263**,573-583

Guharay, F. & Sachs F. (1984). Stretch activated K+ channels in muscle cell membranes as mechanotransducers? *J. Physiol. Lond.* **352**, 685-701.

Hall, A.C. Urban, J.P.G. & Gehl, K.A. (1991). The effects of hydrostatic pressure on matrix synthesis in articular cartilage. *J. Orthop. Res* **9**, 1-10

Self-Organization in Biological Systems as a Result of Interaction between Active and Passive Mechanical Stresses: Mathematical Model

Alexander A. Stein
Institute of Mechanics
Moscow University
Michurinsky prosp. 1
Moscow 117192
Russia

Embryonic tissue is the complex system consisting of spatially distributed cells and extracellular material. Structure and properties including mechanical ones of an individual cell are extremely important for understanding of self-organization principles, but the problem of direct derivation of cell medium characteristics from individual cell model is unresolvable practically. Modeling of multicellular system by continuum is reasonable approach, cell interactions being considered as collective effect.

The most general mechanical peculiarity of embryonic continuum is the presence of active component in mechanical stress which can not be reduced to elastic or viscous effects. Those intrinsic active stresses may have different origins: 1)cell growth processes including mitosis, cell wall growth, etc.; 2)stowage and resorption of extracellular material; 3)osmotic processes; 4)active forces exerted by cells or their components including those producing active cell movements. The said above predetermines following general structure of the main equation:

$$\hat{\sigma} = \hat{N} + \hat{E}\,\hat{\varepsilon} + \hat{\mu}\,\hat{e} \qquad (1)$$

$\hat{\sigma}$ is the total stress tensor, $\hat{\varepsilon}$ is the strain tensor, $\hat{e}$ is the strain rate tensor, $\hat{E},\hat{\mu}$ are tensor coefficients (elastic and viscous), and $\hat{N}$ is active stress tensor. The equation (1) is reasonable for small deformations. Substitution of variables in (1) by their differentials is natural generalization.

In mechanical terms, morphogenesis is deformation accompanied by structural changes. Active stresses may be sometimes predetermined by nonmechanical processes only, but they are nevertheless involved into interaction with the passive stresses due to connections which are determined by the equation (1) and usual dynamics relations. In this case, active stress tensor has to be defined as a function of individual coordinates and time, whereas deformations and internal stresses are found as a result of the problem solving. On the other hand, morphogenetic processes on different scale levels are influenced by mechanical stresses (see e.g. reviews: Regirer & Stein, 1985,1987).

Some partial examples of feedback connections were investigated by Odell et al.(1981), Melikhov et al.(1983), Murray et al.(1983), Belintsev et al. (1987).Physical details of such mechanisms acting in the cell medium are not always crucial for its self-organization principles. The dependence of active stress alteration rate $\hat{N}^{\cdot}$ on the total stress tensor is one of the most elementary and simultaneously general law types. We shall assume that appropriate function also includes active stress and some internal parameters χ_i like chemical concentrations as arguments:

$$\hat{N}^{\cdot} = \hat{R}(\hat{\sigma},\hat{N},\chi_i); \qquad i = 1,2,\ldots n \tag{2}$$

The matter of definition of tensor function time derivatives is out of discussion here. The example under consideration below allows just to use partial ones. We also omit viscous term in (1). Viscosity effects can hardly be determinant factors in realization of organization principles.

The equation (2) is abstract enough and may appear as reasonable simplification for different physical mechanisms.

One of such mechanisms is growth rate dependence on mechanical stress. The latter may be reduced by simple assumptions to dependence of inelastic strain rate on total stress (Stein, 1987). It is easily seen that the equation (1) (with the viscous term omitted) may be rewritten in the form

$$\hat{\varepsilon} = \hat{E}^{-1}\hat{\sigma} - \hat{E}^{-1}\hat{N} \tag{3}$$

where $\hat{E}^{-1}$ is inverse tensor to $\hat{E}$. Both hypotheses are obviously equivalent mathematically if $\hat{E}$ is constant. General case resulting in some complication of the laws (2),(3) is not regarded here.

Dependence of cytosceleton transformation rate on total (or passive) stresses is another possible interpretation of the relation (2). Such remodeling leads to collective cell deformations, cell polarization being a typical example.

Active stress rate dependence on total stress leads to appearance of some opportunities for the control of system as a whole in communication with local cell interactions like transport processes. Some of them are connected with steady solution disappearance due to parameter evolution. An alternative is stability loss. Realization of those schemes includes force interaction between parts of the system which develops in response to deformation. The consideration is confined by thin cell layers in this work. Mathematical analysis may be performed for such objects using equations averaged in short direction but retaining some parameters responsible for their inhomogeneity in this direction.

Cell layer on elastic substratum is probably simplest possible example. We assume that the concentration c of a certain generalized regulator is the only additional internal parameter (χ_1=c, n=1) and this parameter is governed by diffusion equation. Strictly speaking, c may have other interpretations, diffusion-like propagation of this characteristic in cell medium being only essential. The dependence of component c generation rate on active and passive stresses is assumed too.

The system of equations for one-dimensional problem

is following:

$$\frac{\partial\sigma}{\partial x} - \lambda u = 0; \qquad \sigma = N + E\,\frac{\partial u}{\partial x};$$

$$\frac{\partial N}{\partial t} = R(N,N_y,\sigma,c); \qquad \frac{\partial c}{\partial t} = \phi(c,N,N_y,\sigma) + D\,\frac{\partial^2 c}{\partial x^2}; \qquad (4)$$

$$N_y + E_y(h-h_0) = 0; \qquad \frac{\partial N_y}{\partial t} = R_y(N,N_y,\sigma,c)$$

Here σ is the tensile (or compressive if negative) axial stress in X direction (see Fig.1), N and N_y are the tensile (or compressive) active stresses in X and Y directions, u is the averaged displacement of cell medium in X direction, h and h_0 are actual and initial local thicknesses of the layer. The elastic coefficients λ, E, E_y and the diffusion coefficient D are all positive. The equations(4) are true if $h-h_0 \ll L$ and have to be modified when deformation in Y direction becomes large.

Fig.1

Boundary conditions corresponding to fixing of the layer margins and their non-permeability for the component c may be used: $u(0,L)=0$, $\partial c/\partial x(0,L)=0$.

Some important properties of the model are seen from investigation of the system (4) in linearized form. All functions may be found now as trigonometrical series in independent variable x with coefficients dependent of time exponentially. Let the system state be uniform in the initial moment. The structure of functions being present in the equation (4) may allow the uniform solution but may not. In the second case, the non-uniform distribution of variables and the complication of object form may develop in determined way. In the first case, the instability of uniform solution may take place. Dispersion equation connecting exponent increment ω with wave number k is easily investigated for discrete set of wave numbers determined by the boundary problem.

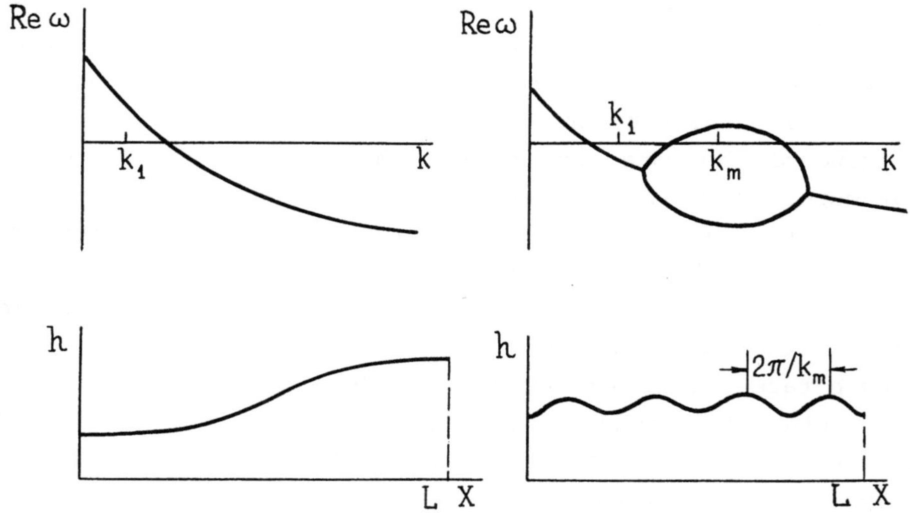

a b

Fig 2

Examples of dispersion curves and stability loss patterns. a) Long wave instability (the simplest case: no diffusion); b) Short wave instability. k_1 is the least wave number, m>1.

Diverse development regimes are possible, most interesting of them taking place for $\beta=\partial R/\partial\sigma>0$. The latter means that tensile stress quickens the development of tensile active stress. Two types of instability are illustrated by dispersion curves and patterns of stability loss for the geometrical variable h (Fig.2). The system behaviour is controled by the parameter β if other parameters are fixed.

The uniform solution is always stable for $\beta=0$ if some reasonable restrictions on other coefficients are imposed; it is also stable for positive β until the latter reaches a certain critical value. Fig. 2a demonstrates instability in the range of small wave numbers including zero. The parameter β evolution to the critical point leads to stability loss of the mode corresponding to the longest wave. If D is small (or equals zero) this type of stability loss is always realized. The other opportunity appears when D is large enough. The range of instability appears as a finite interval of wave numbers not including zero (Fig. 2b). The system may show short wave instability in this case. The critical wave number is proportional to the object length L for long wave instability, but the dependence on L weakens, as this number grows.

This elementary consideration has to be complemented by solving of the more correct problem taking into account slow parameter evolution in time and including nonlinear analysis. Such analysis shows that consolidation of structures developing during stability loss is possible.

There are many other possibilities of self-organization for systems of such kind. Multilayer systems with different growth rates of layers provide for short wave instability due to growth rate dependence on mechanical stresses only (Stein & Logvenkov, 1993). Geometrical form is equal in rights with other characteristics in this model, being the integral part of feedback mechanism. Closed layers with internal pressure manifest wide range of self-organization capabilities due to the bend sensitivity of stress distribution.

REFERENCES

Belintsev BN Beloussov LV Zaraisky AG (1987) Model of pattern formation in epithelial morphogenesis. J Theor Biol 129: 369-394.

Melikhov AV Regirer SA Stein AA (1983) Mechanical stresses as the factor in morphogenesis. Dokl Akad Nauk SSSR (Sov Phys Dokl) 271: 1341-1344.

Murray JD Oster GF Harris AK (1983) A mechanical model for mesenchymal morphogenesis. J Math Biol 17: 125-129.

Odell GM Oster G Alberch P Burnside B (1981) The mechanical basis of morphogenesis. I.Epithelial folding and invagination. Devel Biol 85: 446-462.

Regirer SA Stein AA (1985)Mechanical aspects of growth, development, and remodeling processes in biological tissues Itogi Nauki i Tekhniki VINITI (Advances in Science and Technology of VINITI) (Russ.) Ser. Complex and Special Sections of Mechanics 1: 3-142.

Regirer SA Stein AA (1987) Mechanochemical models of morphogenesis. In Theoretical and Mathematical Aspects of Morphogenesis (Russ) (Presnov YeV Maresin VM Zotin AI, eds) Moscow: Nauka pp 151-161.

Stein AA (1987) A new approach to continual description of volume growth mechanics. The model of growing elastic body. In Biomechanics of Soft Tissues (Cherny GG Ilgamov MA Regirer SA, eds) (Russ.) Kazan pp 90-101.

Stein AA Logvenkov SA (1993) Spatial self-organization of a layer of biological material growing on substratum. Dokl Akad Nauk (Phys Dokl) 328: 443-446.

Biomechanical Events in the Production of Ciliary Movement

Peter Satir
Department of Anatomy and Structural Biology
Albert Einstein College of Medicine
Bronx, New York 10461

Introduction

Ciliary motility consists of a repetitive series of bends that are formed and propagated along the ciliary axoneme. It is clear that the axoneme, the cytoskeletal core of each motile cilium, that consists of the classic 9+2 arrangement of microtubules (*cf.* Sugrue *et al.*, 1991), is responsible for the production of bending and bend propagation. The form of beat in living cilia is matched by that produced by isolated axonemes reactivated by ATP, sometimes with great fidelity. Cilia are found on many different cells of a great variety of single and multicellular organisms. They come in many lengths, ranging from about 5 to over 500 µm and numbers ranging from 1 to about 200 per cell. However, the diameter of the 9+2 axoneme is always the same, about 0.2 µm, suggesting that this is an important parameter for the production of motility. Ciliary bending is also variable, both with regard to amplitude and timing, and the same organelle can exhibit quite different kinds of beat, varying specifically in beat frequency, in overall form or in the spatial direction of net force production. An elegant example is given by sea urchin embryo cilia which spontaneously beat in two different directions - normal and reversed, and whose beat can be recorded and analyzed at high time resolution (Mogami *et al.*, 1993). A second example is given by *Chlamydomonas*, whose so-called 'flagella' beat normally with a ciliary breast stroke, but which reverse, for example during a photophobic response, to beat with what is in effect, undulatory motion. In both of these examples, the motile mechanism of the axoneme remains unchanged, but it is controlled in specific ways, to give the appropriate changes in timing and bend amplitude. The controls work by means of second messengers, notably Ca^{2+} or cAMP, which act directly on axonemal constituents. Any explanation of the mechanism of motility needs to explain how it is possible to change flexural into undulatory motion and back again, or to change the direction of the effective stroke for a short while on a single cilium.

The bend pattern of cilia usually consists of planar bends, which normally alternate in two opposing directions. These are a principal bend, that gives rise to the effective stroke, and a reverse bend that influences the recovery stroke. The bends are separate; as will be discussed, they are probably not part of an oscillatory wave, and in this view it is a major

NATO ASI Series, Vol. H 84
Biomechanics of Active Movement
and Division of Cells
Edited by N. Akkaş
© Springer-Verlag Berlin Heidelberg 1994

misconception to treat them as such. The beat form cannot be fitted by a sine wave; the bends are circular arcs and the regions between them are ruler straight.

Sliding Microtubules

Ciliary motility is driven by microtubule sliding (*cf.* Satir and Sleigh, 1990). The amount of sliding ($\Delta \ell_n$) in micrometers for any doublet is related to the bend produced by the sliding in that:

$$\Delta \ell_n = d_n \Sigma\alpha \qquad (1)$$

where d_n is a constant related to axonemal diameter. The reason that axonemal diameter is invariant for all 9+2 axonemes is that the mechanism by which bending is derived from sliding is evolutionarily conserved. $\Sigma\alpha$ is the amplitude of the bend. For a bend of 100°, a normal principal bend during flexural motion, the displacement of one doublet (N) vs. its neighbor (N+1) is quite small, about 0.1 µm. For a cilium beating at 25 Hz, with the bend formation taking perhaps a quarter of a beat period -ie. 10 msec, the velocity of sliding is approximately 10 μms^{-1}. By labelling doublets with gold beads, Brokaw (1989) has shown that equation (1) holds as axonemes reactivated by ATP actually form bends.

Sliding was confirmed by reactivation of isolated axonemes after a brief treatment with proteolytic enzymes. The digestion destroys constraints such as spoke, central-sheath interactions and regulatory systems that will be discussed shortly. In the presence of Mg^{2+}-ATP, the axoneme telescopes apart. In contrast to the situation in the beating cilium, the sliding of axonemal microtubules on one another continues to completion, producing a displacement of 10 µm or more, so that adjacent doublets end up with only a small overlap zone. The velocity of sliding of a given doublet is constant as the axoneme elongates almost to this point (Takahashi *et al.*, 1982). Velocity depends on ATP concentration with a K_m of 3×10^{-4}M Mg^{2+}-ATP and a Vmax of about 19 μms^{-1}.

Dynein is the Microtubule Motor of the Axoneme

The microtubule motor that powers microtubule sliding during ciliary motility is axonemal dynein. In sliding axonemes, doublet N on whose subfiber A dynein arms are assembled always pushes doublet N+1 (to which those arms transiently attach) toward the (+) end of the axonemal microtubules. All axonemal dyneins and dynein fragments generate force with this polarity. The mechanochemical cycle of axonemal dynein is reminiscent of myosin.

The force produced by one dynein arm is about 1pN (Kamimura and Takahashi, 1981). This is about the same force as is produced by a single kinesin molecule, and it is sufficient to move an axonemal microtubule at maximum velocity for the time that the arm remains attached to the microtubule (the duty cycle: t_S). Vale *et al.* (1992) have shown that the duty cycle of dynein is much shorter than that of kinesin. It lasts perhaps 1-2 msec in an overall mechanochemical cycle of perhaps 33 msec. Conformational changes in a mechanochemical cycle have been modelled by Sugrue *et al.* (1991); a sliding step of 16 nm can be calculated from the differences in position of extended vs. rigor arms.

Continuous sliding of a microtubule for several micrometers at rates up to 19 μms^{-1} requires the action of many different dynein arms, cycling out of phase. Consider an axonemal microtubule moving at 10 μms^{-1} after ATP addition to a protease-treated axoneme. A displacement of one third of a micrometer will take 33 msec, about 1 arm cycle time. Such a displacement would require about 20 steps -ie. 20 independent, asynchronous arm cycles. A typical axonemal doublet possesses 4 outer arms and 3 or 4 inner arms (most people believe 3) per 0.1 µm length, say 400 outer arms that could cycle. These might act cooperatively such that up to 4 outer arms might cycle synchronously as an arm group. Therefore, between 5 and 20% of the arms cycling asynchronously are sufficient to produce the speed and the displacement under discussion. Since sliding continues at the same velocity as the displacement increases to many micrometers, the twenty arms or arm groups could not be localized to any small part of the microtubule except the extreme tip or motion would cease when sliding passed this region. It is highly improbable that the active arms are all at the tip, consistent with *in vitro* motility studies where a single microtubule can glide at constant velocity for hundreds of micrometers over a field of dynein.

The Switch Point Hypothesis

Sliding in normally-bending, intact axonemes occurs at rates that are similar to, or only marginally different from those in the sliding, telescoping axoneme. As discussed further below, it is likely that this sliding is also produced by a small percentage of the dynein arms that are randomly located along the axoneme. However, the duration of unidirectional sliding in the intact axoneme and the displacement during bend formation are only about 1% of the sliding duration and displacement seen in the telescoping axoneme. Further, in a bending axoneme some doublets are moving in the wrong direction, as if they could not be actively translocated by dynein. In these cases, doublet N+1 moves baseward with respect to doublet N. It can be concluded that during bend formation in an intact axoneme, some doublets have active arms, while others do not. Active arms are those that can hydrolyze ATP and complete a mechanochemical cycle; inactive arms are temporarily not cycling. Presumably, when a microtubule is moving in the wrong direction with respect to the polarity of doublet N+1, arm attachment and cycling become negligible.

Now consider the formation of a principal bend by active sliding in a cilium beating at 25 Hz, as before. As $\Sigma\alpha$ changes from 0 to $+100^{o}$ in about 10 msec, some six dynein arm cycle steps will be required for each active doublet. This is consistent with a duty cycle of 1-2 msec, and a maximum of 6% of the arm groups cycling in this stage of beat. When this phase of beat repeats some 40 msec later, it is unlikely that exactly the same arms will be in their duty cycle once again. It seems reasonable to conclude that during each successive active phase on a doublet, a stochastically-determined different set of arms or arm groups is responsible for force generation.

The switch point hypothesis postulates that about half of the axonemal microtubules, ie.- doublets 1-4, have active arms to generate a principal bend, while the other half, doublets 5-9, have active arms to generate a recovery bend. Switching is controlled by a regulatory complex of structures within the axoneme consisting of spokes, links and inner arms, which measures the amount of distortion, turns arm activity off or on and elastically returns the axoneme to a metastable position. Since, according to the hypothesis, opposite halves of the axoneme act independently of one another to form oppositely directed bends, they should be able to be uncoupled experimentally. Lindemann and Goltz (1988) have produced axonemes where principal bend formation is inhibited, but when ATP is added the reverse bend still forms, relaxes and reforms - and vice versa. This is consistent with a doublet activity pattern within the axoneme illustrated in Satir *et al.*(1991). Further evidence in support of the switch point hypothesis is summarized in Satir and Sleigh (1990), where several complications are also mentioned. Earlier versions of the hypothesis do not incorporate the elastic relaxation seen in the Lindemann and Goltz experiments. The switch point hypothesis does not yet specify how bend propagation occurs, but is seems likely that propagation is in part coupled to relaxation.

An attractive feature of the hypothesis is that it solves the problem of how ciliary reversal occurs or how flexural becomes undulatory motion on one axoneme. The amplitude of a new bend depends solely on the 'on-time' of dynein arm activity in the half axoneme after the overall $\Delta \ell_n$ for every doublet returns to zero ($\Sigma\alpha=0$), where:

$$\alpha_p = \nu_p \, t_p \quad \text{and} \quad \alpha_r = \nu_r \, t_r \qquad (2)$$

where α = amplitude of the new principal (p) or reverse (r) bend, ν is the velocity of sliding and t is the 'on-time' of arm activity for the half axoneme. Ciliary reversal is achieved by reversing the 'on-times' of the half axonemes.

To summarize: doublet sliding in the beating axoneme is virtually the same as doublet sliding in an *in vitro* gliding system or in a telescoping, proteolyzed axoneme. All the arms on an active doublet are capable of force generation via ATP hydrolysis and attachment, but in the few msec available for activity, only a few randomly spaced arms or arm groups are actually pushing the microtubule. In the bending axoneme, not all doublets are moving with the proper polarity, so that half of the doublets have inactive arms. A timing mechanism stops the sliding after a few msec and permits a recoil such that the doublets with inactive arms can become active while those previously active are moving with a polarity no longer conducive to arm attachment and force generation. The timing mechanism operates with speeds on the order of an arm cycle, suggesting that different arms actually complete their mechanochemical cycle with each beat. Proteolysis of the axoneme destroys the timing and recoil mechanisms.

Control of Beat Frequency

In the absence of changes in beat form, sliding velocity during bend formation is proportional to beat frequency. At a frequency of 16.7 Hz, where a bend of 100° will be produced in about 15 msec, the sliding velocity will be 6 μms^{-1} vs. 10 μms^{-1} at 25 Hz. At the higher frequency, every phase of beat must be speeded up accordingly. The outer dynein arms primarily regulate beat frequency. Any agent that changes the properties of the mechanochemical cycle of the outer dynein arms will potentially alter beat frequency.

The mechanochemical cycle of an arm can be altered by changing the duty cycle such that the arm remains attached longer and pushes further, changing the step size without changing the duty cycle so that the arm pushes faster, or changing the cycle time, so that the arm pushes more often per unit time. A significant enough population of the 400 outer arms or 100 arm groups must be altered to ensure that the sliding velocity increase is seen with nearly every beat, even though a different set of arms or arm groups normally generates sliding during that beat. Suppose, for example, that the cycle time of an altered arm decreases from 33 msec to 5 msec - then in the 10 msec in which the bend develops at high frequency, this arm will cycle about twice on average, while all the unaltered arms together on average produce four sliding cycles (based on the sliding velocity at low frequency). A small population of altered arms can therefore act as pacemakers.

Many cells control their ciliary beat frequency by increases in intracellular cAMP. In *Paramecium*, monobutyryl-cAMP added to the medium increases forward swimming speed of the living protozoan, while 10 μM cAMP increases the swimming speed of reactivated permeabilized cells (Bonini *et al.*, 1991). In these cells, a 29 kDa axonemal polypeptide (pp 29) is phosphorylated in a cAMP-dependent manner. This polypeptide cosediments with the three-headed 22S dynein that comprises the outer arm of these axonemes, but not with 14S dynein, and it is probably a dynein regulatory light chain.

This enables us to test directly whether cAMP-dependent phosphorylation regulates microtubule sliding velocity as predicted. Hamasaki *et al.* (1991) prepared 22S dynein with and without the phosphorylated 29 kDa light chain (actually thiophosphorylated for technical reasons) and measured sliding velocities in *in vitro* microtubule translocation assays using brain microtubules. When the pp 29 is phosphorylated, sliding is 40% faster. cAMP can be removed and reactivated, permeabilized cells will still swim faster than controls. Only about 2-3% of the 22S dynein has a new experimentally phosphophorylated pp 29 under these conditions; this could be an underestimate, but, if correct, this requires a pacemaker hypothesis such as that discussed above to account for the increase in translocation velocity. A prediction of this analysis is that after cAMP-dependent phosphorylation, pacemaker dyneins are part of the set that is hydrolyzed during arm activity on virtually every active microtubule in every successive beat in an intact axoneme.

Literature references

Bonini NM, Evan TC, Miglietta LAP, Nelson DL (1991) The regulation of ciliary motility in *Paramecium* by Ca^{2+} and cyclic nucleotides. Adv 2nd Messenger and Phosphoprotein Res 23:227-272.

Brokaw CJ (1989) Direct measurements of sliding between outer doublet microtubules in swimming sperm flagella. Science 243:1593-1596.

Hamasaki T, Barkalow K, Richmond J, Satir P (1991) A cAMP-stimulated phosphorylation of an axonemal polypeptide that copurifies with the 22S dynein arm regulates microtubule translocation velocity and swimming speed in *Paramecium*. Proc Nat Acad Sci USA 88:7918-7922.

Kamimura S, Takahashi K (1981) Direct measurement of the force of microtubule sliding in flagella. Nature 293:566-568.

Lindemann CB, Goltz JS (1988) Calcium regulation of flagellar curvature and swimming pattern in Triton X-100 extracted rat sperm. Cell Motil Cytoskel 10:420-431.

Mogami Y, Sekiguchi S, Baba SA (1993) Beating of cilia of sea urchin embryos: A critical comparison of the normal and reversed beating of cilia of isolated cells. J Exp Biol 175:251-266.

Satir P, Goltz JS, Isaac N, Lindemann CB (1991) Switching arrest in cilia and the calcium response of rat sperm: A comparison in *Comparative Spermatology 20 Years After, Vol. 75*, B. Baccetti (ed), Raven Press, N.Y. *p.*377-384.

Satir P, Sleigh MA (1990) The physiology of cilia and mucociliary interactions. Ann Rev Physiol 52:137-155.

Sugrue P, Avolio J, Satir P, Holwill MJ (1991) Computer modelling of *Tetrahymena* axonemes at macromolecular resolution: Interpretation of electron micrographs. J Cell Sci 98:5-16.

Takahashi K, Shingyoji C, Kamimura S (1982) Microtubule sliding in reactivated flagella. Symp Soc Exp Biol 35:159-177.

Vale RD, Malik F, Brown D (1992) Directional instability of microtubule transport in the presence of kinesin and dynein, two opposite polarity motors. J Cell Biol 119:1589-1596.

MICROTUBULE CONVERGING CENTERS: IMPLICATIONS FOR MICROTUBULE DYNAMICS IN HIGHER PLANTS.

Andrew S. Bajer[1]; Elena A. Smirnova[2]; Kolja A. Wawrowsky[3]; Rainer Wolf[3] and Jadwiga Molè-Bajer[1].

[1]Department of Biology, University of Oregon, EUGENE, OR 97403, USA.

INTRODUCTION

The general picture of mitosis is usually based on structure and function of an astral/centrosomal spindle. Indeed, numerous, well documented data demonstrate that the centrosomal complex plays a dominant role in establishing spindle structure, chromosome movement and organization of the interphase cytoskeleton. However, large groups of organisms, notably higher plants, do not have a defined centrosome (rev. Smirnova and Bajer, 1992; Lambert, 1993). Some types of higher plant cells, including endosperm, have an orderly organized cytoskeleton in the form of radiating or/and parallel microtubule (MT) arrays. Moreover, MTs of these cell types have a very pronounced tendency of self-reorganization (Bajer and Molè-Bajer, 1986).

We have studied self-reorganization of MTs during interphase and telophase-interphase transition in endosperm cells and anucleated cell fragments *(Haemanthus katherinae* and *H. nathalensis*). The studies combine analysis of MTs arrangement in immuno-stained cells with observations *in vivo*. MTs in fixed cells were visualized with the immunogold and immunogold-silver methods. In this report, we review the evidence that MT converging centers (MTCCs) are the basic units of cytoskeleton organization in interphase and mitosis in endosperm of the higher plant *Haemanthus* (Smirnova and Bajer, 1993) - i.e. in cell type without the cellulose cell wall and the preprophase band (PPB). Additionally, we provide some new data on MTCCs structure and function. These latest data are based on observations of single MTs in living interphase and mitotic cells of *Haemanthus* endosperm.

HIGHER PLANT MICROTUBULES

We have followed in real time the dynamics and interactions of individual MTs, as well as of MTs within MT arrays, during interphase and mitosis. Video enhanced images (Image I, Universal Imaging System) of MTs *in vivo* were observed with a modified interference contrast microscope (Bajer et al. 1986), recorded on an optical disc at the rate

[2]Moscow State University, Biology Faculty, Department of Cytology and Histology, 119899 MOSCOW, Russia. [3]Biozentrum, University of Würzburg, D-97074 WÜRZBURG, Germany.

NATO ASI Series, Vol. H 84
Biomechanics of Active Movement
and Division of Cells
Edited by N. Akkaş
© Springer-Verlag Berlin Heidelberg 1994

2 frames/s and analyzed frame by frame. MTs *in vivo* were identified by: 1. comparative morphology with immunocytochemistry, 2. chracteristic events of dynamic instability, 3. by their capability to translocate cellular inclusions. All events of dynamic instability like elongation and rapid shortening at constant velocity over a period of time, and transitions between assembly and disassembly, were observed.

Most detailed observations on single MTs in *Haemanthus* were done on interphase cells. Single MTs can be followed at about 10.000x magnification on the TV screen, yet only in very thin (1 μm or less) areas of cytoplasm located at the cell periphery. This limitation is imposed by two facts. The first is the presence of small granular inclusions such as starch grains which obscure detection of MTs in the vast majority of cells. The second is the absence of MTs in many optically favorable areas.

The interphase cytoskeleton is a constantly changing dynamic meshwork, and individual MTs exhibit various types of motility which closely resemble those reported by Cassimeris et al. (1988) in the newt lung epithelium cells. Mechanical deformations of a single MT comprise bending and breaking, resulting either in rapid disassembly or assembly of the free ends. Free MT ends often anneal and form single and stable MTs. Broken fragments are usually very motile and occasionally turn by 180°. We observed such rotation before the fragments fused, similarly as observed *in vitro* (Rothwell et al. 1986). MTs often relocate in a gliding fashion, implying that they slide on a substrate. Parallel MT fragments often slide on each other. Alternatively, only the tip of one MT is translocated along another, stationary MT. *Haemanthus* MTs also have the capability to interact laterally: they may zip together, or unzip and form many unusual morphological structures like loops and rings.

New MTs may form either from unidentified sites or appear at the top or at the side of existing ones, thus giving the impression of "branching". This suggests MT dependent nucleation of new MTs, i.e. the existence of specific nucleating sites associated with MTs. MTs elongate and shorten at very different rates. For example, one MT elongated at 24 μm/min, then elongation slowed down to 18 μm/min, increased up to 30 μm/min, and finally slowed down to 24 μm/min. Another MT elongated at 18 μm/min, then shortened at 30 μm/min, and finally elongated at 30 μm/min.

MTs often associate (zip) laterally as a group, e. g. in the phragmoplast, or individually. The zipping process seems to be responsible for the rearrangement of MTs during the phragmoplast formation. For example, zipping of individual interphase MTs occured at 18 and unzipping at 7 μm/min. The zipping process is often initiated when MTs are about 3 μm apart. This distance is more than 50 times the diameter of two halos seen in electron microscope around single MTs, and about 20 times longer than the resolution of our microscope.

Some rates are faster than those observed in animal cells and in MTs *in vitro*, but somewhat slower than those calculated for mitotic MTs (50 μm/min) on the basis of UV microbeam irradiation (Czaban et al. 1993).

MTCCs: STRUCTURE AND FUNCTION

Multiple centrosome-free asters, bundles or "fan-like" MT configurations may be experimentally induced in animal cells. They are often formed as a result of decrease of critical concentration of tubulin during interphase and mitosis (De Brabander et al. 1986; Harris et al. 1989; Maekawa et al. 1991; Kallajoki et al. 1992). Different patterns of MT arrays (bundles or fan-like structures, as opposed to asters) seem to be dependent on the rates of MT turnover, which differs during interphase and mitosis (Belmont et al. 1990; Verde et al. 1990, 1992). Additionally, in animal cells, the pattern of experimentally induced MT configurations seem to be species-dependent (Harris et al. 1989; Harris and Clason, 1992).

Aster-like MT structures have not been observed in higher plant cells. However, fan-like or cone-like MT configurations, which we define in endosperm of *Haemanthus* as MTCCs, seem to be ubiquitous elements of the higher plant cytoskeleton. Morphologically similar MT foci were described previously as fan-like MT structures or branching MT clusters, which appear abundantly under certain experimental conditions (Falconer et al. 1988; Wasteneys and Williamson, 1989). EM reconstruction of cone-like MT configurations in the cytoplasm adjacent to the cell wall was done in root tips of *Azolla* (Gunning, 1980; Gunning et al. 1978). These foci are the only documented nucleating structures in vegetative cells of higher plants and closely resemble MTCCs in *Haemanthus* endosperm (Bajer and Molè-Bajer, 1982). The latter were found during anaphase-telophase transition in untreated cells (Bajer and Molè-Bajer, 1986), after enhanced MT assembly induced by taxol (Molè-Bajer and Bajer, 1983), during recovery from low temperature shock, and after oryzalin treatment (Molè-Bajer and Bajer, unpubl.).

The organization of fan-like MTCCs on the surface of the nuclear envelope suggests that they initially originate by lateral association, and/or by formation of additional MTs at different sides along one MT, or at one convergent point. This conclusion is based on the observations of MTs *in vivo* and the appearance of MTCCs in various experimental conditions. Thus, any single MT in specific condition may become a starting point for a MTCC.

MTCCs in *Haemanthus* are highly dynamic transitory structures which are instrumental in orderly reorganization of the higher plant cytoskeleton, due to remarkably different dynamics and properties of their converging and diverging ends. These differences seem to be dependent on the intrinsic MT polarity. The converging ends tend to associate laterally, but the diverging ones end-to-end. Cumulative evidence indicates that the converging ends are minus, and divergent ends are plus. This conclusion is based on two sets of data. The first is the direct determination of MT polarity during telophase-interphase transition in *Haemanthus* (Euteneuer et al. 1982). It has been demonstrated that in the phragmoplast the minus ends of MTs point toward the sister nuclei, and the plus ends toward the cell plate. The second is the basic pattern of cytoskeleton organization established during anaphase and telophase.

The pronounced differences of the two ends of MTCCs are of major importance during organization of the interphase and mitotic cytoskeleton. We maintain that MTCCs not only link minus MT ends but they are instrumental in sorting MT polarity and in regulating the direction of their growths and their final length (Smirnova and Bajer, 1992, 1993). The sideways formation of additional MTs within MTCCs is another feature, well expressed during the formation of the phragmoplast and the phragmosome (Sinnott and Bloch, 1941).

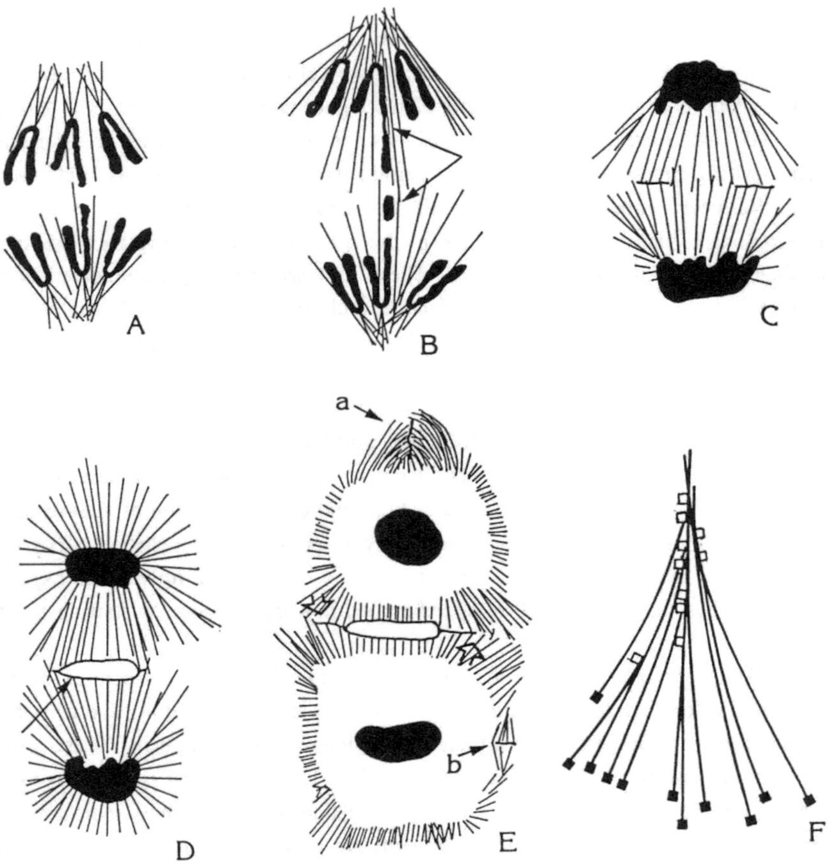

FIG.1. **The formation and development of interphase cytoskeleton**. The onset of this process already takes place in mid-anaphase (A). It begins with the extensive formation of polar MTs. They originate within the two sister half-spindles and between trailing

chromosome arms, but the precise location and mechanism of their formation is not yet known. These MTs elongate toward the equator and increase in number up to late telophase. During their elongation (B-C), in mid-anaphase and early telophase, they exert considerable force directed toward the equator. This force often stretches and may even break chromosome arms (B - forked arrow). In *Haemanthus* endosperm the number of continuous (interzonal) MTs (MTs which are remnants of the metaphase spindle) is extremely variable and often very low. Thus, the phragmoplast is formed predominantly by two sets of polar MTs organized as two inverted cones which meet at the equator (C). A circular cell plate (arrow in D) is formed at the region where these two MT "cones" meet and interdigitate. It begins to form at the phragmoplast edges (arrows in C) and grows centripetally. Occasionally in anaphase but, as a rule, in telophase, the phragmoplast or arrays of polar MTs gradually encircle the whole nucleus (D). Thus these MT radial arrays (radial phragmoplast) are extensions of the primary phragmoplast. These arrays are usually dense and MTs are uniformly distributed (D). During the telophase-interphase transition, the cytoskeleton differentiates: the uniform distribution gradually disappears and the area around the sister nuclei becomes relatively free of MTs which accumulate predominantly at the cell periphery (E). At the same time phragmosome (cell plate extension) forms at the edges of the cell plate (D -hollow arrows) and grows centrifugally until it reaches the cell membrane. Longer MT arrays at the cell periphery often tilt toward each other and form small accessory phragmoplasts (E - arrows a and b). MTCCs (F) which are present during the whole cell cycle, are especially conspicuous during anaphase-interphase transition. They are most pronounced at the edges of the anaphase spindle and phragmoplast. The formation of the phragmosome is the result of lateral propagation of MTCCs. The mechanism of MTCC formation is not well understood. The schematic representation of the structure of MTCC (F) is based on observations *in vivo* and the data of Šmirnova et al. (1993) and Vantard et al. (1990). New MTs form at an angle at the sides of the existing MTs (F- white squares) and elongate at the ends (F - black squares). MTs at the converging end are (-) and are (+) at the diverging end (+).

The development of circular phragmoplasts can be explained by the sideways growth and expansion of MTCCs. Such phragmoplasts are present in syncytial endosperm and sporocytes. Their role is to form cell walls between sister and non-sister nuclei (Brown and Lemmon, 1991). In multinucleate, syncytial and in numerous single cells, MT arrays of each half phragmoplast expand radially and finally encircle the sister nuclei. Consequently, the primary cytoskeleton of telophase and early interphase cells is formed by bi- and monopolar phragmoplasts, and MTs obviously retain their predicted polarity. The circular phragmoplast MTs are directly transformed into interphase cytoskeleton. Thus intrinsic the polarity of interphase MTs is established during anaphase-telophase transition.

MTCCs, MTFTs AND SPINDLE ORGANIZATION

The significance of MT configurations shaped as MTCCs during the cell cycle is clearly revealed in the organization of the mitotic spindle in *Haemanthus* endosperm. During the whole course of mitosis, kinetochore fiber complexes (Bajer and Molè-Bajer,1986) are organized as MT fir trees (MTFTs), which are in fact modified MTCCs. When laterally associated with each other, they form a half-spindle. The kinetochore fiber complex is composed of parallel MTs attached to the kinetochore which represent the trunk of the MTFT, and of skew non-kinetochore MTs, the branches. Such an organization of the kinetochore fiber and the spindle, although still not widely recognized, seems to be an ubiquitous structure of standard higher animal and plant cells. It has been documented not only in *Haemanthus*, but also in other organisms (Fuge and Falke 1991; Inoué, 1988; Palewitz, 1988).

MTFT organization of the kinetochore fiber has been shown to be remarkably functional (Bajer, 1990). In metaphase, skew MTs of the kinetochore fiber of one half-spindle point to the equator where they may interdigitate with MTs of the sister half-spindle. Polar MTs, which represent branches of MTFTs, are continuously formed within the half-spindle from prometaphase through telophase. When chromosomes move, they elongate predominantly toward the equator, but also at the polar region. Consequently, during 2/3 of anaphase, the half-spindle as a whole does not shorten, i.e. anaphase A and B occur at the same time, and disassembly and assembly of the kinetochore fiber also occur simultaneously. Thus, MTFT is supposed to be the basic unit involved both in kinetochore-to-pole movement and separation of the poles (spindle elongation). Polar MTs elongate and increase in number after mid-anaphase, and they become major components of the phragmoplast. Thus, the intrinsic MT polarity of the phragmoplast is already determined in metaphase.

The "fir tree" organization of the kinetochore fiber remained elusive because the present concepts of spindle structure and function are largely based on fragmentary studies with the electron microscope and on observations with the polarizing microscope. Parallel arrays of MTs drew most of the attention, whereas the skew MTs and their massive assembly in late anaphase in *Haemanthus* (Inoué et al. 1985) was not detected, because these MTs are often interlaced under 45° with their optical activities canceling each other. The percentage of skew MTs was also grossly underestimated by electron microscopy. Thus, the generally accepted image of kinetochore fiber and spindle organization based on electron and polarizing microscopy turned out to be highly inadequate.

CONCLUSIONS

Most of the present conceptual understanding of spindle organization and theories of chromosome movements are derived from the studies of spindles of higher and lower organisms where the poles are well-defined due to the presence of a centrosomal complex. All these spindle types are often considered equivalent both structurally and functionally. Such a generalization, however, is not justified. In the widely studied diatoms, for instance, kinetochores reach the poles already in prometaphase, and these lower organisms develop a pronounced central spindle. The latter has never been documented in the typical mitosis of higher organisms (Mazia, 1978). Thus, conclusions derived from one system often are not applicable to another one.

We believe that MTCCs are ubiquitous elements of the higher plant cytoskeleton. They are highly dynamic structural units which regulate the reorganization of the MT cytoskeleton and reflect specific features of MT dynamics. They arise due to lateral association combined with nucleation and attachment of additional MTs. Their development is an expression of a self-reorganizing system of MT turnover (Bajer et al. 1993), which might have been developed in higher plants as a functional substitute for localized centrosome(s). The presence and role of MTCCs in other groups of organisms remains unknown. Supported partly by NIH (GM 37543-01A) to ASB.

ACKNOWLEDGMENTS. We thank Mr. Steve Crowe (Dept. Biol. University of Oregon)for editorial help. Supported partly by NIH (GM 37543-01A) to ASB.

Literature

Bajer AS, Molè-Bajer J (1982) Asters, poles and transport properties within spindle-like microtubule arrays. Cold Spring Harbor Symp 46:263-281

Bajer AS (1990) The elusive organization of the spindle and the kinetochore fiber: a conceptual retrospect. Adv in Cell Biol 3:65-93

Bajer AS, Molè-Bajer J (1986) Reorganization of microtubules on endosperm cells and cell fragments of the higher plant Haemanthus in vivo. J Cell Biol 102:263-281

Bajer AS., Smirnova EA, Molè-Bajer J (1993) Vig BK (ed.) Chromosome segregation and aneuploidy. NATO ASI Series H 72

Bajer AS, Sato H, Molè-Bajer J (1986) Video microscopy of colloidal gold particles and immuno-gold labelled microtubules in improved rectified DIC and epi-illumination. Cell Struc Func 11:317-330

Belmont LD, Hyman AA, Sawin KE, Mitchison TJ (1990) Real-time visualization of cell-cycle dependent changes in microtubule dynamics in cytoplasmic extracts. Cell 62:579-589

Brown, RC, Lemmon, BE (1991) Lloyd CW (ed.) The cytoskeletal basis of plant growth and form. London Acad. Press

Cassimeris L, Pryer NK, Salmon ED (1988) Real-time observations on microtubule dynamic instability in living cells. J Cell Biol 107:2223-2231

Czaban BB, Forer A, Bajer AS (1993) Ultraviolet microbeam irradiation of chromosomal spindle fibers in Haemanthus katherinae endosperm. I. Behavior of the irradiated region. J Cell Sci 105:571-578

De Brabander M, Geuens G, Nuydens R, Willebrords R, Aerts F, De Mey J, McIntosh JR (1986) Microtubule dynamic during the cell cycle: the effect of taxol and nocodazole on the microtubule system of PtK2 cells at different stages of mitotic cycle. Int Rev Cyt 101:215-274

Euteneuer U, Jackson WT, McIntosh JR (1982) Polarity of spindle microtubules in Haemanthus endosperm. J Cell Biol 94:644-653

Falconer MM, Donaldson G, Seagull RW (1988) MTOCs in higher plant cells: an immunofluorescent study of microtubule assembly sites following depolymerization by APM. Protoplasma 144:46-55

Fuge H, Falke D (1991) Morphological aspects of spindle fibers in Mesostoma: "microtubular fir-tree" structures and microtubule association with kinetochores and chromatin. Protoplasma 160:39-48

Gunning BES (1980): Spatial and temporal regulation of nucleating sites for arrays of cortical microtubules in root tip cells of the water fern Azolla pinnata. Eur J Cell Biol 23:53-65

Gunning BES, Hardham AR, Hughes JE (1978) Evidence for initiation of microtubules in descrete regions of the cell cortex in Azolla root-tip cells, and a hypothesis on the development of cortical arrays of microtubules. Planta 143:161-179

Harris PJ, Clason EL, Prier KR (1989) Tubulin polymerization in unfertilized sea-urchin eggs induced by elevated temperature. J Cell Sci 93:9-17

Harris PJ, Clason EL (1992) Conditions for assembly of tubulin-based structures in unfertilized sea urchin eggs. Spirals, monasters and cytasters. J Cell Sci 102:557-567

Inoué S (1988) The living spindle. Zool Sci 5:529-538

Inoué, S., J. Molè-Bajer, and A. S. Bajer (1985) (edts) De Brabander M and De Mey J Microtubules and microtubule inhibitors. Elsevier, Amsterdam

Kallajoki M, Weber K, Osborn M (1992) Ability to organize microtubules in taxol-treated mitotic PtK2 cells goes with the SPN antigen and not with the centrosome. J Cell Sci 102:91-102

Lambert A-M (1993) Microtubule-organizing centers in higher plants. Curr Opin Cell Biol 5:116-122

Maekawa T, Leslie R, Kuriyama R (1991) Identification of a minus end-specific microtubule-associated protein located at the mitotic poles in cultured mammalian cells. Eur J Cell Biol 94:255-267

Mazia D. (1978) Origin of twoness. ICN-UCLA Symp Mol Cell Biol 22:1-14

Molè-Bajer J, Bajer AS (1983) The action of taxol on mitosis. Modification of microtubule arrangements of the mitotic spindle. J Cell Biol 96:527-540

Palevitz BA (1988) Microtubular fir-trees in mitotic spindles of onion root tips. Protoplasma 142:74-78

Rothwell SW, Grasser WA, Murphy DB (1986) End-to-end annealing of microtubules in vitro. J Cell Biol 102:619-627

Sinnott HB, Bloch R (1941) Division in vacuolate plant cells. Am J Bot 28:225-232

Smirnova EA, Bajer AS (1992) Spindle poles in higher plant mitosis. Cell Mot Cytosk23:1-7

Smirnova EA, Bajer AS (1993) Microtubule converging centers and reorganization of the interphase and mitotic cytoskeleton in higher plant Haemanthus. Cell Mot Cytosk (in press)

Verde F, Labbé J-C, Dorée M, Karsenti E (1990) Regulation of microtubule dynamics by cdc2 protein kinase in cell-free extracts of Xenopus eggs. Nature 343:233-238

Verde F, Dogterom M, Stelzer E, Karsenti E, Leibler S (1992) Control of microtubule dynamics and length by cyclin A- and cyclin B-dependent kinases in Xenopus extracts. J Cell Biol 118:1097-1109

Wasteneys GO, Williamson RE (1989) Reassembly of microtubules in Nitella tasmanica: assembly of cortical microtubules in branching clusters and its relevance to steady-state microtubule assembly. J Cell Sci 93:705-714

Elastic Properties of Layered Membranes and Their Role in Transformations of Cellular Shapes

Saša Svetina and Boštjan Žekš
Institute of Biophysics, Medical Faculty and "J.Stefan" Institute
University of Ljubljana
Lipičeva 2, 61105 Ljubljana
Slovenia

Biological membranes are in general layered structures. Phospholipid bilayers as their basic structural unit are composed of two opposing phospholipid monolayers. Many biological membranes have as an additional layer a bilayer linked two-dimensional intracellular network of skeletal proteins. An extracellular matrix can also be considered as a separate layer parallel to other membrane layers. In the work presented here it is taken that membrane layers are in contact but unconnected in the sense that they are free to slide one by the other. Each layer can thus establish its lateral elastic equilibrium in an independent manner. Such physical picture of biological membranes has structural grounds. The two monolayers of a phospholipid bilayer are in contact because of the hydrophobic effect and are unconnected as there are no direct bonds between their molecules. Membrane cytoskeletons are in general in contact with the phospholipid part of the membrane by being attached to it by bonds to the integral membrane proteins. When these proteins can move laterally in the phospholipid milieu sufficiently freely, the cytoskeleton can also be considered as an unconnected membrane layer.

In this paper we will briefly review the work on the effects of the layered structure of membranes on the behavior of vesicular structures such as phospholipid vesicles or cells. We will focus on several unique properties of layered membranes with closed surfaces which we believe represent an important basis for many biological processes. First a general expression for the elastic energy of a closed multilamellar membrane will be presented. Then an overview of some possible shapes which are the result of the tendency of a freely suspended vesicle to attain the state of the minimum elastic energy will be given. We shall discuss the modifications of these shapes due to external forces exerted on a vesicle, and a vesicle (or a cell) squeezed between two plates will be used as an example. A possible chemo-mechanical mechanism of energy transduction will be indicated.

The elastic properties of a closed membrane composed of n layers in

NATO ASI Series, Vol. H 84
Biomechanics of Active Movement
and Division of Cells
Edited by N. Akkaş
© Springer-Verlag Berlin Heidelberg 1994

close contact, but unconnected, are governed by the elastic properties of individual layers and the geometrical constraints which these n layers impose on each other by being in contact. Here we shall consider (Svetina and Žekš 1992) a simple laterally homogeneous membrane of a thickness much smaller than the vesicle dimensions. We shall take that the distances between the neutral surfaces of the layers (h_{jk} for the distance between the j-th and the k-th layer) are the same over all the membrane. We shall also assume that the elastic energy of an individual closed layer (denoted by the index m) is simply given by the sum of the area expansivity term and the bending term, respectively:

$$W_m = \frac{1}{2}\,\frac{K_m}{A_{o,m}}(A_m - A_{o,m})^2 + \frac{1}{2}\,k_{c,m}\int(c_1 + c_2 - c_{o,m})^2\,dA_{o,m} \quad . \tag{1}$$

In the area expansivity term A_m is the area of the neutral surface of the m-th layer, $A_{o,m}$ the equilibrium value of this area, and K_m the area expansivity modulus. In the bending term c_1 and c_2 are the two principal curvatures (reciprocals of the principal radii of curvature), $c_{o,m}$ is the spontaneous curvature (Helfrich 1973) of this layer, and $k_{c,m}$ its bending modulus. Integration is over the whole area of the layer neutral surface. The Gaussian contribution to the bending energy is not included in Eq. 1 because here we are only interested in vesicles of a spherical topology, and for a given topology this contribution is constant.

The total elastic energy for the n-layered membrane is the sum of expressions (1) for all n layers. An expression for this energy (W_{el}) can be obtained by taking into account the constraints on the layer areas due to constant interlayer spacing

$$A_j = A_k + h_{jk}\,C \quad . \tag{2}$$

Here A_j and A_k are areas of the j-th and k-th layer, respectively, and C is the integral of the mean curvature over the whole membrane area:

$$C = \int(c_1 + c_2)\,dA_o \quad . \tag{3}$$

Because there are n - 1 independent constraints defined by a set of equations 2, there are in addition to the membrane bending only two independent elastic deformational modes. An appropriate choice for them are the area of the neutral surface of the membrane A and the integral of the mean curvature C, where the position of the membrane neutral surface is defined in such a way that this independence is fulfilled. Then up to a constant term the elastic energy of the n-layered closed membrane reads

$$W_{el} = \frac{1}{2}\frac{K}{A_o}(A - A_o)^2 + \frac{1}{2}\frac{k_r}{A_o}(C - C_o)^2 + \frac{1}{2}k_c\int (c_1 + c_2 - c_o)^2 dA_o . \quad (4)$$

The three contributions to W_{el} comprise the membrane area expansivity term, the non-local bending (relative stretching) term and the local membrane bending term. The corresponding elastic moduli (K, k_r, k_c) as well as the equilibrium area of the neutral surface (A_o), the equilibrium value of the integral of the mean curvature (C_o), and the spontaneous curvature (c_o) can be expressed in terms of the elastic moduli of the constituent layers, their areas, their spontaneous curvatures, and the distances between their neutral surfaces (Svetina and Žekš 1992). For instance, the area expansivity modulus K and the local bending modulus k_c are simply the sums of the respective moduli of all the layers.

A specific contribution to the elastic energy of closed layered membranes is the non-local bending term in Eq. 4. A notion of this term has been developed for a bilayer membrane by Evans (1974) and Helfrich (1974). The derivation of Eq. 4 for a bilayer has been worked out by Svetina et al. (1985). Due to Eq. 2, the non-local bending term in Eq. 4 can be in this case expressed equivalently in terms of the difference between the two areas ΔA, i.e. $\Delta A = A_2 - A_1 = hC$ with h_{21} denoted by h.

A membrane enclosing a given volume smaller than the volume of the sphere with the same area can be shaped in an infinite number of ways. We assume that a vesicular structure attains the shape at which the membrane elastic energy, given by Eq. 4, has a minimum value. In the following we will first briefly review the present knowledge about the so obtained shapes, and about shape transformations of unsupported flaccid vesicles and cells.

A general shape behavior of closed vesicular systems can be transparently understood on the basis of the results obtained by studying the properties of the bilayer couple system, i.e. the properties of a closed membrane composed of two unconnected layers with uncompressible areas. In view of Eq. 4 the bilayer couple system corresponds to the values of the elastic moduli K and k_r being infinite. In this case the equilibrium shapes correspond to the extrema of the membrane bending energy obtained by the variational procedure where the constraints of constant area of the membrane neutral surface (A), constant difference between the areas of the two layers (ΔA) and constant volume (V) are taken into consideration. The result of such a variation does not depend on the membrane area nor on the bending constant. Therefore the obtained shapes depend only on the values of two geometrical parameters which are the relative volume ($v = V/V_{sph}$ where $V_{sph} = 4\pi R_s^3/3$ with $R_s = (A/4\pi)^{1/2}$) and the relative difference between the two areas ($\Delta a = \Delta A/\Delta A_{sph}$ where $\Delta A_{sph} = 8\pi h R_s$).

It has been shown by the variational analysis of the bilayer couple system (Svetina and Žekš 1989) that at any given v and Δa, there is an infinite number of solutions corresponding to different shapes and different bending energies. By varying v and Δa continuously, each of these shapes changes continuously. However, such continuous shape changes can in general occur only within certain regions in the v/Δa phase space. Accordingly, classes of shapes can be defined, where a class is defined to comprise all shapes of the same symmetry characteristics that can be obtained in the described continuous manner. Recently some of such classes have been characterized in greater detail (Svetina and Žekš 1989, Seifert et al 1991, Heinrich et al 1993). The stability analysis of shapes of these classes indicates (Heinrich et al 1993) that the only stable shape is the shape with the lowest energy at given values of v and Δa.

In order to help visualize the effect of Δa on vesicle shapes, examples of equilibrium shapes of a vesicle with the relative volume v = 0.85 are shown in Fig. 1. These shapes belong to different shape classes which exist within the central region of the Δa values (Δa ≃ 1). At lower Δa values, shapes belong to the class of axisymmetric cup shapes (Fig. 1, example b). By

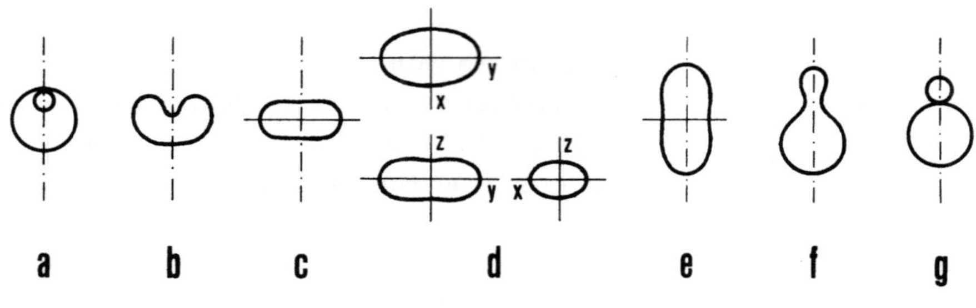

Fig. 1. Cross-sections of stable shapes obtained by the bilayer couple model for a vesicle with the relative volume v = 0.85 and with the relative differences between the areas of the two membrane layers equal to 0.665 (a), 0.93 (b), 1.03 (c), 1.07 (d), 1.085 (e), 1.17 (f) and 1.30 (g). Classes to which these shapes belong are indicated in the text. The axes of the axisymmetric shapes are marked by broken lines and the symmetry planes are marked by full lines. The non-axisymmetric shape (d) is depicted by its front, top and side views.

increasing Δa, at a given value of this area difference these shapes continuously transform into discoidal shapes which are also axisymmetric but have in addition an equatorial mirror symmetry (c). By still increasing Δa, there is a continuous transition into a class of non-axisymmetric shapes involving three mutually perpendicular symmetry planes (d). Then these shapes continuously transform into shapes of a cigar class which are again axisymmetric and have an equatorial mirror symmetry (e). By further increasing Δa, there is a continuous transition into a class of pear (egg) shapes (f). In Fig. 1 two limiting shapes at the chosen relative volume are also shown which are the shape with the lowest Δa value of the cup class (a) and the shape with the largest Δa value of the pear class (g). Such limiting shapes represent one type of class boundaries. Another characteristic class boundary in the bilayer couple model are the lines in the $v/\Delta a$ diagram of the above mentioned continuous transitions between the shapes of classes of different symmetry.

The bilayer couple model has been introduced in a qualitative manner to interpret the drug induced red blood cell shape transformations (Sheetz and Singer 1974). Its theoretical predictions (Svetina and Žekš 1989) have been recently also successfully applied in interpreting temperature (Käs and Sackmann 1992) and phospholipid translocation (Farge and Devaux 1992) induced shape transformations of phospholipid vesicles. The predicted limiting shapes represent a proper basis for the interpretation of physical basis of the budding processes in phospholipid vesicles (Käs et al 1993), and also of some aspects of the biologically important membrane vesiculation and fusion processes (Svetina et al 1988, 1993). A particularly important property which emerged from the analysis of the bilayer couple system is the symmetry breaking behavior of layered membranes. It could serve as the physical basis for the establishment of order in biological systems. An establishment of cellular polarity is a typical example (Svetina and Žekš 1990, 1991).

As already noted the bilayer couple system is the limiting case ($k_r \to \infty$) of a more general situation defined by Eq. 4 and the accompanying requirement of constant vesicle volume. It can be shown that the shapes obtained by variation of Eq. 4 differ from the bilayer couple shapes only in their energies (Svetina and Žekš 1989) and stability characteristics (Heinrich et al 1993), which correspondingly affect the class pattern of equilibrium shapes in the $v/\Delta a$ phase space. At least in phospholipid membranes, the non-local bending contribution to the elastic energy is an important shape determinant. This can be concluded from the measured value for the non-local bending elastic modulus being only about three times larger than the value of the local bend-

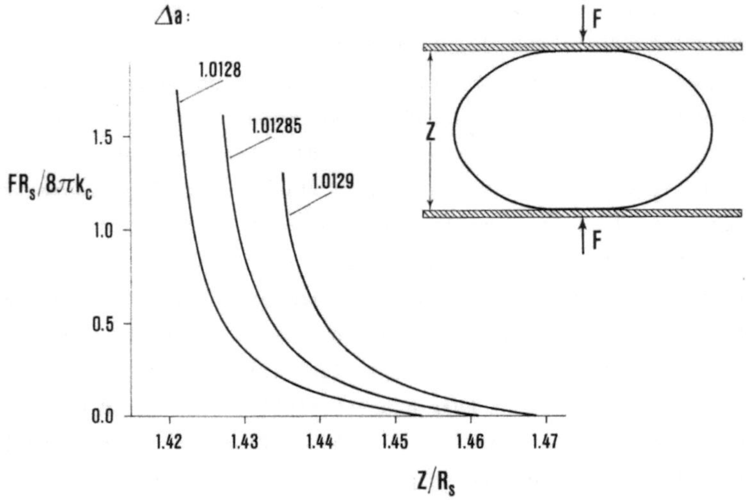

Fig. 2. The dependence of the force squeezing a vesicle (F) on the distance between the plates (Z) at three values of the relative area difference Δa. The system is schematically represented in the inset. Force and distance are given in relative units. The relative volume of the vesicle is 0.95. The vesicle belongs to the class of discoid shapes. Shapes are calculated by solving the Euler-Lagrange equations obtained by varying the free energy of the system at constant vesicle volume, constant membrane area, constant difference between the areas of the membrane layers, and constant distance between the plates.

ing elastic modulus (Waugh et al 1992).

Shapes of freely suspended vesicles and cells are modified under the influence of external forces acting on them. The shape determination in this case is based on varying the free energy of the system which, in general, involves the membrane elastic energy (Eq. 4) and the work performed on a cell by the external forces. Along these lines we treated the shape of the vesicle in an external electric field (Žekš et al 1990), and of a vesicle aspirated partly into a pipette and involving a microtube produced by an external force (Božič et al 1992). A linear response of vesicular structures to mechanical stresses was studied by Peterson (1992). Here, as an example, we determine within the bilayer couple model the shape of a vesicle squeezed between the two plates (Fig. 2, the inset). The equilibrium state of this system can be calculated by varying the sum of the membrane bending energy and the potential energy of the force squeezing the vesicle. The resulting force/thickness dependence for a vesicle with the relative volume $v = 0.95$ and for several values of the relative area difference Δa, is given in Fig. 2. It can be noted that for each Δa value, below a certain thickness the force increases very steeply. This can be related to the decreasing flaccidness of a vesicle.

We conclude by noting that the presented approach represents a link between the cellular processes at the molecular level and the macroscopic cell shape behavior. Essentially this link is represented by the parameters of the elastic state of the individual membrane layers defined by Eq. 1, which are in general expected to be under a metabolic control. For instance, the equilibrium areas of individual phospholipid monolayers are determined by their composition and the areas of layer components, and it is known that in many cells the monolayer compositions are biochemically regulated (Schroit and Zwaal 1991). It is also important to realize that macroscopically, a shape is determined by a relatively small number of membrane parameters which in the described manner depend on the microscopic parameters. Consequently, the same shape can result from many different microscopic mechanisms.

In view of this discussion and the results presented in Fig. 2, a possible chemo-mechanical energy transduction mechanism based on the layered membrane structure can be indicated. The equilibrium area difference ΔA_o (or in a general case the parameter C_o appearing in Eq. 4) could be directly affected by the cell metabolism. On the other hand at a given constant force F, the distance between the two plates depends on ΔA_o. Thus a metabolically driven increase of ΔA_o can perform work by separating the plates.

Acknowledgement

This work was supported by The Ministry of Science and Technology of the Republic of Slovenia grants P3-0244-381 and P1-0108-106.

References

Božič B, Svetina S, Žekš B and Waugh RE (1992) The role of lamellar membrane structure in tether formation from bilayer vesicles. Biophys J 61:963-973

Evans EA (1974) Bending resistance and chemically induced moments in membrane

bilayers. Biophys J 14:923-931
Farge E and Devaux PF (1992) Shape changes of giant liposomes induced by an asymmetric transmembrane distribution of phospholipids. Biophys J 61:347-357
Heinrich V, Svetina S and Žekš B (1993) Nonaxisymmetric shapes in a generalized bilayer-couple model and the transition between oblate and prolate axisymmetric shapes. Phys Rev E 48:3112-3123
Helfrich W (1973) Elastic properties of lipid bilayers: theory and possible experiments. Z Naturforsch 28c:693-703
Helfrich W (1974) Blocked lipid exchange in bilayers and its possible influence on the shape of vesicles. Z Naturforsch 29c:510-515
Käs J and Sackmann E (1992) Shape transitions and shape stability of giant phospholipid vesicles in pure water induced by area-to-volume changes. Biophys J 60:825-844
Käs J, Sackmann E, Podgornik R, Svetina S and Žekš B (1993) Thermally induced budding of phospholipid vesicles - a discontinuous process. J Phys II France 3:631-645
Peterson MA (1992) Linear response of the human erythrocyte to mechanical stress. Phys Rev A 45:4116-4131
Schroit AJ and Zwaal RFA (1991) Transbilayer movement of phospholipids in red cell and platelet membranes. Biochim Biophys Acta 1071:313-329
Seifert U, Berndl K and Lipowsky R (1991) Shape transformations of vesicles: phase diagram for spontaneous-curvature and bilayer-coupling models. Phys Rev A 44:1182-1202
Sheetz MP and Singer SJ (1974) Biological membranes as bilayer couples. A molecular mechanism of drug-induced interactions. Proc Nat Acad Sci USA 71:4457-4461
Svetina S and Žekš B (1989) Membrane bending energy and shape determination of phospholipid vesicles and red blood cells. Eur Biophys J 17:101-111
Svetina S and Žekš B (1990) The mechanical behaviour of cell membranes as a possible physical origin of cell polarity. J theor Biol 146:115-122
Svetina S and Žekš B (1991) Mechanical behavior of closed lamellar membranes as a possible common mechanism for the establishment of developmental shapes. Int J Develop Biol 35:359-365
Svetina S and Žekš B (1992) The elastic deformability of closed multilayered membranes is the same as that of a bilayer membrane. Eur Biophys J 21:251-255
Svetina S, Brumen M and Žekš B (1985) Lipid bilayer elasticity and the bilayer couple interpretation of red cell shape transformations and lysis. Stud Biophys 110:177-184
Svetina S, Gros M, Vrhovec S, Brumen M and Žekš B (1988) Red blood cell membrane vesiculation at low pH and bilayer couple mechanism of red blood cell shape transformations. Stud Biophys 127:193-199
Svetina S, Iglič A and Žekš B (1993) On the role of the elastic properties of closed lamellar membranes in membrane fusion. Ann N Y Acad Sci: in print
Waugh RE, Song J, Svetina S and Žekš B (1992) Local and non-local curvature elasticity in bilayer membranes by tether formation from lecithin vesicles. Biophys J 61:974-982
Žekš B, Svetina S and Pastushenko V (1990) The shapes of phospholipid vesicles in an external electric field - a theoretical analysis. Stud Biophys 138:137-142

CORTICAL FLOW IN FREE-LIVING AMOEBAE

Andrzej Grębecki
Nencki Institute of Experimental Biology
3, Pasteur Str., Warsaw, Poland

INTRODUCTION

The cell cortex of large fresh-water amoebae is composed of the plasma membrane covered with 200 nm thick mucopolysaccharide coat and connected on the inner side with the cytoskeleton. Lateral movements of this complex or of its components are currently called cortical flow (Bray and White, 1988), although the cell cortex cannot be considered as fluid.

The cytoskeleton of Amoeba proteus presents a 1 µm sheet of gel, which is devoid of any structural elements, except the three-dimensional F-actin network connected with plasma membrane by spectrin and α-actinin. This optically empty layer is traditionally called the hyaloplasm (Fig. 1). Deeper layers of the ectoplasm are termed the granuloplasm, because they are packed with organelles, which appear as a myriad of granules in the light microscope. The granuloplasm is gelled as well and bound to the cortical gel layer. Therefore, the granuloplasmic particles behave as natural markers of movements produced by the contractile cell cortex.

LATERAL MOVEMENTS OF THE CORTICAL CYTOSKELETON

In the older literature, the lateral walls of the cortical cylinder of moving amoebae were usually characterized as a static layer. Two apparently independent motor phenomena were however distinguished. At the closed rear end of the cortical cylinder the ectoplasm moves forwards coherently with tail retraction, whereas at its open front on the contrary, the ectoplasm may be pulled backwards from the advancing frontal zone, where it is being continually produced by gelation of the endoplasm (i.e., by actin polymerization and crosslinking). The three-dimensional loop produced in the frontal zone by axial advance of endoplasm and circumferential retreat of ectoplasm looks like a fountain, and is sometimes termed a "fountain streaming".

NATO ASI Series, Vol. H 84
Biomechanics of Active Movement
and Division of Cells
Edited by N. Akkaş
© Springer-Verlag Berlin Heidelberg 1994

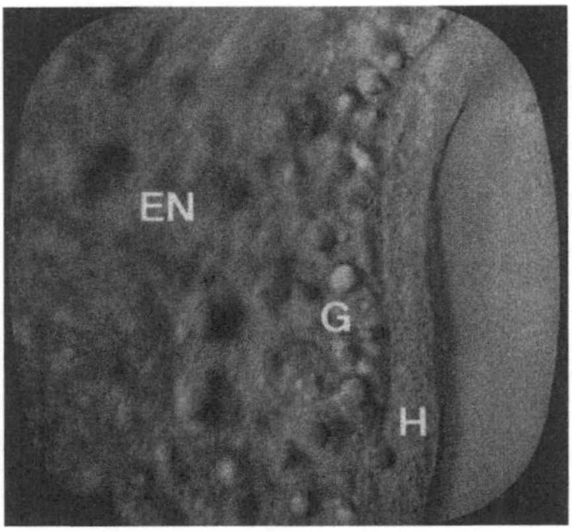

Fig. 1. Cytoplasm layers in a moving A. proteus. Note streaks produced by flowing endoplasm (EN), the granuloplasm (G) and the transparent hyaloplasm (H). Differential interference contrast. Courtesy of L. Grębecka.

We have reexamined by cinematographic analysis the behaviour of the cortical contractile layer in A. proteus (Grębecki, 1984) under different cell-to-substratum attachment conditions, and in suspension.

Without contact with the substratum amoeba is incapable of translatory movements, but the cytoskeletal motor still works, although it is running idle. The axial endoplasm flows forwards as usual, whereas the gelled ectoplasm retreats throughout the full length of amoeba. The fountain movement is no more restricted to the anterior zone of the cell, but extended to its posterior extremity (Fig. 2 A). The velocity profiles of this retrograde motion beneath the cell membrane indicate that no segment of the cytoskeleton is passively pulled as one block, but everywhere a uniform shortening accompanies the retraction. In amoebae which crawl in contact with the substratum the peripheral cyto-

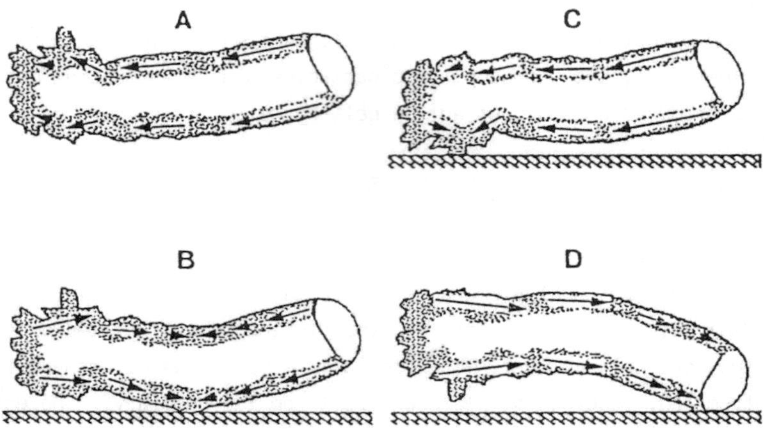

Fig. 2. Backward movement of the cortical gel layer of an amoeba without contact with the substratum (A), and its uniform retraction toward the adhesion sites located either at the middle (B), in the tail region (C), or at the front (D). Velocity of retraction increases with distance from the attachment sites. From Grębecki (1994).

plasm retracts away from unattached extremities toward the attachment sites (Fig. 2 B-D). The velocity analysis proves again that thic centripetal movement is related to the uniform cortical contraction along the whole cell. Therefore, the tail retraction and the fountain movement at the front, which were always separately described, are not two independent phenomena, but merely two local manifestations of the uninterrupted steady retraction of the whole peripheral cytoskeleton.

Steady retrograde transport of structures rich in the cytoskeletal actin was also described in moving fibroblasts, polymorphonuclear leukocytes, lymphocytes, blood platelets, and neuronal growth cones. A large number of these earlier observations have been generalized by Bray and White (1988) in the theory of "cortical flow", a phenomenon manifested by animal cells during locomotion, capping, and cell division.

RELAXATION AND DETACHMENT OF CORTICAL LAYERS AT THE LEADING EDGE

According to the views of Bray and White (1988), Spudich (1989), and ours (Grębecki, 1994), the submembranous actin network is usually drawn from areas of relaxation toward the sites of a more efficient contraction. As a matter of fact, local relaxation of the pseudopodial tips was revealed by application of anaesthetics with a micropipette to the surface of locomoting amoebae (Korohoda, 1977). Stimulation of different cell regions with strictly localized photic stimuli (Grębecki, 1981) demonstrated that the whole peripheral cytoplasm layer is in a state of permanent contraction, with the exception of the advancing edge, which is relaxed. Its relaxation, however, seems to be less a biochemical phenomenon than a biomechanical one. One suspects, rather, not a relaxation of the actomyosin engine, but a deficiency of force transmission within the motor system in the frontal zone, resulting from a local uncoupling between the cytoskeleton which moves backwards, and other cell components that are propelled forwards.

The transmission of force between the contractile layer and the membrane in A. proteus was experimentally blocked by microinjection of a droplet of paraffin oil against the inner side of the plasma membrane (Grębecka, 1977). Vigorous new frontal zones immediately arise at the sites where the connection between the membrane and cytoskeleton has been mechanically broken. It was later shown by electron microscopy that in normal fronts the actin layer is separated from the cell membrane by several µm (Grębecka and Hrebenda, 1979; Stockem et al., 1982).

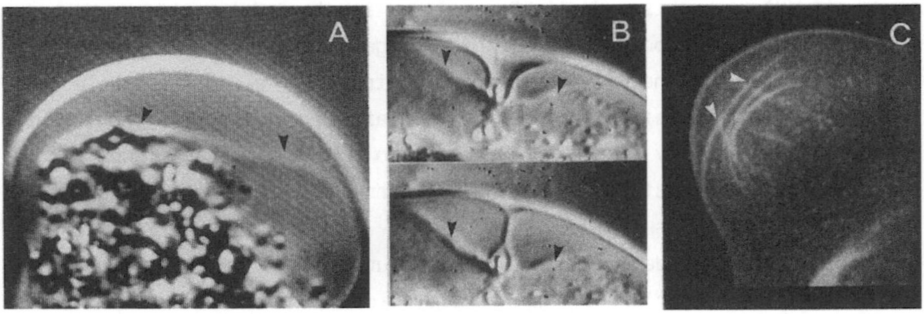

Fig. 3. Detachment of the contractile layer (arrowheads) from the cell membrane seen in the differential interference contrast in the frontal zone of amoeba (A, from Grębecki, 1990), at two stages of endocytosis (B, from Grębecki, 1991), and after actin antibody staining (C, courtesy of P. Pomorski).

It was demonstrated by video-enhancement (Grębecki, 1990, 1991) that the disassembly and reassembly of cortical layers at the frontal margin (Fig. 3 A) incessantly alternate about every 2 s. The detached sheet of actin network moves backwards, whereas the leading edge suddenly expands. The cytoskeleton is then restored by actin polymerization under the frontal membrane, as it was earlier demonstrated by immunofluorescence (Stockem et al., 1983). Reconstruction of the cytoskeleton is followed by successive detachment of the next layer, and next stepwise expansion of the leading edge. The same phenomenon was also seen during the internalization of endosomes (Fig. 3 B). The actin nature of the detached layers has been tested by antibody and phalloidin staining (Fig. 3 C).

Similar phenomena occur in the tissue cells. Concentric arcs composed of a condensed actin network are separated, one after the other, from the leading edges of the lamellae of locomoting fibroblasts (Heath, 1983) and keratinocytes (Theriot and Mitchison, 1991). Local splitting of the membrane-cytoskeleton complex is also necessary before new protrusions can be produced by Dictyostelium (Spudich, 1989), macrophages (Hartwig and Yin, 1988), and astrocytes (Baorto et al., 1992).

MOVEMENTS OVER THE CELL SURFACE

Centripetal character of the cortical flow is locally expressed along the unattached frontal part of amoeba as a retrograde movement of the cytoskeleton. This has further implications on the transport of surface markers on the opposite side of the plasma membrane. The traffic of extracellular material over the surface of amoebae is two-directional (Grębecki, 1986). Latex beads which loosely contact with the cell, move forward as fast as the leading cell margin advances. Even these who freely float a few μm from the membrane, are pulled in the same direction, owing to the viscosity of medium. This extracellular forward flow occurs along the whole length of amoeba (Fig. 4). But the latex beads and glass particles that firmly stuck to the cell surface by adhesion, behave exactly as the cytoskeleton on the opposite side of plasma membrane: they move backward at the front, are stationary in the adhesion zone, and are pulled forward over the tail surface (Fig. 4). Consequently, the adhering and nonadhering particles move in two opposite directions between the leading edge and the substratum attachment sites of amoeba. It was concluded, that the adhering particles are hauled by the retracting actin meshwork, like the receptor-ligand complexes redistributed by the mechanism of capping. The general forward flow of the others is cytoskeleton-independent, but is instead distinctly correlated with the advance and expansion of the frontal edge. Presumably, it is propelled by membrane lipids moving in the direction of cell locomotion (the forward flow of lipids over the cell surface has been directly demonstrated in polymorphonuclear leukocytes by Lee et al., 1990).

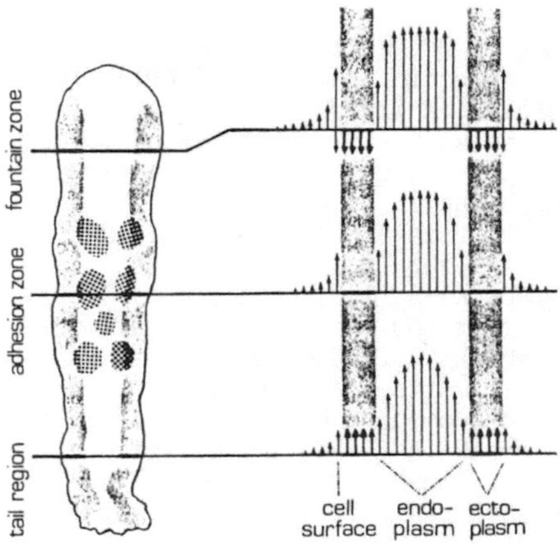

Fig. 4. Transverse velocity profiles of the movements of endoplasm, ectoplasm, cell surface, and the surrounding medium in the anterior, middle, and posterior regions of A. proteus. The relationships between the different moving layers, but not the absolute velocity values, are reflected. From Grębecki (1987).

Also on the surface of macrophages, keratinocytes, and fibroblasts the colloidal gold particles, coated with the concanavalin or poly-L-lysines exhibit the same two types of movements: either random diffusion and passive drift concerted with the cell´s progression, or the cytoskeleton-directed transport (see reviews by Heath and Holifield, 1991 and Grębecki, 1994).

ACKNOWLEDGMENTS. Travel of the author to this meeting was sponsored by Stefan Batory Foundation, Warsaw.

REFERENCES

Baorto DM, Mellado W, Shelanski ML (1992) Astrocyte process growth induction by actin breakdown. J Cell Biol 117: 357-367

Bray D, White JG (1988) Cortical flow in animal cells. Science 239: 881-888

Grębecka L (1977) Changes of motor polarization in Amoeba proteus induced by oil injections. Acta Protozool 16: 107-120

Grębecka L. Hrebenda B (1979) Topography of cortical layer in Amoeba proteus as related to the dynamic morphology of moving cell. Acta Protozool 18: 493-502

Grębecki A (1981) Effects of localized photic stimulation on amoeboid movement and their theoretical implications. Europ J Cell Biol 24: 163-175

Grębecki A (1984) Relative motion in Amoeba proteus in respect to the adhesion sites. I. Behaviour of monotactic forms and the mechanism of fountain phenomenon. Protoplasma 123: 116-134

Grębecki A (1986) Two-directional pattern of movements on the cell surface of Amoeba proteus. J Cell Sci 83: 23-35

Grębecki A (1987) Velocity distribution of the anterograde and retrograde transport of extracellular particles by Amoeba proteus. Protoplasma 141: 126-134

Grębecki A (1990) Dynamics of the contractile system in the pseudopodial tips of normally locomoting amoebae, demonstrated in vivo by video-enhancement. Protoplasma 154: 98-111

Grębecki A (1991) Participation of the contractile system in endocytosis demonstrated in vivo by video-enhancement in heat-pretreated amoebae. Protoplasma 16o: 144-158

Grębecki A (1994) Membrane and cytoskeleton flow in motile cells with emphasis on the contribution of free-living amoebae. Int Rev Cytol 148: 37-80

Hartwig JH, Yin HL (1988) The organization and regulation of the macrophage actin skeleton. Cell Motil Cytosk 10: 117-126

Heath JP (1983) Behaviour and structure of the leading lamella in moving fibroblasts. I. Occurrence and centripetal movement of arc-shaped microfilament bundkes beneath the dorsal cell surface. J Cell Sci 60: 331-354

Heath JP, Holifield BF (1991) Cell locomotion: New research tests old ideas on membrane and cytoskeletal flow. Cell Motil Cytosk 18: 245-257

Korohoda W (1977) Experimental induction of locomotion in enucleated fragments of Amoeba proteus and its bearing on the theories of amoeboid movement. Cytobiologie 14: 338-349

Lee J, Gustafsson M, Magnusson KE, Jacobson K (1990) The direction of membrane lipid flow in locomoting polymorphonuclear leukocytes. Science 247: 1229-1233

Spudich JA (1989) In pursuit of myosin function. Cell Regln 1: 1-11

Stockem W, Hoffmann HU, Gawlitta W (1982) Spatial organization and fine structure of the cortical filament layer in normal locomoting Amoeba proteus. Cell Tiss Res 221: 505-519

Stockem W, Hoffmann HU, Gruber B (1983) Dynamics of the cytoskeleton in Amoeba proteus. I. Redistribution of microinjected fluorescein-labeled actin during locomotion, immobilization and phagocytosis. Cell Tiss Res 232: 79-96

Theriot JA, Mitchison TJ (1991) Actin microfilament dynamics in locomoting cells. Nature 352: 126-131

A Mechanical Model For Growth And Control of the Cell Wall of Rod-shaped Bacteria

John J Thwaites
Engineering Department
University of Cambridge
CB2 1PZ, UK

Unlike animal cells, bacteria have no cytoskeleton. Shape maintenance and, to some extent, cell-cycle regulation are governed by the cell wall. The walls of many rod-shaped bacteria are continuously added to on the inside surface, at the cytoplasmic membrane, and continuously depleted at the outer surface by the action of autolytic enzymes. The walls are stretched longitudinally as the cells grow, but the cell diameter and wall thickness remain remarkably constant during normal growth (Trueba and Woldringh, 1980). This also is a regulatory problem, but one that depends heavily on mechanical behaviour.

The forces involved derive from internal pressure (turgor) and electrostatic repulsion due to a high negative charge which is not completely neutralised *in vivo*. The load-bearing polymer in the wall (peptidoglycan) is clearly highly anisotropically arranged. It has also been shown to be visco-elastic and some of its mechanical properties have been measured (Thwaites and Mendelson, 1991). The model presented here embodies all these features and it is possible to analyse wall stresses in model terms. Because of the growth feature, this is a novel problem of applied mechanics. Its solution yields equations in which rates of deformation (including diametral swelling) are related to the forces and to the rates of material addition and removal. If the wall growth process is taken to be autonomous (Koch, 1983), the equations show it to be unstable, but it can be stabilized by biologically plausible feedback mechanisms. These are examined in model terms.

Mechanical Basis for the Model

The following assumptions are made; they are based mainly on experimental evidence. Where they represent simplifications for ease of analysis, they are not fundamental.
(i) The bacterial wall is taken to be a cylinder of radius a and thickness w . The effects of the polar caps, which close the ends, except for transmitting pressure, are ignored.
(ii) The cylinder is subject to internal pressure p_1 (turgor) and end pressure p_2, representing electrostatic repulsion. In fact electrostatic repulsion contributes a minor part of p_1.
(iii) The wall material is homogeneous and orthotropic, being stiffer in the hoop direction than longitudinal. It is linearly visco-elastic, with a relaxed modulus (longitudinal) E and a

NATO ASI Series, Vol. H 84
Biomechanics of Active Movement
and Division of Cells
Edited by N. Akkaş
© Springer-Verlag Berlin Heidelberg 1994

single creep time constant T. It is also incompressible, from which it follows that the modulus in the hoop direction is $E/2\nu$ where ν is Poisson's ratio.
(iv) Wall material is added uniformly at the inner surface at a volume rate per unit area k_1. It upwells uniformly so that at a given cylinder radius all material is of the same age and it is removed at the outer surface at a volume rate per unit area k_2.

The mechanical significance of wall charge has been shown by Cole *et al.* (1970) in that mutants of *Bacillus subtilis* deficient in teichoic acid are misshapen. For this organism an unneutralised charge of about 3.5% would achieve equality in p_2 and p_1 (Marquis, based on Doyle *et al.* 1980). There is no direct evidence for either homogeneity or orthotropy. Changes in cross-linking as material moves through the wall could alter mechanical properties. Increasing deformation could change the average orientation. For *Bacillus subtilis* the anisotropy could be helical; Thwaites and Mendelson (1991) show that a modification to the following analysis, in which the stiffer direction is inclined at a small angle to the hoop direction, explains the twisting that is observed with growth in these cells (Mendelson, 1976; Fein 1980). The visco-elasticity of cell wall cannot in fact be described by a single time constant (Thwaites and Mendelson 1985; Thwaites and Surana 1991), but the present state of knowledge doesn't justify using more sophisticated visco-elastic relations. Nor is wall material incompressible, but again this simplification makes the analysis possible. Evidence for the uniform upwelling of material through the wall of *Bacillus subtilis* is conclusive (Pooley, 1976b; Merad *et al.* 1989).

Stress Analysis

As new material moves through the wall it undergoes increasing deformation. It is nonetheless useful to write the basic equations in terms of average (through the wall) stress and strain. The stress components σ_ℓ (longitudinal) and σ_h (hoop) are simply related to the pressures by consideration of equilibrium (Fig. 1). The corresponding average strain components are given by

$$E\bar{\varepsilon}_\ell = \sigma_\ell - \nu\sigma_h = \{(1-2\nu)p_1 + p_2\}a/2w \qquad (1)$$
$$E\bar{\varepsilon}_h/\nu = 2\sigma_h - \sigma_\ell = (3p_1 - p_2)a/2w \qquad (2)$$

Because of the visco-elasticity the average strains involve weighted integrals of the corresponding strain rates over the time since the material was newly inserted. This time depends on position in the wall, so the averages are by no means simple. The strain rates involved are quite straightforward. In the longitudinal direction the strain-rate η is the same at all points. The hoop strain-rate at radius r, $\dot{r}/r$, is found by considering mass balance for the material within that radius, as it changes with time. The averaging results in the following expressions for the *viscous terms only* in the average strains

$$\bar{\varepsilon}_\ell = \eta T \qquad (3)$$
$$\bar{\varepsilon}_h = (\dot{a} + k_1 - \tfrac{1}{2}\eta w)T/a \qquad (4)$$

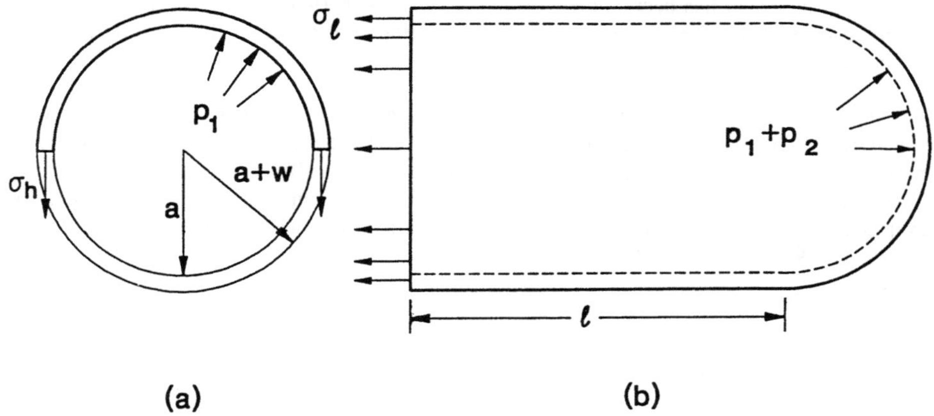

Figure 1. Average longitudinal and hoop stress σ_l and σ_h in a cylinder subject to internal and end pressures p_1 and p_2. (a) Cross-section; (b) longitudinal view.

The elastic terms are, for the non-steady state, given by complicated expressions involving finite delays in several of the variables. In the steady state they take the simple forms $1-k_2T_w/w$ (longitudinal) and $w/2a$ (hoop) where T_w is the time taken for material to move from the inside to the outside of the wall (Thwaites 1993).

Substitution for $\bar{\varepsilon}_\ell$ and $\bar{\varepsilon}_h$ yields two equations linking wall radius a, thickness w and elongation rate η with the 'inputs' p_1, p_2, k_1 and k_2. A third, which is derived from mass balance for the whole wall, is

$$(\dot{a}/a+\eta)w+\dot{w}=k_1-k_2 \tag{5}$$

Steady Growth

Deferring the question of stability, the three equations give the geometry of steady growth in terms of pressures, wall-insertion and -loss rates and material properties. In practice the geometry and some of the properties are known. For *Bacillus subtilis* for example at room temperature, the cell diameter is about 0.8 μ,m the wall thickness about 40 nm, the modulus E about 10 MPa (Thwaites and Mendelson 1991) and the wall loss rate is about 0.1 times the insertion rate (Pooley, 1976a). The equations are satisfied when p_1 is 1 MPa, p_2 is 1.5 MPa and ν is 0.1. This value of Poisson's ratio implies a modulus in the hoop direction (yet to be measured) of $5E$ which is comparable to that of other biopolymers. The value for turgor pressure is less than has been quoted (Mitchell and Moyle 1956) but, given the difficulties involved in measurement, is not unreasonable. Some confidence in the usefulness of the model is thus engendered.

The model yields other steady-state results such as the stress distribution through the wall (Thwaites 1993) and, using the helical anisotropy variation referred to above, the angle of twisting-with-growth. The latter is shown to depend significantly on the ratio p_2/p_1 (Thwaites and Mendelson 1991) so that small changes in unneutralised charge due, for example, to the presence of ions in the growth medium, affect twist, as observed (Mendelson and Favre, 1987).

Stability of Wall Growth

Stability analysis for the full visco-elastic model is, because of the finite delays, very complicated. However, in practice the viscous terms dominate, so that an analysis of the system for which the average strains are given by equations (3,4) should be sufficiently representative of the real situation. If the wall-growth process is taken to be autonomous, i.e. the pressures and material insertion/loss rates are independent variables, it is highly unstable. This is most readily seen by linearising equations (1-3, 2-4 and 5) for small changes from the steady state. The matrix of the coefficients of the proportional variations $\delta\eta/\eta$, $\delta a/a$ and $\delta w/w$ is then, using the Laplace operator s,

$$\begin{pmatrix} 1 & -1 & 1 \\ -0.5 & (a/w)s - 1 - 2k_2/k_1 & k_2/k_1 \\ 1 & s & s+1 \end{pmatrix} \tag{6}$$

where a, w, k_1 and k_2 are the appropriate steady-state values and the unit of time is $1/\eta$. The determinant of the coefficients is a quadratic in s which has a positive root whatever the steady-state conditions.

But the process is not autonomous, for irrespective of any active controls, and neglecting for the moment osmotic flow, if the cytoplasm volume increases the turgor pressure must decrease, i.e. if η or $\dot{a}$ increase, p_1 decreases. This represents a negative feedback. The full relation, including osmotic flow, is found by consideration of cytoplasm mass balance. Taking the cytoplasm to be incompressible, this is described by

$$(1-\rho)(\eta + 2\dot{a}/a) = \dot{\rho} + (\pi_0 - p_1)\frac{2c}{a} \tag{7}$$

where ρ is the proportion of the cytoplasm that is non-aqueous (i.e. ribozomes etc). π_0 is the difference in osmotic potential between the growth medium and the cytoplasm and c is the wall permeability coefficient, i.e. flow rate per unit area per unit pressure.

On the reasonable assumption that the non-aqueous content of the cytoplasm increases at a rate proportional to its amount, ρ can be eliminated from equation (7) and p_1 can then be substituted from it in the three wall-growth equations. When linearised for small changes the result is to add to the matrix of coefficients (6) the terms

$$\begin{pmatrix} e_1\tau(1+\rho/s) & e_1\tau(1+\rho+2\rho s) & 0 \\ e_2\tau(1+\rho/s) & e_2\tau(1+\rho+2\rho s) & 0 \\ 0 & 0 & 0 \end{pmatrix}$$

where e_1 and e_2 are functions of ν and the steady-state value of p_2/p_1 and τ is $a\eta/2cp_1$ (steady-state values). τ has the following physical meaning; it is the time taken (in $1/\eta$ units) for an empty cell to fill due to flow through the wall/membrane complex under pressure p_1. Insufficient data appear to exist for its computation, but it is a significant parameter, for the model predicts the wall-cytoplasm-growth process to be conditionally stable - provided τ is large enough. This appears to be highly unlikely.

Control of Wall Growth

The achievement of stability in a process is not of itself sufficient to ensure a satisfactory response to external, or even internal, disturbances. The changes in steady state due to disturbance must not be too great and the fluctuations during the transient between one steady state and another must be relatively small. Thus it is almost certain that there are active control mechanisms for the wall-cytoplasm growth process. The model suggests that changes in turgor (basically p_1) or the rate of new material insertion (k_1) could be used to effect control, on the basis of information about stress or dimensional changes, using negative feedback. Turgor is known to be controlled by altering the osmotic potential by means of the concentration in the cytoplasm of certain ions. But it is not clear what the measured 'signal' can be. It cannot be the turgor pressure itself, since even if the bacterium possesses a stress-sensing 'transducer' it can measure only its *internal* pressure. There are also difficulties in using wall stress, at the cytoplasmic membrane, say. Dimensional changes in the cytoplasm appear to be more likely candidates. These are of course the same as two of the wall dimensions. They are represented in the model by the elongational strain rate η and the strain rate $\dot{a}/a$ of diametral swelling. These are the variables affected by the coupling between cytoplasm and wall growth, with beneficial effect on stability.

Although $\dot{a}/a$ might seem the obvious choice, it has some disadvantages. Firstly it must be made zero in any steady state. This involves the application of 'integral' control, i.e. essentially a memory term. There is no objection to this in principle, integral control is used in higher organisms, but it involves more sophistication and greater energy input than straightforward 'proportional' control. Secondly, it is difficult to see how it could be sensed, because the strain-rate at no part of the wall equals $\dot{a}/a$. At the cytoplasmic membrane, which is the obvious location for a transducer, the hoop strain rate is $\dot{a}/a + k_1/a$ which has a substantial steady component during growth. In model terms η is the most effective feedback loop in order to achieve reasonable steady-state changes, i.e. less energy is required of the bacterium. It is also 'measurable' at any point of the cell wall.

The model suggests that altering the osmotic potential in response to changes in elongational strain rate is the most effective method of simultaneously controlling turgor pressure and wall growth. This doesn't mean that it is the method actually employed. Experimental confirmation is required, first of all by comparing transient responses to various disturbances, ultimately perhaps by breaking the feedback link using mutants. There is also the possibility of more than one control mechanism being used simultaneously, providing robustness against potentially damaging environmental changes. The model has been used in this way, by adding control of new material insertion (k_1) and using both η and $\dot{a} / a$ as control signals. This leads to no significant improvement over the simpler single loop mechanism described.

References

Cole RM, Popkin TJ, Boylan RJ and Mendelson NH (1970) Ultrastructure of a temperature-sensitive *rod*-mutant of *Bacillus subtilis.* J Bacteriol. 103, 793-810

Doyle RJ, Matthews TH and Streips UN (1980) Chemical basis for selectivity of metal ions by the *Bacillus subtilis* cell wall. J Bacteriol. 143, 471-480

Fein, JE (1980) Helical growth and macrofiber formation of *Bacillus subtilis 168* autolytic enzyme deficiet mutants. Can J Microbiol 26, 330-337

Koch AL (1983) The surface stress theory of microbial morphogenesis. Adv Microbial Physiol 24, 301-366

Mendelson NH (1976) Helical growth of *Bacillus subtilis.*: a new model for cell growth. Proc Natl Acad Sci USA 73, 1740-1744

Mendelson NH and Favre D (1987) Regulation of *Bacillus subtilis.* macrofiber twist development by ions. J Bacteriol 169, 519-525

Merad T, Archibald AR, Hancock JC, Harwood CR and Hobot J (1989) Cell wall assembly in *Bacillus subtilis.*: vizualisation of old and new wall material by electron-microscopic examination of samples stained selectively for teichoic acid and teichuronic acid. J Gen Microbiol 135, 645-655

Mitchell P and Moyle J (1956) Osmotic function and structure in bacteria. Symp Soc Gen Microbiol 6, 150-180

Poole HM (1976a) turnover and spreading of old wall during surface growth of *Bacillus subtilis..* J Bacteriol 125, 1127-1138

Pooley HM (1976b) Layered distribution according to age within the cell wall of *Bacillus subtilis..* J Bacteriol 125, 1139-1147

Thwaites JJ (1993) Growth and control of the cell wall: a mechanical model for *Bacillus subtilis..* In Bacterial growth and lysis (de Pedro MA, ed) Plenum, New York

Thwaites JJ and Mendelson NH (1985) Biomechanics of bacterial walls: studies of bacterial thread made from *Bacillus subtilis..* Proc Natl Acad Sci USA 82, 2163-2167

Thwaites JJ and Mendelson NH (1991) Mechanical behaviour of bacterial cell walls. Adv Microbial Physiol 32, 173-222

Thwaites JJ and Surana UC (1991) Mechanical properties of *Bacillus subtilis.* cell walls: effects of removing residual culture medium. J Bacteriol 173, 197-203

Trueba FJ and Woldringh CL (1980) Changes in cell diameter during the division cycle of *Escherichia coli.* J Bacteriol 142, 869-879

Measurement of passive and active force in single isolated smooth muscle cells.

R van Mastrigt, JJ Glerum, R Schot

dept.Urology-urodynamics, Room EE1630
Erasmus University Rotterdam
PO Box 1738
3000 DR Rotterdam
the Netherlands

fax : 31-10-4366428
internet : vanmastrigt@udn.fgg.eur.nl

Introduction

In urodynamics, a subspecialism of urology, the function of the urinary tract is studied from a physical point of view, and methods are developed to objectively (and eventually automatically) diagnose the condition of the urinary tract. An important aspect of the function of the lower urinary tract is the contractility of the urinary bladder. Methods to measure this contractility are based on models developed using mechanical measurements on samples of bladder tissue. Such methods do not lead to a complete understanding of contractile properties of the (smooth muscle) tissue as the organization and structure of this tissue are irregular. Therefore measurements on single smooth muscle cells are necessary. Such measurements would also enable the use of smooth muscle biopsies for diagnostic reasons. In this chapter the development and first results of a practical method for measuring active and passive mechanical forces in single smooth muscle cells of the urinary bladder are described. Basically, such a method requires that viable and contractile cells can be isolated from the surrounding tissue and be attached to a transducer with sufficient resolution and stability to measure the very small forces

NATO ASI Series, Vol. H 84
Biomechanics of Active Movement
and Division of Cells
Edited by N. Akkaş
© Springer-Verlag Berlin Heidelberg 1994

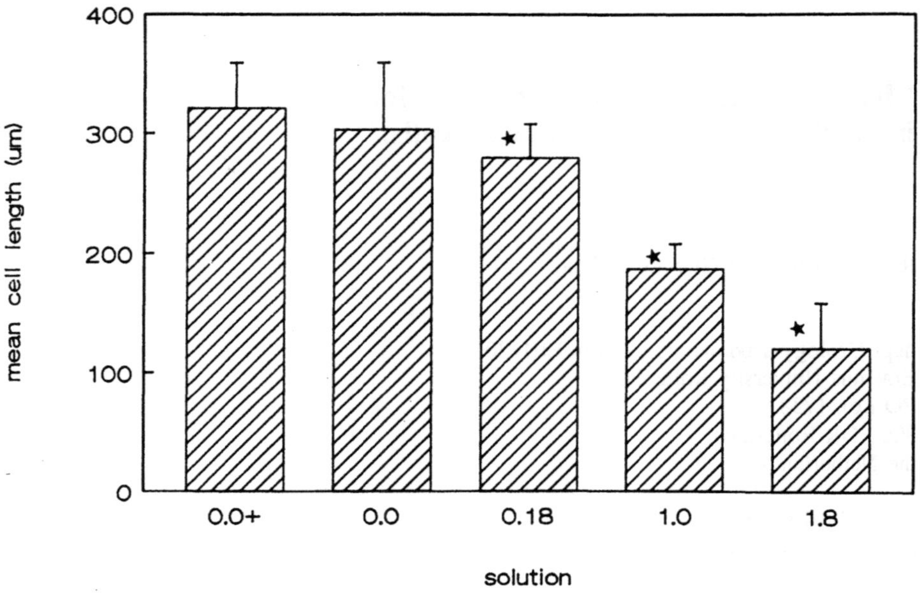

Figure 1. Mean ± standard deviation of pig urinary smooth muscle cell lengths in five different physiological solutions containing 0.0, 0.18, 1.0 and 1.8 mmol/l of calcium. Figure reproduced from Schot et al [1993] with permission.

involved. The three main issues, isolation, attachment and force measurement will be separately discussed, and some first results will be shown.

Isolation of viable contractile cells from pig urinary bladder

Pig bladders were chosen for our experiments as these proved to be a good model of the human urinary bladder, both mechanically and anatomically [Douglas, 1972]. In our first experiments cells were isolated using a solution of collagenase, DNAase and antitrypsine in a modified Krebs Hepes buffer [Glerum et al, 1987]. In later experiments only collagenase and papain were used, in combination with the application of an especially developed mechanical device for continous resuspensation of muscle/cell suspensions through a wide bore 5 ml pipette [Schot et al,1993]. As compared to isolation of other muscle cells, isolation of urinary bladder smooth muscle cells proved to be extremely difficult, probably as a result of both the high concentration of collagen between and around the cells and the many close contacts between the cells [Gosling et al, 1983]. Using the second method, two out of three isolation procedures resulted in a high yield of smooth muscle cells which remained viable for about 3-4 hours as tested with fluorescein diacetate. When trying to measure mechanical responses from these cells it was found that "viability" tested in this way does not guarantee that contractile responses can be evoked, so that the ultimate test for the physiological condition of isolated cells must involve a measurement of contractility. It

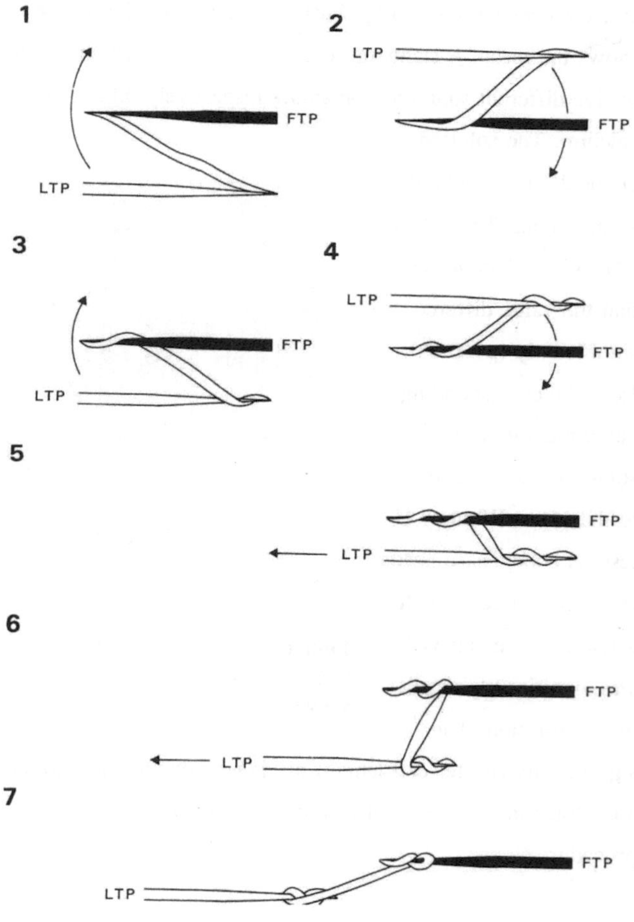

Figure 2. A schematic representation of the knotting procedure used for attaching isolated mammalian single smooth muscle cells to two axially aligned micropipettes. Figure reproduced from Glerum et al [1990b] with permission.

was further found that the length and volume of the isolated cells depended on the calcium concentration in the buffer solution used during the isolation procedure. Fig. 1. shows the mean ± standard deviation of the cell lengths of three batches of cells, isolated in five different solutions containing respectively 0.0, 0.18, 1.0 and 1.8 mmol/l of calcium. The solution denoted 0.0+ contained an additional 2 mmol/l of EGTA. An * indicates that the value differed significantly ($p<0.05$) from the corresponding value measured in solution 0.0+ according to the Mann-Whitney U-test.As collagenase needs calcium as a cofactor for activation, the cell yield was significantly smaller in this solution. The significantly shorter cell lengths at higher calcium concentrations are probably caused by a contraction as a result of mechanical stimulation of the cells during the isolation procedure.

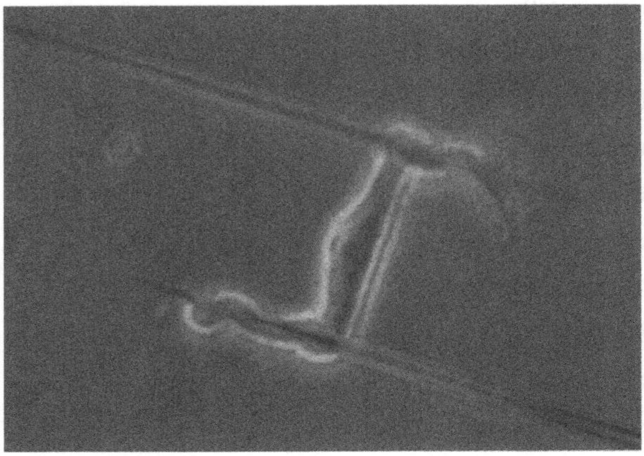

Figure 3. A pig urinary bladder smooth muscle cell attached to two glass micropipettes, halfway the knotting procedure shown in Fig. 2.

Attachment of isolated smooth muscle cells

The isolated cells were attached to glass micropipettes using a modified version [Glerum et al, 1990b] of the technique described by Fay [1977]. Fig. 2. illustrates this technique. A selected cell approached by two micropipettes did generally slightly stick to it, probably as a result of static forces. By rotating the two pipettes around each other, and subsequently pulling these apart, a knot was formed that was often strong enough to enable the measurement of passive and active forces. Fig. 3. shows an example of a pig urinary bladder cell halfway the knotting procedure. The reliability of the knots was

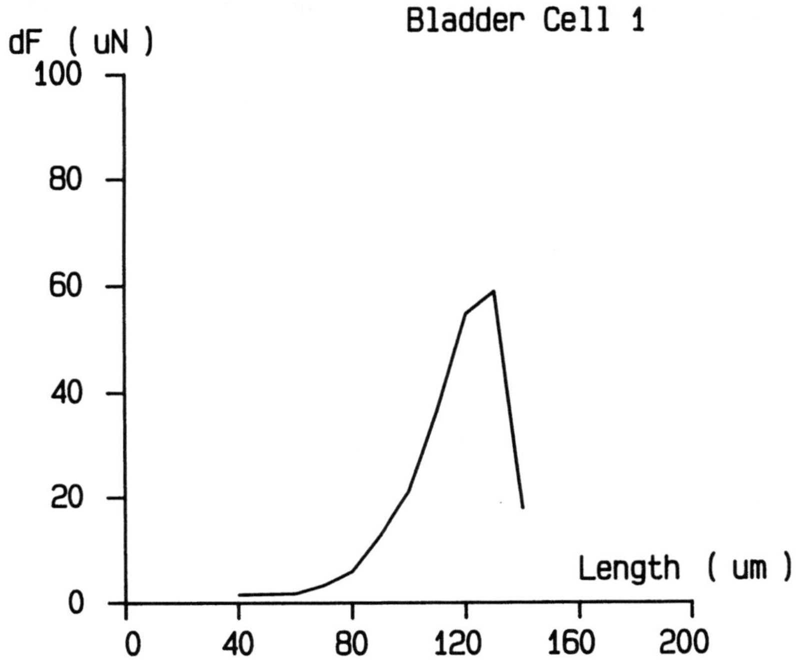

Figure 4. A typical example of the force increments in response to successive 10 μm stretches of a smooth muscle cell as a function of the cell length. Figure reproduced from Glerum et al [1990b] with permission.

evaluated by repeatedly stretching cells by length increments of 10 μm. The resulting passive force increments were measured using the microforce transducer described in the next section. Fig. 4 shows an example of the resulting force increments as a function of cell length. It was observed that frequently the cell would break or tear without detaching from the micropipettes. It should be noted that the forces necessary to break or tear the cells significantly exceeded the active forces developed by contracting cells (see next sections). Although the method allowed relatively short cells to be attached (length $>$ 80 μm) alternative methods are necessary to attach the shortest smooth muscle cells.

Transducers for measuring passive and active force in isolated smooth muscle cells

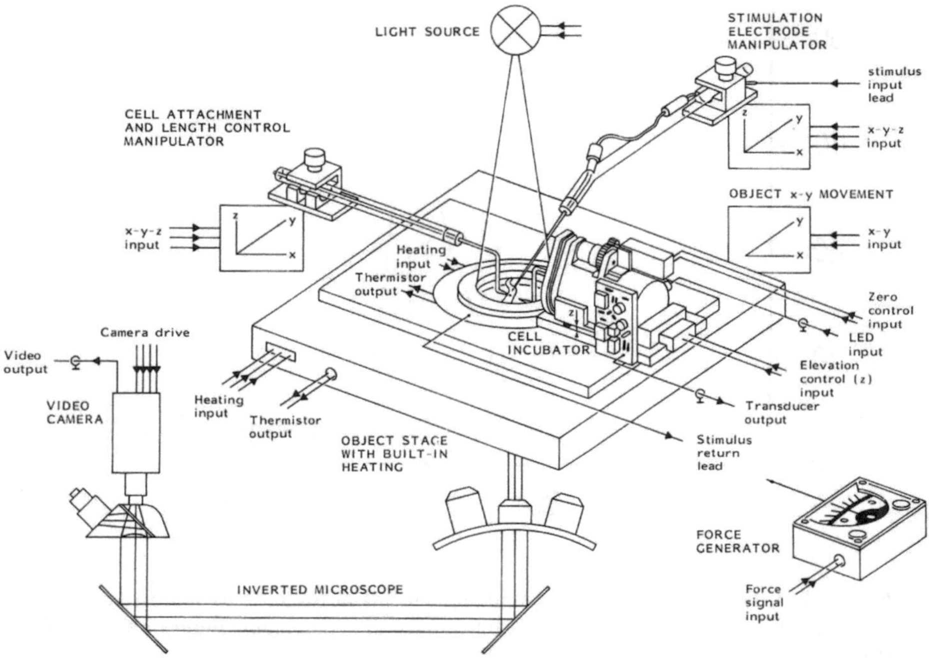

Figure 5. Schematic diagram of cell incubator and microforce transducer used in our first experiments. Figure reproduced from Glerum et al [1990c] with permission.

Fig. 5. shows the set-up in our first series of experiments. Cells were incubated in a heated bath mounted on the heated object table of an inverted microscope. Cells were knotted to micropipettes bent in a Z-like form, the left one was attached to a micromanipulator to control the length of the cell. The right pipette was suspended from two quartz beams in such a way that only movement in the longitudinal direction was possible. This movement was detected by a small vane at the end of the pipette that intercepted an infrared beam directed at a bi-cellular photo diode [Glerum et al, 1990a]. Fig. 6. shows the excellent short term stability and sensitivity of this transducer.

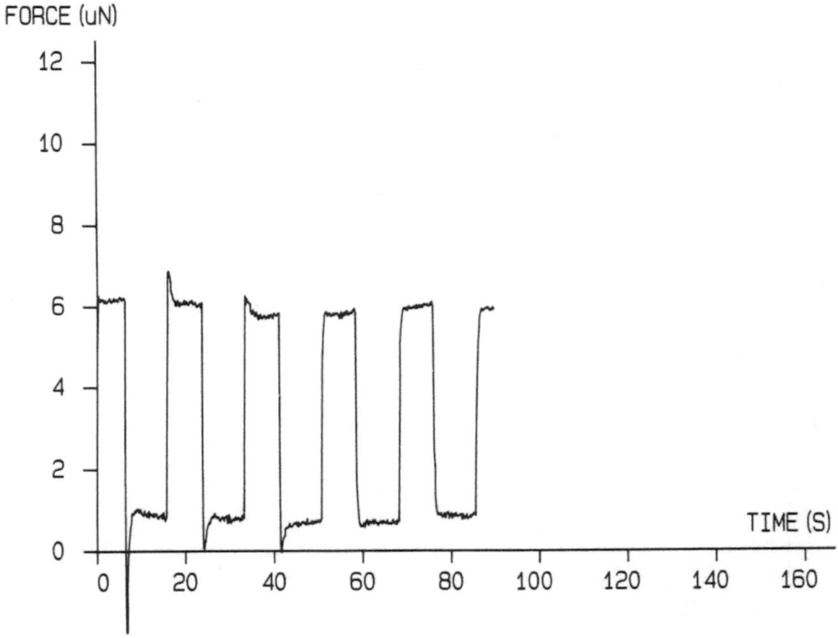

Figure 6. Response of the force transducer shown in Fig. 4. to a square wave calibration force of 5 μN p-p. Figure reproduced from Glerum et al [1990a] with permission.

In later experiments, cells were incubated between two microscope coverglasses about 2.5 mm apart. A thin fluid film between the glasses was continuously refreshed and heated by infrared irradiation. Very short, straight micropipettes were used that entered the fluid film from both sides, the right pipette was attached to a Cambridge Technologies 406A transducer with a full range of 500 μN, 10 nN resolution and an unloaded resonance frequency of 100 Hz. As compared to the earlier set-up, the lower mass as a result of the shorter pipettes resulted in a higher frequency response and a better long term stability.

First results

In a first series of measurements passive responses of pig urinary bladder cells to stretching were measured. Fig. 7 shows a typical example of the resulting stress

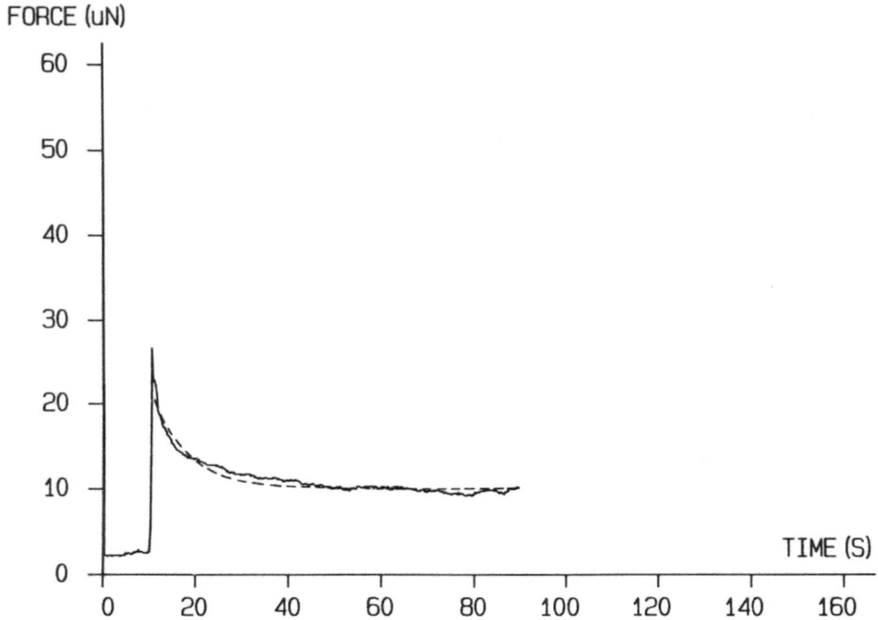

Figure 7. A typical example of a stress relaxation curve measured in response to a 10μm stepwise length increase of a single smooth muscle cell. Figure reproduced from Glerum et al [1990c] with permission.

relaxation curves. These could adequately be described with a mono exponential curve. It was found that the cells relaxed considerably less than intact bladder tissue, and that relaxation did not continue to zero force, as published by other authors [van Dijk et al, 1984], but leveled off at a definite constant force [Glerum et al, 1990c]. These passive properties were ascribed to the cytoskeleton and cell membrane. In more than half of the measurements wavelike excursions from the mono exponential decay were observed, as illustrated in Fig. 8. These were ascribed to stretch induced contractions caused by membrane depolarization. In only a fraction of the cells that showed these "contractions" active force development could also be evoked by direct electrical stimulation. Fig. 9. shows the best example of these. The cells were stimulated by applying positive, mono-

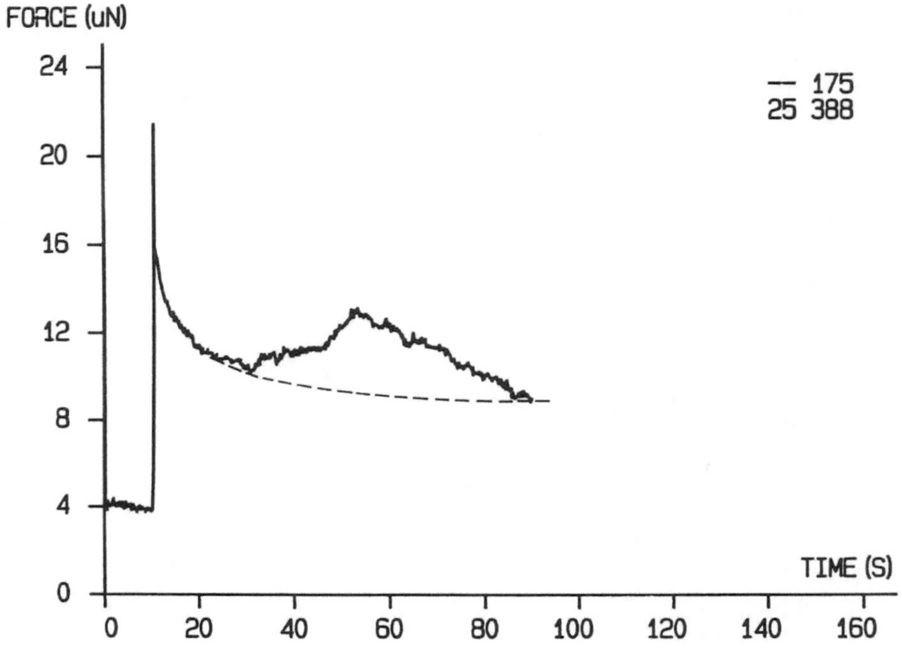

Figure 8. A typical example of a stress relaxation curve measured in response to a 10 μm stepwise length increase of a single smooth muscle cell with a superimposed wavelike force deviation. Figure reproduced from Glerum [1991] with permission.

phasic, rectangular pulses of 1.8 V amplitude, 10 ms duration and a repetition frequency of 50 Hz to a 50 μm platinum wire electrode close to the centre of the cell. Surprisingly in many cases a negative response to such stimulation was observed, i.e. force suddenly decreased. Such decreases were not observed in the second series of measurements, using the different cell isolation method and measurement technique where cells were incubated in a thin fluid film between two microscope cover glasses and attached to short straight micropipettes mounted to a commercially available force transducer. Fig. 10 shows the best measurement from this series. Again stimulation was switched on and off at the drawn lines. The second wave of force development at t=40 s is probably again a stretch induced contraction. It is concluded that using the described techniques and apparatus it is possible to measure passive responses to stretch and isometric contractions in response to

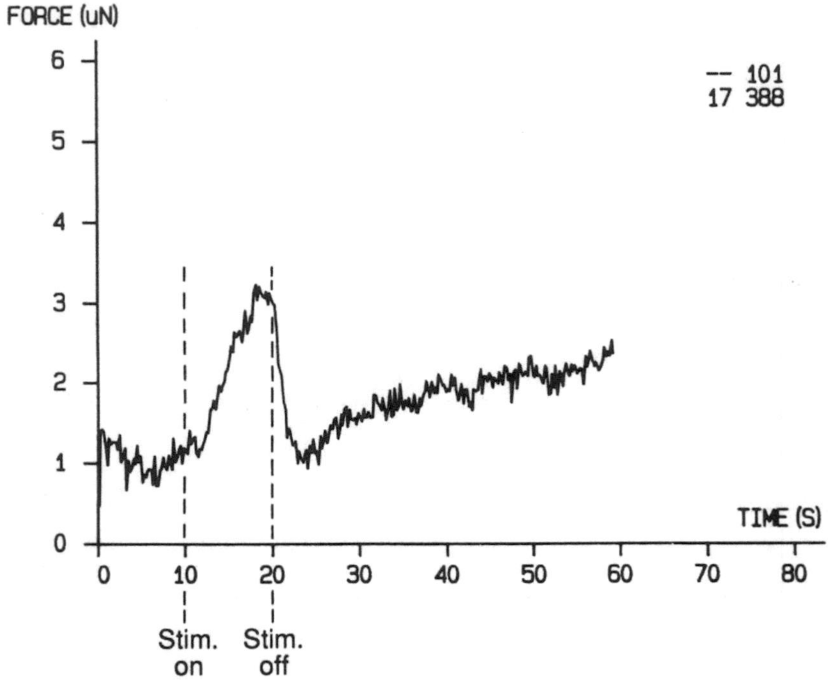

Figure 9. Typical example of an active response to electrical stimulation of a smooth muscle cell. Dotted lines indicate the moments at which stimulation was switched on and off. Figure reproduced from Glerum [1991] with permission.

electrical field stimulation of single isolated smooth muscle cells of the pig urinary bladder. Routine application of these methods is still hindered by the difficult attachment procedure and the fact that only a relatively small percentage of the cells responds to stimulation.

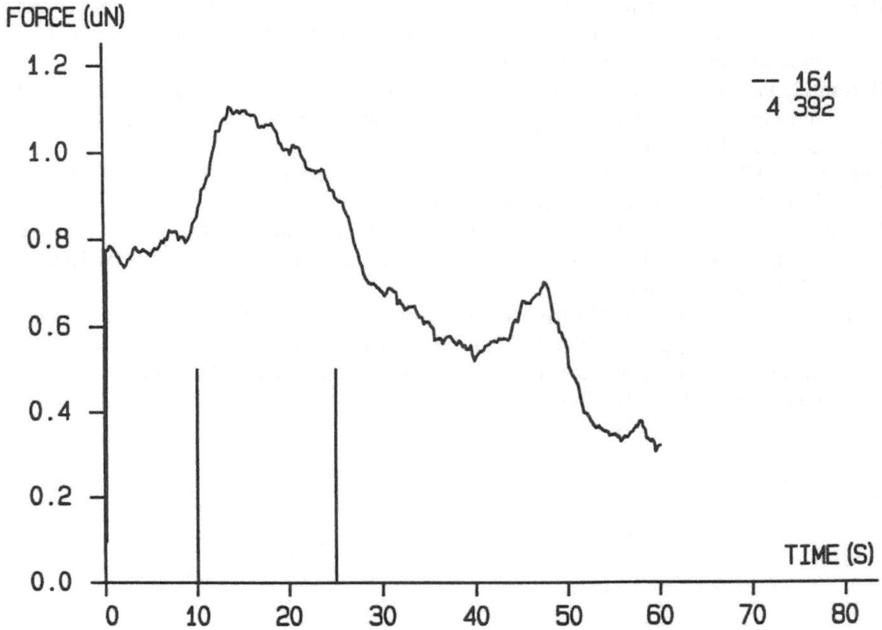

Figure 10. An example of an active contraction of a pig urinary bladder smooth muscle cell from the second series of measurements.

References

Douglas WR (1972). Of pigs and men and research. Space Life Sci 3 : 226-234.

van Dijk AM, Wieringa PA, van der Meer M, Laird JD (1984). Mechanics of resting isolated single vascular smooth muscle cells from bovine coronary artery. Am.J.Physiol. 246:C277-C287.

Fay FS (1977). Isometric contractile properties of single isolated smooth muscle cells. Nature 265:553-556.

Glerum JJ, van Mastrigt T, Romijn JC, Griffiths DJ (1987). Isolation and individual electrical stimulation of single smooth-muscle cells from the urinary bladder of the pig. J.M.Res.Cell Mot. 8 : 125-134.

Glerum JJ, van Mastrigt R (1990a). Mechanical properties of mammalian single smooth muscle cells. I. A low cost large range microforce transducer.J.M.Res.Cell Mot. 11:331-337.

Glerum JJ, van Mastrigt R (1990b). Mechanical properties of mammalian single smooth muscle cells. II. Evaluation of a modified technique for attachment of cells to the measurement apparatus. J.M.Res.Cell Mot. 11:338-343.
Glerum JJ, van Mastrigt R (1990c). Mechanical properties of mammalian single smooth muscle cells. III. Passive properties of pig detrusor and human a terme uterus cells. J.M.Res.Cell Mot. 11:453-462.
Glerum JJ (1991). Mechanical properties of mammalian single smooth muscle cells. PhD Thesis Erasmus University Rotterdam.
Gosling JA, Dixon JS, Humpherson JR (1983). Gross and microscopic anatomy of the urinary bladder. In : Functional anatomy of the urinary tract. Gower Medical Publishing, London.
Schot R, van Asselt E, van Mastrigt R (1993). A method for isolating smooth muscle cells from pig urinary bladder with low concentrations of collagenase and papain : The relation between calcium concentration and isolated cell length. Urol.Res. 21:49-53.

Molecular Mechanisms of Neutrophil Adhesion Studied by Inducing a High Avidity State at β2-Integrin.

Scott I. Simon Ph.D.
Department of Pediatrics
Section of Leukocyte BIology
Baylor School of Medicine
Houston Texas
USA, 77030

INTRODUCTION

The immune host response of neutrophils is dependent upon adhesion to endothelium, chemotaxis into the periferral tissue, and phagocytosis of microorganisms. Essential to these adhesion dependent functions are the Leu-CAM family of β2-integrin adhesive molecules which consist of 3 distinct alpha subunits (LFA-1 CD11a, Mac-1 CD11b, and P150.95 CD11c), noncovalently linked to a common beta subunit (CD18) (for review see ref. 1). The physiologic role of these adhesion molecules is elucidated in patients presenting with a deficiency in leukocyte integrins. These patients suffer from opportunistic infections due to an inability of their neutrophils to firmly adhere and migrate across the endothelium and fight tissue borne infections (2). Normal neutrophils isolated from blood respond to the binding of chemotactic formyl peptide to its surface receptor by an increase in the adhesive avidity of CD18 to bind a multitude of ligands including ICAM-1, fibrinogen, and an as yet uncharacterized ligand on the neutrophil plasma membrane (3). At saturating doses of chemotactic peptide (>500 nM) Mac-1 is rapidly mobilized to the plasma membrane from intracellular granular stores, increasing the surface expression ≈10 fold over constitutive levels (3). However, recent studies from our laboratory and others indicate that only a portion of these sites participate in adhesion. Stimulation of neutrophils in a step-wise fashion by binding nanomolar concentrations of formyl

NATO ASI Series, Vol. H 84
Biomechanics of Active Movement
and Division of Cells
Edited by N. Akkaş
© Springer-Verlag Berlin Heidelberg 1994

peptide results in newly mobilized Mac-1 which participate in adherence-dependent locomotion (4). It has been recently reported that as little as 10% of the surface Mac-1 on activated cells participate in binding to fibrinogen and ICAM-1 (5). The high avidity state of Mac-1 appears to be mediated by a change in the structure or conformation of the epitope. This adhesive state of the molecule is transient and requires continued binding of stimulatory ligand since most adhesion dependent functions are reversible within minutes following stimulation (6). The adhesive function of integrins is also a temperature and energy dependent process which requires an intact cytoskeleton and the presence of divalent cations including Ca^{2+} and Mg^{2+}. Monoclonal antibodies produced against active forms of the adhesion molecules are able to mimic intercellular activation and induce a high avidity state. We present data here on one such monoclonal antibody KIM-127 which recognizes an epitope on the ß-subunit of neutrophils and lymphocytes and promotes homotypic aggregation and other adhesion dependent functions (7). Morphologic evidence indicates that CD18 can cluster and redistribute upon stimulation and binding of adhesive ligand, in fact locomotion is dependent on this process (4). Neutrophil adhesive dependent functions rely on reversible adhesion of the β2-integrin. This along with coordinated cytoskeletal deformation and perhaps molecular redistribution enables chemotaxis and phagocytosis. Key questions which remain unknown is the relationship between cell adhesivity, the mobility of bound receptor, and the dissociation or off rate of the Mac-1-ligand bond. In this report we present data on homotypic aggregation induced by the binding of KIM-127 monoclonal antibody which induces a high avidity state in β2-integrins. The results shed light on the relative importance of the number of activated Mac-1 and their mobility in the membrane, the area of cell contact, and cell shape change on neutrophil adhesive function.

MATERIALS AND METHODS

Blood and neutrophil isolation. Human blood was collected by venipuncture into a sterile syringe containing 10 Units/ml of heparin. Neutrophils were isolated using a ficoll hypaque gradient. The neutrophil band was collected then washed and resuspended in Hepes buffer at 4°C . In aggregation assays, 1.5 mM Ca^{++} was added to the cells ($5x10^6$ cells/ml) in a ≈2 min incubation at 37°C before stimulation with 1μM formyl peptide (4-pep, CHO-Nle-Leu-Phe

Bachem BioScience Inc., Torrance, CA). Some cell preparations were preincubated with KIM-127 at a saturating concentration of 20 μg/ml at room temperature for 10 min (generous gift from Dr. M.K. Robinson, Celltech Inc., Slough, UK.). For the aggregation assay cell suspensions (0.5 ml) were placed in 12 x 75 mm polystyrene tubes and stirred with a small bar magnet (7mm x 2mm VWR Scientific, Media PA) at stir speeds of ≈500 rpm for optimum neutrophil aggregation (6). Analysis of particle and molecular events were carried out on a Facscan flow cytometer (Becton Dickinson, Mountain View, CA).

RESULTS

Flow cytometric detection and kinetics of formyl peptide stimulated aggregation. Neutrophil aggregate formation in a shear stirred suspension in response to chemotactic stimulation was quantitated on the basis of LDS-751 fluorescence and cell autofluorescence as shown in Figure 1. At maximum aggregate formation (t = 2 min) the scattergram (Fig 1a) depicts resolution of aggregates up to pentuplets. The kinetics of singlet recruitment to form doublets-quartets reveals a reversible process characterized by three distinct phases. Following stimulation with formyl peptide (4-pep) neutrophil aggregation proceeds within seconds at the highest rate. Within the first 2 minutes aggregate formation plateaus and formed aggregates remain attached for several minutes until disaggregation ensues. By 6-7 minutes more than 90% of the formed aggregates have dissociated back into single cells. We have previously formulated a model which predicts the kinetics of aggregate formation within a single population of stimulated cells (6). Particles up to pentuplets are modeled based on first order reaction kinetics where the encounter frequency is proportional to the square of the particle concentrations and a probability constant describing the formation of each aggregate geometry. The time course of stimulated aggregation was modeled as a three phase process, each phase with distinct rate constants. A fit of the kinetic data is shown by the solid line of Fig 1b. The first phase was modeled over the initial 90 sec of peptide addition with the most rapid rate of aggregation, and virtually no disaggregation. The rate of .6/P/min corresponds to ≈ 4/10 cell collisions per/minute remaining adherent. The second phase was characterized by a plateau in the rate of aggregation and a sustained low disaggregation rate. Disaggregation proceeded rapidly by 4 minutes as cells became refractory to continued peptide stimulation.

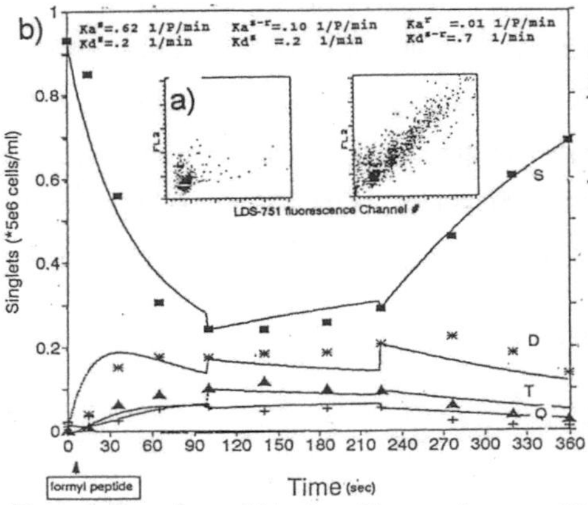

Figure 1. Detection and kinetics of homotypic neutrophil aggregation fit with a mathematical model.

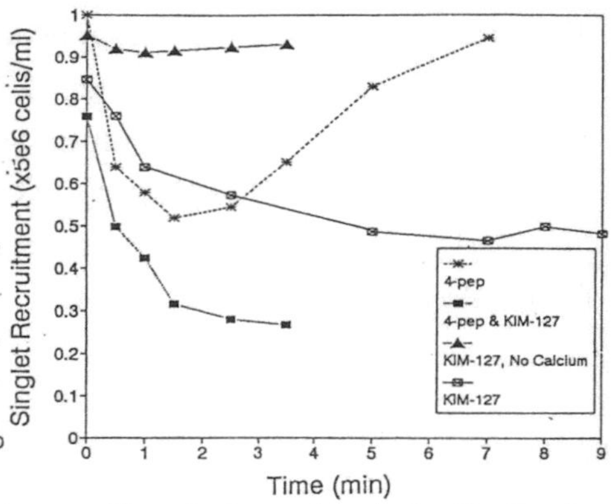

Figure 2. Aggregation stimulated by formyl peptide and mAb KIM-127.

Stimulation of cell aggregation by the binding of mAb KIM-127. Aggregation was also found to be stimulated by the binding of mAb KIM-127 to the ß2-Integrin. The rate of KIM-127 stimulated aggregate formation under conditions where mAb was prebound at room temperature was one half that found for 4-pep stimulation (K_{agg} = .3/P/min). However, the maximum extent attained with either 4-pep or KIM-127 antibody stimulation was comparable, 50-60% of singlets were recruited into aggregates. A remarkable difference was found in the time required for maximum aggregate recruitment, 4-pep required ≈ 2 min as compared to KIM-127 in which > 7 min were required. Addition of 4-pep to cells treated with mAb KIM-127 did not potentiate the rate or extent of aggregate formation over that with 4-pep alone, but did have the effect of sustaining the formed aggregates. It was confirmed that aggregation with KIM-127 required the presence of calcium ions, a requirement for the conformational change and integrin function.

Dose dependence of mAb KIM-127 binding and neutrophil aggregation. Using fluorescently labeled KIM-127 the amount of mAb bound to CD18 was determined along with the maximum extent of aggregation achieved. At binding concentrations as low as 20% of saturation with KIM-127 an aggregation response was measured. There was a linear relationship between amount bound and aggregates formed up to a plateau at 80% KIM binding (≈ 50,000 sites). The number of KIM-127 sites appears to be constant at equilibrium binding and represents only about 25% of the total number of CD18 sites which can be upregulated to the surface, estimated at ≈250,000 sites (3). However, the extent of aggregation achieved with KIM-127 was equivalent to that stimulated by 4-pep or the combination of both activators. This suggested that a subpopulation of cells were unable to participate in aggregation.

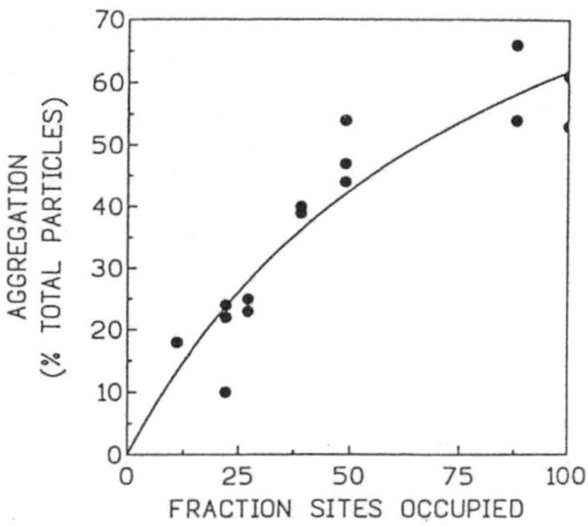

Figure 3. Dose response of KIM-127 binding and the extent of aggregate formation.

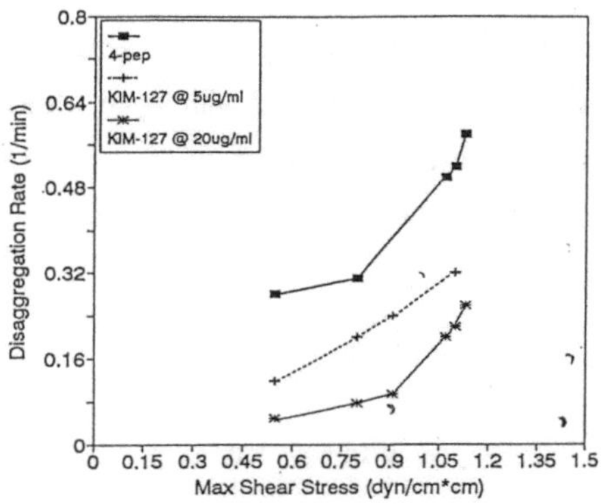

Figure 4. The rate of aggregate dissociation a a function of shear stress.

The Critical shear stress for aggregate breakup. We explored the level of shear stress required to induce aggregate breakup over the plateau phase of stable aggregate formation (Fig 4). Cells were stimulated with 4-pep or KIM-127 at an optimal shear rate ($\approx 15\ sec^{-1}$) and stable aggregates allowed to form. The shear stress in the mixing chamber was estimated assuming Newtonian laminar flow conditions. Aggregates were exposed to a step increase in the shear rate and the disaggregation rate was estimated from the particle dissociation kinetics using the mathematical model. Stimulation with 4-pep produced aggregates which were dissociated rapidly at shear stresses in excess of 1 dyne/cm^2. At the maximum dose of KIM-127 a similiar critical shear stress to induce aggregate breakup was found. KIM-127 activated adhesion sites exhibited a 4 fold slower rate of dissociation than for 4-pep stimulated adhesion sites. At KIM-127 concentrations corresponding to ≈25% of available sites aggregate breakup occured at a higher rate and did not exhibit a critical threshold, but disaggregated at a rate linearly proportional to the applied shear stress.

Adhesion and motility of albumin coated latex microspheres. Neutorphils in suspension stimulated with low doses of 4-pep (2 nM) will capture albumin coated latex beads by a ß2-Integrin adhesion receptor dependent mechanism (4). These beads which are initially homogeneously distributed along the cell body are transported backwards toward the uropod within minutes of binding to the plasma membrane (Fig 5). For cells stimulated with 4-pep alone

75% of the total number of microspheres attatched exhibit this retrograde motion and by 30 minutes are located at the uropod. A remarkable difference is observed for cells incubated with KIM-127 and then stimulated with 4-pep. These cells show an equivalent capture efficiency of beads, but virtually no motion of these adherent beads. The adherent microspheres appear to be immobile in the plasma membrane where they were bound to KIM-127 activated ß2-Integrin sites.

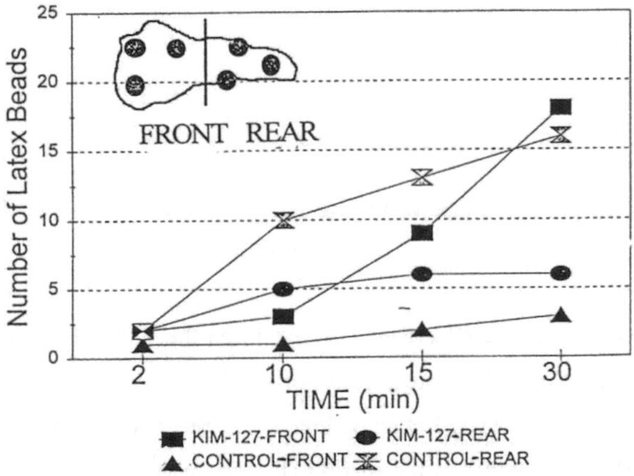

Figure 5. Adherence and motility of albumin coated latex microspheres.

DISCUSSION

Molecular Mechanisms of increased cell adhesive avidity. The onset of adhesion has been attributed to a change in the structural or chemical conformation of the ß2-integrin from one in which it has a low adhesivity to one in which it can resist external forces and mediate cell attachment. This high avidity state is reversible over the time course of chemotactic peptide stimulated aggregation, as are adhesive contacts with the endothelium and extracellular matrix in marginating neutrophils. A multitude of cell responses are measured within seconds of 4-pep stimulation including microvilli and pseudopod formation and upregulation and activation of the ß2-integrin. Neutrophils treated with mAb KIM-127 did not exhibit these rapid changes but did exhibit a homotypic aggregation response. The rate of aggregation was only 1/2 that of 4-pep stimulation but the extent was similiar and the formed aggregates remained attatched for at least 20 min.

Insight into the mechanical requirements of neutrophil adhesion. Several factors appear to be important in the initiation and strengthening of adhesive bonds these include: 1) The number of adhesive receptors and their rate of recruitment; 2) mobility of the free and bound adhesive sites; 3) the lifetime of the formed adhesive bonds; 4) recruitment of contact area and active deformation of plasma membrane. This report has begun to identify the role each of these play in neutrophil adhesion. In response to formyl peptide or other activating stimuli it appears that only a portion of the total number of sites expressed on the surface participate in adhesive function, by our estimates ≤ 20%. In addition these adhesive sites have an adhesive lifetime estimated at less than a minute, a fraction of the time in which cells remain attatched to each other or albumin coated beads (4,6). Activation with KIM-127 binding to Mac-1 sites results in intercellular bonds with a sustained lifetime. Since upregulation of the integrin from intercellular stores in response to treatment with KIM-127 alone is minimal (less than 10% that with 4-pep stimulation) it is unlikely that the maintenance of aggregation is achieved by newly upregulated sites, but rather by the same sites having a sustained adhesive conformation. Moreover, these sites appear to have a limited motility in the plane of the membrane.compared to 4-pep activation as indicated by the bead adhesion assay. This result was further confirmed by electron micrographs of cell aggregates in which the contact area of KIM-127 stimulated cells was 4 times larger than that formed by formyl peptide stimulation (data not shown). The strength of adhesion appeared to be greater for KIM-127 treated cells which showed a slower rate of breakup even at the critical shear stress for cell dissociation. This rate was dependent on the number of KIM-127 mediated adhesive sites indicating a strong dependence of strength on bond number. We assume that the bond density in the contact region of KIM-127 treated cells is at least 4 fold smaller than for formyl peptide stimulation. However, the KIM-127 bonds are less mobile and presumably have a decreased intrinsic dissociation rate of the receptor-counterstructure complex. These results shed light on likely mechanisms for the transient nature of the formyl peptide stimulated adhesive bonds. We have previously measured specific endocytosis of ß2-Integirn at a rate as high as 20%/min (9). This along with an intrinsic off rate of the adhesive conformation could account for a decrease in bond number and adhesive strength with time. We are extending this study to focus on the role of active cell deformation in the aggregation and disaggregation phases.

REFERENCES

1. Arnout MA (1990) Structure and Function of the leukocyte adhesion molecules CD11/CD18. Blood 75: 1037.

2. Anderson DC, et al (1984) Abnormality of polymorphonuclear leukocyte function associated with a heritable deficiency of high molecular weight surface glycoprotein (GP138): common relationship to diminished cell adherence. J Clin Invest 74: 536.

3. Simon SI, Rochon YP, Lynam EB, Smith CW, Anderson DC, Sklar LA. (1993) ß2-Integrin and L-Selectin are obligatory receptors in neutrophil aggregation. Blood 82: 1097.

4. Hughes BJ, Hollers JC, Crockett-Torabi E, Smith CW. (1992) Recruitment of CD11b/CD18 to the neutrophil surface and adherence-dependent cell locomotion. J Clin Invest 90: 1687.

5. Diamond MS, Springer TA. (1993) A subpopulation of Mac-1 (CD11b/CD18) molecules mediates neutrophil adhesion to ICAM-1 and Fibrinogen. J Cell Biol 120:545.

6. Simon SI, Chambers DC, Sklar LA. (1990) Flow cytometric analysis and modeling of cell-cell adhesive interactions: The neutrophil as a model. J Cell Biol 111: 322.

7. Robinson MK, Andrew D, Rosen H, Brown D, Ortlepp S, Stephens P, Butcher EC. (1992) Antibody against the Leu-CAM ß chain (CD18) promotes both LFA-1 and CR3-dependent adhesion events. J Immun. 148:1080.

8. Simon SI, Schmid-Schoenbein GW. (1990) Cytoplasmic strains and strain rates in motile polymorphonuclear leukocytes. Biophys. J. 58:319.

9. Chambers DC, Simon SI, Berger EM, Sklar LA, Arfors K-E. (1993) Endocytosis of ß2-integrins by stimulated human neutrophils analyzed by flow cytometry. J. Leuk. Biol. 53:462.

Mechanical Tension Regulates the Phenotype of Cells Cultured in a Collagen Gel.

Charles A. Lambert, Pascal Y. Lefebvre, Christophe Deroanne, Betty V. Nusgens and Charles M. Lapière.

Laboratory of Experimental Dermatology
University of Liège
CHU Sart Tilman
4000 –Liège
Belgium.

In vivo, the mechanical activity of cells expressed by migration and traction plays a major role in physiological processes such as embryogenesis, growth and development. It is equally significant in pathological conditions responsible of stenosis, sclerosis and fibrocontractile diseases. This mechanical activity of cells *in vitro* was evidenced by the wrinkling of a rubber support (Harris et al., 1980) and by the retraction of a floating collagen gel (FG) populated by cells (Bell et al., 1979). The development of mechanical forces by cells is dependent upon an efficient cytoskeleton and a functional system of connection to the extracellular matrix (ECM, Paye et al., 1987). The integrins are involved in this process (Gullberg et al., 1990; Klein et al., 1991).

In order to investigate the effect of mechanical forces on the phenotype of cells we have created a model composed of a floating collagen gel containing cells restrained in its retraction by a stainless steel wire and compared to an otherwise completely identical gel but allowed to freely contract. In our model of restrained collagen gel (BG, Lambert et al, 1992), cells tracting on the fibrous support are submitted in return to a self-generated tensional force that can be measured *in vitro.* Mechanical forces transmitted to cells by collagen polymers regulate their replicative and biosynthetic phenotype.

Measurement of mechanical forces developed by cells.

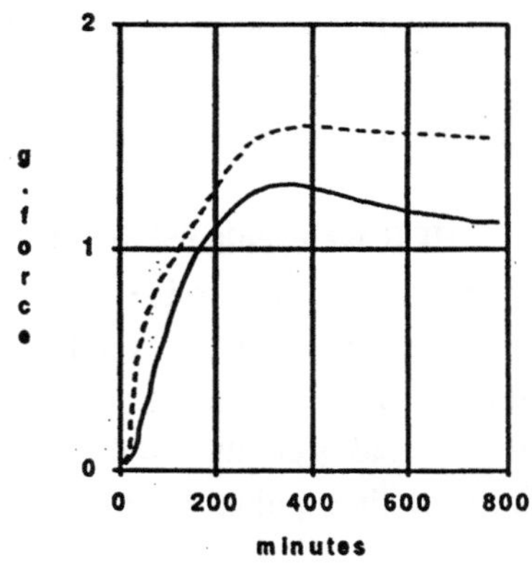

The technique used is essentially that described by Delvoye et al. (1991). Briefly, cells were suspended in a solution of collagen in culture medium and casted in a rectangular dish containing captors positionned at each

NATO ASI Series, Vol. H 84
Biomechanics of Active Movement
and Division of Cells
Edited by N. Akkaş
© Springer-Verlag Berlin Heidelberg 1994

extremities and connected to strain gauges. The collagen was allowed to polymerize. Cells attached to the collagen polymers generate a force inducing an electrical signal in the strain gauges recorded as a function of time. A typical recording for human dermal fibroblasts (HDF, 2.10^6 cells, solid line) and vascular human smooth muscle cells (HSMC, 10^6 cells, dotted line) is shown in figure 1. After a short lag time, the tension progressively developed attaining a maximal value within 5 to 6 hours. It persisted for the next 10 hours.

Effects of the mechanical tension on the cell phenotype.

1. Proliferation rate.

The proliferation rate of HDF was measured by incorporation of (^{3}H)-thymidine during a 5 days culture period in **FG** and **BG**. The incorporation was low in **FG** already at day 1 as previously reported (Sarber et al., 1981), and much higher in **BG**, indicating an increase in the rate of multiplication in response to tension (table I).

Proliferation rate of HDF
Thymidine incorporation
(cpm x 10^{-4}/dish)

Day	FG	BG
1	0.19±0.07	30.4±7.2
3	0.21±0.11	14.8±0.5
5	0.08±0.01	1.6±0.4

Table I: Incorporation of (^{3}H)-thymidine was measured as trichloroacetic acid-precipitable radioactivity. Values are mean ±SD calculated on three samples.

2. Biosynthetic activity.

The effects of tensional forces on protein synthesis was investigated both at a post and a pretranslational level (Lambert et al, 1992).

a. Proteins and collagen synthesis.

HDF were cultured in **FG** or **BG** for up to 5 days and then metabolically labelled with (^{3}H)-proline. The radioactivity incorporated in the non-dialyzable fraction, representing the overall protein synthesis, and the radioactivity associated to hydroxyproline, an index of the collagen synthesis, were measured. The non-dialysable radioactivity and the (^{3}H)-hydroxyproline were found higher in **BG** than in **FG** at all time points tested (table II). Tensional forces promote the synthesis of proteins, and more specifically that of collagen. The level of mRNA coding for the different polypeptides of the major collagens produced by fibroblasts (types I, III and VI) was measured in similar cultures. At day 2 the mRNA steady-state level of α1(I) (figure 2) and α2(I) (not shown) was 5 times higher in **BG** than in **FG** and 22 times at day 5 (not shown). A similar, although lower increase was observed on the level of mRNA coding for the three chains of collagen type VI and the collagen type III chain (not shown). The tension-induced stimulation of collagen synthesis is therefore due, at least in part, to a pretranslational regulation in the expression of the genes.

Total protein and collagen synthesis

Day	Total protein FG (cpm x 10^{-3}/µg DNA)	Total protein BG	Collagen FG	Collagen BG
1	4.2±2.8	16.1±1.6	0.4±0.3	5.5±0.8
2	6.6±1.5	27.1±3.3	1.5±0.7	7.9±1.5
3	6.2±1.7	31.1±7.8	0.2±0.6	6.1±2.9
5	6.5±3.1	25.3±7.7	1.9±0.5	6.7±0.7

Table II: Total protein and collagen synthesis were assayed as described (Juva and Prockop, 1966). The radioactivity of the cell layer and the medium were summed. Values are mean ± SD calculated for three samples.

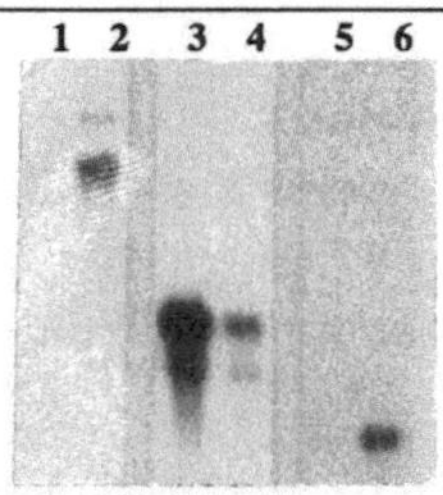

Figure 2: α1(I) (1,2), collagenase (3,4) and β-actin (5,6) mRNA levels in **FG** (1,3,5) and **BG** (2,4,6). Northern blots were probed with ^{32}P-labelled cDNA, washed and autoradiographed.

b. Interstitial collagenase.

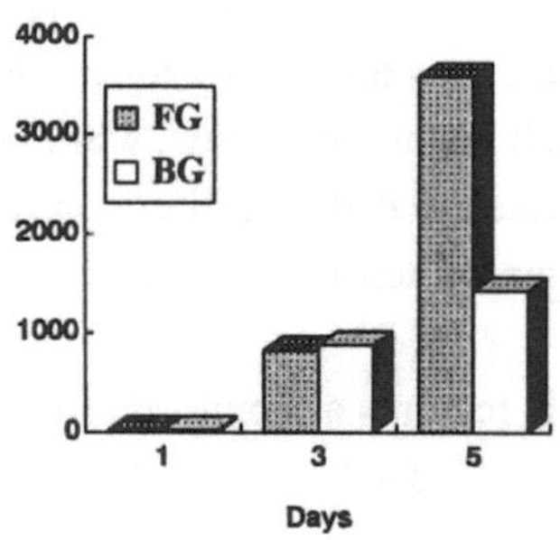

Figure 3: Collagenase activity accumulated in the medium of FG and BG as a function of time.

Interstitial collagenase is the enzyme required for initiating the degradation of the collagen types I,II and III. Its expression is often inversely coordinated with that of collagen type I. The collagenase activity secreted in the medium by the cells cultured in **FG** and **BG** measured as described by Bailly et al. (1989) similarly increased as a function of time in both **FG** and **BG** during the first 3 days of culture, but was significantly lower in **BG** as compared to **FG** after five days of culture. A reduction of the same amplitude in the level of collagenase mRNA was observed in HDF cultured in **BG** as compared to **FG** (figure 2). The expression of collagenase is therefore also modulated by mechanical forces.

β–actin concentration.

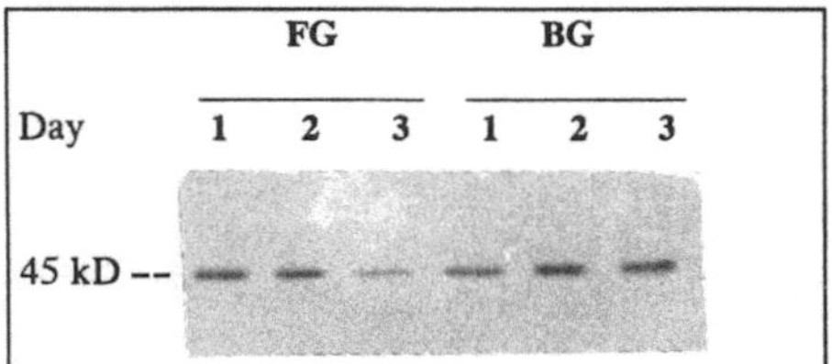

Figure 4: Proteins were separated by SDS–PAGE and transfered onto nitrocellulose membrane. The membrane was sequentialy probed with a monoclonal anti β–actin and goat anti–mouse IgG antibodies conjugated to peroxidase. Signal was detected using an ECL kit (Amersham).

The level of β–actin in cells cultured in **FG** and **BG** for one to three days was investigated by western blotting using specific antibodies. It decreased as a function of time in cells cultured in **FG**, and remained constant in cells cultured in **BG** (figure 4). A similar tension–induced increase was observed at the level of the mRNA in HDF (figure 2).

Discussion

The proliferative and biosynthetic phenotype of cultured cells is submitted to multiple regulatory mechanisms arising mainly from growth factors present in the medium and from interactions with their support through specific receptors. By comparing cells embedded in floating collagen gels allowed to freely retract (**FG**) or mechanically restrained (**BG**), the main variable parameter is the tension to which cells are submitted. In **FG**, the mechanical tension on the cell membrane is low, while in **BG** it is much higher. By using this model we demonstrated that the proliferation of the cells and their biosynthetic phenotype were modulated by tension. These regulations operate, at least in part, at a pretranslational level.

The development of mechanical forces by cells requires a connection between the cytoskeleton and the extracellular substrate. Cells interact with their protein substratum through specific transmembrane receptors, the integrins (Hynes, 1992). Their participation in the mechanical activity of the cells (attachment, migration, contraction) is clearly supported by the elegant studies of Chan et al. (1992) demonstrating that the cytoplasmic domain of the α sub–unit is involved in its specification. Actin actively participates in the generation of the mechanical forces (Bell et al, 1979). The cytoplasmic domain of the integrins is connected to actin microfilaments at the level of focal adhesion plaques through a series of interlinked proteins (Burridge and Fath, 1989) providing a continuum between the ECM and the contractile apparatus of the cell. The cytoskeleton might be involved in the perception of the tension by the cell since modifications of its organisation responsible for changes in shape often parallels modulation of specific genes (Unemori and Werb, 1986). The high level of collagenase synthesis by cells under low mechanical tension as in **FG** and its reduction in **BG** might depend on a similar process. The regulation of the collagenase gene is largely under the control of a cis regulatory element binding the AP1 complex. An overexpression of c–fos and c–jun has been observed in **FG** (unpublished). The lower expression of collagen and β–actin in the same **FG** seems to be controlled by other transacting factors.

Although numerous studies have dealt with the transduction pathway of the mechanical signal to the cell machinery , the nature of the prime event converting a physical stimulus into a chemical signal remains essentially unknown. Tension in the plasma membrane could direct conformational and functional changes of membrane associated molecules such as integrins, growth factor receptors, phospholipases, G proteins, Na^{+}–K^{+}–ATPases or ion channels. In our model, gadolinium (Gd^{3+}, up to 100μM) did not affect the force developed by HDF nor the levels of mRNA for α1(I), collagenase and β–actin (data not shown) suggesting that Gd^{3+}–sensitive stretch–activated ion channels are unlikely to be the mechanotransducer. A similar observation was made by Sadoshima et al. (1992) on cardiac myocytes. The cytoplasmic domain of the integrins might operate in the control of metabolic functions by its association with the tyrosine kinase c–src. A cascade of phosphorylations catalysed by kinases of various types participate in the process since their inhibition can suppress the overexpression of collagenase and down regulation of α1(I) collagen gene in **FG** (Lambert et al., 1993). The control of activity of the kinases by cytoskeleton–associated molecules can be suggested on the basis of the multiple interactions existing between cytoskeletal proteins and elements of the signaling machinery. Integrins and phospholiapase C are linked to microfilaments (Payrastre et al, 1991). Profilin binds to phosphatidylinositol bisphosphate and blocks its susceptibility to phospholipase C , this binding being displaced by actin monomers (Lassing and Lindberg, 1985). Even though we have observed that phospholipase C and tyrosine kinases participate in the control of collagenase and α1(I) collagen gene expression in the **FG** (Lambert et al., 1993), further work is required to understand how they are involved in the transduction of a mechanical signal.

Bailly C, Dreze S, Asselineau D, Nusgens BV, Lapière ChM, Darmon M (1990) Retinoic acid inhibits the production of collagenase by human dermal keratinocytes. J Invest Dermatol 94:47–50.

Bell E, Ivarsson B, Merrill C (1979) Production of a tissue–like structure by contraction of collagen lattices by human fibroblasts of different proliferative potentials *in vitro*. Proc Natl Acad Sci USA 76:1274–1278.

Burridge K, Fath K (1989) Focal contacts – transmembrane links between the extracellular matrix and the cytoskeleton. Bioessays 10: 104–108.

Chan BMC, Kassner PD, Schiro JA, Byers HR, Kupper TS, Hemler ME (1992) Distinct cellular functions mediated by different VLA α subunit cytoplasmic domains. Cell 68: 1051–1060.

Delvoye P, Wiliquet P, Lévêque JL, Nusgens BV, Lapière ChM (1991) Measurement of mechanical forces generated by skin fibroblasts embedded in a three–dimensional collagen gel. J Invest Dermatol 97:898–904.

Gullberg D, Tungstrom A, Thuresson, AC, Olsson L, Terracio L, Borg TK, Rubin K (1990) β1 integrin–mediated collagen gel contraction is stimulated by PDGF. Exp Cell Res 186:264–272.

Harris AK, Wild P, Stopak D (1980) Silicone rubber substrata: a new wrinkle in the study of cell locomotion. Science 208: 177–179

Hynes RO (1992) Integrins – versatility, modulation, and signaling in cell adhesion. Cell 69:11–25.

Juva K, Prockop DJ (1966) Modified procedure for the assay of ^{3}H– or ^{14}C–labeled hydroxyproline. Anal Biochem 15:77–82.

Klein LE, Dressel D, Steinmayer T, Mauch C, Krieg T, Weber L (1991) Integrin alpha 2 beta 1: evidence for a pivotal role in the reorganization of collagen I by human fibroblasts. J Invest Dermatol 96: 1015–1022.

Lambert ChA, Soudan EP, Nusgens BV, Lapière ChM (1992) Pretranslational regulation of extracellular matrix macromolecules and collagenase expression in fibroblasts by mechanical forces. Lab Invest 66: 444–451.

Lambert ChA, Martens H, Nusgens BV, Lapière ChM (1993) Regulation of collagenase and Col1A1 genes, but not of β–actin gene, in fibroblasts populated collagen gel is mediated via tyrosine kinases and phospjolipase C. J Invest Dermatol 100:436 (abstract)

Lassing I, Lindberg U (1985) Specific interaction between phosphatidylinositol 4,5–bisphosphate and profilactin. Nature 314:472–474.

Payrastre, B, van Dergen en Henegouwen PMP, Breton M, den Hartig JC, Plantavid M, Verkleij AJ, Boonstra J (1991) Phosphoinosotide kinase, diacylglycerol kinase, and phospholipase C activities associated to the cytoskeleton: effect of epidermal growth factor. J Cell Biol 115: 121–128

Sadoshima J–I, Takahashi T, Jahn L, Izumo S (1992) Role of mechano–sensitive ion channels, cytoskeleton,and contractile activity in stretch–induced immediate–early gene expression and hypertrophy of cardiac myocytes. Proc Natl Acad Sci USA 89:9905–9909.

Sarber R, Hull B, Merrill C, Soranno T, Bell E (1981) Regulation of proliferation of fibroblasts of low and high population doubling level grown in collagen lattices. Mech Ageing Dev 17:107–112.

Unemori EN and Werb Z (1986) Reorganisation of polymerised actin: a possible trigger for induction of procollagenase in fibroblasts cultured in and on collagen gels. J Cell Biol 103:1021–1031.

The Mechanism of Actomyosin and Microtubule-Kinesin Systems

Yong-Ze Ma and Edwin W. Taylor
Dept. of Mol. Gen. and Cell Biol.
Universiy of Chicago
Chicago, IL,60637
U. S. A.

INTRODUCTION

Force and motion in cells are generally produced by the action of three kinds of protein complexes, actin-myosin, microtubule-dynein and microtubule-kinesin. The actomyosin system is responsible for muscle contraction, cell cleavage, organelle movement in plant cells, amoeboid movement and some aspects of fibroblast locomotion. The kinesin family of proteins is involved in the movement of membrane bounded vesicles and possibly chromosomes.

The interesting problems are how the energy of ATP is used to produce force and motion and whether the different protein systems work by the same underlying mechanism. A mechanism in the sense of a kinetic scheme which specifies the intermediates in the hydrolylsis cycle and the rate and equilibrium constants of each step, provides the basis for a mechano-chemical model. The problem to be discussed is whether the same kinetic mechanism is used by the actomyosin and microtubule-kinesin systems.

The assumption of most mechanochemical models is that the motor protein-filament complex undergoes a structural change in one or more

NATO ASI Series, Vol. H 84
Biomechanics of Active Movement
and Division of Cells
Edited by N. Akkaş
© Springer-Verlag Berlin Heidelberg 1994

steps of the cycle which generates a strain force. This step can be thought of as a rotation or deformation of the motor protein which stretches an elastic element in the globular head region or a connecting link (the coiled-coil region of myosin or kinesin). The increase in strain along the direction of the motion by a distance *d* produces a force F= ad where *a* is the stiffness of the spring. The force produces relative motion of the actin or microtubule or if motion is prevented an external force is generated. The distance *d* is refered to as the step distance and is expected to be the order of 5 to 10 nm. Recent evidence on the structure of the myosin head and the actomyosin complex (Rayment et. al 1993) provides support for a large deformation of the motor. The force produced per molecule is a few picoNewtons and the stiffness is approximately 10^{-3} Nm^{-1} (Huxley and Simmonds,1971).

THE ACTOMYOSIN MECHANISM

The kinetic scheme for actomyosin in solution has been studied in several laboratories (reviewed by Hibberd and Trentham,1986). The scheme has been extended to myofibrils to determine the effect of shortening velocity on the rate constants of the mechanism (Ma and Taylor,1994). The scheme is summarized in the table. The important features are the following. The binding of ATP induces a very fast structural change in the actomyosin complex which greatly weakens the binding of actin (factor of 10^4) and the complex dissociates at a rate too fast to measure. M·T can rebind at a high rate in the lattice where the effective concentration is much higher than in solution. Both the M·T and M·D·P are essentially in equilibrium with the corresponding bound states. This feature of weak binding and very fast attachment and detachment means that these states do not contribute to active force and do not impede the motion by generating a strain force in the opposing direction.Dissociation of phosphate and isomerization to a more strongly bound state (MC·D) is the step in which force is generated. In muscle increasing the phosphate concentration decreases the isometric force (Hibberd et. al.,1985) which supports the assignment of force generation to step three. The dissociation of ADP is a fast step which will generally occur before the protein complex can dissociate. The strain force, in the positive or negative direction is exerted over the time interval from formation of MC·D to detachment of the complex in the MC·T state. Does the kinetic scheme account for the mechanical properties? In general rate and equilibrium constants are expected to be altered for transitions in which strain increases (or decreases). For myofibrils which are shortening while developing a net zero external force the effect is small since values of the four rate constants measured in solution and in myofibrils differ by less than a factor of two.The constant velocity of shortening (U) requires internal strain forces to cancel on average. The force developing state MC·D decays with rate constant $k = k_1k_4/(k_1 + k_4)$ therefore the condition of zero force is $\int(d-Ut) \exp(-kt)\, dt = 0$ which gives $U = dk$. The simple model is approximately correct but we have neglected the lag in the decay of the force developing state since the process occurs in two steps. U is 6 micrometers sec^{-1} for myofibrils and from k of 400 sec^{-1} the step distance is 15 nm which is in the range expected.

The kinetic scheme is consistent with the mechanical behavior. The motor exerts a force for only about 5% of the ATPase cycle time ($V/k_4 = 0.05$). Thus a high velocity can be produced by summing the contributions of a large number of motors acting independently.

ACTOMYOSIN AND KINESIN MICROTUBULE MECHANISMS

$$
\begin{array}{ccccccccc}
 & k_1 & & k_2 & & k_3 & & k_4 & \\
MC & \longrightarrow & MC{\cdot}T & \longrightarrow & MC{\cdot}D{\cdot}P & \longrightarrow & MC{\cdot}D & \longrightarrow & MC \\
\updownarrow & +T & \updownarrow & & \updownarrow & -P & \updownarrow & -D & \updownarrow \\
M & \longrightarrow & M{\cdot}T & \longrightarrow & M{\cdot}D{\cdot}P & \longrightarrow & M{\cdot}D & \longrightarrow & M
\end{array}
$$

STATE	TRANSITION TO STATE		RATE OF DISSOCIATION	
	ACTOMYO	MT-KINESIN	ACTOMYO	MT-KINESIN
MC·T	$k_1 \geq 1500$	$k_1=200$	>1500	<<50
MC·D·P	$k_2=100$	$k_2=50$	>1500	?(slow)
MC·D	$K_3=100$	$k_3 > 30$	50-100	?30-50
MC	$k_4= 400$ $k_4/V=20$	$k_4=30$ $k_4/V=2$	0.2	0.2

MC is motor protein complex, M is motor protein, T, D, and P are ATP, ADP and phosphate, all rate constants in sec-1, V is steady state rate (sec^{-1}), 20°, pH 7, 100mM salt for actomyosin and myofibrils, 50 mM salt for microtubule kinesin. Rate constants for transitions: k_1, maximum rate of ATP binding, apparent rate constant for step 1 is $K_0k_1[T]/(K_0[T] + 1)$ where K_0 is rapid equilibrium complex; k_2, phosphate burst transient rate; k_3, calculated from burst and steady state rate; k_4, maximum rate of ADP dissociation in reaction of M·D with actin or microtubules; for myofibrils k_4 calculated from K_M for shortening, k_1 and K_0. Dissociation rate constants for actomyosin measured directly or from rate of association and equilibrium constant; for kinesin rates measured indirectly from rate of association and equilibrium constant. References, Taylor(1991), Sadhu and Taylor(1992), Ma and Taylor(1994) and Ma and Taylor(in prep).

MICROTUBULE-KINESIN MECHANISM

The motion of vesicles along microtubules and the motion of microtubules over a kinesin coated surface is strikingly different from the actomyosin system. A single (two headed) kinesin molecule produces the same or a slightly larger velocity as many kinesins and the motion can continue for several micrometers before detachment of the motor protein (Howard et. al.,1989,Block et. al.,1990).Is this difference explained by the kinetic mechanism?

The mechanism has not been worked out in the same detail as the actomyosin mechanism but there are clear differences. Although the same set of steps occurs and there is a rapid phosphate burst the rate constants of some steps are different(see table). For kinesin, phosphate dissociation has a much larger rate than ADP dissociation consequently the major steady state intermediate is M·D and its decay is the rate limiting step (Hackney, 1988,Sadhu and Taylor,1992).The relative rates of these steps is reversed for myosin. The rate of ADP dissociation from the MC·D complex is 500 to 1000 times larger than for M·D as is the case for actomyosin versus myosin. However the rate constant is at most twice as large as the steady state turnover rate while for actomyosin the ratio is twenty. It is this large ratio and the weak binding and rapid dissociation of substrate and product states which allows the summation of the contributions of myosin heads to produce the large velocity.

This difference is part of the explanation of why a single motor is as fast as many. But what about the rate of detachment of the motor? The binding of single or double headed kinesin to microtubules is only slightly weaker in the presence of ATP compared to ADP although both nucleotides reduce the binding at least one hundred fold. The lack of a suitable signal has so far prevented a reliable measurement of the rate of dissociation of the motor protein complexes.The larger rate of the hydrolysis step for the motor protein complex suggests that this step must occur before dissociation and therefore the rate of dissociation of the MC·T is much less than 50 sec^{-1}. From the rate constant of binding of M·D and the equilibrium constant the rate constant of dissociation is calculated to be 30 to 50 sec^{-1}. From the very preliminary evidence it appears that detachment of the motor is slow and probably slower than the transitions between the motor protein complexes in the hydrolysis pathway. The evidence supports a model in which the heads remain attached for most of the cycle.

Which transition corresponds to the force step? In actomyosin this step must be post hydrolysis because earlier states detach very rapidly. In the kinesin system there is no such restriction. It has been proposed(Romberg and Vale,1993) that force is produced in the isomerization induced by ATP(step 1). This possibility is not contradicted by the available evidence.It is harder to explain why kinesin has a low probability of detachment per cycle. Models in which one head detaches when the other steps are easy to propose but difficult to test or even to formulate in a consistent way.

REFERENCES

Block, S. M.,Goldstein, L. S. B., and Schnapp, B. J. (1990) Bead movement by single kinesin molecules studied wilth optical tweezersNature **348**, 348- 352.

Hackney, D. D.,(1988) Kinesin ATPase; Rate limiting ADP release. Proc. Nat. Acad. Sci. U. S. A. **85**, 6314-6318.

Hibberd,M. G., and Trentham, D. R.,(1986) Relationships between chemical and mechanical events during muscle contraction. Ann.

Rev. Biophys. Chem. **15**, 119-161.

Hibberd, M. G., Dantzig, J. A., Trentham, D. R., and Goldman, Y. E. (1985) Phosphate release and force generation in skeletal muscle fibers. Science **228**, 1317-1319.

Howard, J., Hudspeth, A. J. and Vale, R. D. (1989) Movement of microtubules by single kinesin molecules. Nature **342,** 154-158.

Huxley, A. F., and Simmonds, R. M. (1971) Proposed mechanism of force generation in striated muscle. Nature **233**, 533-538.

Ma, Y. Z. and Taylor, E. W. (1994) Kinetic Mechanism of Myofibril ATPase. Biophys. J. in press.

Rayment, I., Holden, H. M., Whittaker, M., Yohn, C. B., Lorenz, K. C., Holmes, K.C.and Milligan, R. A. (1993) Structure of the Actin-Myosin complex and its implications for muscle contraction. Science **261**,58-65.

Romberg, L. and Vale, R. D. (1993) Chemomechanical cycle of kinesin differs from that of myosin. Nature **361**, 168-170.

Sadhu A. and Taylor, E. W. (1992) Kinetic study of the kinesin ATPase. J. Biol. Chem. **267**, 11352-11359.

Taylor, E. W. (1991) Kinetic studies on the association and dissociation of myosin subfragment 1 and actin. J. Biol. Chem. **266**, 294-302.

Immunolocalization of acetylated microtubules in germ cells of insects

Klaus Werner Wolf
Institut für Biologie
Medizinischen Universität zu Lübeck
Ratzeburger Allee 160
23538 Lübeck
Germany

Summary

The acetylation status of microtubules (MTs) has been assessed in dividing germ cells of males of four insect species using an antibody specific for acetylation in positon 40 of the amino-terminal domain of α-tubulin and indirect immunofluorescence. As a rule, spermatogonial spindles were highly acetylated. Acetylation in spermatocytes of two Lepidoptera species, the Mediterranean mealmoth and the large white, differed. In the relatively small spindles of the moth, only the MTs of the midbodies, an array of interdigitated MTs located between the daughter nuclei in telophase, were consistently acetylated. In the larger spindles of spermatocytes of the large white, however, kinetochore-MTs were acetylated in addition to midbody MTs. The finding that the degree of acetylation increases with spindle size was confirmed by comparing the small meiotic spindles of a beetle with the larger spindle apparatuses in male meiosis of a locust. I conclude that MTs of spermatogonia and spermatocytes differ significantly in their ability to serve as a substrate for acetylases, which, in turn, may be interpreted to indicate higher stability of MTs in spermatogonia. In male meiosis, MTs of large spindles tend to be acetylated to a higher degree.

NATO ASI Series, Vol. H 84
Biomechanics of Active Movement
and Division of Cells
Edited by N. Akkaş
© Springer-Verlag Berlin Heidelberg 1994

Introduction

Tubulin is the substrate for various reversible post-translational modifications such as phosphorylation, glutamylation, tyrosination and acetylation. Addition of an acetyl residue to lysine 40 in the N-terminal domain of α-tubulin is widespread (for a review, see LeDizet and Piperno 1991). The functional meaning of α-tubulin acetylation is not understood. Possibly it is not essential, since it is missing in specific tissue culture cells. However, evidence from a variety of systems indicates that acetylated microtubules (MTs) have a lower turnover than non-acetylated MTs (e.g. Webster and Borisy 1989). Thus, the degree of α-tubulin acetylation serves as a marker to assessing MT dynamics.

Using a monclonal antibody, 6-11B-1 (Piperno and Fuller 1985), and indirect immunofluorescence, the acetylation status of MTs in mitotic and meiotic spindles of a series of insect species was determined. For comparison, a monoclonal antibody against tyrosinated α-tubulin, YL1/2 (Kilmartin et al. 1982), and a commercially available monclonal antibody against ß-tubulin were used. The animals were the Mediterranean mealmoth, Ephestia kuehniella (Pyralidae, Lepidoptera), the large white, Pieris brassicae (Pieridae, Lepidoptera), a beetle, Tenebrio molitor (Tenebrionidae, Coleoptera), and the desert locust, Schistocerca gregaria (Acrididae, Orthoptera).

Material and Methods

Larvae of wild type laboratory strains of the Mediterranean mealmoth, E.kuehniella and T. molitor were reared on rolled oates. S.gregaria was obtained from the Institute of Zoology (Univ. Hamburg, Germany). Larvae of P.brassicae were collected in the field.

The preparation of the cytoskeletons was performed as described previously (Wolf and Bastmeyer 1991). The specimens were stained with one of three different monoclonal anti-tubulin antibodies. These were 1) a mouse antibody against ß-tubulin (Sigma), 2) a rat antibody, YL 1/2 (Kilmartin et al. 1982), against tyrosinated α-tubulin, and 3) a mouse antibody, 6-11B-1 (Piperno and Fuller 1985), against acetylated α-tubulin. The chromatin was stained with a DNA-specific fluorescent dye, DAPI (for details, see Wolf and Bastmayer 1991).

Results

Use of an antibody against ß-tubulin, tyrosinated α-tubulin, and acetylated α-tubulin produced virtually the same images in spermatogonia. As exemplified in E.kuehniella using the 6-11B-1 antibody (Fig. 1a, b, 2a, b), the entire spindles were stained in all 4 species. The reaction pattern in spermatocytes was more complex (Figs. 3 - 13). Although the images obtained with an antibody against ß-tubulin or tyrosinated α-tubulin did not differ much - the entire microtubular cytoskeleton was stained in all stages of both meiotic divisions in all 4 systems studied - there was differential α-tubulin acetylation (Tab.1).

Tab. 1: α-tubulin acetylation in meiosis of 4 insect species.

species	spindle size	spindle component stained in	
		metaphase, prometaphase anaphase	late telophase
E.kuehniella	small	- (Fig. 5a)	midbody MTs (Fig. 6a)
P.brassicae	large	kinetochore MTs (Fig. 8a)	midbody MTs
T.molitor	small	- (Fig. 10a)	midbody MTs
S.gregaria	large	kinetochore MTs (Fig. 13a)	midbody MTs

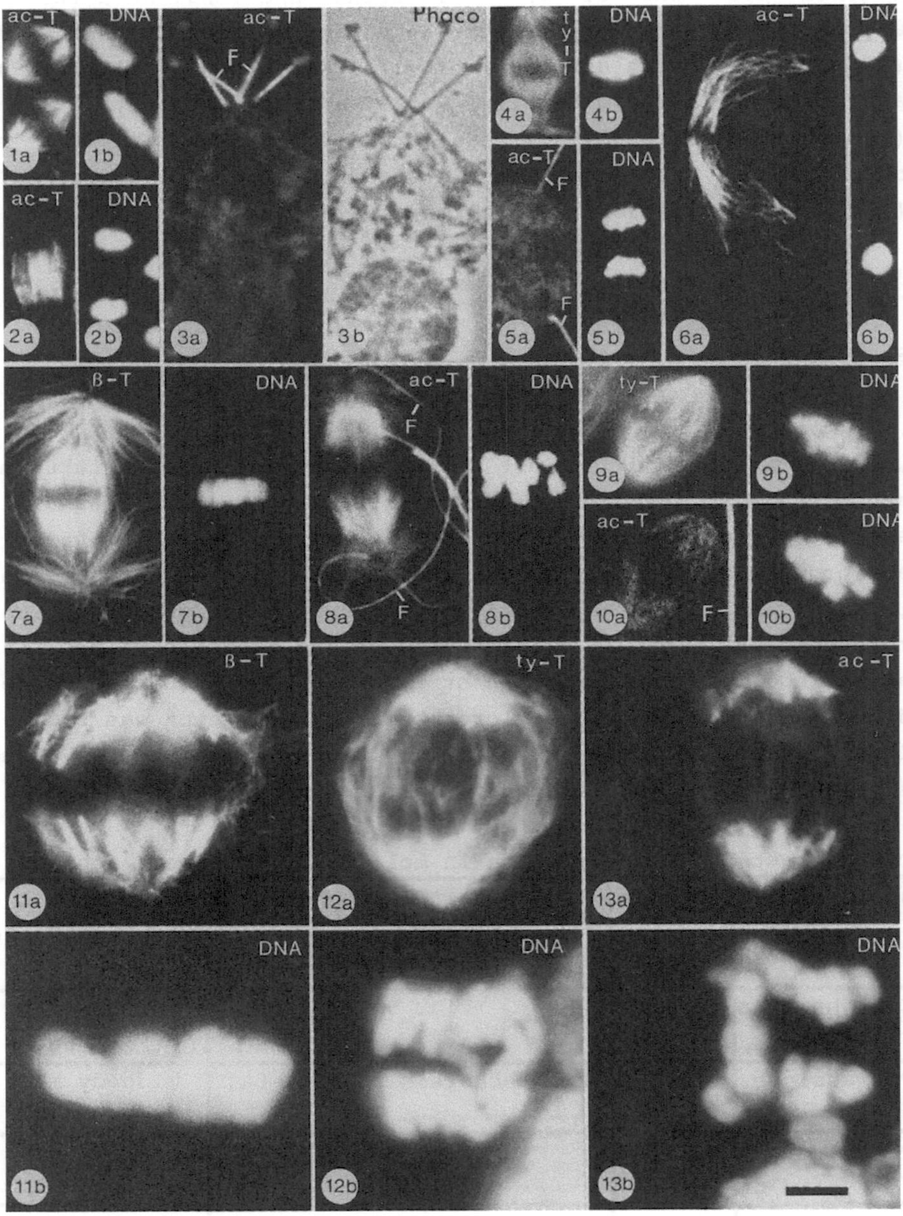
ac-T
DNA
ac-T
Phaco
ty-T
DNA
ac-T
DNA
F
1a
1b
ac-T
DNA
2a
2b
3a
3b
4a
4b
ac-T
DNA
5a
5b
6a
6b
β-T
DNA
ac-T
DNA
ty-T
DNA
7a
7b
8a
8b
9a
9b
ac-T
DNA
10a
10b
β-T
ty-T
ac-T
11a
12a
13a
DNA
DNA
DNA
11b
12b
13b

Discussion

In a series of 4 insect species, MTs were highly acetylated throughout spermatogonial mitosis. In contrast, acetylation was confined to midbody MTs in meiotic telophase spindles of male animals. When the meiotic spindles were large, also kinetochore MTs were acetylated. In light of the finding that stable MTs tend to be acetylated (e.g. Webster and Borisy 1989), the differences between mitosis and male meiosis in the acetylation status of spindle MTs point at a generally higher turnover within the latter. The conclusion that acetylation indicates more stable MTs is corroborated by the observations in meiotic spindles. Midbody MTs are considered relatively stable entities (Shelden and Wadsworth 1990) and were acetylated in male meiosis of all 4 species examined.

Errors during meiotic chromosome segregation have direct consequences on the propagation of the species, since aneuploid sperm arise which may fertilze eggs. It is, therefore, tempting to speculate that nature achieves greater fidelity in chromosome segregation in male meiosis through a higher turnover of the microtubular cytoskeleton.

Figs. 1a - 13a display immunolabelled cytoskeletons of mitotic and male meiotic cells of 4 different insect species printed at the same magnification. Figs. (b) show a phase contrast (Phaco) image of the same cells or the DNA stained with DAPI (4',6-Diamidino-2-phenylindole·2 HCl). Bar in Fig. 13b represents 5μm. Elements stained: ß-tubulin: ß-T, tyrosinated α-tubulin: ty-T, acetylated α-tubulin: ac-T, flagella: F.

Fig. 1 E.kuehniella, metaphase spermatogonium.
Fig. 2 E.kuehniella, telophase spermatogonium.
Fig. 3 E.kuehniella, late prophase I spermatocyte.
Fig. 4 E.kuehniella, metaphase II spermatocyte.
Fig. 5 E.kuehniella, anaphase II spermatocyte.
Fig. 6 E.kuehniella, telophase II spermatocyte.
Fig. 7 P.brassicae, metaphase II spermatocyte.
Fig. 8 P.brassicae, prometaphase II spermatocyte.
Fig. 9 T.molitor, metaphase II spermatocyte.
Fig. 10 T.molitor, metaphase II spermatocyte.
Fig. 11 S.gregaria, metaphase I spermatocyte.
Fig. 12 S.gregaria, anaphase I spermatocyte.
Fig. 13 S.gregaria, anaphase I spermatocyte

Acknowledgements

The author is indebted to Prof. W. Traut (Lübeck) for critically reading the manuscript. I would like to thank Dr. J.V. Kilmartin (MRC, Lab. Molec. Biol. Cambridge, U.K.) and Dr. G. Piperno (Rockefeller Univ., New York, U.S.) for the generous supply of YL1/2 and 6-11B-1 antibody respectively.

References

Kilmartin JV, Wright, B Milstein, C (1982) Rat monoclonal antitubulin antibodies derived by using a new nonsecreting rat cell line. J Cell Biol 93:576-582.

LeDized, M, Piperno, G (1991) Detection of acetylated α-tubulin by specific antibodies. In Methods in Enzymology Vol. 196, Molecular Motors and the Cytoskeleton, Vallee RB (ed), pp.264-274, San Diego: Acad. Press.

Piperno G, Fuller MT (1985) Monoclonal antibodies specific for an acetylated form of α-tubulin recognize the antigen in cilia and flagella from a variety of organisms. J Cell Biol 101: 2085-2094.

Shelden E, Wadsworth P (1990) Interzonal microtubules are dynamic during spindle elongation. J Cell Sci 97:273-281.

Webster DL, Borisy GG (1989) Microtubules are acetylated in domains that turn over slowly. J Cell Sci 92:57-65.

Wolf KW, Bastmeyer M (1991) Cytology of Lepidoptera. V. The microtubule cytoskeleton in eupyrene spermatocytes of Ephestia kuehniella (Pyralidae), Inachis io (Nymphalidae), and Orgyia antiqua (Lymantriidae). Eur J Cell Biol 55:225-237.

Mechanisms of α_1-adrenergic induced positive inotropism in cardiac cells

G. Vassort, C. Vannier, V. Lakomkine, M. Pucéat, O. Clément-Chomienne
Laboratoire de Physiologie Cellulaire Cardiaque
I.N.S.E.R.M. U-241
Centre Universitaire, Bât. 443
F-91405 Orsay cedex
France

Key Words : Muscle, Heart cell, Neurohormone, Phenylephrine, Contraction, Crossbridge cycling.

Force development and shortening in muscle result from cycling interaction of the contractile proteins, myosin-heads and actin forming crossbridges. In all muscle cells, the extent of interaction between actin and myosin is regulated by the cytosolic Ca concentration whatever the specific protein that confers Ca sensitivity of striated muscles is: troponin and tropomyosin associated with the thin filament in Vertebrates, or directly myosin in Invertebrates. Crossbridges cycle between force generating and non-force generating states; the force generating states are believed to be strongly attached to actin (Huxley, 1957). According to this two states model, the force that a muscle develops can be characterized by the apparent rate constants for the formation of force generating states (f_{app}) and the dissociation of force generating states (g_{app}). Assuming that the apparent rate constants for the reverse transitions can be neglected, the steady-state fraction of turning over crossbridges in the force generating states (α_{Fs}) is: $\alpha_{Fs} = f_{app}/(f_{app} + g_{app})$ and the isometric force $F = n\, F\, \alpha_{Fs}$, where n is the number of turning over crossbridges per half sarcomere and F is the mean force produced by one crossbridge in the force generating state.

On increasing Ca concentration, force increases due to an increase in n and/or to an increase in α_{Fs}; indeed, an increase in f_{app} has been reported occurring with or without a decrease in g_{app} (Brenner, 1988; Kerrick et al., 1991). According to Brenner (1988), isometric force, stiffness and ATPase activity are regulated by

NATO ASI Series, Vol. H 84
Biomechanics of Active Movement
and Division of Cells
Edited by N. Akkaş
© Springer-Verlag Berlin Heidelberg 1994

Ca^{2+} through changes in crossbridge turnover kinetics, whereas the total number of crossbridges involved in active cycling appears unchanged.

It has long been considered that neuromodulation of cardiac inotropism is mediated by variations in the intracellular Ca concentration. Even if in general Ca is the major controlling factor, there are now a few, well-defined situations in which the force developed by the contractile proteins at a given Ca concentration is altered. Thus, contractile proteins appear less sensitive to Ca after ß-adrenergic stimulation so that the force-pCa relation is shifted to the right. The opposite occurs following an increased myofilament Ca sensitivity after α_1-adrenergic stimulation or when applying the so-called "Ca-sensitizers" such as bepridil, sulmazol, EMD... (Terzic et al., 1993).

To investigate in greater details the latter effects and to be able to analyse the underlying mechanisms, was a technique developed allowing to measure the force on a single cardiac cell whose contractile proteins are made accessible to the bathing fluid after chemical skinning of the cell with Triton X-100 (Pucéat et al., 1990). Moreover, kinetic properties of the cardiac muscle cell were studied by a method of analysis in the frequency domain. This method overcomes the problem of resolving whether differences arise from variations in the number of active crossbridges or from the rate at which they cycle (Kawai & Brandt, 1980).

The system for recording tension consisted of a transducer (model AE 801, SensoNor a.s., Horten, Norway) with a thin 3-cm-long glass rod. It was connected to an amplifier to yield a sensitivity of 12.5 mV/μN with a noise level below 2 μN and a bandwith of 200 Hz. A suspension of skinned cells in the relaxing solution was placed in a Petri dish on the stage of an inverted microscope. The tip of the glass rod, first soaked in a small drop of optical adhesive (Norland Products, Inc., North Brunswick, NJ) was gently put on one end of a myocyte. This end of the cell was then exposed to long wavelength ultraviolet light for 3 minutes to cure the adhesive. This adhesive had the advantage of remaining in the form of a drop at the tip of the glass rod when going through the air-water interface. The other end of the cell was similarly glued to the tip of a high voltage piezoelectric translator including a position controller (model P 172.07, Physik Instrumente, Waldbronn, Germany). This allowed to subject the cell to small-amplitude sinusoidal length oscillations (peak to peak amplitude 1-5 % L_{max}) at 9 discrete frequencies ranging from 0.2 to 10 Hz software generated and defined by 400 steps (0.2, 0.5, 0.7 Hz) or by 100 steps (1, 1.3, 1.7, 3, 5, 10 Hz). Tension and length signals were digitized and processed to yield a frequency response function. The time required to collect one spectrum was approximatively 90 s; the sequence being repeated 4 fold in a given condition to increase signal-noise ratio. The frequency at which dynamic stiffness is minimum was termed f_{min} (Kawai & Brandt, 1980; Rossmanith et al., 1986). The modulus of dynamic stiffness was defined as the ratio $\Delta F/\Delta L$ in which ΔF is the amplitude of the force oscillations and ΔL is the amplitude of muscle length oscillation. Sarcomere length was adjusted under the microscope and was estimated by counting the number of sarcomeres in two different 25-μm-long areas of the cell (average length, 2.08 ± 0.04 μm; n=33 cells). The attached myocyte was then

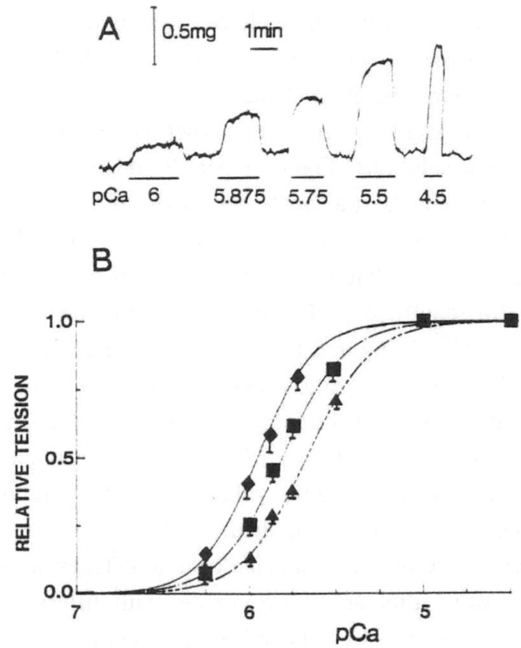

Figure 1: Mechanical activity recorded on an isolated chemically skinned cardiac cells.

A) Original recording of tension elicited by applying solutions containing different Ca concentrations (pCa = -Log Ca). The relaxing solution has a pCa below 8. In each case Ca concentration is controlled by EGTA buffer and calculated according to Fabiato (1981).

B) Tension/pCa relationships established on isolated skinned cells. Control (■). Before the skinning procedure some cells were submitted for 5 min to either 1 μM isoproterenol a ß-adrenergic agonist (▲) or to 10 μM phenylephrine, an α_1-adrenergic agonist (♦). Values are expressed relative to the maximal tension elicited in each cell (n≥5; SEM).

positioned at the tip of a conical micropapillary that received the outlets of several microcapillaries connected to 10-ml syringes. Outflow (20 μl/min) was obtained by elevating a syringe 60 cm above the level of the dish. The cell was superfused with solutions at 20-22°C. The dish was continuously superfused with the relaxing solution, and the drain for the bath perfusion was glued 2 mm to the rear of the opening of the conical microcapillary. The delay in response after elevating a syringe was below ten seconds.

Continuous records of the force developed by a single skinned cell submitted to EGTA-buffered solutions of different Ca concentrations are shown in figure 1. Under control conditions the threshold pCa was 6 and the maximal force obtained at pCa 4.5 was between 1 and 1.5 mg (i.e. about 180 mN/mm^2). The effects were reproducible although maximal Ca-activated force was generally less on subsequent applications. Also in about half of the experiments, cells detached upon application of the lower pCa solutions. Figure 1 part B shows an averaged force-pCa relation obtained for control cells at a sarcomere length of 2.1 μm. In these preparations this relation is well described by the Hill equation: $\%F=S^{n}H/(K+S^{n}H)$, where %F is the relative force, S is the calcium concentration, K is the apparent affinity constant, and n_H is the Hill coefficient. K and n_H permit the calculation of pCa_{50}, the concentration of Ca^{2+} required to elicit half-maximal force since a pCa of 50 equals $(-\log K)/n_H$. Use of this equation and a linear regression fit provides a convenient and objective description of the force-pCa relation. pCa_{50} and n_H were 5.80 ± 0.01 and 2.78 ± 0.11, respectively (n=25 cells). There was only a very small scatter of the pCa_{50} when particular care was taken to estimate sarcomere length and when the same sequence of solutions was applied.

The sensitivity of single skinned cells to calcium ions was markedly modified when the cells were stimulated for 5 minutes with different agonists before skinning. Figure 1 summarizes results obtained on different cells exposed to isoprenaline and to phenylephrine in the presence of propranolol. The ß-adrenergic agonist clearly reduced the calcium sensitivity so that the threshold pCa was around 5.87. On the other hand, the α_1-adrenergic agonist increased the myofibrillar calcium sensitivity; noticeable force was recorded at pCa 6.25. The use of skinned cells at variable times (up to 8 hours) after the skinning procedure confirms that the contractile proteins in relaxing solution maintained their previously acquired stimulated state. However, these effects (in average a 0.17 pCa unit rightward shift with isoprenaline and a 0.13 pCa unit leftward shift with phenylephrine) were abolished after treating the skinned cells with alkaline phosphatase. The latter indicates that both shifts are due to phosphorylation of some contractile proteins (the inhibitory protein troponin I and the myosin light chain 2 respectively). Phosphorylation of troponin I accounts for the rightward shift of the tension/pCa curves; and we recently got evidence that phosphorylation of the myosin light chain 2 is responsible for the leftward shift (Terzic et al., 1993).

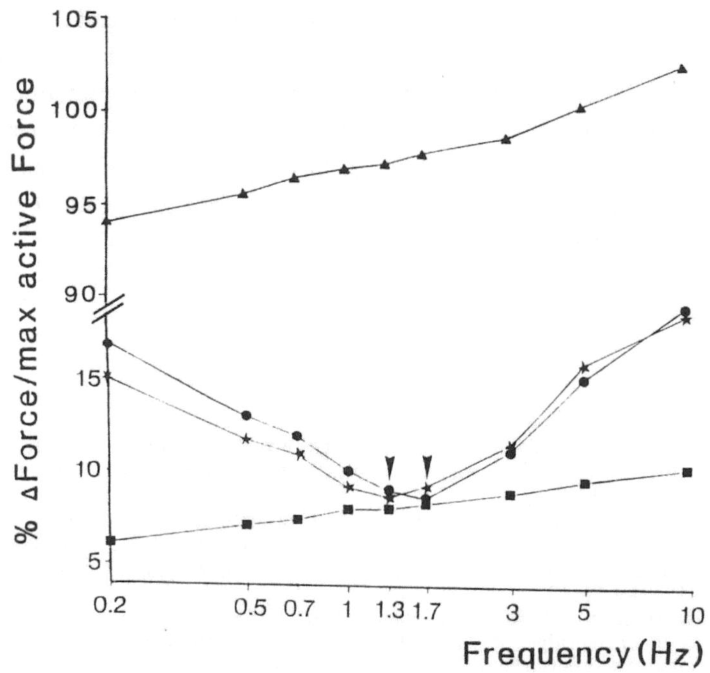

Figure 2: Variation in force elicited by sinusoidal length perturbations at different frequencies. The variation length in was 5% of active cell length (initial sarcomere length 2.1 μm). Length perturbations were applied on a control cell in the relaxing solution (pCa=9; ■) in a rigor-inducing solution (no ATP and no phosphocreatine: ▲) and at pCa 5.87 (●). Similarly an α_1-adrenergic stimulated cell previously to skinning was submitted to the same length perturbation at pCa 5.87 (★). The minimum resonnance frequencies (arrows) were 1.7 and 1.3 in control and after α_1-adrenergic stimulation respectively. The variation in free Ca^{++} is expressed relative to maximal active force elicited at pCa 4.5; note that rigor force was about 45% of maximal active force and that the length-induced in force of very large amplitude as expected from attached rigor crossbridges.

To get more insight into the mechanisms of Ca sensitization, sinusoidal length perturbations were applied on cells made to develop about half-maximal force. In control conditions, the minimum resonance frequency, f_{min}, was 1.7 Hz. In agreement with previous results, cells which have been ß-adrenergic stimulated before skinning, or submitted to cyclic AMP-dependent protein kinase exhibit a significant higher f_{min}. To account for this observation, it had been suggested that crossbridge turnover is accelerated with less force developed at a given pCa due to a shorter time in the force-generating state (Hoh et al., 1988). More original is our observation that f_{min} was significantly reduced after applying Ca sensitizers (EMD) or after α_1-adrenergic stimulation (f_{min}=1.3 Hz) as illustrated in Fig. 2. Consequently, it can be proposed that the α_1-adrenoceptor-mediated increase in force developed at a given pCa, is consequent to a higher proportion of crossbridges in the attached state. This result is of importance since it accounts for the improvement of cardiac energetics under this experimental condition, a situation which is tentatively mimicked by the use of Ca-sensitizing agents to cure cardiac mechanical deficiency.

REFERENCES

Brenner B (1988) Effect of Ca^{2+} on cross-bridge turnover kinetics in skinned single rabbit psoas fibers: Implications for regulation of muscle contraction. Proc Natl Acad Sci USA 85:3265-3269

Fabiato AJ (1981) Myoplasmic free calcium concentration reached during the twitch of an intact isolated cardiac cell and during calcium-induced release of calcium from the sarcoplasmic reticulum of a skinned cardiac cell from the adult rat or rabbit ventricle. J Gen Physiol 78:457-497

Hoh JFY, Rossmanith GH, Kwan LJ, Hamilton AM (1988) Adrenaline increases the rate of cycling of crossbridges in rat cardiac muscle as measured by pseudo-random binary noise-modulated perturbation analysis. Circ Res 62:452-461

Huxley AF (1957) Muscle structure and theories of contraction. Prog Biophys 7:255-318

Kawai M, Brandt PW (1980) Sinusoidal analysis: a high resolution method for correlating biochemical reactions with physiological processes in activated skeletal muscles of rabbit, frog and crayfish. J Muscle Res Cell Motil 1:279-303

Kerrick WGL, Potter JD, Hoar PE (1991) The apparent rate constant for the dissociation of force generating myosin crossbridges from actin decreases during Ca^{2+} activation of skinned muscle fibres. J Muscle Res Cell Motil 12:53-60

Puceat M, Clement O, Lechene P, Pelosin JM, Ventura-Clapier R, Vassort G (1990) Neurohormonal control of calcium sensitivity of myofilaments in rat single heart cells. Circ Res 67:517-524

Rossmanith GH, Hoh JFY, Kirman A, Kwan LJ (1986) Influence of V_1 and V_3 isomyosins on the mechanical behaviour of rat papillary muscle as studied by pseudo-random binary noise modulated length perturbations. J Muscle Res Cell Motil 7:307-319

Terzic A, Puceat M, Vassort G, Vogel SM (1993) Cardiac α_1-adrenoceptors: an overview. Pharmacol Rev 45:147-175

KINETOCHORE DIRECTIONAL INSTABILITY IN VERTEBRATE MITOTIC CELLS

Robert V. Skibbens and E. D. Salmon
C B # 3280, Department of Biology
University of North Carolina
Chapel Hill, NC 27599-3280

Introduction:

Vertebrate mitotic spindles are formed from oppositely oriented polar microtubule (MT) arrays nucleated from centrosomes. MTs mostly grow and shorten by the addition and loss of tubulin subunits from the MT plus-ends distal from the pole. Chromosomes become attached to the spindle when kinetochores capture and stabilize the dynamically instable MT plus-ends (reviewed in Salmon, 1989).

There are typically 5 stages of chromosome movement in spindles of vertebrate cells (reviewed in Skibbens et al., 1993). Initially, kinetochores occasionally associate to the lateral face of polar MTs and rapidly glide poleward. However, the bulk of chromosome movement occurs after the end-on association of MTs with kinetochore attachment sites (reviewed in Salmon, 1989). Usually, the sister kinetochore facing the closest pole first acquires kinetochore MTs (kMTs) and the chromosome moves toward that pole (mono-orientation). Chromosomes congress to the metaphase plate at the spindle equator after the untethered sister kinetochore captures MTs emanating from the opposite spindle pole (bi-orientation). After all chromosomes have congressed to the metaphase plate, sister chromosomes separate and then move toward their associated pole (Anaphase A). In addition to the net displacements during congression and segregation, kinetochores of both mono- and bi-oriented chromosomes also exhibit oscillations over several microns poleward (P) and away from the pole (AP) (reviewed in Skibbens et al., 1993).

Recent kMT labeling and severing experiments have shown that tethered chromosomes move within the spindle by forces produced at or near kinetochores (reviewed in Salmon, 1989; Gorbsky, 1992 and in Mitchison and Salmon, 1992) rather than by forces generated along the lengths of kMTs (Ostergren, 1950; Hays *et al.*, 1982). These results have shown that kinetochores move P and AP at 1-3µm/min, relative to the pole, along relatively stationary kMTs. Though there is evidence that kMTs move slowly poleward (flux) at 0.5µm/min (Mitchison, 1989; Mitchison and Salmon, 1992), the majority of kinetochore movement is tightly coupled to kMT depolymerization and growth at plus-end attachment sites (Mitchison *et al.*, 1986; Gorbsky *et al.*, 1987; Nicklas, 1989; Mitchison and Salmon, 1992). Kinetochore motility could be driven by kMT plus-end growth and shortening (Inoue and Sato, 1967; Koshland *et al.*, 1988; Hotani and Miyamoto, 1990;

NATO ASI Series, Vol. H 84
Biomechanics of Active Movement
and Division of Cells
Edited by N. Akkaş
© Springer-Verlag Berlin Heidelberg 1994

Coue *et al.*, 1991) or by MT plus- and minus-end directed motors located at kinetochores (Steuer *et al.*, 1991; Pfarr *et al.*, 1990; Hyman and Mitchison, 1991; Wordeman *et al.*, 1991; Sawin *et al.*, 1992; Yen *et al.*, 1992).

Given the finding that kinetochores move toward and away from mitotic spindle poles along relatively stationary kMTs, it became important to determine the nature of kinetochore motility at the plus-ends of kMTs and whether kinetochore motility pushes as well as pulls on the chromosome. To answer these questions, we developed video microscopy methods and a semi-automatic tracking technique to obtain high resolution (50 - 200 nm, 2 sec) kinetic measurements of kinetochore motility and centromere deformation in mitotic newt cells (Fig. 1)(Skibbens et al., 1993). Newt lung epithelial cells provided an excellent model system because of their large size, optical clarity and because the rate of kMT poleward flux had previously been measured (Mitchison and Salmon, 1992). Thus, kinetochore motion at the plus-ends of kMTs could be calculated from the rates of kinetochore motility relative to the pole and flux. We discovered in this study that tethered kinetochores primarily exhibit two functional states, abruptly switching between persistent constant velocity phases of P and AP motion (kinetochore directional instability). Kinetochore switching on adjacent mono-oriented chromosomes is uncorrelated while sister kinetochore switching on bi-oriented chromosomes is highly cooperative. The relative durations of P and AP phases govern net chromosome displacements during congression and segregation. Also, kinetochore P motion pulls on the centromere while AP motion can push.

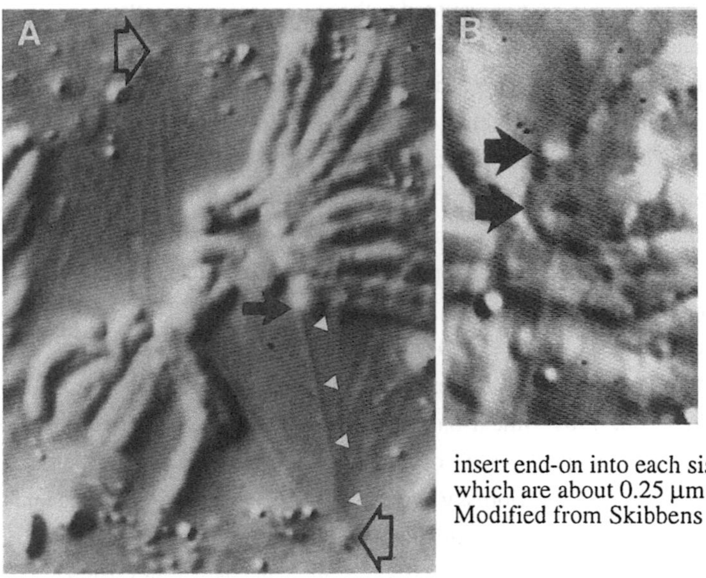

Figure 1: (A) Image obtained by VE-DIC microscopy shows kinetochores (filled arrows) tethered to spindle poles (open arrows) by kinetochore fibers (arrowheads) in mitotic newt lung cell. (B) At high resolution, kinetochore fibers, composed of approximately 20 microtubules in newt lung cells (Rieder, 1990), are seen to insert end-on into each sister kinetochore, which are about 0.25 µm in diameter (arrows). Modified from Skibbens et al., 1993.

Results and Discussion:

We tracked the positions over time of both spindle poles and kinetochores and analyzed the movement of kinetochores along kMT plus-ends (Fig. 1). Kinetic plots of kinetochore to pole distance over time showed that tethered kinetochores on both mono- and bi-oriented chromosomes moved P and AP in a saw-toothed fashion comprised of extended P and AP constant velocity domains interrupted by abrupt switches (Fig. 2). Abrupt switching between directions of motion occurred below our sampling resolution but typically occurred within 6 sec. Velocities of kinetochore P and AP motion relative to the pole were about 1.5 to 2 μm/min, on average. Since kMTs move only poleward, kinetochore motility is a property of kinetochore motion at kMT plus ends which we termed directional instability. We calculated that the velocity of kinetochore movement relative to the kMT lattice was about 1 - 1.5 μm/min for P motion and 2 - 2.5 μm/min for AP motion for a flux velocity of 0.5 μm/min (Mitchison and Salmon, 1992).

We analyzed the movements of kinetochores on adjacent mono-oriented chromosomes to test if kinetochore directional instability was regulated by fluctuations in some gradient or soluble factor within the spindle. The results showed that kinetochores on mono-oriented chromosomes switch independent of adjacent kinetochores 77% of the time. Furthermore, the duration of P and AP movements of adjacent kinetochores was random; approximately 50% of the time they were in-phase and 50% of the time they were out-of-phase. These results strongly argue that kinetochores represent autonomous motile entities that both switch and move in an independent fashion.

In contrast, the incidence of independent kinetochore switching decreased for sister kinetochores on bi-oriented chromosomes. We found that sister kinetochores on bi-oriented chromosomes primarily exhibited 75% out-of-phase movement, relative to their associated poles (Fig. 2). This out-of-phase sister kinetochore movement was required for net chromosome movement (Fig. 2) and showed that chromosome bi-orientation modifies sister kinetochore movements.

We focused on the extent of centromere deformation to infer the state of force exerted during kinetochore oscillations. Kinetochore P movement stretched the centromere poleward as expected from previous reports. However, we found that this centromere stretch disappeared during kinetochore AP movement and the kinetochore often punched into and deformed the centromere (Fig. 2). This is the first clear demonstration which confirms earlier speculations (Bajer, 1982; Mitchison, 1988) that kinetochores can produce both pushing and pulling forces in vivo.

Of key interest is how kinetochore directional instability is regulated to produce chromosome congression and segregation. Factors located at kinetochores which regulate kMT assembly or the activity of MT motors located at the kinetochore are likely to be

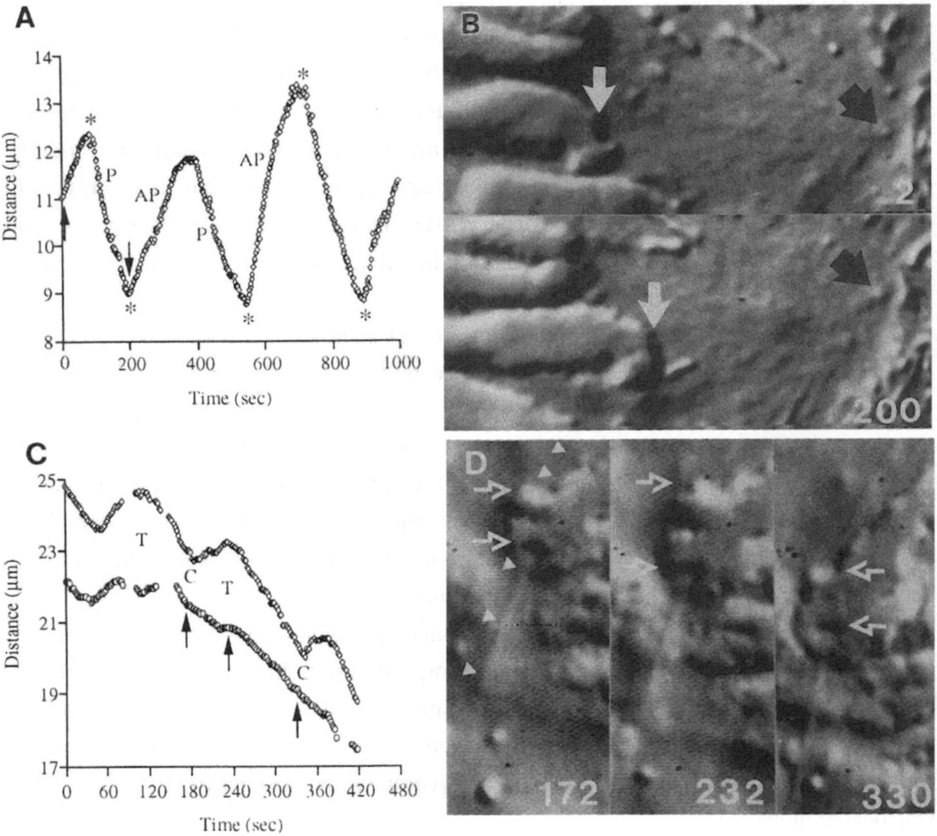

Figure 2: Distance-versus-time plots of kinetochore movement relative to centrosomes at spindle poles for both mono- (A) and bi-oriented chromosomes (C). Arrows in (A) and (C) correspond to micrograph images in (B) and (D), respectively. P and AP denote phases of motion toward and away from the spindle pole. Asterisks indicate abrupt transitions. T and C in lower panel denote periods of tension and compression across the centromere, respectively, due to sister kinetochore activity. Note that net displacement of bi-oriented chromosomes only occurs during coordinated sister kinetochore movement (one sister in P and the other in AP motion). Micrographs in (B) and (D) also show centromere stretching during P motion and compression during AP motion for both mono- (B) and bi-oriented chromosomes (D). White arrows = kinetochores; black arrows = poles; Arrowheads = bundle of kinetochore microtubules. Modified from Skibbens et al., 1993.

involved. Mitchison (1989) initially suggested that kinetochores are "smart" motile organelles capable of switching direction in response to positional cues in the spindle. We have found that kinetochores are also capable of sensing vectorial information within the spindle. Even though sister kinetochores are very close together, sister kinetochores must sense chromosome bi-orientation so that one sister kinetochore can persist in P pulling motion while the other persists in AP pushing motion during congression. Sister kinetochores must also sense chromosome disjunction at anaphase onset so that each kinetochore persists in P motion. Tension across the centromere seems a likely mechanism of inducing sister kinetochore cooperativity based on the extensive micromanipulation studies performed by Nicklas (1977, 1988). We have proposed that tension at the kMT attachment site is a vectorial regulator of kinetochore directional instability (Figure 3). Increased tension increases the probability of switching to AP motion while compression or decreased tension increases the probability of switching to P motion. Polar ejection forces on chromosomes may also provide vectorial cues in the spindle that modulate tension at kinetochores. Polar ejection forces have previously been proposed to be oriented in the direction of MT growth and increase in strength as chromosomes move into regions of higher MT density nearer the pole (Rieder et al., 1986; Salmon, 1989; Theurkauf and Hawley, 1992; Leslie, 1992; Cassimeris et al., 1993). In this model, tension is the upstream regulator of molecular mechanisms, like phosphorylation, which could control the direction of motor activity or MT assembly dynamics at the kinetochore (Hyman and Mitchison, 1991; Skibbens et al., 1993).

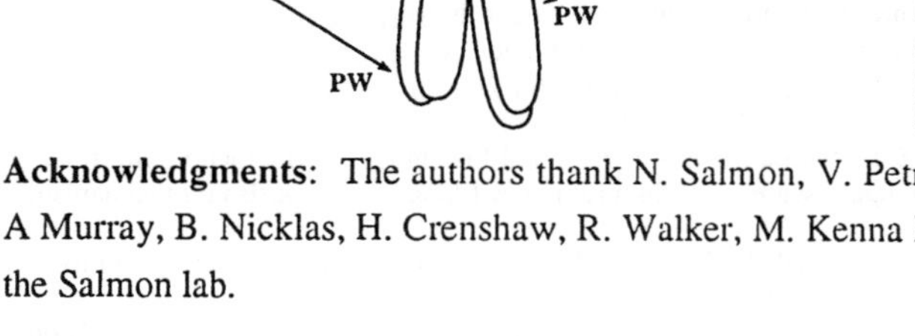

Figure 3: Schematic shows summary of potential forces within the mitotic spindle that could exert tension or compression at a kinetochore (F = flux, PW = polar ejection forces or "winds", K_P, K_{AP} = forces produced by P and AP kinetochore movement. Curved arrows indicate addition and loss of tubulin subunits in kMTs.

Acknowledgments: The authors thank N. Salmon, V. Petrie Skeen, S. Magers and Drs. A Murray, B. Nicklas, H. Crenshaw, R. Walker, M. Kenna N. Gliksman and members of the Salmon lab.

References:

Bajer, A.S. 1982. Functional autonomy of monopolar spindle and evidence for oscillatory movement in mitosis. *J. Cell Biol.* 93:33-48.

Cassimeris, L., C.L. Rieder and E.D. Salmon. 1993. Microtubule assembly and kinetochore directional instability in vertebrate monopolar spindles: implications for the mechanism of chromosome congression. *J. Cell Science*, submitted.

Coue, M., V.A. Lombillo and J.R. McIntosh. 1991. Microtubule depolymerization promotes particle and chromosome movement in vitro. *J. Cell Biol.* 112:1165-1175.
Gorbsky, G.H., P.J. Sammak, and G.G. Borisy. 1987. Chromosomes move poleward in anaphase along stationary microtubules that coordinately disassemble from their kinetochores ends *J. Cell Biol.* 104:9-18.
Gorbsky, G. 1992. Chromosome motion in mitosis. *BioEssays* 14:73-80.
Hays, T.S., D. Wise, and E.D. Salmon. 1982. Traction force on a kinetochore at metaphase acts as a linear function of kinetochore fiber length. *J. Cell Biol.* 93:374-382.
Hotani, H. and H. Miyamoto. 1990. Dynamic features of microtubules as visualized by dark-field microscopy. *Adv. Biophysics* 26:135-156.
Hyman, A.A. and T.J. Mitchison. 1991. Two different microtubule-based motor activities with opposite polarities in kinetochores. *Nature* 351:206-211.
Inoue, S. and H. Sato. 1967. Cell motility by labile association of molecules. The nature of mitotic spindle fibers and their role in chromosome movement. *J. General Phys.* 50:259-292.
Koshland, D.E., T.J. Mitchison and Marc W. Kirschner. 1988. Poleward chromosome movement driven by microtubule depolymerization in vitro. *Nature* 331:499-504.
Leslie, R.J. 1992. Chromosomes attain a metaphase position on half-spindles in the absence of an opposing spindle pole. *J. Cell Science* 103:125-130.
Mitchison, T.J., L. Evans, E. Schulze, and M. Kirschner. 1986. Sites of microtubule assembly and disassembly in the mitotic spindle. *Cell* 45:515-527.
Mitchison, T.J. 1988. Microtubule dynamics and kinetochore function in mitosis. *Annu. Rev. Cell Biol.* 4:527-549.
Mitchison, T.J. 1989. Chromosome alignment at mitotic metaphase: balanced forces or smart kinetochores? Cell Movement, Volume 2: Kinesin, Dynein, and Microtubule dynamics, 421-430.
Mitchison, T.J. 1989. Polewards microtubule flux in the mitotic spindle: evidence from photoactivation of fluorescence. *J. Cell Biol.* 109:637-652.
Mitchison, T.J. and E.D. Salmon. 1992. Poleward kinetochore fiber movement occurs during both metaphase and anaphase-A in newt lung cell mitosis. *J. Cell Biol.* 119:569-582.
Nicklas, R.B. 1977. Chromosome movement: facts and hypotheses. In Mitosis Facts and Questions, eds. Little, Paweletz, Petzelt, Ponstingl, Schroeter and Zimmermann. Springer-Verlag, 150-155.
Nicklas, R.B. 1988. The forces that move chromosomes in mitosis. *Annu. Rev. Biophys. Chem.* 17:431-449.
Nicklas, R.B. 1989. The motor for poleward chromosome movement in anaphase is in or near the kinetochore. *J. Cell Biol.* 109:2245-2255.
Ostergren, G. 1950. Considerations on some elementary features of mitosis. *Hereditas* 36:1-18.
Pfarr, C.M., M. Coue, P.M. Grissom, T.S. Hays, M.E. Porter and J.R. McIntosh. 1990. Cytoplasmic dynein is localized to kinetochores during mitosis. *Nature* 345:263-265.
Rieder, C.L., E.A. Davison, L.C.W. Jensen, L. Cassimeris and E.D. Salmon. 1986. Oscillatory movements of mono-oriented chromosomes and their position relative to the spindle pole result from the ejection properties of the aster and half-spindle. *J. Cell Biol.* 103:581-591.
Rieder, C.L. 1990. Formation of the astral mitotic spindle: ultrastructural basis for the centrosome-kinetochore interaction. *Electron Micros. Rev.* 3(2):269-300.
Salmon, E.D. 1989. Microtubule dynamics and chromosome movement. In Mitosis: Molecules and Mechanisms. eds. Hyam and Brinkley. Academic Press Limited, 119-181.
Sawin, K.E., T.J. Mitchison, and L.G. Wordeman. 1992. Evidence for kinesin-related proteins in the mitotic apparatus using peptide antibodies. *J. Cell Science* 101:303-313.
Skibbens, R. V., V. Petrie Skeen and E. D. Salmon. 1993. Directional Instability of Kinetochore motility during chromosome congression and segregation in mitotic newt lung cells: a push-pull mechanism. J. Cell Biol. 122(4): 859-875.
Steuer, E.R., L. Wordeman, T.A. Schroer, and M.P. Sheetz. 1990. Localization of cytoplasmic dynein to mitotic spindles and kinetochores. *Nature* 345:266-268.
Theurkauf, W.E. and R.S. Hawley. 1992. Meiotic spindle assembly in Drosophila females: behavior of non exchange chromosomes and the effects of mutations in the nod kinesin-like protein. *J. Cell Biol.* 116:1167-1180.
Wordeman, L., Eric R. Steuer, Michael P. Sheetz and Tim Mitchison. 1991. Chemical subdomains within the kinetochore domain of isolated CHO mitotic chromosomes. *J. Cell Biol.* 114:285-294.
Yen, T.J., Gang Li, Bruce T. Schaar, Illya Szilak and Don W. Cleveland. 1992. CENP-E is a putative kinetochore motor that accumulates just before mitosis. *Nature* 359:536-539.

Signal Transduction and Myofibrillogenesis in Isolated Neonatal Heart Myocytes In Vitro

Will Sharp, David G. Simpson, Louis Terracio, and Thomas K. Borg
Department of Developmental Biology and Anatomy
University of South Carolina
Columbia, South Carolina 29208
USA

Introduction A precise sequence of morphological events is associated with the process of myofibrillogenesis in vitro in neonatal cardiac myocytes (Hilenski et al 1991; Sharp et al 1993; Sanger et al 1984). A similar set of sequences probably occurs in vivo; however, the rate of formation is more gradual, thus making investigation of precise events more difficult. In vitro analyses of morphological and biochemical events associated with myofibrillogenesis can provide an understanding of these processes in normal heart development, response to stimulation, and cardiac hypertrophy.

The rate of attachment and subsequent myofibril formation in vitro is determined by the age of the isolated myocytes and the extracellular matrix (ECM) substrate upon which they are plated (Borg et al 1984; Terracio et al 1990). As the age of the isolated myocytes increase, they become more difficult to culture. In part this is due to the organization of the cytoskeleton which is intimately associated with myofibrillar structure. The cytoskeleton links the myofibrillar apparatus to adjacent myofibrils and to the sarcolemma (Terracio et al 1991). In rodent embryonic and fetal myocytes, this cytoskeletal connection is not well formed; however, these cytoskeletal connections become more complex by neonatal and adult stages where the isolation and establishment in culture of these cells becomes more difficult. It is presumed that the increasing cytoskeletal connections are a reflection of increased mechanical tension on the cell as well as an increased organization of the myocytes to become more efficient in force generation.

In vivo, cardiac myocytes at all stages are surrounded by a complex ECM (Borg and Caulfield 1981). This 3-dimensional network is in intimate contact with the cellular components of the heart and plays a functional role in cardiac development and physiology (Robinson et al 1983). While it is clear that the ECM is essential for the form and function of the heart, the mechanisms underlying these processes are not well understood. These processes involve a combination of both chemical and mechanical signalling from the ECM to the inside of the cell as well as from the cell to the ECM.
To investigate these mechanisms of signalling, isolated neonatal cardiac myocytes have been grown on different ECM substrates and various processes including myofibrillogenesis have been investigated (Hilenski, 1991).

NATO ASI Series, Vol. H 84
Biomechanics of Active Movement
and Division of Cells
Edited by N. Akkaş
© Springer-Verlag Berlin Heidelberg 1994

By plating isolated cells at different stages on different ECM components it is possible to show that: 1) myocytes from different developmental stages attach at different rates to the substrate; and 2) recognition of the ECM component by the myocytes is receptor mediated. The role of the ECM substrate in myofibrillogenesis is less well understood; however, the ECM clearly plays a significant role in myocyte differentiation in vitro (Hilenski et al 1991). Cells plated on different ECM components, the cells show differences in cytoskeletal and myofibrillar patterns.

Characteristic of the interaction with the ECM substrate is the formation of a structure termed a focal adhesion (Burridge et al 1988). Focal adhesions represent sites of contact of the sarcolemma with the substrate and are composed of a clustering of both membrane and cytoskeletal proteins (Hilenski et al 1992). In myocytes, the focal adhesions contain receptors for the ECM components and cytoskeletal components that are associated with the myofibrils. The principal receptors for the ECM components are members of the integrin superfamily (Hynes 1992). Integrins are heterodimers of variable α chains and a common β chain. While there is a large variety of potential combinations of these chains, only a relatively few are involved in the recognition of ECM components in cardiac myocytes (Terracio et al 1991). The α chains 1,2,3,5,6 and 7 are associated with the β_1 chain and are involved in recognition of the ECM components of fibronectin ($\alpha 3, \alpha 5$), interstitial collagens ($\alpha 1$, $\alpha 3$) and laminin ($\alpha 1$, $\alpha 3$, $\alpha 6$, $\alpha 7$).

The integrin receptors are transmembrane and have been proposed to function in the translation of chemical and mechanical signals from the ECM into the cell (Hynes 1992). How these signals were processed within the cell has been the subject of much speculation; however, recent data seems to support the fact that both chemical and mechanical stimulation is important in the signal transduction process. The purpose of this report is to describe the potential role of the focal adhesions in transmitting both chemical and mechanical signals that result in the formation of myofibrils.

<u>Materials and Methods</u> Myocytes were isolated from 4-5 day old neonatal Sprague-Dawley rat hearts as previously described (Borg et al 1984). Myocytes were plated on laminin (20 μg/ml of culture media) and allowed to adapt to culture conditions over 72 hrs. Cells were treated with 10 μM verapamil for 72 hr followed by a washout procedure. The verpamil media was replaced and the phenotypic changes of the cells observed by laser scanning confocal microscopy (LSCM) (Sharp et al 1993). The temporal and spatial expression and patterns of: 1) myofibrillar assembly (phalloidin); 2) integrin localization (antiβ1-integrin); 3) anti-FAK localization; 4) anti-phosphotryosine; 5) costamere formation (anti-vinculin); and 6) reflection interference for focal adhesion were examined by LSCM at 4 hr intervals following the washout of the verpamil. Controls consisted of untreated cells and verapamil-vehicle.

Results and Discussion The myofibrils were apparent in isolated cells by 72 hrs of culture (fig.) and the cells exhibited synchronous contraction. Focal adhesions were located at the terminal ends of the individual myofibrils as were the antibodies for the integrins, FAK and phosphotryosine (figs 1-3). Following treatment with verapamil for 72 hrs, the contraction of the myocytes was inhibited. The focal adhesions showed a dispersed pattern although the cell shape was similar to the controls. The pattern of FAK, phosphotryosine and integrin also gave a similar pattern (fig 4-6). Vinculin staining also show a diffuse sarcolemma and cytoplasmic pattern. The staining with phalloidin showed that few myofibrils were apparent; however, stress fibers were apparent in some cells.

Following the washout procedure the myofibrils began to reassembly and were similar to the controls by 72 hrs. The contractile activity was also similar by this time point. The focal adhesions began to cluster in association with the FAK and phosphotryosine patterns within 2 hrs. Antiβ1-integrin initially stained the focal adhesion in a rapid (2 hr) manner followed by increased staining in the costameres by 24 hrs. The pattern for vinculin staining of the costameres showed reassembly by 24 hrs. While patterns similar to controls were attained by 48 hrs, individual cellular variation was apparent making quantitation difficult.

Taken together these data indicate that mechanical tension, as exhibited by contractile activity, is necessary for the maintanence of the myocyte phenotype. When contraction was inhibited with the calcium channel blocker verpamil, the myofibrils underwent disassembly and only reformed when the verapamil was removed. The rapid formation of the focal adhesions following the removal of the verapamil suggests that they are critical for the formation of myofibrils. Data from these preliminary observations and other laboratories indicates that the presence of FAK and phosphotryosine antibodies togther with the β1-integrin indicates that these molecules appear to be essential in the formation of the focal adhesion (Juliano and Haskill 1993). The focal adhesion is the critical site in the formation of the myofibrils. However, further experimentation is necessary to determine the specific roles of mechanical and chemical signalling at this particular site. These data are consistent with the hypothesis that the integrins are necessary in signal transduction from the ECM to the cytoskeleton and subsequent myofibril formation. Inhibition of the focal adhesion by blocking integrins also showed an aberrant myofibrillar pattern (Hilenski et al 1992).

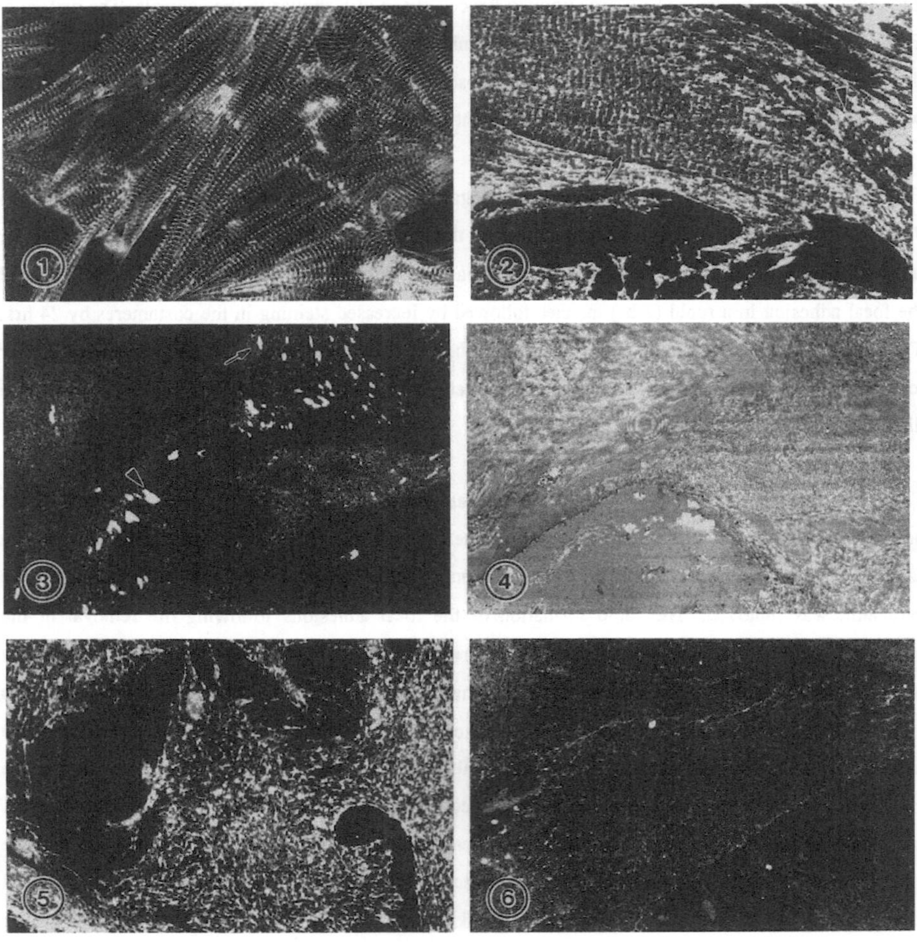

Figure 1. LSCM of phalloidin stained neonatal cardiac myocytes after 72 hrs of culture. The myofibrillar structure is evident in all cells.

Figure 2. LSCM of the antiβ1-integrin which is localized to the costameres (C) as well as the focal adhesions (arrows)

Figure 3 LSCM of anti-FAK showing the localization is confined to the focal adhesions (arrows).

Figure 4. Reflection interference microscopy showing the dispersed focal adhesion pattern after 72 hr of verapamil treatment.

Figure 5. LSCM of antiβ1-integrin showing the wide spread distribution and the absence of focal adhesions and myofibrils after verapamil treatment.

Figure 6. LSCM of phosphotryosine staining of verapamil treated cells showing a pattern similar to figs 4 and 5.

Two possible pathways for integrin signalling have been proposed (): 1) organization of the cytoplasmic cytoskeleton into a defined network; and 2) direct chemical signalling by FAK and/or other tryosine kinases (Juliano and Haskill 1993). Previous data supports the former pathway because in the formation process there is no new proteins synthesis, only the reassembly of existing proteins and the scaffolding to form the myofibril in composed of critical cytoplasmic components organized in a particular manner. The chemical signalling as indicated by the FAK and phosphotryosine may play a significant role in focal adhesion assemby (Hilenski et al 1991; Bockholt and Burridge 1993); however, its potential role in myofibril formation remains to be shown.

Acknowledgements

The authors wish to thank the sponsor of the meeting for the opportunity to present these data and to Margaret Currie for her technical assistance in cell culture procedures. These studies were sponsored in part by grants HL-37669, HL-40424 and HL-42249.

References

Bockholt SM and Burridge K (1993) Cell spreading on extracellular matrix proteins induces tryosine phosphorylation of tensin. J Biol Chem 268:14565-14567.

Borg TK and Caulfield JB (1981) The collagen materix of the heart. Fed Proc 40:2037-2042.

Borg TK, Rubin K, Lundgren E, Borg K, and Obrink B (1984) Recognition of extracellular matrix components by neonatal and adult cardiac myocytes. Dev Biol 104:86-96

Burridge K, Fath K, Kelly G, Nucknolls G, and Turner C (1988) Focal adhesions:transmembrane junctions between the extracellular matrix and the cytoskeleton. Ann Rev Cell Biol 4:487-525.

Hilenski LL, Terracio L and Borg TK (1991) Myofibrillar and cytoskeletal assembly in neonatal rat cardiac myocytes cultured on laminin and collagen. Cell Tissue Res 264:577-587

Hilenski LL, Xuehui M, Vinson N, Terracio L and Borg TK (1992) The role of β1 integrin in spreading and myofibrillogenesis in neonatal rat cardiomyocytes in vitro. Cell Motility and Cytoskel 21:87-100

Hynes RO (1992) Integrins:versatility, modulation and signaling in cell adhesion. Cell 69:11-25.

Juliano RL and Haskill S (1993) Signal transduction from the extracellular matrix. J Cell Biol 120:577-585.

Robinson TF, Cohen-Gould L, and Factor SM (1983) Skeletal framwork of mammalian heart muscle. Arrangement of inter and pericellular connective tissue structures. Lab Invest 49:482-495.

Sanger JW, Mittal B, and Sanger JM (1984) Analysis of myofibrillar structure and assembly using fluorescently labeled contractile proteins. J Cell Biol 98:825-834

Sharp WW, Terracio L, Borg TK and Samarel AM (1993) Contractile activity modulates actin synthesis and turnover in cultured neonatal rat heart cells. Circ Res 73:172-183

Terracio L, Rubin K, Jyring R, Gullberg D, Carver W and Borg TK (1991) Expression of collagen binding integrins during cardiac development and hypertrophy. Circ Res 68:734-744.

Terracio L, Simpson DG, Hilenski L, Carver W, Decker RS, Vinson N and Borg TK (1990) Distribution of vinculin in the Z-disk of striated muscle: Analysis by laser scanning confocal microscopy. J Cell Physiol 145:78-87.

Quantitative EMG and Biomechanics

George V Dimitrov and Nonna A Dimitrova
Central Laboratory of Biomedical Engineering
Bulgarian Academy of Sciences
Acad.G.Bontchev Str. Bl.105
BG-1113 Sofia
Bulgaria

Lippold (1952) was a pioneer in studying relationships between the surface EMG and muscle force. Since then, an easy recording of surface EMG and close relation between muscle force and integrated EMG under isometric contraction, attract investigators to searching for force-EMG quantitative relationships. The general force-EMG relationships, however, are still not obtained probably due to incomplete knowledge of EMG quantitative characteristics. The greater the number of active fibres the greater both the integrated EMG and muscle force. The opposite statement is, however, not true because not only the number of fibres producing extracellular potentials affects the surface EMG, but so does the profile of intracellular action potential (IAP) along each fibre. Could variations in the IAP profile induce significant changes in the motor unit potentials (MUPs) and thus in the EMG signals?

The power spectrum of the EMG signal shifts toward to lower frequency while the amplitude, *rms* and rectified integral of the surface EMG signal, increase during fatigue (Piper 1912, Cobb and Forbes 1923). Three main explanations have been proposed to account for these changes. They are: motor unit (MU) recruitment, MU synchronisation and variations in the muscle fibre propagation velocity (MFPV). On the other hand, fatigue is accompanied by increasing in duration of IAP as well as in the

NATO ASI Series, Vol. H 84
Biomechanics of Active Movement
and Division of Cells
Edited by N. Akkaş
© Springer-Verlag Berlin Heidelberg 1994

amplitude of the early negative afterpotential (Persson, 1963; Lüttgau, 1965).

The aim of the present study is analysis of effects of the IAP profile on quantitative characteristics of MUPs recorded at skin surface.

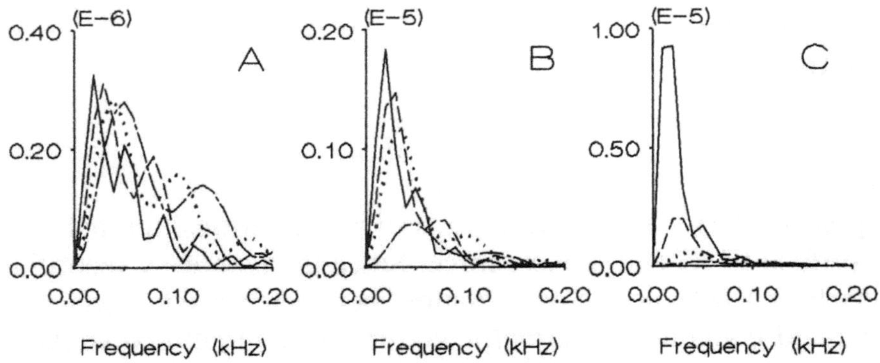

Fig.1. Profiles of IAP.

Method

The extracellular potentials of muscle fibres were calculated basing on the model suggested by Dimitrova (1974) by means of the method described by Dimitrov (1987). The IAP profile (Fig.1) included negative afterpotential falling exponentially (Frank, 1957). Three groups of numerical experiments were performed. The first group included IAPs without afterpotentials (A), duration of IAP was maintained constant (1.5ms) and MFPV was 2(——) ,3(- - - -), 4(········) or 5(— · —) m/s and thus the wavelength changed in the 3-7.5mm range. The second group was identical to the first one but the IAPs had negative afterpotentials (B). In the third group of experiments (C), variations of the MFPV (2-5m/s) and IAP duration (3-1.2ms) were performed in such a way that the wavelength was maintained unchanged (6mm). The amplitude of the early negative afterpotential increased with slowing of the membrane processes typical for fatigue, when the MFPV decreasing was accompanied by the IAP duration increasing (Persson, 1963; Lüttgau, 1965).

The MUPs were calculated as a linear summation of a temporally and spatially dispersed potentials generated by individual muscle fibres. Diameter of the MU territory was 8mm, end-plates were dispersed within 6mm, and desynchronization in the fibres' activation was up to 5ms. We supposed that the muscle fibre was cylindrical with length L=123mm, the fibre ends were insulated (due to the tendons), the end-plates were not in the middle of the fibres (L_1=60mm, L_2=63mm). The distance from the territory centre to the electrode was 10mm, and the anisotropy ratio of the volume conductor was 5.

Results

The variations in the IAP profile affected amplitude and time course of the surface MUPs (Fig.2). The amplitude and *rms* of

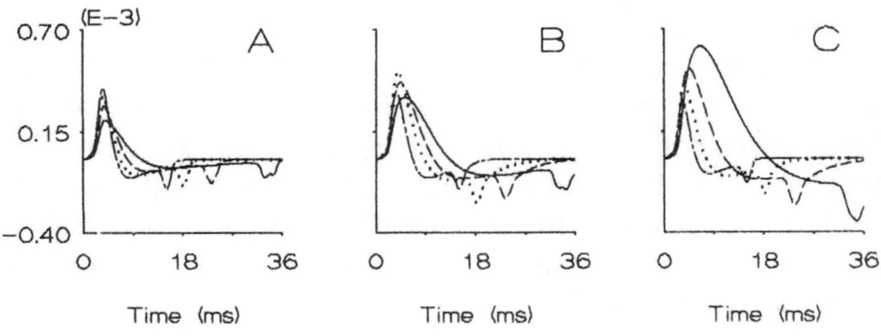

Fig.2. Time course of MUPs

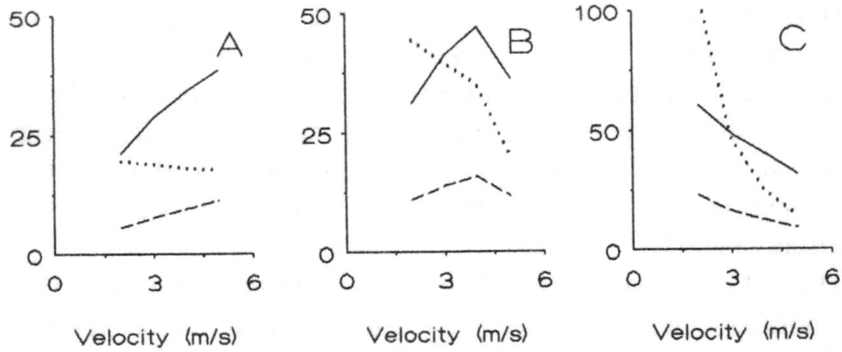

Fig.3. Amplitude (——), *rms* (- - - -) and rectified integral (········) of MUPs

MUPs increased with lengthening of profile of the IAP without afterpotential (Fig.3A). When the IAP had afterpotential, both amplitude and *rms* changed non monotonously (Fig.3B). When MFPV and IAP duration were varied simultaneously and the magnitude of negative after-potentials increased with the slowing of the membrane processes, the amplitude and *rms* of the MUPs were higher for slower processes (Fig.3C). MFPV was the main factor affecting character of changing of the rectified integral when the IAP had afterpotential. The rectified integral decreased under velocity increasing in the three groups of experiments. The degree of the changes, however, was the highest in the third group.

Power spectra are also sensitive to variations in the profile (Fig.4). Power spectra shifted toward to lower frequency with MFPV decreasing (Fig.4A). The spectral compressing was higher when the IAP had afterpotential (Fig.4B) and especially when the decreasing in the MFPV was accompanied by simultaneous increasing in the IAP duration and amplitude of negative afterpotential (Fig.4C).

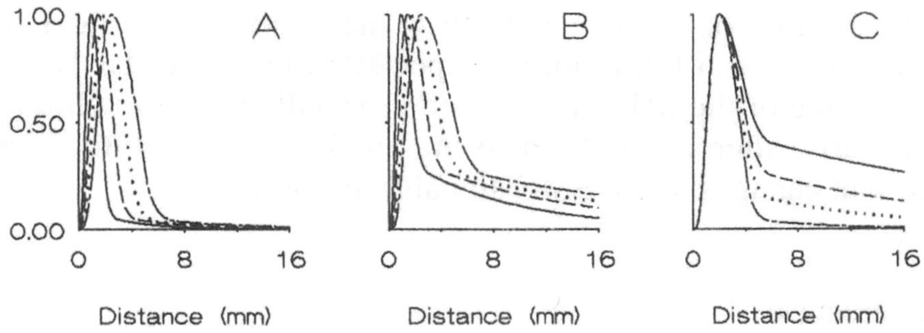

Fig.4. Power spectra of MUPs

Discussion

The results obtained show that variations in the IAP profile can significantly change the MUPs and EMG signal, especially due to lengthening of MUPs negative phase with increasing of afterpotentials. Thus, upon the force - EMG analysis, one must have in mind effects of variations in the IAP profile on the quantitative characteristics of MUPs.

After Lindstrom (1970), the spectral shift of the power spectrum has been generally tied with reduction in the propagation velocity in result of fatigue. However, Bigland - Ritchie et al. (1981) have shown that a similar variation in MFPV induced by temperature reduction, resulted in a much less shift of the power spectrum. Since then, additional factors affecting spectrum have been looking for. Increasing in the amplitude, *rms* and rectified integral accompanying fatigue, puzzle the investigators. By intuition, one could expect their reduction with fatigue and correspondingly with the force reduction. The explanations for these phenomena are looking for in recruitment of fresh MUs and/or in synchronisation of MUs.

The results obtained show that the well-known (and non conflicting with the intuitive expections) experimental facts on increasing of the IAP duration and especially of the amplitude of the early afterpotential (Persson, 1963, Luttgau, 1968) could explain changes in the EMG signals with fatigue.

References

Bigland-Ritchie B, Donovan EF, Roussos CS (1981) Conduction velocity and EMG power spectrum changes in fatigue of sustained maximal efforts. J Appl Physiol 51:1300-1305

Cobb S, Forbes A (1923) Electromyographic studies of muscle fatigue in man. Am J Physiol 65:234-251

Dimitrov GV (1987) Changes in the extracellular potentials produced by unmyelinated nerve fibre resulting from alterations in the propagation velocity or the duration of the action potential. Electromyogr Clin Neurophysiol 27:243-249

Dimitrova NA (1974) Model of the extracellular potential field of a single striated muscle fibre. Electromyogr Clin. Neurophysiol 14:53-66

Frank GB (1957) Negative afterpotential of frog's skeletal muscle. J Neurophysiol 20:602-614

Lippold OCJ (1952) The relation between integrated action potentials in a human muscle and its isometric tension. J Physiol (London) 117:492-499

Lüttgau HC (1965) The effect of metabolic inhibitors on the fatigue of the action potential in single muscle fibers. J Physiol (London) 178:45-67

Persson A (1963) The negative after-potential of frog skeletal muscle fibers Acta Physiol Scand 58 Suppl 205:1-32

Piper H (1912) Electrophysiologie Muschliche Muskein. Verlag von Julius Springer Basel

The Anaphase Movement
and its relashionship with cell cleavage

Miguel Mota
Department of Genetics,
Estação Agronómica Nacional,
Oeiras,
Portugal

All steps of the mitotic and meiotic processes have given rise to a large number of theories and much discussion but none has been so prolific in this respect as the anaphase movement, when daughter chromosomes leave the equatorial plate and move their centromeres up to the poles of the achromatic spindle.

The anaphase movement is a high precision mechanism that assures that any somatic cell receives an exact copy of the genetic material present in the chromosomes, that is, almost the total genetic information of the organism.

The similarity of the anaphase movement in plants and animals, in mitosis and meiosis, permits us to consider that the basic mechanism is the same in all cases.

Van Beneden (1883) proposed the pulling fibers theory to explain the anaphase movement and this has been for many years the most accepted theory. Another theory, which several authors considered acceptable, was the repulsion theory, suggested by Watase (1891). Many other theories were suggested but did not last. Good reviews of early work can be found in Schrader (1944, 1954) and Hughes (1952). Van Beneden's pulling fibers theory has been, until recently, the most generally accepted explanation.

I started to be interested in the anaphase movement a long time ago, during a study of inducing mitotic disturbances (Mota, 1952). Aberrations were described that contradicted several of the most accepted theories of such movement. One of them was the presence of chromosomes dividing as in c-mitosis but arranged in an equatorial plate, with a spindle present. It was possible to confirm the presence of a spindle, suggested by the arrangement of the chromosomes, which defined a clear equatorial plate, by looking at the slides (obtained by a different method) used in the work reported in the only paper in the literature reporting identical phenomenon, as caused by nickel

NATO ASI Series, Vol. H 84
Biomechanics of Active Movement
and Division of Cells
Edited by N. Akkaş
© Springer-Verlag Berlin Heidelberg 1994

nitrate (Levan, 1945). Such metaphases resulted, at least in some cases, in what I called "diffuse anaphases".

It is obvious that the presence of the spindle fiber is necessary for the anaphase movement. However, the possibility of a totally or partially inactivated centromere could not be ruled out and led to a detailed analysis of the arguments for and against the traction fibers theory. The argument in favor is essentially the general appearance, while the arguments against it were far more serious. It is difficult to conceive of an elastic fiber contracting practically to zero length to pull a heavy body such as the chromosome. Furthermore, such a fiber does not have an anchorage point on the other end.

Another aberration described (Mota,1952) consisted in the inactivation of one chromatid, which was left in the equatorial plate, while its partner moved normally to the pole. This case ruled out the theory of repulsion between daughter centromeres.

Belar (1929a, b) by studying living animal cells (long before phase contrast microscopy) gave an excellent description of anaphase, considering the traction fibers and the simultaneous action of an expansion body ("Stemmkorper"). Morphologically everything appears to be according to Belar's description. However, we have to find a mechanism producing the same effect without traction fibers or expansion bodies. (The so-called "anaphase-B", an elongation of the distance between the spindle poles, which occurs in many animal cells, is not considered in the present study, which deals only with true anaphase or "anaphase-A")

A few authors (Metz, 1933 and even Watase, citing Strasburger) had proposed the authonomy of the chromosomes (as opposed to the pulling and the pushing theories) but no explanation was offered, as far as I know, on the type of motor.

Following previous experimental work and the analysis of the abundant literature on the subject, "A new hypothesis of the anaphase movement" was presented (Mota, 1957) in which the chromosome is not considered as a passive body, pushed or pulled by external forces, but that moves autonomously, with the centromere acting as its "motor". The type of "motor" proposed, on the basis of existing evidence, was a "jet mechanism", in which the centromere expelled some substance in the direction of the equatorial plate. This creates a reaction that propels the chromosome to the pole.

Releasing of substances by the chromosomes during anaphase has been reported or suggested by several authors. Very good morphological evidence is given by Ris (1949) in a study of meiosis in grasshopper spermatocytes. Anaphases in which one or more bivlentes are sticky and unable to complete separation show highly distorted spindles, as could be caused by the "jet engines" working in opposite directions. The spindle, as oulined by the mitochondria that surround it, is projected laterally. In one case it even induces a partial cleavage, despite the fact that no chromosomal material exists in that part of the cell. The "concave" aspect of the projeted spindle material is exactly what could be expected. Such a deformation of the spindle can even be seen in time lapse movies of normal meiosis, just before the onset of anaphase.

Mota's new hypothesis, although largely cited in papers and text books (as, for instance, Mazia, 1961), has not, until recently, been generally accepted. More experimental work by several authors, published since 1987 (Gorbsky et al. 1987; Nicklas, 1988; Brinkley et al. 1988, etc.), has supported this theory. The most interesting evidence in its favor was the discovery of a specific centromere protein (Yen et al. 1991) which is released to the equatorial plate during anaphase.

This theory of anaphase may have some implications in cell cleavage.

Mota (1959) proved the existance of a contractile ring (postulated by some authors to explain the mechanism of cell cleavage) in grasshopper spermatocytes. In abnormal cells the ring had detached from the cell membrane and contracted alone and was visible in optical phase contrast microscopy.

Cleavage (cytokinesis) occurs after nuclear division (karyokinesis) in animal cells. Its plane is located where the plane of the equatorial plate was. In most cells this occurs at the equator of the cell but there are many cases (grasshopper neuroblasts, oocytes, etc.) of assymetrical division. In such cases the chromosomes arrange in an equatorial plate on one side of the cell and this plane determines the position of the cleavage furrow. Therefore, as pointed out by Mota (1959), the mechanism that triggers the action of the contractile ring may originate in the interaction of the material expelled from the two anaphase groups.

In one of the cases described by Ris (1949, Fig. 12) it is possible to see the formation of a cleavage furrow, away from the chromosome groups, where the material expelled by the chromosomes mixes.

References

Belar, K. (1929a) Beitrage zur Kausanalyse der Mitose. II. *Ach. Entwmk..* 118:359-480

Belar, K. (1929b) Beitrage zur Kausanalyse der Mitose. III. *Z. Zellf.* 10:73-134

Beneden, E. van (1883) Recherches sur la maturation de l'oeuf, la fécondation et la division cellulaire. *Arch.Biol.* 4:265-641

Brinkley, B. R.; Zinkowsky, R. P.; Mollon, W. L.; Davis, F. M.; Pisegna, M. A.; Pershouse, M. and Rao, P. N. (1988) Movement and segregation of kinetochores experimentally detached from mammalian chromosomes. *Nature* 336:251-254

Gorbsky, G. J.; Sammak, P. J. and Borisy, G. G. (1987) Chromosomes move poleward in anaphase along stationary microtubules that coordinately disassemble from their kinetochore ends. *J . Cell Biol.* 104:9-18

Hughes, A. F, (1952) *The mitotic cycle.* New York

Levan, A. (1945) Cytological reactions induced by inorganic salt solutions. *Nature* 153:751

Mazia, D. (1961) Mitosis and the physiology of cell division. *The Cell* (J. Brachet and A. E. Mirsky, editors), Academic Press, vol. 3:77-412

Metz, C. W. (1933) Monocentric mitosis with segregation of chromosomes in Sciara and its bearing on the mechanism of mitosis. *Biol. Bull.* 64:333-347

Mota, M. (1952) The action of seed extracts on chromosomes. *Arquivo de Patologia* 24 (3):336-357

Mota, M. (1957) A new hypothesis on the anaphase movement. *Proceedings of the International Genetics Symposium, l956, Supplement volume of Cytologia* pp.113-116

Mota, M. (1959) Karyokinesis without cytokinesis in the grasshopper. *Experimental Cell Research* 17:76-83

Nicklas, B. (1988) Chromosomes and kinetochores do more in mitosis than previously thought. *Chromosome structure and function* (J. P. Gustafson and R. Appels, editors), Plenum Press: 53-74

Ris, H. (1949) The anaphase movement of chromosomes in the spermatocytes of the grasshopper. *Biol. Bull.* 96:90-106.

Schrader, F. (1944, 1954) *Mitosis* (1st and 2nd editions), New York

Watase, S. (1891) Studies on cephalopods I. *J. Morph.* 4:247-303.

Yen, T. J.; Compton, D. A.; Wise, D.; Zinkowsky, R. P.; Brinkley, B. R.; Earnshaw, W. C. and Cleveland, D. W. (1991) CENP-E, a novel human centromere-associated protein required for progression from metaphase to anaphase. *EMBO J.* 10:1245-1254

Mechanism of Induction of Microspore Embryogenesis in *Brassica napus*: Significance of the Preprophase Band of Microtubules in the First Sporophytic Division.

Daina H. Simmonds
Plant Research Centre
Agriculture Canada
Ottawa, Ontario K1A 0C6
Canada

SUMMARY. A heat treatment interrupts pollen development in isolated uninucleate microspore cultures of *Brassica napus* cv. Topas inducing embryogenesis. The mechanisms involved in the switch from gametogenesis to sporogenesis has been studied. The first pollen division occurs without the participation of a preprophase band of microtubules (PPB) which results in a cell plate which degenerates. The cultured microspores subjected to a heat treatment undergo a change in cellular organization and form PPBs in the first microspore division. It is suggested that this PPB formation ensures the stability of the future cell plate a critical step in the initiation of future multicellular structures which lead to embryogenesis.

INTRODUCTION

Cell movement distinguishes animal morphogenesis from morphogenesis in higher plants. The somatic cells of higher plants are not capable of movement as are the animal cells. Therefore, morphogenesis in plants is carried out through precise spatial and temporal control of planes of cell division and cell shape change. Microtubules and other cytoskeletal elements participate in both of these processes (Gunning and Hardham 1982, Simmonds 1992). Cortical microtubules (MTs) and/or microfilaments interact to determine cell shape through controlling the orientation of cellulose microfibrils. The preprophase band of microtubules (PPB), a cortical ring of microtubules (MTs) circumscribing the nucleus, predicts the position and orientation of the future cell plate. The PPB may also have a role in consolidating the immature cell plate following cytokinesis (Mineyuki and Gunning 1990). PPBs are found in almost all sporophytic cell divisions of higher plants

NATO ASI Series, Vol. H 84
Biomechanics of Active Movement
and Division of Cells
Edited by N. Akkaş
© Springer-Verlag Berlin Heidelberg 1994

and play a very important role in determining where new cell walls will be inserted during morphogenesis. However, PPBs do not participate in the two terminal divisions during pollen development. The uninucleate microspores of *Brassica napus* undergoes one division to produce a vegetative and a generative cell. The generative cell later divides to form two sperm cells. Both of these divisions take place without PPBs.

An *in vitro* system has been optimized to study the mechanism(s) involved in changing gametogenesis to sporogenesis. Immature microspores isolated from *B. napus* cv. Topas continue to develop as pollen when cultured at 25°C. Embryo development is induced when the microspores are subjected to a short heat treatment (32.5°C for 18-24 h) prior to culture at 25°C. By culturing the microspores near the late uninucleate stage of development, embryo yields averaging 20% are routinely obtained (Telmer et al. 1992). This developmental change from gametogenesis to sporogenesis requires a major change in microspore morphogenesis which must involve a reorganization of the cytoskeleton. This study shows that the most dramatic change is that of the appearance of (PPBs) after 6 h of heat induction; this key structural element identifies future sporophytic divisions and thus predicts embryogenesis.

MATERIALS AND METHODS

Buds were harvested from *B. napus* plants grown at 5°C, 8 h dark/10°C, 16 h light in growth cabinets. Microspores from individual buds of *B. napus* cv. Topas (4079), a highly embryogenic line, were isolated and cultured as previously described (Telmer et al. 1992). Embryogenesis was induced by subjecting the cultures to 32.5°C for 24 h followed by 3-4 weeks at 25°C. Microspores cultured continuously at 25°C developed as pollen. The non-embryogenic *B. napus* line 0025 (Allelix Inc.) was treated in the same manner as Topas. This line does not produce embryos when subjected to the heat treatment.

To examine the effect of MT depolymerization, microspores were cultured at 25°C in the presence of 10 ug/ml colchicine.

MT organization was analyzed by using indirect

immunofluorescence labelling on microspores fixed immediately after isolation and during the first 24 h of culture at 32.5°C or 25°C. The cells were simultaneously labelled for DNA with Hoechst 33258 and MTs with the primary monoclonal rat anti-yeast tubulin (MAS 078, clone YOL 1/34, Cedarlane Laboratories, Hornby, Ont. Canada) followed by the secondary antibody, fluorescein-conjugated goat anti-rat immunoglobulin IgG (Sigma), (Simmonds et al. 1985, Simmonds & Setterfield 1986).

RESULTS

Immunofluorescence labelling of MTs in developing microspores of both cv. Topas and line 0025 shows that the pattern of MT organization remains the same from the late uninucleate to the late prophase stages (Figs. 1 and 2). During these developmental stages, the nucleus is acentric in one of the microspore lobes with a segment of the nuclear membrane closely appressed to the plasma membrane. The nucleus shows MTs radiating from the nuclear envelope into the cytoplasm. There is no PPB present prior to or during prophase in pollen development (Fig. 2). Isolated microspores, cultured at 25°C, show similar MT organization, i.e. they do not form PPBs and they develop as pollen.

Heat treated microspores, cultured at 32.5°C for 6-8 h show changes in nuclear position and microtubule organization. The nucleus moves to a more central position and PPBs are formed prior to and during prophase (Fig. 3). This treatment induces 10-40% of the microspores to develop as embryos (Telmer et al. 1992).

Microspores from the non-embryogenic line 0025 do not show nuclear relocation and MT reorganization when cultured at 32.5°C. The acentric position of the nucleus is maintained and PPBs are not formed (Fig. 4).

After 8 h of culture at 25°C in the presence of colchicine, the MTs are depolymerized and the nuclei are positioned more centrally in the microspores (Fig. 5).

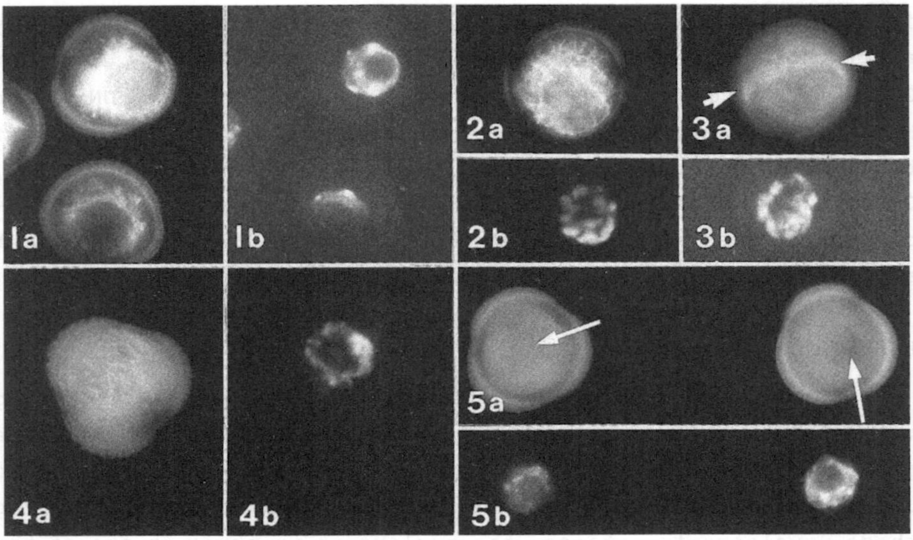

Micrographs showing indirect immunofluorescence labelling of MTs (1a, 2a, 3a, 4a, 5a) and DNA staining with Hoechst 33258 (1b, 2b, 3b, 4b, 5b). Fig. 1. Late uninucleate microspores showing acentric nuclei (1b) and MTs radiating from the nuclear envelope (1a). Fig. 2. Prophase of pollen development or non-induced microspores (2b) with MTs radiating from the nuclear envelope (2a). Fig. 3. Prophase of induced microspores of cv. Topas (3b) showing a PPB (3a, arrows). Fig. 4. Prophase of induced non-embryogenic 0025 (4b) showing cortical MTs but no PPB (4a). Fig. 5. Colchicine treated micropsores of Topas in non-inductive conditions showing nuclei (5b) but no MTs (5a); the darker area in each microspore indicates the nuclear location (arrows).

DISCUSSION

Microspores of *B. napus* cv. Topas continue pollen development when isolated and cultured at 25°C. However, if the cultures are subjected to a heat treatment of 32.5°C for 18-24 h prior to being cultured at 25°C their development changes to embryogenesis. The first structural changes indicating that gametogenesis is altered to sporogenesis is the movement of the nucleus from its acentric position and the appearance of a PPB. If the microspores are very close to mitosis when they are subjected to the high temperature, the PPB will appear before the nucleus leaves the edge of the cell. The *B. napus* line 0025 does not undergo these initial structural changes when subjected to the high temperature treatment and it is not embryogenic. This suggests that the nuclear movement and PPB formation are key factors in triggering sporophytic development.

It is evident that MTs play a critical role in maintaining the nucleus appressed to the edge of the microspore because a short treatment with colchicine which depolymerizes the MTs also frees the nucleus from its acentric position.

The heat treatment of 32.5°C is within the physiological growth range of *B. napus*, but the large temperature difference between the plant growth temperatures of 5°C/10°C and the induction temperature of 32.5°C probably induces a heat shock response (Pechan et al. 1991). Heat shock normally inhibits the synthesis of housekeeping proteins and induces the synthesis of heat shock proteins (Vierling 1991). This would result in the inhibition of synthesis of pollen specific proteins (Mascarenhas 1990, Carpenter et al. 1992) including the proteins responsible for the maintenance of the MT organization found in microspores at the late uninucleate and prophase stages (Figs. 1 and 2) which appears to be critical for continuation of pollen development. Thus, induction of embryogenesis may be the result of a default mechanism caused by the suppression of pollen development.

It seems evident that the formation of PPBs, after suppression of pollen development, signals the onset of multicellular structures. Multicellular structures are formed when cell division occurs by cell plate formation. Cell division which occurs without cell plate development (e.g. ingrowing cleavage furrows found in animal and lower plant cytokinesis) leads to single cell formation. It follows that when cell plates degenerate (Telmer et al. 1993) or are malformed (Simmonds and Setterfield 1986), multicellular structures cannot develop. In the case of pollen development, a cell plate forms after the first pollen mitosis separating the generative cell from the vegetative cell in the early bicellular structure. However, as the bicellular structure matures, the newly formed cell wall degenerates allowing the generative cell to migrate into the cytoplasm of the vegetative cell (Telmer et al. 1993). It has been suggested that PPBs nct only predict the plane and location of the future cell plates but also have a role in consolidating them at the end of cytokinesis (Mineyuki and Gunning 1990). When the cell plate is first formed it is fluid and convoluted but after reaching the site previously occupied by the PPB it becomes rigid and flat. Mineyuki and Gunning (1990) suggested that a cell wall maturation factor(s) may be deposited at the site occupied by the PPB thus allowing the future cell plate to become consolidated. This hypothesis fits the situation found in the first pollen mitosis which takes place without a PPB and results in the formation of an unstable cell plate which eventually degenerates. It seems reasonable to suggest that the PPBs, induced in uninucleate microspores by the heat treatment, play a critical role in consolidating the cell plate formed after cytokinesis. This stable cell wall is essential for initiating the development of multicellular structures.

REFERENCES

Carpenter JL, Ploense SE, Snustad DP, Silflow CD (1992) Preferential expression of an -tubulin gene of Arabidopsis in pollen. Plant Cell 4:557-571

Gunning BES, Hardham AR (1982) Microtubules. Annu. Rev. Plant Physiol. 33:651-698

Mascarenhas JP (1990) Gene activity during pollen development. Annu. Rev. Plant Physiol. Plant Mol. Biol. 41:317-338

Mineyuki Y, Gunning BES (1990) A role for preprophase bands of microtubules in maturation of new cell walls, and a general proposal on the function of preprophase band sites in cell division in higher plants. J. Cell Sci. 97:527-537

Pechan PM, Bartels D, Brown DCW and Schell J (1991) Messenger-RNA and protein changes associated with induction of *Brassica* microspore embryogenesis. Planta 184:161-165

Simmonds DH (1991) Microtubules in cultured plant protoplasts. Acta Bot. Neerl. 40:183-195

Simmonds DH, Setterfield G (1986) Aberrant microtubule organization can result in genetic abnormalities in protoplast cultures of Vicia hajastana Grossh. Planta 167:460-468

Simmonds DH, Seagull RW, Setterfield G (1985) Evaluation of techniques for immunofluorescent staining of microtubules in cultured plant cells. J. Hist. and Cytochem. 33:345-352

Telmer CA, Simmonds DH, Newcomb W (1992) Determination of developmental stage to obtain high frequencies of embryogenic microspores in *Brassica napus*. Physiol Plant 84:417-424

Telmer CA, Newcomb W, Simmonds DH (1993) Microspore development in *Brassica napus* and the effect of high temperature on division *in vivo* and *in vitro*. Protoplasma 172:154-165

Vierling E (1991) The roles of heat shock proteins in plants. Annu. Rev. Plant Physiol. Plant Mol. Biol. 42:579-620

SUBJECT INDEX

acetylation 533,535,537
acrosome reaction 184
actin 56,192,245,431
 beta-actin 522,523
 depolymerization 195,215
 F-actin 287,293,302,376, 439,445
 G-actin 287,293,302,446, 451
 in cell motility 189,193
 network 287,312
 polymerization 184,192, 200,210,215,432
 sliding 196
actomyosin 525
adherence 516
adhesion 184,199,330,437, 440,442,511
 receptor 286,303
 regulation 286,306
aggregation 513
Allomyces 396
alpha helices 4
alpha-tubulin 534
amoeba 215,447
amoeboid motion 231
anaphase 563
arc-line 398
arcs 192
atomic force microscopy 185
ATP 417,419,467,525
atrial natriuretic peptide 421
auto-correlation 437,438
axonemal
 splitting 407
 twisting 406
axoneme 403,466

bacteria 493
beat frequency 469
beetle 534
bend shape 398
bending
 modulus 417,420
 moment 400,407,409
biased-baseline 407
biopolymer 17
Blastocladiella 396,397
blastocoel 134,158,164
blastopore 141,159
blastula stage 146,158,160
blebbing 434
boundary condition 249
buckling 167
BWO cells 190

calcium 208
 effect on elasticity 185, 187
capillary 420
cardiac cell 539
cell
 collision 191,206,212
 contact 214
 migration 180,218
 poking 185
 shape 184,188,201
 immigration 144,156
 intercalation 142,153,160
 polarization 138,150
 movement 133,164
cGMP 421
chemotaxis 442
Chlamydomonas 393,397
choanocyte 100
chromosome 1
cilia 393,465
Ciona sperm 407
circumferential links 403, 406
cleavage 38,50,58,67,81, 565
collagen gel
 bound 523
 free floating 523
collagenase 521
compliance 247
constitutive laws 243
continuum 15,238,259,459
contractility 45,61,251
contraction 204,209,293, 298,314,338,451,509
control 397
cortex 42,67,439,445,446, 451,
cortical
 contraction 121
 flow 439,440,487
Crithidia 393,396
curvature 12,43
cytochalasin 208,211
cytogel expansion 200

cytokinesis 37,68
cytoplasm 48,68,238,488
cytoskeleton 183,245,287, 307,372,461,474,487,519, 523,535
 components 426
cytosol 446

deformation 454
depolymerization 292,294, 304
diffusion 268
discrete analysis 25
DNA 1,2,18,30
double helix 7,9
DPG 419
drag
 membrane 296
dynamic reciprocity 181
dynein 404,409,466
dynein arms 403,406,409

echinoderm egg 57
eigenvalue 23
E.kuehniella 534
elasticity 197,226,402,407
 measurement 185
embryo 425,427
 virtual 427
embryogenesis 569
EMG-force 557,561
endocytosis 141
endothelial cell 186,332, 369
energy
 elastic 480
 expenditure 401,402
 helix 14
 minimization 113
epidermis cell 190,198, 212,216
equatorial force 75
erythrocyte 415,422
 ghost 420
Euglena 394,396
Eulerian instability 159
exocytosis 141
extracellular matrix 181, 377,551
eye rudiment 135

feedback 460,464
fibroblast 215
filterability 421

finite element method 425, 430
flagella 393
flickering 417
flow chamber 371
fluctuation
 local 416
 mechanical 417,422
 membrane 415,421
focal contact 189
force 117,288,294,309,484, 502,543
 adhesive 354
frontal contraction 196
furrow 40

gastrocoel 134
gastrulation 141,146,149, 174
gelation 241
gene expression 182
germ cells 533
growth 461,464,495
growth pulsation 136,167, 171

homeostasis 112
Hydra 103
hydraulic see hydrostatic
hydrodynamics 399,403
hydrostatic pressure 184, 198,201,205,432

immunofluorescence 571
incompressibility 239
inductive substance 133
inertia 399
inotropism 540
insect 533
instability 462,464,545
 morphogenic 448
integrin 511
intermediate filament 188, 194
 arrangement 184
intracellular action potential
  afterpotential 558,561
  duration 557,561
  profile 557,560
invagination 428

keratinocyte 442,445,447, 451

kinesin 529
kinetochore 38,545

lamella 201,212
lamellipod 438,445
lanthanum action 193
Laplace's law 44
leading edge 192
lepidoptera 534
leukocyte 215,437,442
light scattering 415
linear momentum 240
locomotion
 model 288,292,315
locust 534

magnesium 420
mass balance 239,337
mastigonemes 394
meander 398
mechanical
 properties 183,345
 strength 184
 stresses
  pattern 132,144
  hyperrestoration 152,176
 transduction 453
mechanics
 cellular 359
 intracellular 358
 multicellular 87
mechanotransduction 522,
 456
meiosis 537
membrane 479
 actin binding 192
 displacement 416,421
 extension 200
 turnover 200
metaphase plate 39
microcolliculi 214
microfilament 425,428
micropipette 501
microtubule 188,193,425,
 428,466,471,534,546,569
 kinetochore 535
 sliding 404,409
midbody 535
migration 333
mitosis 58,471
mitotic apparatus 38
mitotic spindle 549
model
 biomechanical 69
 cell crawling 214,217
 cell locomotion 182
 fluid-mechanical 72
 mathematical 46,71,76,383
 mechanical 493
 rotor 31
modelling 407,409
modulus of elasticity 59
morphogenesis 98,460
motility 252,515
motive force 195,200
motor unit potential
 amplitude 559,561
 rectified integral 560,
  561
 rms 557,559
 time course 557,559
movement
 morphogenetic 426,427
 patterns 394,397,407
muscle
 cell 499
 fatigue 557,561
 fiber propagation velocity
  557,561
mussel 406,407
Mutilus 397
myocyte 551
myosin 56,192,196,289,299,
 313,

Na/K-ATPase 183
neural plate 140,166,429
neurulation 427,428
neurits 158
neurons 215,216
neurula stage 146,160
neutrophil 432,511

Ochromonas 394,402
osmotic pressure 134,151,
 183,200,420
 see also hydrostatic
oxygenation 419

Paramecium 396
pattern formation 428
P.brassicae 534
Pelomyxa 234
Peranema 396
phalloidin 208,211
Physarum cytoplasmic
 strands 150,153

Poisson's deformation 134, 164
polar body formation 53
polarization 461
polymerization 253,287, 291,295,
power 401,402
 spectrum 416,418,561
pressure
 intracellular 80
proliferation 182,520
propagation 405
protein 1,5,528,552
 kinase 212
protrusion 447
 wave 450
pseudopod 237
 protrusion 284,307,316

rat sperm 407
receptor 336,437,441
reflection interference microscopy 203,217
reorientation 449
retraction 447
Reynolds number 398
ring force 80
rigor 406
rolling 288,309
ruffles 199,434

Sabellaria 397
scanning acoustic microscopy 185,206,212, 218
sea urchin sperm 394,401, 405,407
self-organization 92,459, 464
S.gregaria 534
signal 47
 generation 300,302
 furrow-inducing 54
 recognition 379
 transduction 379,551
simulation
 computer 426,427
 numerical 20
 stochastic 441
solation 241
soliton 27
spermatocyte 537
spermatogonia 537
spermatozoa 393,404
spindle 475,537
spokes 403,406
sponge 95,98
spreading of cells 205
stability 496
stiffness 77,405,408,339, 541
strain 455
stress 459,464,494
 active 313,460,463
 elastic 312
 internal 460
 relaxation 506
 shear 515
stress fibers 189,199,334
structure 184,403
surface angle 147
switch-point 406,407

tension 195,519,523
 surface 439,446,447
Tetrahymena 393,404
T.molitor 534
torque 21,440
traction 342
transducer 504
translocation 440
trypanosomes 394
tug of war 197
tyrosinated 535

urodynamics 499

velocity 400,402,491
vesicle 482
viscoelasticity 74,494
viscosity 184,188,399,240,
 intermediate filaments 188

wave 394,397,404,407
 nonlinear 27
wave propagation 428
wavelength 400

Xenopus see epidermis cell

Young's modulus 346

zygote 141,158,174

NATO ASI Series H

Vol. 21: **Cellular and Molecular Basis of Synaptic Transmission.** Edited by H. Zimmermann. 547 pages. 1988.

Vol. 22: **Neural Development and Regeneration. Cellular and Molecular Aspects.** Edited by A. Gorio, J. R. Perez-Polo, J. de Vellis, and B. Haber. 711 pages. 1988.

Vol. 23: **The Semiotics of Cellular Communication in the Immune System.** Edited by E.E. Sercarz, F. Celada, N.A. Mitchison, and T. Tada. 326 pages. 1988.

Vol. 24: **Bacteria, Complement and the Phagocytic Cell.** Edited by F. C. Cabello und C. Pruzzo. 372 pages. 1988.

Vol. 25: **Nicotinic Acetylcholine Receptors in the Nervous System.** Edited by F. Clementi, C. Gotti, and E. Sher. 424 pages. 1988.

Vol. 26: **Cell to Cell Signals in Mammalian Development.** Edited by S.W. de Laat, J.G. Bluemink, and C.L. Mummery. 322 pages. 1989.

Vol. 27: **Phytotoxins and Plant Pathogenesis.** Edited by A. Graniti, R. D. Durbin, and A. Ballio. 508 pages. 1989.

Vol. 28: **Vascular Wilt Diseases of Plants. Basic Studies and Control.** Edited by E. C. Tjamos and C. H. Beckman. 590 pages. 1989.

Vol. 29: **Receptors, Membrane Transport and Signal Transduction.** Edited by A. E. Evangelopoulos, J. P. Changeux, L. Packer, T. G. Sotiroudis, and K.W.A. Wirtz. 387 pages. 1989.

Vol. 30: **Effects of Mineral Dusts on Cells.** Edited by B.T. Mossman and R.O. Begin. 470 pages. 1989.

Vol. 31: **Neurobiology of the Inner Retina.** Edited by R. Weiler and N.N. Osborne. 529 pages. 1989.

Vol. 32: **Molecular Biology of Neuroreceptors and Ion Channels.** Edited by A. Maelicke. 675 pages. 1989.

Vol. 33: **Regulatory Mechanisms of Neuron to Vessel Communication in Brain.** Edited by F. Battaini, S. Govoni, M.S. Magnoni, and M. Trabucchi. 416 pages. 1989.

Vol. 34: **Vectors asTools for the Study of Normal and Abnormal Growth and Differentiation.** Edited by H. Lother, R. Dernick, and W. Ostertag. 477 pages. 1989.

Vol. 35: **Cell Separation in Plants: Physiology, Biochemistry and Molecular Biology.** Edited by D. J. Osborne and M. B. Jackson. 449 pages. 1989.

Vol. 36: **Signal Molecules in Plants and Plant-Microbe Interactions.** Edited by B.J.J. Lugtenberg. 425 pages. 1989.

Vol. 37: **Tin-Based Antitumour Drugs.** Edited by M. Gielen. 226 pages. 1990.

Vol. 38: **The Molecular Biology of Autoimmune Disease.** Edited by A.G. Demaine, J-P. Banga, and A.M. McGregor. 404 pages. 1990.

NATO ASI Series H

Vol. 39: **Chemosensory Information Processing.**
Edited by D. Schild. 403 pages. 1990.

Vol. 40: **Dynamics and Biogenesis of Membranes.**
Edited by J. A. F. Op den Kamp. 367 pages. 1990.

Vol. 41: **Recognition and Response in Plant-Virus Interactions.**
Edited by R. S. S. Fraser. 467 pages. 1990.

Vol. 42: **Biomechanics of Active Movement and Deformation of Cells.**
Edited by N. Akkas. 524 pages. 1990.

Vol. 43: **Cellular and Molecular Biology of Myelination.**
Edited by G. Jeserich, H. H. Althaus, and T. V. Waehneldt. 565 pages. 1990.

Vol. 44: **Activation and Desensitization of Transducing Pathways.**
Edited by T. M. Konijn, M. D. Houslay, and P. J. M. Van Haastert. 336 pages. 1990.

Vol. 45: **Mechanism of Fertilization: Plants to Humans.**
Edited by B. Dale. 710 pages. 1990.

Vol .46: **Parallels in Cell to Cell Junctions in Plants and Animals.**
Edited by A. W Robards, W. J . Lucas, J . D. Pitts, H . J . Jongsma, and D. C. Spray. 296 pages. 1990.

Vol. 47: **Signal Perception and Transduction in Higher Plants.**
Edited by R. Ranjeva and A. M. Boudet. 357 pages. 1990.

Vol. 48: **Calcium Transport and Intracellular Calcium Homeostasis.**
Edited by D. Pansu and F. Bronner. 456 pages. 1990.

Vol. 49: **Post-Transcriptional Control of Gene Expression.**
Edited by J. E. G. McCarthy and M. F. Tuite. 671 pages. 1990.

Vol. 50: **Phytochrome Properties and Biological Action.**
Edited by B. Thomas and C. B. Johnson. 337 pages. 1991.

Vol. 51: **Cell to Cell Signals in Plants and Animals.**
Edited by V. Neuhoff and J. Friend. 404 pages. 1991.

Vol. 52: **Biological Signal Transduction.**
Edited by E. M . Ross and K . W. A. Wirtz. 560 pages. 1991.

Vol. 53: **Fungal Cell Wall and Immune Response.**
Edited by J. P. Latge and D. Boucias. 472 pages. 1991.

Vol. 54: **The Early Effects of Radiation on DNA.**
Edited by E. M. Fielden and P. O'Neill. 448 pages. 1991.

Vol. 55: **The Translational Apparatus of Photosynthetic Organelles.**
Edited by R. Mache, E. Stutz, and A. R. Subramanian. 260 pages. 1991.

Vol. 56: **Cellular Regulation by Protein Phosphorylation.**
Edited by L. M. G. Heilmeyer, Jr. 520 pages. 1991.

NATO ASI Series H

Vol. 57: **Molecular Techniques in Taxonomy.**
Edited by G . M . Hewitt, A. W. B. Johnston, and J. P. W. Young .
420 pages. 1991.

Vol. 58: **Neurocytochemical Methods.**
Edited by A. Calas and D. Eugene. 352 pages. 1991.

Vol. 59: **Molecular Evolution of the Major Histocompatibility Complex.**
Edited by J. Klein and D. Klein. 522 pages. 1991.

Vol. 60: **Intracellular Regulation of Ion Channels.**
Edited by M. Morad and Z. Agus. 261 pages. 1992.

Vol. 61: **Prader-Willi Syndrome and Other Chromosome 15q Deletion Disorders.**
Edited by S. B. Cassidy. 277 pages. 1992.

Vol. 62: **Endocytosis. From Cell Biology to Health, Disease and Therapie.**
Edited by P. J. Courtoy. 547 pages. 1992.

Vol. 63: **Dynamics of Membrane Assembly.**
Edited by J. A. F. Op den Kamp. 402 pages. 1992.

Vol. 64: **Mechanics of Swelling. From Clays to Living Cells and Tissues.**
Edited by T. K. Karalis. 802 pages. 1992.

Vol. 65: **Bacteriocins, Microcins and Lantibiotics.**
Edited by R. James, C. Lazdunski, and F. Pattus. 530 pages. 1992.

Vol. 66: **Theoretical and Experimental Insights into Immunology.**
Edited by A. S. Perelson and G. Weisbuch. 497 pages. 1992.

Vol. 67: **Flow Cytometry. New Developments.**
Edited by A. Jacquemin-Sablon. 1993.

Vol. 68: **Biomarkers. Research and Application in the Assessment of Environmental Health.** Edited by D. B. Peakall and L. R. Shugart.
138 pages. 1993.

Vol. 69: **Molecular Biology and its Application to Medical Mycology.**
Edited by B. Maresca, G. S. Kobayashi, and H. Yamaguchi. 271 pages.
1993.

Vol. 70: **Phospholipids and Signal Transmission.**
Edited by R. Massarelli, L. A. Horrocks, J. N. Kanfer, and K. Löffelholz.
448 pages. 1993.

Vol. 71: **Protein Synthesis and Targeting in Yeast.**
Edited by A. J. P. Brown, M. F. Tuite, and J. E. G. McCarthy. 425 pages.
1993.

Vol. 72: **Chromosome Segregation and Aneuploidy.**
Edited by B. K. Vig. 425 pages. 1993.

Vol. 73: **Human Apolipoprotein Mutants III. In Diagnosis and Treatment.**
Edited by C. R. Sirtori, G. Franceschini, B. H. Brewer Jr. 302 pages. 1993.

NATO ASI Series H

Vol. 74: **Molecular Mechanisms of Membrane Traffic.**
Edited by D. J. Morré, K. E. Howell, and J. J. M. Bergeron. 429 pages. 1993.

Vol. 75: **Cancer Therapy. Differentiation, Immunomodulation and Angiogenesis.**
Edited by N. D'Alessandro, E. Mihich, L. Rausa, H. Tapiero, and T. R. Tritton. 299 pages. 1993.

Vol. 76: **Tyrosine Phosphorylation/Dephosphorylation and Downstream Signalling.**
Edited by L. M. G. Heilmeyer Jr. 388 pages. 1993.

Vol. 77: **Ataxia-Telangiectasia.** Edited by R. A. Gatti, R. B. Painter. 306 pages. 1993.

Vol. 78: **Toxoplasmosis.** Edited by J. E. Smith. 272 pages. 1993.

Vol. 79: **Cellular Mechanisms of Sensory Processing. The Somatosensory System.**
Edited by L. Urban. 514 pages. 1994.

Vol. 80: **Autoimmunity: Experimental Aspects.**
Edited by M. Zouali. 318 pages. 1994.

Vol. 81: **Plant Molecular Biology. Molecular Genetic Analysis of Plant Development and Metabolism.**
Edited by G. Coruzzi, P. Puigdomènech. 579 pages. 1994.

Vol. 82: **Biological Membranes: Structure, Biogenesis and Dynamics.**
Edited by Jos A. F. Op den Kamp. 367 pages. 1994.

Vol. 83: **Molecular Biology of Mitochondrial Transport Systems.**
Edited by M. Forte, M. Colombini. 420 pages. 1994.

Vol. 84: **Biomechanics of Active Movement and Division of Cells.**
Edited by N. Akkaş. 587 pages. 1994.

MIX
Papier aus verantwortungsvollen Quellen
Paper from responsible sources
FSC® C105338

If you have any concerns about our products,
you can contact us on
ProductSafety@springernature.com

In case Publisher is established outside the EU,
the EU authorized representative is:
Springer Nature Customer Service Center GmbH
Europaplatz 3, 69115 Heidelberg, Germany

Printed by Libri Plureos GmbH
in Hamburg, Germany